Advanced Structured Materials

Volume 150

Common engineering materials reach in many applications their limits and new developments are required to fulfil increasing demands on engineering materials. The performance of materials can be increased by combining different materials to achieve better properties than a single constituent or by shaping the material or constituents in a specific structure. The interaction between material and structure may arise on different length scales, such as micro-, meso- or macroscale, and offers possible applications in quite diverse fields.

This book series addresses the fundamental relationship between materials and their structure on the overall properties (e.g. mechanical, thermal, chemical or magnetic etc.) and applications.

The topics of *Advanced Structured Materials* include but are not limited to

- classical fibre-reinforced composites (e.g. glass, carbon or Aramid reinforced plastics)
- metal matrix composites (MMCs)
- micro porous composites
- micro channel materials
- multilayered materials
- cellular materials (e.g., metallic or polymer foams, sponges, hollow sphere structures)
- porous materials
- truss structures
- nanocomposite materials
- biomaterials
- nanoporous metals
- concrete
- coated materials
- smart materials

Advanced Structured Materials is indexed in Google Scholar and Scopus.

More information about this series at http://www.springer.com/series/8611

Leonid Borisovich Getsov

Materials and Strength of Gas Turbine Parts

Volume 1: Materials, Properties, Damage, Deformation and Fracture Models

Translators and Editors
Holm Altenbach and Konstantin Naumenko

Springer

Leonid Borisovich Getsov
Joint-Stock Company "I.I.Polzunov
Scientific and Development Association on
Research and Design of Power Equipments"
(NPO CKTI)
St. Petersburg, Russia

Translators and Editors
Holm Altenbach
Institut für Mechanik, Fakultät für Maschinenbau
Otto-von-Guericke-Universität Magdeburg
Magdeburg, Sachsen-Anhalt, Germany

Konstantin Naumenko
Institut für Mechanik, Fakultät für Maschinenbau
Otto-von-Guericke-Universität Magdeburg
Magdeburg, Sachsen-Anhalt, Germany

ISSN 1869-8433 ISSN 1869-8441 (electronic)
Advanced Structured Materials
ISBN 978-981-16-0533-8 ISBN 978-981-16-0534-5 (eBook)
https://doi.org/10.1007/978-981-16-0534-5

This Springer imprint is published by the registered company Springer Nature Singapore Pte Ltd.
The registered company address is: 152 Beach Road, #21-01/04 Gateway East, Singapore 189721, Singapore

In memory of the parents B.B. Getsov and S.Ya. Getsova and my academic teachers G.S. Pisarenko, A.V. Stanyukovich, and E.L. Greenzaid.

Foreword

The two-volume book “Materials and Strength of Elements in Gas Turbines” has appeared in Russian four times since the 1970s. The translation of parts of the 4th Russian edition into English is intended to fill a gap, as Russian original sources have so far not been explored abroad.

The author of the monograph has worked for many years in the field of power engineering, and especially in the gas turbine industry in Russia. He is the author/coauther of more of 500 scientific publications in Russian and foreign scientific journals. The first edition of the book "Materials and Strength of Gas Turbine Parts" was published in 1973. In it, he published the results of his research for the first

Fig. 1 Cover sheets of volume 1 (left) and 2 (right)

L.B. Getsov	**Л.Б. Гецов**
MATERIALS AND STRENGTH OF GAS TURBINE PARTS	**МАТЕРИАЛЫ И ПРОЧНОСТЬ ДЕТАЛЕЙ ГАЗОВЫХ ТУРБИН**
IN TWO BOOKS	**В ДВУХ КНИГАХ**
BOOK 1	**КНИГА 1**
Materials, properties, damages, models of deformation and fracture	Материалы, свойства, повреждения, модели деформирования и разрушения
Rybinsk Publishing House Gas Turbo Technology 2010	Рыбинск ООО «Издательский дом «Газотурбинные технологии» 2010

Fig. 2 Cover sheets of Volume 1

two decades of work. With the accumulation of experience in further developments, L.B. Getsov prepared and published new editions of his book (1982, 1996, 2010-2011). In the second edition, the author, while remaining within the same volume, placed new materials and excluded some of the earlier ones. In 1996, the third edition of a markedly increased volume was published (591 pages instead of 295). And finally, in the 4th edition, his book was published in two volumes with a total volume of 1100 pages. The current edition in English 2020 will contain the text of the 4th edition and include lists of publications of articles from 2010-2020. Many issues and results concerning the behavior of materials at high temperatures, covered in the monograph, are the original contributions of the author.

The book itself contains essential information from turbine design and construction, with material science and mechanical aspects being the focus. Turbine construction had a major impact on energy supply over the past hundred years. Safe operation is a basic requirement for energy security in national economies and thus for progress and economic growth.

At the end of the last century and the beginning of the present, scientists from the former Soviet Union (mainly now from Russia and Ukraine) and France were intensively involved in the creation of models of visco-elasto-plastic media, their experimental study and verification of models. A number of scientific schools were created, concentrated in various institutes and universities. Moreover, the work was carried out not in isolation, but collegially.

Such organizations include: *Saint Petersburg University* (L.M. Kachanov, Yu.I. Kadashevich, R.A. Aratyunyan, V.A. Likhachev), *Peter the Great Saint Petersburg Polytechnic University* (A.I. Lurie, V.A. Palmov, P.A. Zhilin, A.M. Krivtsov, A.S. Semenov, B.E. Melnikov), *Lomonosov Moscow State University* (A.Yu. Ishlinsky, A.A. Ilyushin, Yu.N. Rabotnov, G.P. Cherepanov, V.S. Lensky, S.A. Shesterikov, R.V. Goldstein, A.M. Lokoshchenko, R.A. Vasin, G.L. Brovko, V.S. Bondar), *Novosibirsk State University* (Yu.N. Rabotnov, V.V. Bolotin), *Lobachevsky State University of Nizhny Novgorod* (A.G. Ugodchikov, V.G. Korotkikh V.G.), *Central Institute of Aviation Motors "P.I. Baranov" - CIAM* (I.A. Birger, B.F. Shorr, E.R. Golubovsky, R.N. Sizova, R.A. Dul'nev), *Joint-Stock Company "I.I. Polzunov Scientific and Development Association on Research and Design of Power Equipments" - CKTI* (A.V. Stanyukovich, A.A. Chizhik, L.B. Getsov, V.K. Adamovich, Yu.K. Petrenya), *Central Research Institute of Mechanical Engineering Technology - CNIITMASH* (I.I. Trunin, M.G. Kabelevsky), *Mechanical Engineering Research Institute "A.N. Blagonravov" of the Russian Academy of Sciences - IMASH* (S.V. Serensen, R.M. Shneiderovich, N.A. Makhutov, A.P. Gusenkov), *A. F. Ioffe Physico-Technical Institute of the Russian Academy of Sciences* (S.N. Zhurkov, N.N. Davidenkov, V.I. Betekhtin), *Tver State Technical University* (V.G. Zubchaninov), *South Ural State University, former Chelyabinsk Polytechnic Institute* (D.A. Gokhfeld, O.S. Sadakov, O.F. Chernyavsky, K.M. Kononov), *Perm National Research Polytechnic University* (P.V. Trusov), *Institute of Mechanics "S.P. Timoshenko" of the Academy of Sciences of Ukraine* (Yu.N. Shevchenko) and *Institute of Strength Problems of the Academy of Sciences of Ukraine* (G.S. Pisarenko, V.T. Troshchenko, A.A. Lebedev, V.V. Pokrovsky, A.Ya. Krasovsky).

In addition, it should be mentioned the *Federal State Scientific Center "All-Russian Institute of Aircraft Materials" (VIAM)*, which was and is the leading Russian institute of materials for aviation technology. Metallic materials developed at VIAM based on iron, nickel, titanium and aluminum, as well as non-metallic materials are widely used in various industries in Russia. Leading scientists of VIAM, such as S.T. Kishkin, I.N. Frindlyander, Ya.B. Fridman, F.F. Khimushin, R.E. Shalin, N.M. Sklyarov, I.P. Bulygin, I.L. Svetlov, E.R. Golubovsky, N.I. Kolobnev, S.Z. Bokshtein and now E.N. Kablov, N.V. Petrushin, O. G. Ospennikov, B.S. Lomberg, I.M. Demonis and many others are widely known throughout the world for major developments in new materials and new technologies. In the monograph by L.B. Getsov, who worked and works in stationary and ship power engineering, references to the published works of VIAM employees occupy a significant place. A number of works by L.B. Getsov have been and are being carried out jointly with VIAM employees.

The French school is associated with the names of J.-L. Chaboche and J. Lemaitre and their students J. Besson, G. Cailletaud, S. Forest and E. Busso (*ONERA* and *Center de Matériaux at MINES Paris Tech*).

In contrast to the Russian edition, this book only contains the first four chapters of Volume 1 and an appendix that was originally included in Volume 2. This material was selected after discussions with the author because it deals with a certain aspect and can be published independently. The aim was to present the book to the readers as quickly as possible, as turbine design and construction is still developing very dynamically. The existing literature is very extensive. A complete overview cannot be given. With the present book, however, an insight into Russian literature is given, so that the reader at least partially gets an overview of the research work in the former Soviet Union (now Commonwealth of Independent States).

The English edition differs from the Russian 4th edition not only in scope. Literature sources from the period 2010-2020 were added by the author and the translators. At the same time, printing errors from the previous editions were eliminated and the literature references made more precise. A further improvement of the source situation was not possible because not all sources were publicly accessible or were victims of the restrictive publication policy in the former Soviet Union. The author was therefore forced to link some research results only with the names of scientists without citing sources.

In the hope that the book will give readers a first glimpse of the Russian research, the translators would like to invite them to take part in a disputation of further developments in the research area.

With respect to content of Volume 1, the author L.B. Getsov and the translators will prepare Volume 2 devoted to the operating experience and technology, and with the focus on new methods for calculating the stress-strain state and strength of gas turbine parts.

Magdeburg, November 2020 *Holm Altenbach*

Preface

Nowadays, gas turbines are widely used as power plants for various purposes. Examples are power machines, engines of aircraft, ships, automobile engines, as part of gas turbine generators for various purposes. However, the largest number of different gas turbine units manufactured at various enterprises in Russia is applied at gas compressor stations of gas main pipelines for gas transportation. The elements of these units are made of different materials. The most important issues in design, manufacturing and maintenance of gas turbines are the selection of optimal component materials and solving strength problems, which allow ensuring reliability of their long-term operation. These issues are especially important when solving problems of resource extension.

In general, the process of creating a gas turbine unit (GTU), which has a given resource, consists of two main stages: design, an integral part of which are strength analysis and choice of materials, and manufacturing and finishing of prototypes. Duration and labor intensity of the second stage is usually significantly higher than that of the first stage, since the reliability issues of GTU are solved mainly at this stage on the basis of the results of experimental analysis of various kinds (by means of resource, equivalent and cyclic tests), as well as data of pilot operation. At this stage the designer is often forced to change the design and material of individual parts, to change the requirements to the technology of manufacturing of blanks, technology of finishing operations and the volume of control.

For the majority of GTUs the resource is determined by the state of its elements during operation (resource as it is). Therefore, the work to increase the life of GTU elements is the work to increase the life of the GTU as a whole. If at a design stage the designer had the exhaustive information on conditions of operation of all components, about properties of materials of components in these conditions, and also about regularities of deformation and destruction of materials at any programs of loading, then, using perfect design methods of definition of kinetics of stress and strain states of components for all service life, corresponding algorithms and programs of calculation on the computer, he could solve questions of durability and reliability of GTU components. As a result the time of finishing and testing of GTU prototypes could be significantly reduced.

All said above testifies to the fact that solving reliability issues requires knowledge in various fields: material science, strength of materials, design, manufacturing and assembly technology, calculation of thermal fields. The author has set a task to consider on the basis of available in the literature data and results of own researches the basic aspects of a problem of a choice of materials and durability of GTU components. As a whole idea of the book consists not in illumination of two themes "materials" and "durability of details", and in consideration of the questions which are on a joint of these themes. The book does not intend to compete neither with manuals for designers and calculators and books for mechanics, nor with books on theories of heat resistance and alloying of materials, as well as manuals on the properties of heat-resistant alloys.

Nevertheless, it tries to show that the traditional method of selecting materials of components by the characteristics of long-term durability, given in the standards, does not allow to adequately assess their resource both because of the differences between the real stress-strain state of the parts from the true, and because of the dependence of the characteristics of the material from the used method of statistical processing and extrapolation of the test results, from the method of assessing the impact of the loading program, the type of stress state, from the temperature conditions and the corrosive environment.

In the book, on the one hand, for materials widely used in the GTU, the number of which increases every year, the regularities of their mechanical behavior at various loading programs, the impact on them technological features of blank production, as well as fundamental aspects of the influence of their structure on the strength and operational reliability, which, according to the author, are necessary to assess the strength of parts and their residual life. On the other hand, the book covers those strength issues, the knowledge of which is necessary for the correct formulation of metallurgical research; in some cases, the focus is on those problems that have been studied only in recent years. Special attention is paid to the ideology of new methods for calculating the strength of blades, disks and a number of stator parts of turbomachinery operating in unsteady conditions, including in contact with corrosive environments.

The translation is based on the fourth, considerably revised edition of the same title book published earlier (Getsov, 1973, 1982, 1996, 2010). Its appearance is connected with significant progress achieved in recent years in the field of materials science of heat-resistant alloys and strength of structural elements, including the research with the participation of the author. A large number of studies has been carried out to improve the feasibility of decisions on the operational reliability of GTU and mainly GTU, operated on gas main pipelines. They include, in particular, works on examination of the structure state and metal properties of blades and disks of various GTU after long-term operation, development of new approaches to the solution of strength issues based on calculations by the finite element method, development of technology and solution of issues of operational reliability of centrifugal compressor wheels made of high-strength steels, expansion of knowledge about properties of single crystal materials used for blades manufacturing, creation of methods for prediction of residual life of blades.

The analysis of damage to gas turbine parts during operation has shown that the most common are damage from thermal fatigue of materials and gas turbine parts. Therefore, a significant place in the book is occupied by the issues of resistance of materials to thermal fatigue, the development of criteria for their destruction under thermal cyclic loading, the development of modern methods for calculating the stress-strain state and the strength of the GTU elements under these conditions.

In the author's opinion, the book is rather broad, sanctifying the experience of operation of the main parts of gas turbine units used in RAO Gazprom, analyzing the damage that occurred during operation, considering the issues of methodology to extend their service life, paying attention to the control of the structure and properties of metal parts during long-term operation.

Striving to cover as fully as possible various aspects of materials behavior under complex operating conditions of GTU parts for various purposes, the author was forced to pay relatively little attention to a number of issues covered in detail in the literature, for example, they include various theories of heat resistance, fatigue and methods of heat treatment and welding of parts made of heat-resistant alloys. Questions related to the calculation of temperature fields in the details are also considered only in perspective. It should be noted that it is the unsteady temperature fields that cause the accumulation of thermal fatigue damage to the parts of gas turbines.

For the convenience of practical use of the book in the appendices, the designations of Russian grades of materials according to the standards are given, as well as those available to the author for a number of materials: isochronous creep curves, cyclic creep curves, dependence of the crack growth rate on the values of the stress intensity factor. They are also given references to the respective table and figure numbers.

On the progress in the field of creation of new heat-resistant and heat-resistant alloys for GTU, both in Russia and abroad, is reported in Khimushin (1969); Sims and Hagel (1972); Sahm and Speidel (1974); Metals Society (1978); Sims et al. (1982); Betz et al. (1986); Paton et al. (1987); Maslenkov and Maslenkova (1991); Kablov et al. (2002); Kablov (2006), as well as conference proceedings (Coutsouradis et al., 1994; Lecomte-Beckers et al., 1998; Lecomte-Beckers, 2002; Lecomte-Beckers et al., 2006, 2010, 2014; Green, 2004, and others).

The first part describes materials used in gas turbine industry, including single crystal alloys, coatings, their basic properties, regularities of behavior in different media, damages, deformation and fracture models. In Appendix A are given a list of Russian steel and alloys discussed in the book.

The author expresses his deep gratitude to his colleagues, without whose help most of the studies underlying the book could not have been carried out. First of all, this applies to my colleagues A.I. Rybnikov, I.S. Malashenko, A.S. Semenov, G.D. Pigrova, M.G. Kabelevsky, N.V. Mozhaiskaya and A. Staroselsky for many years of joint work in the directions laid down in the basis of various sections of the book (participation is reflected in joint publications), as well as A.A. Lanin (Subsect. 3.2.4) and B.E. Melnikov (Sect. 2.5), The author also expresses his heartfelt gratitude to V.T. Troshchenko and B.Z. Margolin, and V.K. Dondoshansky, D.A. Gokhfeld, G.S. Pisarenko and A.V. Stanyukovich for their help in organizing the work used to write the book and in discussing their results. Special thanks to Holm Altenbach coordinating the English version and organzing the contacts to Springer publisher.

St. Petersburg,
November 2020

Leonid Getsov

References

Betz W, Brunetaud R, Coutsouradis D, Fischmeister H, Gibbons TB, Kvernes I, Lindblom Y, Marriott JB, Meadowcroft DB (eds) (1986) High Temperature Alloys for Gas Turbines and Other Applications 1986. Springer Netherlands

Coutsouradis D, Davidson JH, Ewald J, Greenfield P, Khan T, Malik M, Meadowcroft DB, Regis V, Scarlin RB, Schubert F, Thornton DV (eds) (1994) Materials for Advanced Power Engineering 1994. Springer Netherlands, Proceedings of a Conference held in Liège, Belgium, 3–6 October 1994

Getsov LB (1973) Materialy i Prochnost' Detalei Gazovykh Turbin (in Russ., Materials and Strength of Gas Turbine Parts). Mashinostroenie, Leningrad

Getsov LB (1982) Detalei Gazovykh Turbin - Materialy i Prochnost' (in Russ., Gas Turbine Parts - Materials and Strength). Mashinostroenie, Leningrad

Getsov LB (1996) Materialy i Prochnost' Detalei Gazovykh Turbin (in Russ., Materials and Strength of Gas Turbine Parts). Nedra, Moscow

Getsov LB (2010) Materialy i Prochnost' Detalei Gazovykh Turbin (in Russ., Materials and Strength of Gas Turbine Parts), vol 1 & 2. Publishing house "Gazoturbinnye tekhnologii", Rybinsk

Green KA (ed) (2004) Superalloys 2004. The Minerals, Metals & Materials Society, Warrendale, PA, Proceedings of the Tenth International Symposium on Superalloys, held September 19 - 23, 2004

Kablov EN (ed) (2006) Litejnye Zharoprochnye Splavy. Effekt Kishkina (in Russ., Casting Heat-resistant Alloys. Kishkin's Effect). Nauka, Moscow

Kablov EN, Toloraiya VN, Orekhov NG (2002) Single-crystal rhenium-bearing nickel alloys for turbine blades of GTE. Metal Science and Heat Treatment (7-8):274–278

Khimushin FF (1969) Zharoprochnye Stali i Splavy (in Russ., Heat-resistant Steels and Alloys). Metallurgiya, Moscow

Lecomte-Beckers J (ed) (2002) Materials for Advanced Power Engineering 2002 (Pt. I-III), Schriften des FZ Jülich, Reihe Energietechnik / Energy Technology, vol 21. FZ Jülich Zentralbibliothek, Jülich

Lecomte-Beckers J, Schubert F, Ennis PJ (eds) (1998) Materials for Advanced Power Engineering 1998 (Pt. I-III), Schriften des FZ Jülich, Reihe Energietechnik / Energy Technology, vol 4. FZ Jülich Zentralbibliothek, Jülich

Lecomte-Beckers J, Carton M, Schubert F, Ennis PJ (eds) (2006) Materials for Advanced Power Engineering 2006 (Pt. I-III), Schriften des FZ Jülich, Reihe Energietechnik / Energy Technology, vol 53. FZ Jülich Zentralbibliothek, Jülich

Lecomte-Beckers J, Contrepois Q, Beck T, Kuhn B (eds) (2010) Materials for Advanced Power Engineering 2010, Schriften des FZ Jülich, Reihe Energie & Umwelt / Energy & Environment, vol 94. FZ Jülich Zentralbibliothek, Jülich

Lecomte-Beckers J, Dedry O, Oakey J, Kuhn B (eds) (2014) Materials for Advanced Power Engineering 2014, Schriften des FZ Jülich, Reihe Energie & Umwelt / Energy & Environment, vol 234. FZ Jülich Zentralbibliothek, Jülich

Maslenkov SB, Maslenkova EA (1991) Spravochnik Stali i Splavy dlya Vysokikh Temperatur (in Russ., Handbook Steels and Alloys for Advanced Temperatures), vol 1 & 2. Metallurgiya, Moscow

Metals Society (ed) (1978) Forging and Properties of Aerospace Materials, Metals Society Series, vol 188, Metals Society, London, Proceedings of an International Conference Organized by Activity Group Committee 2 (Applied Metallurgy and Metals Technology) of the Metals Society, in Association with the Leeds and Bradford Metallurgical Societies, held in the Houldsworth School of Applied Science, University of Leeds, on 5-7 January 1977

Paton BE, Stroganov GB, Kishkin ST (eds) (1987) Zharoprochnost' Litejnykh Nikelevykh Splavov i Zashchita ikh ot Okisleniya (in Russ., Heat Resistance or Casted Nickel Alloys and Protection Against Oxydation). Naukova Dumka, Kiev

Sahm PR, Speidel MO (eds) (1974) High Temperature Materials in Gas Turbines. Elsevier, Amsterdam, Proceedings of the Symposium on High-temperature Materials in Gas Turbines, Brown, Boveri & Company Limited, Baden, Switzerland, 1973

Sims CT, Hagel WC (eds) (1972) The Superalloys. Wiley-Interscience, New York

Sims CT, Stoloff NS, Hagel WC (eds) (1982) Supperalloys, vol II. Wiley-Interscience, New York et al.

Volume 1: Materials, Properties, Damage, Deformation and Fracture Models

Problems concerning selection of materials for gas turbine parts, mechanism of deformation and destruction at stationary and nonstationary loading, mechanism of material damage including corrosive damage at wide temperature range typical for gas turbine operation are discussed. Accelerated tests and other methods of increasing the reliability of gas turbine engines are considered. Attention is given to nontraditional methods of calculating the strength characteristics as well as longevity of main parts. Influence on the strength value of the following factors: structure, methods of blank and part manufacturing as weil as protecting coats is considered. Unlike the previous edition of this book the special attention is given to operating experience GTU (gas turbine unit) at compressor stations, to studying of features of behaviour of single-crystal materials and modern methods of calculation of durability turbine blades.

This first volume focuses on the selection of materials, deformation and destruction mechanisms in connection with stationary and non-stationary loading, and types of material damage such as the thermal fatigue. Particular attention is paid to the issues of the properties of single crystal alloys, the relationship between structure and properties, the influence of technological factors and long-term operation. The characteristics of creep resistance, crack resistance, and resistance to cyclic deformation of different alloys are given.

Contents

1 **Operation Conditions of Gas Turbine Parts and Materials used for them** . . . 1
1.1 Operation Conditions of High-temperature Components of Gas Turbine Units and Damage Mechanisms During Long-term Operation and Laboratory Testing . . . 1
1.1.1 Turbine Guide Blades . . . 3
1.1.2 Turbine Working Blades . . . 6
1.1.3 Turbine Disks . . . 14
1.1.4 Stator Components . . . 24
1.1.5 Compressor Blades, Disks, and Housings . . . 26
1.1.6 Reduction Gears . . . 28
1.2 Requirements for the Materials of Gas Turbine Units Components . . 30
1.2.1 Turbine Disks . . . 31
1.2.2 Turbine Working Blades . . . 33
1.2.3 Turbine Guide Blades . . . 34
1.2.4 Flame Tubes . . . 34
1.2.5 Fasteners . . . 34
1.2.6 Regenerators . . . 35
1.2.7 Blades, Disks and Compressor Housings . . . 35
1.3 Materials of Gas Turbine Unite Components . . . 36
1.3.1 Perlitic Steels . . . 40
1.3.2 Ferritic Steels . . . 40
1.3.3 Austenitic Steels . . . 43
1.3.3.1 Group I . . . 45
1.3.3.2 Group II . . . 45
1.3.3.3 Group III . . . 45
1.3.3.4 Group IV . . . 46
1.3.4 Nickel, Nickel-Cobalt, and Cobalt Based Alloys . . . 46
1.3.4.1 Cast Heat-resistent Nickel Basis Alloys . . . 55

1.3.5 Composite Materials ... 56
1.3.6 Ceramic Materials ... 59
1.3.7 Titanium Alloys ... 61
1.4 Concluding Remarks ... 64

References ... 65

2 Deformation and Strength of Heat-resistant Materials under Static and Cyclic Loading ... 67
2.1 Mechanical Behavior under Uni-axial Stress State ... 67
2.1.1 Elasticity and Plasticity ... 67
2.1.2 Young's Modulus ... 69
2.1.3 Heterogeneity of Plastic Deformations ... 72
2.1.4 Scale Factor ... 72
2.2 Creep and Stress Relaxation ... 73
2.2.1 Creep Phenomenon ... 73
2.2.2 Creep Theories ... 79
2.2.3 Isochronous Creep Curves ... 84
2.2.4 Stress Relaxation ... 88
2.3 Multi-axial Strain and Stress States ... 93
2.3.1 Elasticity ... 93
2.3.2 Stress Concentrations ... 94
2.3.3 Simplest Models of Ductility and Creep ... 95
2.4 Resistance of Materials to Cyclic Deformation ... 98
2.4.1 Cyclic Elasto-plastic Deformation ... 98
2.4.2 Cyclic Elasticity Limit ... 101
2.4.3 Cyclic Creep and Stress relaxation Under Constant Sign Loading ... 104
2.4.4 Cyclic Creep Under Alternating Stress ... 108
2.4.5 Cyclic Creep Under Varying Temperature ... 109
2.4.6 Cyclic Elasto-Plasticity Under Varying Temperature ... 109
2.5 Strain Theories at Complex Loading ... 115
2.5.1 Effects of the Behavior of Metallic Materials ... 116
2.5.2 Models of Visco-Elasto-Plasticity ... 122
2.5.2.1 Model 1. Theory of Kinematic Hardening ... 126
2.5.2.2 Model 2. Multisurface Theory with one Active Surface ... 127
2.5.2.3 Model 3. Model of Isotropic Thermo-visco-elasto-plasticity ... 127
2.5.2.4 Model 4. Model of Cyclic Plasticity and Creep Under Proportional Loading ... 129
2.5.2.5 Model 5. Microstructural Model ... 131
2.5.2.6 Methods for Implementing a Multi-model Approach ... 137
2.5.2.7 Database of Material Properties ... 139

2.5.3 Implementation of Cyclic Plasticity and Creep Models under Proportional Loading . . . 143
2.5.3.1 Methodology for Determining the Material Parameters . . . 144
2.5.3.2 Calculation Methodology . . . 146
2.5.3.3 Comparison of Calculated and Experimental Data . 147
2.5.3.4 Criteria for Choosing a Material Model . . . 147
2.5.3.5 Example of the Structural Analysis . . . 148
2.6 Concluding Remarks . . . 149

References . . . 153

3 Deformation Response of Heat Resistant Materials at Static and Cyclic Loading . . . 167
3.1 Ductile and Brittle Fracture Criteria . . . 167
3.1.1 Fracture Criteria . . . 167
3.1.2 Conditions for Ductile and Brittle Fracture . . . 171
3.1.3 Bearing Capacity of Structures . . . 180
3.2 Long-term Strength at Constant and Variable Temperatures and Stresses . . . 180
3.2.1 Time Dependency . . . 182
3.2.2 Influence of Temperature . . . 188
3.2.3 Damage Mechanisms and Deformation Capacity of Materials During Creep . . . 195
3.2.4 Fracture Under Stress Relaxation Conditions . . . 199
3.2.4.1 Fracture Models Under Stress Relaxation Conditions . . . 202
3.2.4.2 Methods for Predicting the Conditions of Fracture . 203
3.2.4.3 Development of Cracks During Stress Relaxation. Influence of Crack Resistance . . . 209
3.2.4.4 Influence of the Rheological Properties of the Material . . . 210
3.2.4.5 Influence of the Supply of Elastic Energy . . . 210
3.2.4.6 Fracture Under Conditions of Relaxation and Repeated Loading . . . 211
3.2.5 Long-term Strength Under Complex Stress State . . . 215
3.2.5.1 Fracture Criteria . . . 215
3.2.5.2 Effect of Stress Concentration . . . 217
3.2.6 Long-term Strength at Variable Temperatures and Stresses . . 224
3.3 Resistance to Fracture for Cyclically Varying Stresses . . . 229
3.3.1 High Frequency Fatigue . . . 229
3.3.1.1 Influence of Cycle Asymmetry . . . 230
3.3.1.2 Statistical Processing of Test Results . . . 231
3.3.1.3 Effect of Stress Concentration . . . 232
3.3.1.4 Effect of Temperature . . . 233

3.3.1.5 Influence of the Type of Loading 235
3.3.1.6 Fatigue Under Complex Stress State 236
3.3.1.7 Complex Influence of Various Factors 237
3.3.2 Low Cycle Fatigue 239
3.3.2.1 Basic Features 239
3.3.2.2 Influence of Frequency 240
3.3.2.3 Influence of Temperature, Cycle Asymmetry, Stress Concentration and Type of Loading 242
3.3.3 Thermal Fatigue 244
3.3.3.1 Test Methods 244
3.3.3.2 Test Procedure for Installations of the Coffin Type . 246
3.3.3.3 Test Procedure for Installations of the ALATOO Type (IMASH-5S-65) 249
3.3.3.4 Test Procedure for Wedge-shaped Specimens 254
3.3.3.5 Features of Thermal Fatigue Failure............. 254
3.3.3.6 Influence of Maximum Temperature and Cycle Period 255
3.3.3.7 Influence of Non-metallic Inclusions 258
3.3.3.8 Thermal Fatigue of High Temperature Alloys..... 259
3.3.3.9 Progressive Deformation of Materials Under Thermal Cyclic Loading 265
3.4 Failure Criteria for Complex Loading Programs 265
3.4.1 Damage Under Static and Low-cycle Loading 265
3.4.2 Fracture Criteria for Elasto-plastic Deformation........... 266
3.4.3 Adaptability Theory 270
3.4.4 Failure Criterion for Alternating Cyclic Creep 272
3.4.5 Modified Fracture Criteria for Materials Under Cyclic Loading 274
3.4.6 Experimental Verification of Criteria................... 281
3.4.7 Fracture Criteria for Complex Stress State 284
3.4.7.1 First Approach 284
3.4.7.2 Second Approach 286
3.5 Features of the Formation and Propagation of Cracks in Heat-resistant Alloys under Static and Dynamic Loading.......... 287
3.5.1 Conditions and Nature of Cracking 287
3.5.2 Crack Propagation Rate 292
3.6 Concluding Remarks... 305

References 307

4 Influence of Technological Factors and Long-term Operation on the Microstructure and Properties of Heat-resistant Materials 321
4.1 Dependence of Properties on Metallurgical Factors, Size, and Orientation of Grains 321
4.1.1 Influence of Melting and Casting Methods 321

4.1.2 Influence of Deformation Conditions of Workpiece 325
4.1.3 Effect of Grain Size of Deformed Alloys 326
4.2 Influence of Crystal Orientation on the Properties of Cast Alloys ... 329
4.2.1 Microstructure of Monocrystalline Materials 329
4.2.2 Anisotropy of Elastic Moduli 331
4.2.3 Poisson's Ratio Anisotropy 332
4.2.4 Anisotropy of the Coefficient of Thermal Expansion 332
4.2.5 Short-term Mechanical Properties 335
4.2.6 Resistance to High Frequency Fatigue 338
4.2.7 Long-term Strength and Creep Resistance 341
4.2.8 Thermal and Low-cycle Fatigue 347
4.2.8.1 Influence of Maximum Cycle Temperature and Temperature Range 355
4.2.8.2 Influence of Orientation on Durability Under Thermocyclic Loading 356
4.2.8.3 Effect of Holding at Maximum Cycle Temperature 357
4.2.8.4 Effect of Stress Concentration 357
4.2.8.5 Features of the Deformation Relief 357
4.2.8.6 Features of the Nature of Fracture 361
4.2.8.7 Processes of Unilateral Progressive Deformation .. 362
4.2.8.8 Fractographic Features of Thermal Fatigue Cracks 362
4.2.8.9 Analysis of the Stress-strain State of the Samples .. 363
4.2.8.10 Analysis of Fracture Mechanisms 368
4.2.9 Fracture Criteria for Monocrystalline Materials under Thermal Cyclic Loading 372
4.2.10 Crystallographic Features of High-cycle Fatigue Fracture of Single Crystals 376
4.2.11 Oxidation Resistance 382
4.3 Relationship of Structure and Properties 382
4.3.1 Influence of Heat Treatment Regime 383
4.3.2 Effects of Long-term Aging 385
4.3.2.1 Carbide Transformations 386
4.3.2.2 Changes in the Number, Composition and Size of Particles of γ'-phase 388
4.3.2.3 Formation of Intermetallic Phases like Ni_3Ti, R-Phases and Laves Phases 390
4.3.2.4 Formation of Topologically Close-packed Phases . 390
4.3.2.5 Boundary Structure Changes 392
4.3.2.6 Changes in Properties During Long-term Operation 393
4.3.3 Reductive Heat Treatment 407
4.4 Dependence of Properties on the State of the Surface 409
4.4.1 Effect of Mechanical and Heat Treatment on the State of the Surface Layer 410
4.4.2 Dependence of Properties on Surface Condition 412
4.4.3 Methods of Surface Hardening of Gas Turbine Parts 414

4.5 Concluding Remarks 418

References 421

A Russian Steels and Alloys for Gas Turbine Units 429

B Isochronous Creep Curves 437
B.1 Materials and Test Conditions 437
B.2 Diagrams 438

C Cyclic Creep Curves 449
C.1 Materials and Test Conditions 449
C.2 Diagrams 450

D Fracture Toughness Characteristics at Cyclic Load 455
D.1 Materials and Their Parameters 455
D.2 Cyclic Fracture Toughness Curves 458

References 469

Acronyms

Abbreviations

NPO	Scientific Production Association
PO	Production Associationitem[JSC]Joint Stock Company
CJSC	Closed Joint Stock Company
OJSC	Open Joint Stock Company
CIAM	Central Institute of Aviation Motors "P.I. Baranov"
CNIITMASH	Central Research Institute of Mechanical Engineering Technology
VIAM	Federal State Scientific Center *All-Russian Institute of Aircraft Materials*
CKTI	Joint-Stock Company "I.I.Polzunov Scientific and Development Association on Research and Design of Power Equipments" (NPO CKTI)
CNIIChermet	State Scientific Center of the Russian Federation Federal State Unitary Enterprise "Central Scientific Research Institute of Ferrous Metallurgy *I.P. Bardin*
IMASH	Mechanical Engineering Research Institute of the Russian Academy of Sciences
SPbSPU	Peter the Great St. Petersburg Polytechnic University
n.s.	not standardized
GTU	gas turbine unit
GPA	gas pumping unit
GTE	gas turbine engines
HPT	high pressure turbine
LPT	low pressure turbine
ICC	intercrystalline corrosion
PC	pitting corrosion
GOST	state standard
OST	industry standard
HB	Brinell hardness
HRC	Rockwell hardness scale C
HV	Vickers hardness

HPC	heat-protection coating
TCP	topologically close-packed
ISM	impact strength on Menage specimen with circular notch
KP	strength category
LSD	least squares deviation
SEM	Scanning Electron Microscope
FCGR	fatigue crack growth rate
AY	alternating yielding
PD	progressive deformation
ChS	steel brand Chelyabinskaya stal' (Chelyabinsk steel)
L	cast steel

Symbols

a_H	impact toughness
dl/dN	fatigue crack growth rate
E	Young's modulus
E^T_{st}	static Young's modulus at temperature T
E^T_{dyn}	dynamic Young's modulus at temperature T
N	number of cycles
N_f	number of cycles to failure
N_{ci}	number of cycles to crack initiation
t_{expl}	exploitation temperature
α	thermal expansion coefficient
δ	relative extension
δ_ε	hysteresis loop width
ε_f	creep strain accumulated before fracture
ε_{pl}	plastic strain
ε_{stru}	structural deformation
p	creep strain
σ_i	principal stress ($i = 1, 2, 3$)
σ_b	bending stress
σ_{lim}	strength limit
$\sigma_{0,2}$	yield stress when 0.2 plastic strain occurs
σ_{el}	elastic limit (lowest stress point at which permanent deformation can be measured)
σ_{prop}	proportionality limit (up to this amount of stress, stress is proportional to strain)
σ_{ps}	static proportionality limit
σ_B	ultimate tensile strength (maximum of the engineering stress – engineering strain curve under tension conditions)
σ^T_B	ultimate tensile strength at temperature T
σ_{100}	long-term stress resulting in fracture after 100 h
σ^T_{100}	long-term stress resulting in fracture after 100 h at temperature T in °C

σ_{500}	long-term stress resulting in fracture after 500 h
σ_{1000}^{T}	long-term stress resulting in fracture after 1000 h at temperature T in °C
σ_{1000}	long-term stress resulting in fracture after 1000 h
σ_{-1}	fatigue limit in the case of stress ratio $R = -1 = \sigma_{min}/\sigma_{max}$
σ_{-1}^{T}	fatigue limit in the case of stress ratio $R = -1 = \sigma_{min}/\sigma_{max}$ and temperature T in °C
$\sigma_{0,2/100}^{T}$	stress at temperature T for creep 0.2 % after 100 h in °C
$\sigma_{0,5/1000}^{T}$	stress at temperature T for creep 0.5 % after 1000 h in °C
$\sigma_{10^{-5}}^{600°C}$	stress at temperature 600°C for creep rate of 10^{-5}%/h
σ_0	low cycle fatigue stress with non-changing stress ratio and $N = 10^4$ cycles
$\sigma_{vM} = \sigma_i$	equivalent stress (according to von Mises)
σ_{eq}	equivalent stress
σ_{res}	residual stress
σ_{limit}	limit stress
$\sigma_{lt\,s}$	long term strength
$\sigma_{tension}$	limit stress at tension
$\sigma_{compression}$	limit stress at compression
$\tau_{torsion}$	limit stress at torsion
$\sigma_{inner\ pressure}$	limit stress in the test of thin-walled tube under inner pressure
σ_m	mean stress during the cycle
σ_a	stress amplitude during the cycle
K_I	stress intensity factor at normal separation
K_{II}	stress intensity factor at transverse shear
K_{III}	stress intensity factor at longitudinal shear
K_{th}	threshold value of the stress intensity factor (at K less than K_{th}, the fatigue crack does not grow)
K_c	critical value of the stress intensity factor for brittle fracture
K_{1c}	critical value of the stress intensity factor for brittle fracture corresponding to the start of a crack in the presence of plane deformation at its tip
λ	coefficient of thermal conductivity
ρ	density
ψ	relative narrowing
K_{Ic}	stress intensity coefficient
K_t	theoretical stress concentration coefficient (according to Neuber)
S_s	true separation resistance
S_k	true break strength
T	temperature
T_K^{50}	critical temperature critical temperature at which fractured toughness samples have 50% fractured fibers

T_M	exploitation temperature limit for a metal under the following conditions: long-term strength during 1000 h not lower 250 MPa
T_{melt}	melting temperature
t_f	time to fracture
(hkl)	Miller indices (family of crystal lattice planes orthogonal to h$\boldsymbol{b}_1$ + k$\boldsymbol{b}_2$ + l$\boldsymbol{b}_3$, where $\boldsymbol{b}_i$ are the basis of the reciprocal lattice vectors
{hkl}	notation for Miller indices which is the set of all planes that are equivalent to (hkl) by the symmetry of the lattice
[hkl]	Miller indices which denotes a direction in the basis of the direct lattice vectors instead of the reciprocal lattice
<hkl>	Miller indices which denotes the set of all directions that are equivalent to [hkl] by symmetry

During the translation, the author and the translators encountered difficulties associated with the need to change the designations of elements, types of blanks and the casting method of the Russian steels and alloys. Below are given some designations - the first letter is the Russian symbol in Latin, the second one the international symbol.

C, C	carbon
Ch, Cr	chromium
N, Ni	nickel
K, Co	cobalt
Yu, Al	aluminium
T, Ti	titanium
B, Nb	niobium
W, W	tungsten
M, Mo	molybdenum
F, V	vanadium
R, B	boron
D, Cu	copper
G, Mn	manganese
S, Si	silicon
A	reduced impurity content
A or B	modification
VAR	vacuum arc remelting
VA	vacuum arc
IAR	induction-arc remelting
IA	induction arc
OI	open induction smelting
DC	directional crystallization
C	cast
SR	slag remelting
ESR	electroslag remelting

About the Author

Leonid Borisovich Getsov is Chief Researcher at the Central Boiler and Turbine Institute (at present Joint-Stock Company "I.I .Polzunov Scientific and Development Association on Research and Design of Power Equipments" - NPO CKTI). In 1953, he graduated from the Leningrad Polytechnic Institute and at the same time Faculty of Physics, Leningrad University. He defended his Ph.D. thesis in 1962. Then he was appointed as a senior researcher in the specialty "metallurgy and heat treatment of metals" in 1966. In 1979, he defended his doctoral dissertation (Dr. of Engineering Sciences). He has 65 years of experience in production engineering, research and teaching, including 20 years in higher educational institutions. He is the author/co-author of about 500 articles in Russian and foreign journals and several monographs.

Chapter 1
Operation Conditions of Gas Turbine Parts and Materials used for them

1.1 Operation Conditions of High-temperature Components of Gas Turbine Units and Damage Mechanisms During Long-term Operation and Laboratory Testing

Modern gas turbine units (GTUs) are distinguished by a variety of designs and types (open and closed cycles), working fluids (working by burning fuel or by an external source of heat, such as the heat of flue gases produced in industrial processes, etc.) and applications (stationary, transport, air craft, ship). Over the last decades, a large number of gas turbines have been installed in Russia, which are used to drive compressor superchargers of main gas pipelines and to cover the peaks of electrical loads. Some of them have an operating time that has reached 100000 h or more. Aircraft engines of industrial versions have been widely used, converting them allows to get GTU with high efficiency. The design features of aviation and transport gas turbine engines (GTE) are caused by high demands on both their mass and dimensions, and cost-effectiveness. For stationary turbines, the dominant requirement is the efficiency. A feature of stationary gas turbine engine designs with a gas temperature above 820°C is the introduction of cooling system for turbine blades. In the manufacture of ship gas turbine generators, high efficiency is achieved without cooling blade system, which is partly due to the limitation of the resource at nominal loads and loads exceeding nominal values. Some of them have an operating time that has reached 40000 h or more. Atomic gas turbines of a closed cycle use air, nitrogen, helium, and carbon dioxide as the working medium heated in the reactor.

The stress state of the main parts of gas turbines for various purposes is determined by their structural differences and the specifics of operating conditions. Most modern GTUs during the service life undergo a large number of relatively quick start-ups and operate on variable regimes, which causes special requirements. The structure of the open cycle GTU may include the following main components: combustion chambers, compressors, turbines, a regenerator, various stator components, control and regulation units. For parts of gas turbines operating at elevated temperatures (flame tubes of combustion chambers and other stator parts, turbine

L. B. Getsov, *Materials and Strength of Gas Turbine Parts*, Advanced Structured Materials 150, https://doi.org/10.1007/978-981-16-0534-5_1

blades, turbine disks, regenerator elements, fasteners, compressor blades of the last stages of compressors with high compression ratios), when choosing which to take the specifics of their stress state and features of operation of turbines for different purposes.

The parameters affecting the choice of material related to the operating conditions should include the working temperature of parts, the degree of temperature non-uniformity, the type and composition of fuel, and the degree of maneuverability of the installation. The choice of gas temperature is determined largely by whether the GTU is used as a base, peak or backup.

For stationary, ship and transport gas turbines, various fuels are used, differing in the content of impurities that adversely affect the resistance materials against corrosion. In addition to the main ingredients of the fuel entering into the combustion products (Table 1.1), the following components may fall into the flow-through part of a gas turbine plant: black carbon formed due to incomplete combustion of fuel; impurities contained in groundwater, sometimes entering fuel storage tanks or combusted with natural gas containing them; salt water from the spray, sucked into the compressor with the air in the ship gas turbines; sand - getting into the air, including during sandstorms, etc.

Table 1.1 Percentage of chemical additives in liquid fuels (n.s. - not standardized)

fuel	GOST	percentage, % (no more)			
		V	Na	S	ash
diesel fuel	305-82	-	-	0.2-0.5	0.01
engine fuel:					
- low sulfurous	1667-68	n.s.	n.s.	0.5	0.02-0.5
- sulfurous		(~0.01)	(~0.01)	1.5-3.0	0.02-0.15
oil (masut)	14298-79	n.s.	n.s.	0.5	0.3
		(0.017)	(0.013)	1.0-2.5	0.3
	10585-75	n.s.	n.s.	0.6	0.1
		(0.0005)	(0.013)	2.0	0.05
jet engine fuel	10227-86			0.1-0.25	0.03
oil for gas turbines	10433-75	0.00002	0.00005	1	0.01
	10433-75	0.00002	0.00005	2.5	0.01
destillate fuels					
- from sulfurous oil	-	0.0001	0.0006	2.5	0.01
- from low sulfurous oil	-	0.0002	0.0004	0.4	0.01

As gases of stationary GTU, various gases are also used that differ in the content of H_2S, CO_2, H_2, C_3H_8, C_2H_6, CH_4: natural deposits, artificial (from anthracite, underground gasification, semi-coked, blast-furnace, coke-refined) and associated, obtained from various oil fields. Natural and associated gases, as a rule, contain only traces of sulfurous compounds, artificial - from 0.2 to 58% in semi-coked gas (Starozhuk, 1978).

In Russia and abroad, considerable experience has been gained in the operation of aviation, transport and stationary gas turbines, whose parts were made of various materials.

1.1.1 Turbine Guide Blades

The guide (nozzle) blades are the most heated parts of the turbine with the exception of the flame tube of the combustion chamber and the gas collector. Unlike working blades, the temperature of which is determined by the average temperature of the gas, the individual guide blades of the annular grid can have a temperature increase of 50°C, and in some cases even 100-150°C above the average gas temperature. Since the cross-sectional area of the guide blades is usually larger than that of the working blades, gas temperature changes that occur during start-ups, shut-downs and variable modes, lead to thermal stresses in them greater than in the working blades. Depending on the design of the guide vane, blades are subjected to tensile stresses from the mass of the stator parts, and, most importantly, to bending stresses from the action of gas forces. However, stresses due to centrifugal forces in the guide blades are usually much less than those in the working blades.

The most typical types of damage in guide blades during long-term operation are:

- corrosive, leading to thinning of the edges of the blades, the appearance of ulcers and surface roughness (Fig. 1.1 b), reducing the aerodynamic parameters of the guide vane,

Fig. 1.1 Damage after operation in guide vane made of the alloy: a) ZhS6K, b) EI826, c) cobalt-based.

- thermal fatigue, as a result of which cracks are formed on the edges and shelves (Figs. 1.1-1.4), and in some cases, pieces of metal that fall into the flow part of the turbine are torn off,
- changes in blade geometry due to loss of stability and progressive deformation under thermo-cyclic loading (Fig. 1.5 a, b), and
- brittle fracture of blades from materials with low ductility, when foreign objects get into the flow part of the turbine (Fig. 1.1c).

The segments of the guide blades are often susceptible to fatigue damage (Fig. 1.4). Corrosive damage of the guide blades is caused by the processes of interaction of the metal of the blades with gas flowing around them and ash deposits containing various components of the combustion products of the fuel (Fig. 1.6). Aggressive corrosion processes of the blades particularly occur when using heavy fuels containing vana-

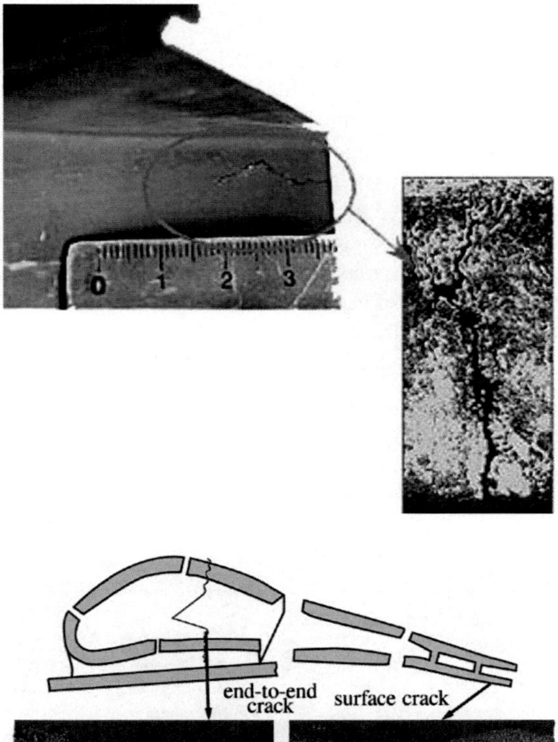

Fig. 1.2 Thermal fatigue crack in a guide vane made of the casted alloy EI893(L) (CrNi65WMoTiAl) after 22098 h exploitation in a compressor station.

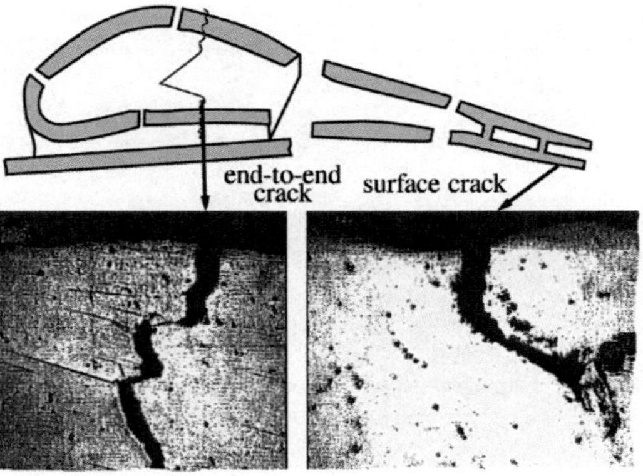

Fig. 1.3 Cracks in the guide blade in the first high pressure turbine stage made of alloy ZMI3U after 23500 h.

Fig. 1.4 Part of a guide vane of gas turbine GTK25-I made of the alloy FSX-414 with thermal fatigue cracks

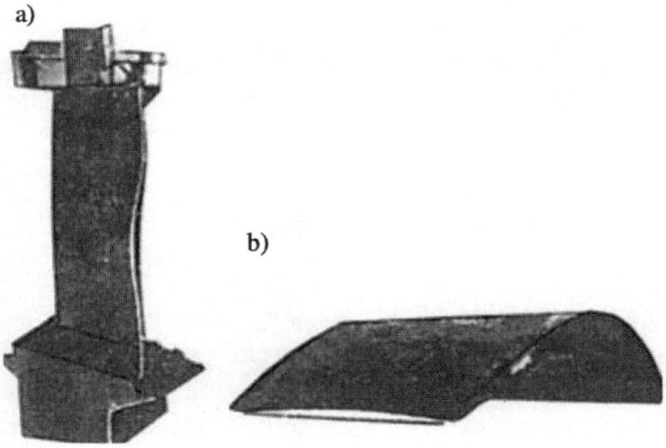

Fig. 1.5 Deflection of the boundaries of guide vanes made of the alloy: a) EI787, b) EI868

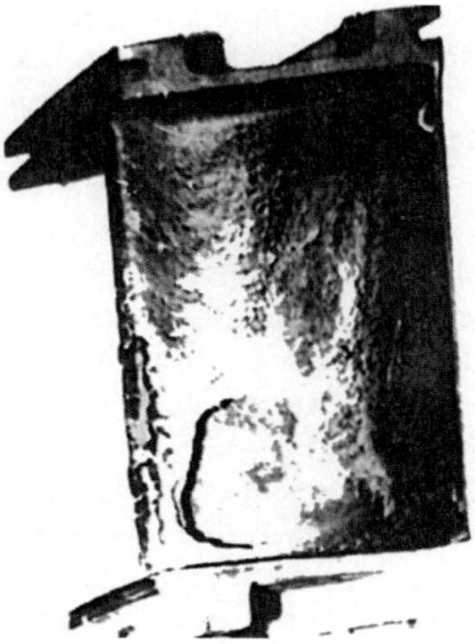

Fig. 1.6 Corrosion damage of a turbine blade made of ZhS6K as the result of overheating and corrosion-active deposits

dium, sulfur, and when seawater salts enter the turbine. The formation of thermal fatigue cracks on the edges of the guide blades during the operation of a gas turbine is observed relatively often. The reason for this is the high thermo-mechanical load of these blades, which are directly subjected to all changes in gas temperature during start-ups and changes in operating modes, and temperature gradients around the circumference and height of the channel of the flow part. For example, on the guide blades of the stage I of high-pressure turbines GT-750-100 of the production association of turbine construction "Leningrad Metal Plant" (LMP) from ZhS6K alloy, intense corrosion damage (peptic corrosion)[1] was observed, which contributed to the formation of thermal fatigue cracks on the output edges (Korsov, 1978; Nikitin, 1987).

In cases, when during the start of a gas turbine unit, the temperature of the middle sections of the blades is significantly higher than the peripheral ones, and especially when the blades have shelves that impede the free expansion of the blades, laboratory tests of the gas turbine units show the bending of output edges leading to a decrease in the flow section of the guide blades and reduction of efficiency.

1.1.2 Turbine Working Blades

Working turbine blades are subjected to centrifugal forces, which create in them variable stress distribution over the hight. The maximum stress is usually located at the root of the blade, but in the presence of a bandage it may also be in another place, for example, directly under the bandage or in the bandage itself. Under the action of gas forces, bending stresses in the blades occur, having the maximum at the blade root. Blades have a temperature variations over the height and cross section. The value of temperature gradients depends on the flow parameters and the geometric dimensions of the blades. Temperature gradients across the section increase sharply during start-ups and shut-downs, creating thermal stresses. In twisted blades, which provide better gas flow conditions and, consequently, higher efficiency than cylindrical blades, tensile stresses from centrifugal forces are distributed non-uniformly over the cross section: in the central part, stresses are higher and at the edges lower than average stresses. In the case of blades with shifted center of mass of the feather part, tensile stresses arise at the edges, caused by bending, which significantly exceed the allowable values. These local stresses due to difficulties in redistribution can cause premature failure.

The edges of the cooled blades are practically not cooled, however, since considerable thermal compressive stresses are superposed on the tensile stresses from the centrifugal forces, the total stresses are close to zero. The middle zone of such blades has a lower temperature, while the thermal stresses there are relatively small (about 10-30 MPa). The lock of the blade (fir-tree or T-shaped tail) has significant stress concentrators. Each tooth is subjected to stresses of shear, crushing and bending. A

[1] The cause of corrosion was the aggressive elements contained in the fuel (0.2-0.8 g/t V and 0.4-2 g/t Na).

great influence on the distribution of stresses on the teeth has an accuracy of their manufacturing (accepted tolerances).

In addition to the static stresses in the blades, cyclic stresses occur, under the influence of which most of the failures of the blades are caused by fatigue damage processes. This is due to the fact that the level of dynamic stresses in the blades of the designed turbine, as a rule, is unknown (there are no corresponding calculation methods). The oscillations of the blades can have relatively large amplitudes, when the frequency corresponding to some principal form of free vibrations of the blades coincides with the frequency of the disturbing force, a multiple of the number of revolutions, i.e. there is a resonance. The level of dynamic stresses under resonance conditions depends on the stiffness of the blade itself or the blade pack, on the energy losses due to oscillations (damping) and, finally, on how close the nature of the distribution of disturbing forces to the resonating form of free oscillations. It is usually assumed that there is a correlation relationship between the vibration stresses and the bending stresses σ_b in the blade, because the level of σ_b characterizes, on the one hand, the stiffness of the blade, and on the other hand, the average value of the forces of the gas flow, a certain proportion of which are the pertubation forces. Therefore, despite of the fact that bending stresses usually do not determine the strength of the blades, the value of σ_b should be reduced to a minimum. The level of dynamic stresses can be changed significantly only by the designer. The maximum possible decrease in stress concentration, increase in surface cleanness, as well as increase in structural damping by appropriately designing a locking joint, applying bandages, thickening the shelf and choosing the optimal pairing with the locking part are the most effective methods to increase the strength of the blades (Dondoshanskiy, 1978; Borovkov and Getsov, 2008).

By applying materials with a high fatigue limit or using one or another strengthening treatment for the same purpose, it is possible to increase the permissible level of vibration stresses. Some reduction in the level of vibration stresses can also be achieved by choosing materials with a high damping capacity (high vibration decrement), which significantly depends on the chemical composition: chromium steels (see Subsect. 1.3.2) have the greatest decrement, austenitic ones have the lowest vibration ratio. However, one should not overestimate the role of damping in the material as compared with structural and aerodynamic damping.

Damage of turbine blades is the most common type of failure of high-temperature parts of GTU. The causes of blades failures are foreign particles caught in the flow part, increased vibration stress levels, high cyclically acting thermal stresses, significant corrosion, changing the stress state of the material, eccentricity of the center of mass caused by manufacturing of blades and, finally, overheating, in which the metal of the blades has insufficient heat resistance.

Consider some examples of failure of the working blades when testing GTU in laboratory conditions and during the operation. Damage statistics show that the blades are often damaged as a result of strikes by external objects and fragments of damaged parts. In most cases, this leads to the formation of nicks, dents and scratches that act as cuts, causing stress concentrations and thereby reducing the fatigue resistance.

Failure of the blades from alloy EI765 in the pen under the action of centrifugal forces, as shown in Fig. 1.7, occurred due to an emergency metal temperature rise of approximately 300°C above the nominal value. The reason for this temperature increase was the incomplete combustion of fuel in the chamber and its burning on the blades. The operating time at elevated temperatures was sufficient to warm the blades and soften them due to the dissolution of the intermetallic phase, as evidenced by microstructure studies and a sharp decrease in hardness in the blade section from the base of the root to the location of failure.

Static damage of the edges and failures of the blades due to the displacement of the center of mass was observed during the operation of ship gas turbines after 3000 - 17000 h and power plant turbines after 370 - 5132 h (Fig. 1.8, Getsov et al., 1993). The blades of these GTUs were distinguished by the presence of a bandage in the first case and by the material: EI826 and EI893. In addition, the cracks on the blades of shipboard gas turbines were observed mainly on one of the kit, while cracks on the blades of the power plant GTUs were massive, which is associated with a greater frequency of deviations from the manufacturing technology. Unfortunately, during the manufacturing of these blades, there was no system for monitoring the displacement of the center of mass.

The high concentration of stresses in the T-shaped tail of EI612 alloy blades caused the formation of cracks (Fig. 1.9) during long-term operation due to the low

Fig. 1.7 Damage of blades made of alloy EI765 as a result of overheating

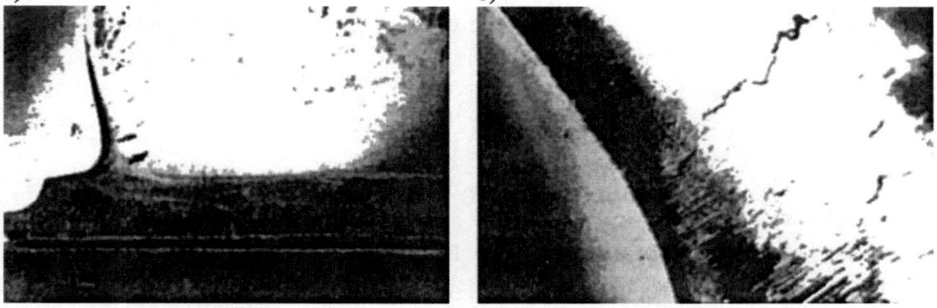

a) b)

Fig. 1.8 Cracks in the working blades of the HPT GT-100 made of EI893 alloy: a) magnification 1x, b) magnification 13x

Fig. 1.9 Cracks at the end of blades made from steel EI612

ductility of the material of the workpieces, its sensitivity to notch and high relaxation resistance of steel at operating temperature.

The operating experience of 134 GTUs (General Electric Company, USA) for more than 20000 hours showed that most of the failures are due to the performances of the material. For example, from 15 failures of blades of step I made from nimonic 80 alloy, five occurred due to chipping of the material from the feather surface due to aging at working temperatures (insufficient stability of the microstructure), and 10 due to thermal fatigue. High compressive stresses in the blades (due to the difference in the rate of temperature change on the central and outer parts of the blades) with minor scratches on the input edges led to the formation of cracks. A special case of damage in the blades is the static failure from the hole under the retaining wire (Fig. 1.10).

During the operation of gas turbines CJSC "Nevsky Plant" (NZL) on the main gas pipelines, there were also some cases of failure of rotor blades. So, after 2500-3000 h of operation of the GT-750-6 turbine, chipped shelves of working blades made from EI893 alloy were found on several units at the point of contact with the inserts, as a result of increased material brittleness.

Due to the relatively low fatigue strength of the blades of stage II GT-700-5, made of steel EI726 (fatigue strength 65 MPa for 10^8 cycles at a temperature of 600°C), some working blades failed after 600-19000 h (in 34 from 90 machines in operation).

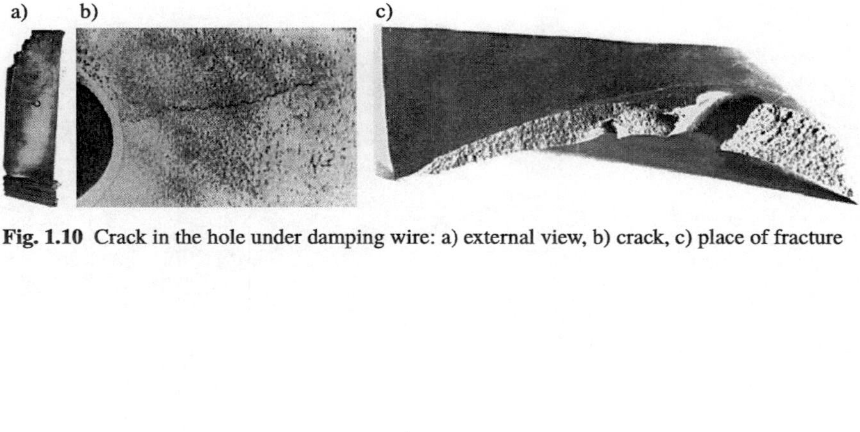

Fig. 1.10 Crack in the hole under damping wire: a) external view, b) crack, c) place of fracture

Fatigue failures are observed on the working blades of the TLP GTK-10-4, made of alloy EI893.

Operation of GT-750-100 units (JSC Leningrad Mechanical Plant) working on liquid gas turbine or diesel fuel under conditions of relatively frequent start-ups (every 3-4 h) was also accompanied by the appearance of individual damage of working blades (Korsov, 1978). On the blades of step I of EI893 alloy, fatigue cracks along the neck of the T-shaped tail were found, on the blades of step II - brittle fractures at the hole for the damper connection. Another example of fatigue damage is the failure of the blades of stage II from the alloy EI765 of the gas turbine GTU-15 after 320 h of laboratory testing (Fig. 1.11). Inspection of the rotor made it possible to establish that the cracks in the root of the first tooth of the fir-tree shaped shank were found on four blades of stage II, while no cracks were observed in blades of the stage I made of alloy EI893. An analysis of the accident showed that the blades failed by following reasons: the type of conjugation of the feather blades of stage II (trough) with the shank (this design was chosen in contrast to the design of the blades of stage I for technological reasons); shutting down the five nozzle channels while finalizing the installation in order to coordinate the operation of the turbine and compressor; the latter created a partial gas supply, which led to an increased vibration of the blades. The reason that only the blades of stage II were failed was due to the fact that the frequency diagram for them was less favorable (the reserves between the working and resonant speeds were less) than for the blades of stage I.

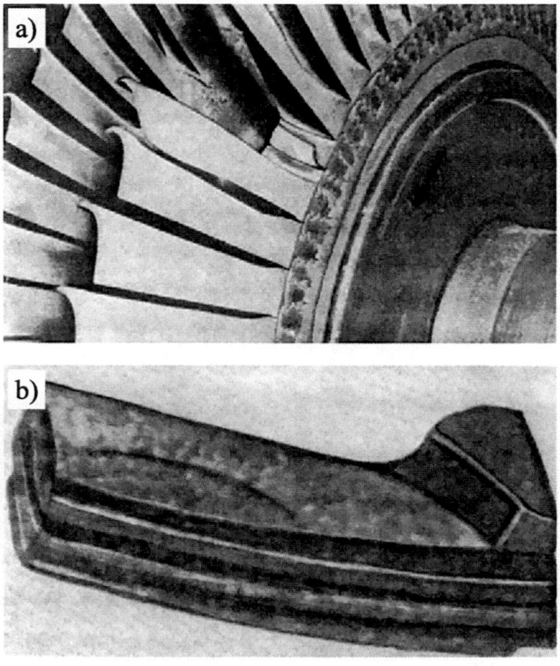

Fig. 1.11 Fatigue damage of blade made of EI765 alloy: a) exterior view of the rotor, b) break of the blade

Thermal fatigue damage of the blades, shown in Figs. 1.12 and 1.13, have a different character due to the different microstructure of the metal. In the blades of alloy EI765 with a fine-grained microstructure, a crack was formed (see Fig. 1.12), the development of which would eventually lead to a failure in the blade pen. In the blades of steel EI726 with a coarse-grained microstructure cracks formed at the edge of its chipping (Fig. 1.13).

After 20000 h of operation of gas turbines in marine conditions, corrosion damage to the output edges was detected (Fig. 1.14). There have been cases when corrosive damage in bandage shelves of blades due to the ingress of sodium salts into the fuel weakened the section so much that cracks formed in the shelves (Fig. 1.15). Corrosive damage due to sulfide oxide corrosion is often observed in in the blade pens of stationary turbines, especially in the absence of protective coatings (Fig. 1.16).

Fig. 1.12 Thermal fatigue damage of a blade made from the alloy EI765

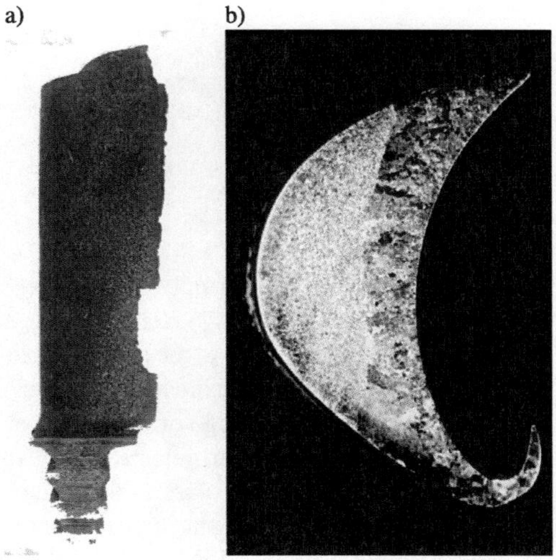

Fig. 1.13 Thermal fatigue burning of a part of the blade made from steel EI726: a) exterior view, b) macrostructure of the cross section

Fig. 1.14 Corrosion-erosion damage of the end of the blade made from alloy EI826

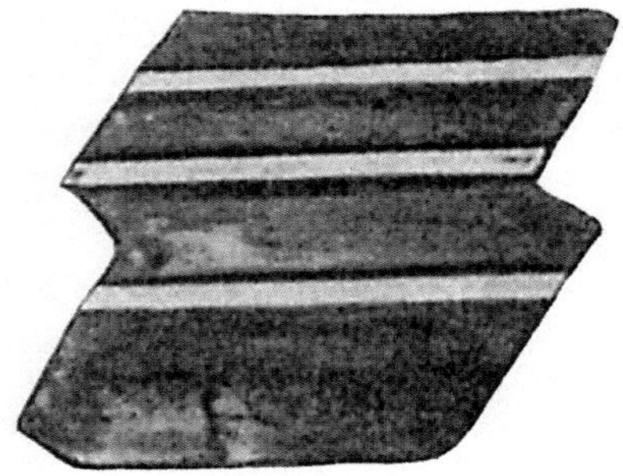

Fig. 1.15 Cracks of the blade tracking made from alloy EP220 as the result of static loading under corrosion damage

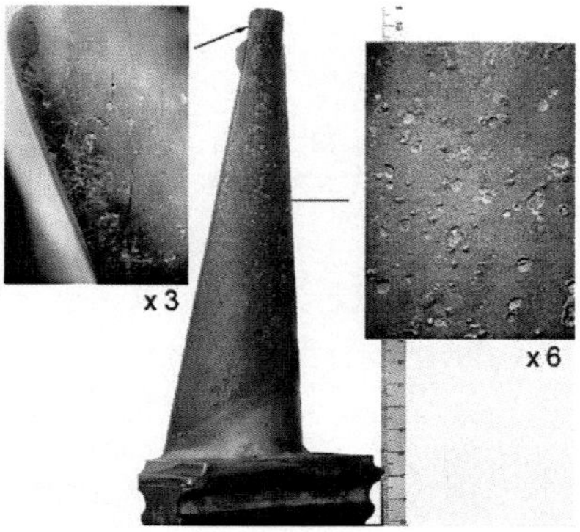

Fig. 1.16 Ulcerated damage of the turbine blade surface

In the cooled blades, thermal fatigue cracks are fairly often formed at the holes on the input edges (Fig. 1.17). Failures are observed also in the blades of high-temperature gas turbines due to the exhaustion of long-term strength under conditions of corrosive environmental exposure and overheating (Fig. 1.18). In Figs. 1.19-1.21 photographs of fatigue damage of the blades are presented.

Analysis of failures in various stator and rotor blades during operation allowed different damage types to be classified for machine-based purposes (Table 1.2). This classification of damage in the blade apparatus is used in the development of a database for reliability. In addition to the classifier indices A1-5, B1-4, C1-3, D1-3,

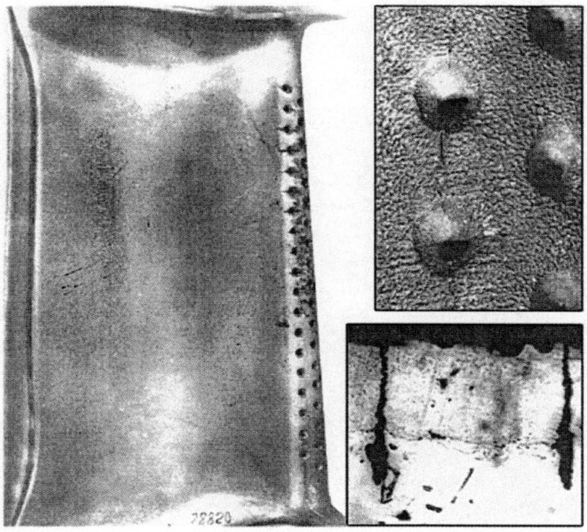

Fig. 1.17 Thermal fatigue cracks near the perforation on the inlet edges of the cooled turbine blades (on the inlet edge and on the trough, traces of a burned surface are visible from the effects of a high-temperature flow of products of combustion of fuel)

Fig. 1.18 Photographs of the whole (left) and destroyed (right) working blades GTK-25-I from the alloy IN738 after 22759 h of operation at the compressor station

Fig. 1.19 Fatigue damage of the blades from torsional vibrations

Fig. 1.20 Fatigue failure of the working blade of the turbine from the alloy EI893 during long-term operation. 1–failed part, 2–fatigue line, 3–start of failure

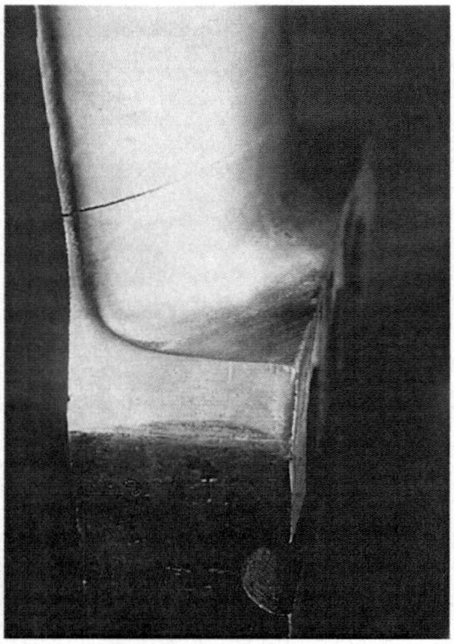

Fig. 1.21 Fatigue failure of the blade model

E1,2, F1-5, G1-4, H1-3 the serial number of the machine, the operation life at the time of detection of damage, the installation site, the type and composition of the fuel used, the date and nature of the change to the drawing are included (Getsov and Rybnikov, 1993).

1.1.3 Turbine Disks

When considering the stress state of the disks, one should distinguish between the web of disks and the protrusions of the locking part. The complex stress state of the blade is caused by the presence of centrifugal forces and temperature gradients (along the radius and axis), resulting in radial, tangential and axial stresses (in the hub), which are usually much smaller. Temperature gradients along the radius in some disks with stationary modes can reach 400°C, and during start-up 600°C.

Table 1.2 Classification of blade damage

Feature	Variants
A. Version of damage	1. Corrosion 2. Erosion 3. Static 4. Thermal fatigue 5. Fatigue
B. Type of damage	1. Complete destroyed 2. Macrocracks 3. Microcracks (on and under surface, fragmentation) 4. Change of the geometry (intensive corrosion)
C. Location of defect	1. Input edge, output edge, through 2. Bandage shelf (abutment surface, outer tootth) 3. Tree connection (tooth base, tooth, stem)
D. Number of blades with damage of the given GTU	1. Individual (less then 5%) 2. Multiple (10-20%) 3. Mass (50-100%)
E. Methods of damage observation	1. Destructive (fractography, metallograhy, mechanical testing) 2. Nondestructive (visual, luminescence, weight, hardness)
F. Specific design and type of the blade material showing damage	1. Stator (a) or rotor (b) 2. Cooled (a) or uncooled (b) 3. With bandage shelf (a) or without (b) 4. Twisted (a) or untwisted (b) 5. Made from casted (with equiaxed (a) or oriented (b) crystallization) or strained (c) alloy
G. Reasons for damage	1. Structural 2. Technological 3. Manufactural 4. Exploitation
H. Conclusions for further exploitation	1. No influence on the lifetime prediction 2. Limitation on the exploitation time 3. Immediate change of the blades

The nature of the stress state of the disk during operation varies both as a result of the redistribution of stresses due to creep, and during start-ups and shut-downs, which cause sudden changes in thermal stresses. Difficulties arise in assessing the stress state of the disks at the roots of the grooves intended for connecting blades and located along the rim of the disk, as well as near the holes located on a certain radius for connecting the disk.

Stress analysis in the protrusions of the disk is usually limited to determining the average tensile stresses from centrifugal forces (including those from the blade load), and in some cases the analysis is supplemented with an estimate of the strung stresses on the contact surfaces and the bending stresses in the teeth, assuming the uniform distribution of stresses over teeth. However, in reality, the stress state in the protrusions of the disk is much more complicated. Significant complications of the stress state occur due to stress concentration in the grooves, uneven clogging of the installation gaps, differences in the linear expansion coefficients of the metal of the blades and disks, axial bending under the action of the axial temperature gradient and cyclic stresses generated from oscillating blades. Additional stresses in the disk can occur in the case of anisotropy of the mechanical and physical properties of the material. Depending on the design of the disk and the conditions of the GTU operation, the relations between the stresses acting in different places of the disk vary significantly: in some places the radial stresses take maximum values while in others the tangential stresses dominate. Rapid start-ups and shut-downs of the machine sometimes cause so significant thermal stresses that they prevail over the stresses from centrifugal forces. When working on pulverized fuel, the installation clearances in the slots of the disks become clogged, as a result of which the conditions of heat transfer from the blades to the disks, as well as the rigidity of the blades. All this affects the stress state of the disk.

During the start-up performed in a relatively short time, in the disks and rotors of turbines the most loaded area adjacent to the axial hole is practically cold. Metallurgical defects, usually found in the center of forgings, can play the role of stress concentrators. During operation of the turbine, as well as compressors of aviation gas turbine engines, oscillations of the bladed disks occur, which are transmitted from blades clamped during rotation of the rotor in the rim of the disk.

In contrast to the disks of aircraft and transport gas turbines, disks of stationary turbines of high power have very large diameters and thicknesses. Therefore, in them, as a rule, there are no significant vibration stresses and they are usually made of iron-based alloys that are easier to process by pressure. This was possible by the use of blades with long shanks, which allow to reduce the temperature of the periphery of the disks in the area of the locks.

Failures of turbine disks, deflectors and impellers during operation can be caused by static loadings (Figs. 1.22 - 1.25), thermal cycles (Figs 1.26 - 1.28) and vibrations (Figs. 1.29, 1.30). Table 1.3 summarizes the working conditions of several disks and the causes of their failure.

Other types of damage are also known:

1. breaks along a cylindrical section (from radial stresses);
2. damage caused by material defects in the central part of the disk that were not detected during the inspection and significantly reduced the ductility;
3. failure of the weld seam of composite disks due to welding defects;
4. formation of cracks in the holes for the locking pins of the working blades (due to intergranular corrosion and thermal fatigue);
5. failures of gear projections due to uneven application of load (due to errors in the manufacture of grooves on machines, insufficient compensation of the bending of the blades by gas forces and distortion of the core part of the blades, leading to a shift in their center of gravity);

a) b)

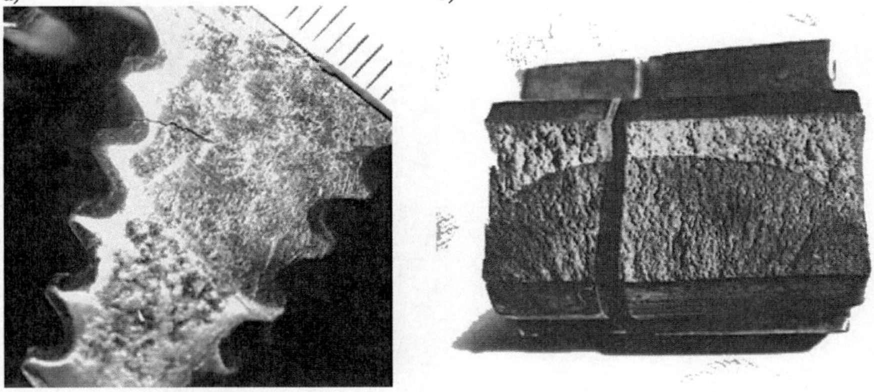

Fig. 1.22 Damage of a disk made of steel EI481 at protrusion groove base: a) view of the crack, b) place of fracture

a) b) c)

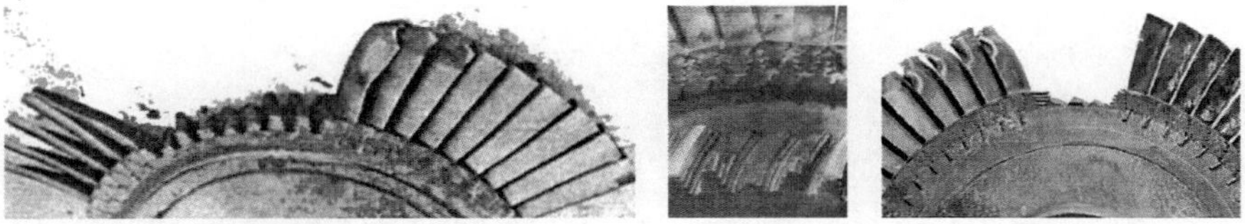

Fig. 1.23 Destruction of several disks made of EI612 (a, b) and EI481 (c)

Fig. 1.24 Failure of a disk with a forged crack

a) b)

Fig. 1.25 Failure of deflectors. a) alloy EI698, cut under the action of thermal stresses, b) steel EI481, rupture under the action of centrifugal load

Fig. 1.26 Intergranular cracks in a disk made of steel EI612 in the grooves for locking plates at the tearing out of a piece of the disk

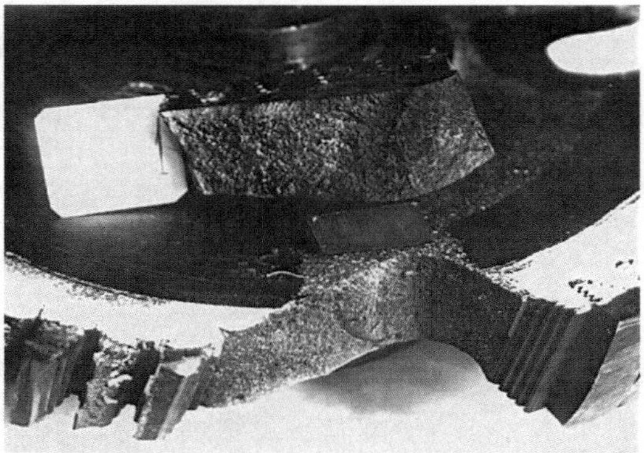

Fig. 1.27 Failure of a full-scale disk of a shipboard gas turbine made of steel EI612 under test conditions for thermal fatigue with rotation at working revolutions

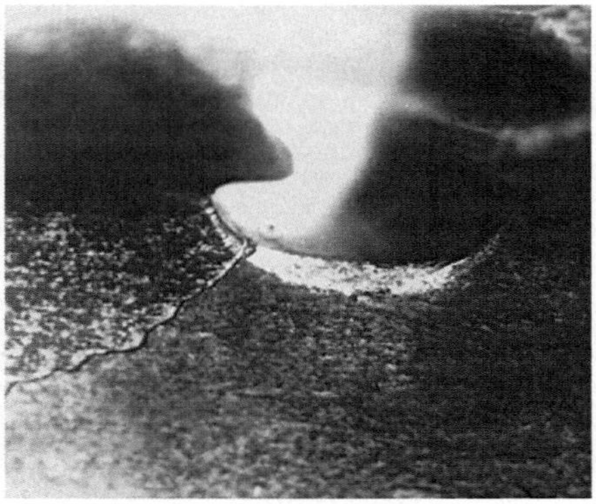

Fig. 1.28 Formation of thermal fatigue cracks on the rim of the disk

Table 1.3 Working conditions and reasons of damage of disks

Figure	Location and type of damage	Material	Working condition	Reason of damage
1.22	Intergranular macrocrack formation at the base of the groove of the protrusion of the disk	EI481	Exploitation at elevated temperature	High static stresses at elevated temperature
1.23a	Break of three ledges of a disk	EI612	Exploitation during 9797 h at temperature 700°C of the protrusion metal	Destruction of disk made of notch sensitive material as a result of superposition of bending stresses and tension in stress concentration zones
1.23b	Break of three ledges of a disk of second stage at the base of the grooves with intergranular distribution of kink at middle protrusion	EI612	Exploitation during 1250 h at temperature 650°C of band, 300 launches took place	Thermal stresses in the lock connection with wedged locking plate, pull out of the middle protrusion leading to nonsymmetric loading and destruction of neighboring protrusion
1.23c	Break of three ledges of a disk	EI481	Exploitation during 3000 h at temperature 460-550°C of band	Not established
1.24	Destruction of disk	EI698	Bench test during 200 h at temperature 600°C of band	Forging crack not established by ultrasound control
1.25a	Crack in the protrusion of the deflector	EI698	Bench test, to reduce leakage of cooling air the annular protrusion of the deflector fits tightly into the groove of the disk	High temperature stress cut as result of different temperature of disk and deflector during lunch
1.25b	Intergranular cracks in the canvas of the deflector and his destruction	EI481	Exploitation during 3500 h, maximum temperature of deflector 600°C, 800 launches took place	Not established
1.26	Intergranular cracks in the grooves for the locking plates and the tearing of a piece of disk	EI612	Multiple lunches and stops	Thermal-fatigue destruction during long-term exploitation
1.29a	Impeller destruction	EI572(L)	-	Fatigue destruction
1.29b	Impeller destruction	EI787	Exploitation during 507 h at temperature 550-600°C	Fatigue edge crack developing during exploitation under high static stresses

a) b)

Fig. 1.29 Failure of centrifugal wheels of turbines from steels EI572(L) (a) and EI787 (b)

Fig. 1.30 Failure of the impeller turbine aviation GTU from the combined action of repeated static and vibration loads

6. the formation of cracks in eccentric holes for mounting disks;
7. fatigue damage of the teeth protrusions due to vibrations transmitted from the blades;
8. a cut of the annular protrusions and the formation of cracks in the grooves mating with them under the action of thermal stresses.

Disk failures may also occur due to the formation of increased deformations, even if they do not lead to excessive damage. For example, noticeable residual deformations were found after long-term operation in GTU 700-4 disks, as well as GTU-3 disks made of EI726 and EI415 steels, which have a lower resistance to deformation at working temperatures. During manufacturing of disks from blanks that have non-uniform mechanical characteristics for several reasons, the formation of asymmetric deformations is possible, leading to increased rotor vibration. This was observed on a disk of the stage III of the alloy EI607A with bench tests of GTU-6A.

Failures of deflectors (Fig. 1.31) were observed in cases of increased static and thermal cyclic stresses. Stress concentrators are possible causes of disk damage under low cycle fatigue loading conditions (Fig. 1.32). It was found that the use of substandard corrosion inhibitors led to corrosion cracking of the disk with residual stresses under long-term storage conditions in a corrosive-active environment (Fig. 1.33).

One of the specific types of damage are cracks found in the rotor disks of high pressure turbines (HPT) and low pressure turbines (LPT) made of GTK-10-4 gas turbine units from EP428 steel with a life of 100000 h or more. By the middle of 2008, 113 gas pumping units GTK-10-4 were diagnosed directly at the compressor stations during the overhauls, i.e. about 16% of the fleet of GTK-10-4, which are in operation at OJSC "Gazprom" (Kryukov et al., 2009). When inspecting by color flaw

a)
b)

Fig. 1.31 Failure of the deflector on the disk of the ship GTU: a) deflector disk destruction, b) destroyed deflector element

a)
b)

Fig. 1.32 Low-cycle fatigue cracks in the zone of stress concentration of the turbine disk of an aircraft engine: a) general view, b) view of crack

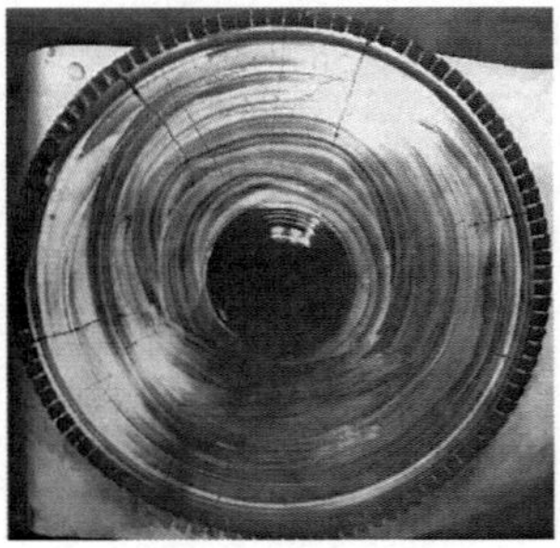

Fig. 1.33 Stress corrosion cracks in the disk of steel EI481 after long-term storage

detection, 22% of the disks (27 HPT, 21 LPT) from 221 examined with different operating time, more than 200 cracks were found on the inner surface of the fir groove in the cavity under the first tooth of the interphase protrusion and on the surface of the trapezoidal groove under the lock plate (Fig. 1.34). Several cases of cracks on the end surfaces of the disks and one case of through cracks around the second tooth with access to the end surface were recorded (Fig. 1.34f).

In Fig. 1.34b it is visible, as the crack comes through, coming to the surface of the trapezoidal groove under the locking plate. First it starts at the corners of the trapezoid, then it occupies the entire surface of the groove (Fig. 1.34c). In Fig. 1.34d) the indicator trace of the crack is visible with the appearance of cracks on the disk end surface. Figure 1.34e shows the development of cracks and chipping of the metal

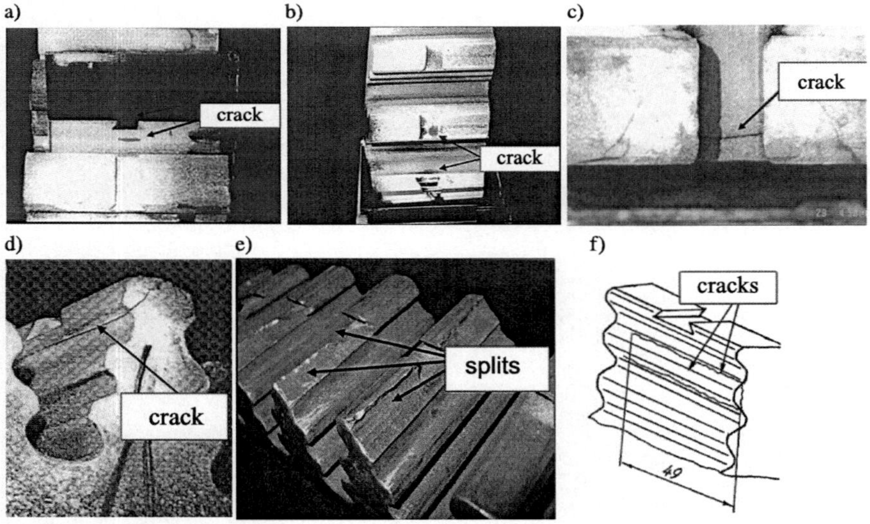

Fig. 1.34 Appearance of indicator traces of cracks of various disks of the HPT and LPT GTK-10-4 gas turbine after long-term operation, detected by color flaw detection

of the comb from two sides on one interphase protrusion (left and right along the gas combustion products).

The occurrence of cracks on the end surfaces of HPT disks (for LPT disks such cases are extremely rare) was observed both from the entrance side and from the exit side of the gas combustion products. Moreover, the development of cracks was sometimes observed from two sides on one interphase overhang (to the left and to the right along the gas flow). Perhaps this is due to the design features of the installation of blades in the HPT and LPT disks. In HPT disks, blade shanks are located in fir-tree shaped grooves at a certain angle to the axis, in LPT disks - parallel to the axis of the turbine.

Fractographic analysis revealed the fatigue nature of the initiation of all cracks. The initial nucleation of a crack in the LPT disks occurs in the area under the first tooth of the interphase protrusion (Fig. 1.35a), then it goes to the surface of the trapezoidal groove 3-4 mm from the edge just at the place of installation of the stopper (Fig. 1.35b). The crack under the first tooth is about 5 mm below the groove plane under the stopper, it is located in the middle, symmetrically from the edges, and has as a rule dimensions of 5-20 mm and the maximum opening width in its central part. Crack nucleation in HPT disks is the same as in LPT disks, i.e. under the first tooth with the exit then to the surface of the trapezoidal groove (Fig. 1.35a, b). But at the same time there are following features. First, the cracks are located

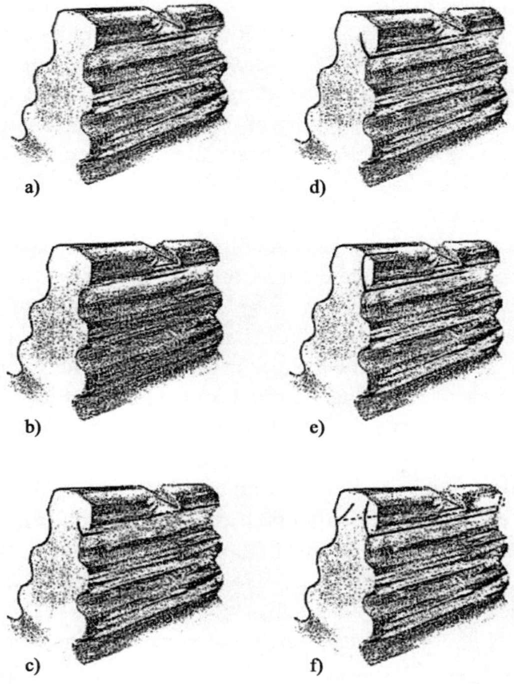

Fig. 1.35 Schematic representation of the type of detected cracks in the disks of various specimens of gas turbines GTK-10-4

both on the left and right along the gas flow. As already mentioned, the development of cracks was sometimes observed from two sides on the same interphase overhang, to the left and to the right along the gas flow (Fig. 1.35f). Second, the crack can be located symmetrically and asymmetrically, move both towards the gas inlet and towards the gas outlet, can have an exit to one of the end surfaces along the inlet or outlet (Fig. 1.35d, e) or both (Fig. 1.35f), i.e., have a length of from 5-10 to 62 mm. Third, at the end surfaces, through cracking can also be observed along the upper surface of the ridge in the direction of the trapezoidal groove (Fig. 1.35e). Fourth, the initiation of a crack can begin under the first tooth from the end surface (Fig. 1.35c) with further development towards the trapezoidal groove, including the upper part of the ridge (Fig. 1.35d, e). Fifth, with a further increase in through cracks, the upper part of the disk ridge may be chipped, as shown in Fig. 1.35e).

1.1.4 Stator Components

One of the most important stator parts of GTU are combustion chambers. The greatest number of problems with the combustion chambers during operation is associated with the flame tubes. During the operation process of GTU the flame tubes of the combustion chambers are exposed to very high temperatures, the distribution of which is strongly non-uniform. Air cooling of the flame tubes leads to a significant decrease in the temperature of the metal, keeping the temperature field non-uniform (some parts of the combustion chambers in modern gas turbines may have temperatures up to 1000-1200°C and even more). The rate of the metal temperature decrease depends not only on the parameters of the cooling air, but also on the wall thickness, the thermal conductivity of the metal and the amount of deposited soot. Therefore, the combustion chamber is usually made from sheet metals. Non-uniform and cyclic changing temperature leads to the occurrence of significant thermal stresses causing the formation of thermal fatigue cracks and warping, and the degree of warping for materials with a higher coefficient of thermal expansion is greater than for materials with a lower coefficient of thermal expansion. In addition, various stress concentrators (holes, defects in welds) can contribute to the formation of cracks during long-term operation of the flame tubes under variable loading conditions.

Microstructural changes in the course of long-term operation can lead to embrittlement of individual materials of the flame tubes, which contributes to their cracking under the action of various loads. Fatigue failures of the metal were observed in the transition tubes of power machines and gas collectors of gas turbine generators. For example, during the trial operation of a GTU with a capacity of 150 MW at the transition nozzles, after 63 h cracks were found on the flange with teeth, along which the nozzle was attached to the guide vane, Fig. 1.36. The length of individual cracks reached 230 mm (Fig. 1.37). Fractographic analysis revealed the fatigue nature of the initiation of all cracks.

The GTU regenerators serve to increase the efficiency of the installation by using the heat of the exhaust gas by heating the air supplied from the compressor to the

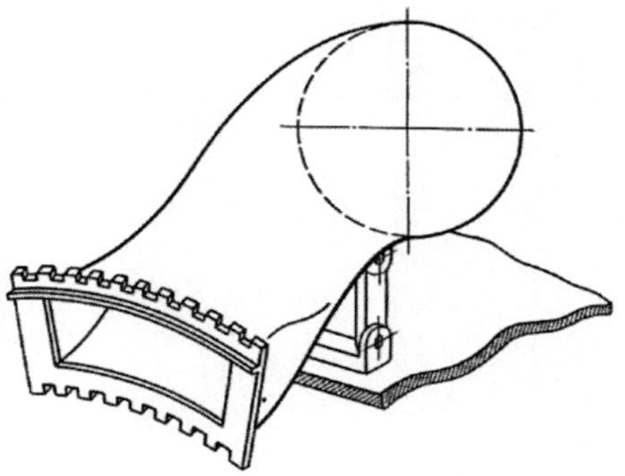

Fig. 1.36 Design of the suction inlet

Fig. 1.37 Cracks in branch pipe

combustion chamber of the turbine. Ensuring high heat transfer from gas to air is achieved in the most common designs of regenerators using thin-walled (sheet or tubular) components through which the heated air and exhaust gases move in countercurrent.

Air regenerators of gas turbines operating in the marine environment, contains salts of sea water. Saltwater salts deposited in the regenerator are moistened during the stand-by regimes due to the hygroscopicity of magnesium chloride and calcium chloride. As a result, the following types of corrosion damage of the metal of the regenerator are possible: intergranular corrosion (MCC) (during stand-by modes), for example in nickel-chrome steels during prolonged exposures at temperatures above 500°C; ulcerative, crevice and contact corrosion (during parking) and corrosion cracking. Combustion products containing vanadium and sulfur are also corrosive.

Various housing components, including rims of guide vanes and gas collectors undergo thermal loads during operation, which in some cases lead to the appearance of irreversible deformations, and sometimes to the initiation of cracks (Figs. 1.38 and 1.39). The consequences of irreversible deformation of stator parts are sometimes disastrous. The formation of thermal fatigue cracks in the nozzles of the combustion chamber and in the flame section of the combustion chambers is shown in Figs. 1.40 and 1.41.

Fig. 1.38 Thermal fatigue crack in the rim of GTU turbine guide made of alloy EI868

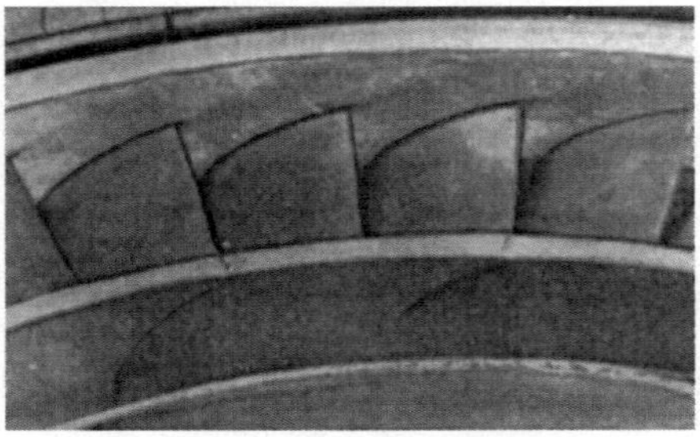

Fig. 1.39 Thermal fatigue crack in a guide vane

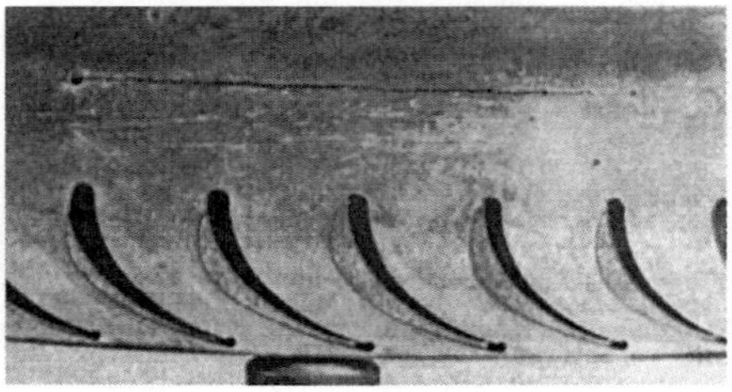

Fig. 1.40 Thermal fatigue cracks of the combustion chamber nozzles (Anurov and Fedorchenko, 2004)

1.1.5 Compressor Blades, Disks, and Housings

The level of static stresses in the blades of axial compressors of the GTE is usually relatively small. Significantly large stresses arise in the blades due to the bending moment generated by the air flow forces. In some cases, to compensate for these bending stresses, the blades with offset of the center of gravity are designed. The stress level

Fig. 1.41 Thermal fatigue cracks of the flame part of the combustion chamber (Anurov and Fedorchenko, 2004)

in the locking connections of the fir-tree and dovetail shape is also rather low. The level of vibration stress can reach such large values that the crucial characteristic is fatigue resistance. It should also be taken into account that compressor blades, operating at moderate temperatures (up to 300-500°C) may be exposed to corrosive and erosive effects from impurities contained in the air drawn in by the compressor (sand, dust and spray of seawater). In some cases, the erosion of the blades reduces the chord by 5-10 mm, which dramatically changes the characteristics of the compressor (Fig. 1.42). An attention should be paid to the fact that corrosion damage accumulates not only during operation, but also in parking modes, during which moisture condenses on the blades.

In compressors of machines in which there are no anti-icing devices, fatigue failures of the blades are often observed during periods when the outside air temperature is 0. . .+5°C. At these temperatures there are ice deposits on the inlet guide vane, which causes disturbances - the source of failure of the blades.

Typical damage forms of GTE compressor blades are nicks from external objects falling into the flow part. These nicks, as well as corrosion damage, can significantly reduce the vibration strength of the blades. The nature of the main oscillatory phenomena causing vibration of compressor blades, and ways to reduce the level of dynamic stresses in them are discussed in Dondoshanskiy (1978); Borovkov and Getsov (2008).

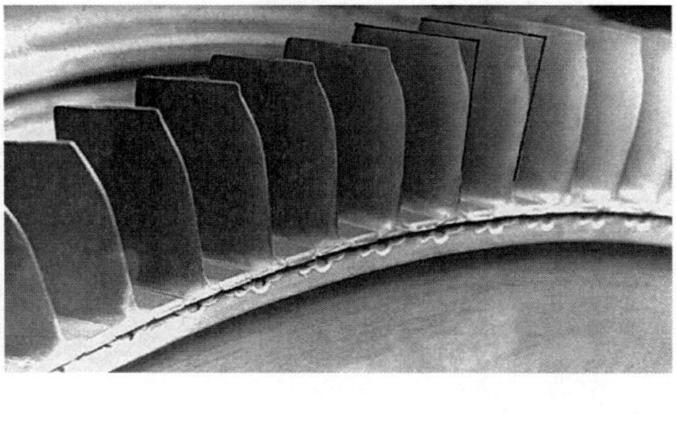

Fig. 1.42 Abrasive wear of compressor blades after long-term operation in a dusty atmosphere

The conditions of static loading of the blades of centrifugal compressors differ significantly from each other for wheels of different designs (half-open or with a covering disk, with thin or thickened disks, with weakly and strongly curved disks). The problems of corrosion damage and dynamic loading also exist for centrifugal compressor blades.

Fatigue failures of blades of axial compressors during laboratory tests of gas turbines during their fine-tuning and accelerated testing, as well as in some cases and during long-term operation are the most common type of damage. A typical fracture of blades of steel 20Ch13 due to corrosion fatigue, which occurred under the action of torsional vibrations, is shown in Fig. 1.43.

Figures 1.44-1.46 show various types of damage in compressor blades, and Figs. 1.47-1.49 illustrate damage in compressor disks according to Anurov and Fedorchenko (2004). Figure 1.50 presents a photograph of the consequences of a titanium fire resulting from friction of blades and a body made of titanium alloys.

1.1.6 Reduction Gears

The most common damage in gears are wear, fretting-corrosion, fretting-fatigue, chipping and tooth chipping (see, for example, Fig. 1.51). In some cases, fatigue

a)

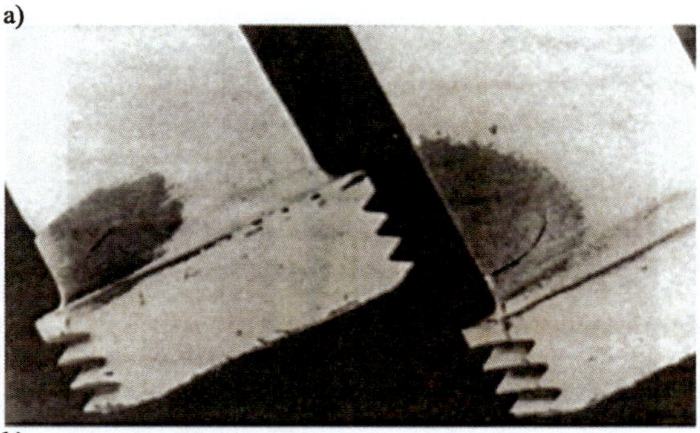

b)

Fig. 1.43 Crack (a) and kink (b) in compressor blades

Fig. 1.44 Fatigue failure of blades of the compressor guide device, caused by excitation from the blades of the impeller

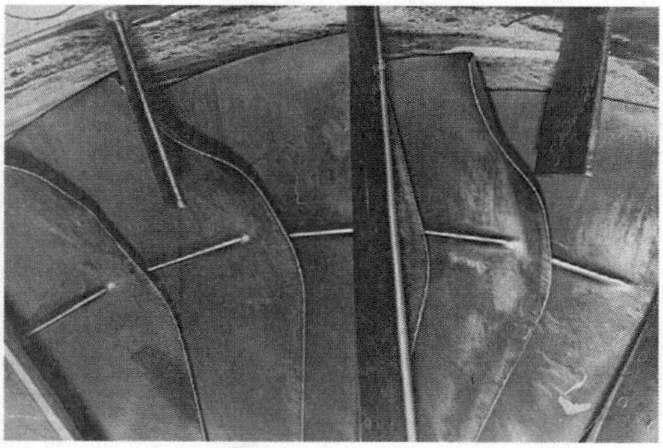

Fig. 1.45 Bending of compressor blades after entering the inlet poultry engines

Fig. 1.46 Fatigue damage of a compressor blade made from the alloy VT-3-1

Fig. 1.47 Fatigue failure of the compressor casing with the onset of the development of fatigue cracks from the stress concentrator in the form of rivets (Anurov and Fedorchenko, 2004)

Fig. 1.48 Fatigue failure of a compressor disk caused by increased circumferential non-uniform pressure field (Anurov and Fedorchenko, 2004)

Fig. 1.49 Crack at the base of the interphase protrusion of a titanium alloy compressor disk (Anurov and Fedorchenko, 2004)

Fig. 1.50 Consequences of a titanium fire in a compressor (Anurov and Fedorchenko, 2004)

failures of gears are also encountered (see Fig. 1.52, Anurov and Fedorchenko, 2004). Typical damage forms in components of superchargers are described in Kryukov et al. (2008).

1.2 Requirements for the Materials of Gas Turbine Units Components

Analysis of the stress states in GTU parts and their operating conditions allow us to identify a set of necessary requirements for the materials. It is known that the best performance of loaded parts is ensured with a combination of high yield strength,

Fig. 1.51 Chipping contact surfaces of gears

Fig. 1.52 Example of fatigue damage of a gear (Anurov and Fedorchenko, 2004)

long-term strength and high deformation capacity (plasticity) of the material. However, in most cases, such a combination of strength and plastic properties cannot be achieved. In this regard, in each particular case, the minimum permissible level of plasticity should be established, which guarantees reliable work of the part in the machine. Unfortunately, the question of the minimum allowable levels of plasticity is still debatable. The advantages for ensuring high performance of parts are those materials that have the greatest durability at the stage of crack growth, i.e. have higher crack resistance characteristics.

1.2.1 Turbine Disks

The performance of locking protrusions of turbine disks depends on such material characteristics as elastic limit, plasticity, creep resistance, long-term strength, sensitivity to stress concentration, Young's modulus and thermal expansion coefficient (compared to thermal expansion coefficient of metal of blades), as well as heat resistance in contact with fuel combustion products. To ensure high strength of the lock part, it is desirable that the materials along with high long-term durability have a high long-term ductility, which allows to prevent brittle fracture. With sufficient ductility, creep of the material plays a positive role, since it leads to the redistribution of stresses, as a result of which the level of maximum stresses decreases and the efficiency of the lock joint increases. The fatigue strength of the material may play a role in the strength of the interphase protrusions of the disk operating under the vibration loads transmitted from the blades.

To ensure reliable operation of the disk, the material must resist the action of heat changes, i.e. have a certain heat resistance. The values of its long-term strength and creep resistance should be such as to prevent the failures of the disk and the appearance of excessive deformation. The requirement for long-term plasticity is caused by the need to ensure the strength near the stress concentrators (holes, fillets) located in the heated part of the disk blade. For disks made of pearlitic steels, the requirement of increased resistance to brittle fracture is also made.

For long-life turbine disks, an important requirement is the stability of the structure and properties of the material over the entire range of operating temperatures. The softening of the material can lead to the appearance of plastic deformations in the disk, which, by changing its stress state, will lead to a noticeable decrease in working capacity. Additional hardening phase precipitations can lead to a decrease in the volume of the material (negative creep), as a result of which the stresses in the disk are redistributed and can increase in some positions.

The requirements for relaxation resistance are also sometimes imposed on the material of the turbine disk. These requirements depend on the working conditions. Low relaxation resistance of the metal can cause the appearance of permanent deformations even at stresses lower than the elastic limit. With repeated start-ups characteristic for transport and aviation gas turbine engines, this can lead to thermal fatigue failure of the disk. Therefore it is desirable that the material has a high level of relaxation resistance. In the disks of stationary turbines there are only few start-ups and shut-downs during the operation. A decrease in the level of thermal stresses as a result of the relaxation process leads to an increase in working capacity. In this case, it is desirable that the relaxation resistance should be reduced.

For multi-mode GTUs, the requirements for heat resistance, creep resistance, and relaxation resistance depend on the nature of temperature and stress changes during operation. When choosing a particular grade of alloy for the disks, the designer has to consider their manufacturing and cost. In those cases when mastering the manufacturing technology for disks made from a specific alloy requires a long time and high costs, the decision of its use should be made depending on the expected serial production of the designed GTU.

For turbine disks of various purposes, the above material requirements cannot be formulated as a list of specific values for yield strength, creep, long-term strength, ductility, resistance to thermal and mechanical fatigue, relaxation, tendency to brittle fractures, the number and size of acceptable metallurgical defects, critical stress intensity factor values under cyclic loading, etc. Nevertheless, some ideas about the mechanical properties that developed materials of GTU disks of various types must possess, have been established.

In recent years, the technology of hot isostatic pressing (HIP) of powders of hard-to-deform materials is used for the manufacture of high-loaded disks operating at high temperatures. The quality of powders, their purity, the composition of the protective atmosphere, the modes of pressing and heat treatment of the disks after processing should ensure, on the one hand, the properties of the material of the disks at the level of the properties of samples from the alloy of the same composition, and on the other hand - the absence of defects capable to cause failures.

1.2.2 Turbine Working Blades

The choice of material for blades is usually made according to the characteristics of long-term strength at operating temperatures, which should provide the necessary margin of safety in relation to the maximum tensile stresses. In cooled blades, it is usually not possible to significantly reduce the temperature of the edges. Therefore, for these blades, in addition to heat resistance, one of the important requirements for metal is heat resistance in the combustion products of the corresponding fuel at temperatures close to the gas temperature. The requirement of heat resistance is given, of course, also for the metal of non-cooled blades. It is precisely in connection with this requirement that considerable difficulties arise when choosing the material for the blades of shipboard gas turbines working in contact with sediments, which include aggressive salts of sea water. A similar situation arises with energy GTU operating in conditions of polluted air and using contaminated fuel.

Certain requirements are also imposed on the resistance of the metal to thermal fatigue for gas turbines operating under conditions of fast start-ups and shut-downs. As noted above, damage statistics show that the blades are often damaged as a result of strikes by foreign objects and fragments of damaged parts. In most cases, this leads to the formation of nicks, dents and scratches that act as cuts, causing stress concentration and thereby reducing fatigue resistance. The high probability of impact damage of the blades causes increased demands on the sensitivity of blades to notches under the conditions of static and fatigue loads. The material of the blades with bandages must have a certain relaxation resistance. This requirement is caused by the need to maintain a certain tension in the bandage created when assembling the blades and increasing when the swirling blades work under the action of centrifugal forces, which provides damping of the oscillations of the blades and a decrease in the amplitude of oscillations.

1.2.3 Turbine Guide Blades

The material of the guide blades should have the following characteristics: heat resistance, heat resistance under conditions of abrupt temperature changes corresponding to the change in gas temperature, which is observed during start-ups, shut-downs and changes in the GTU operating mode; long-term strength; creep resistance, ensuring their operability at operating stress levels during the entire service life at temperatures, sometimes significantly exceeding the average temperature of the gas (due to its non-uniformity around the circumference).

1.2.4 Flame Tubes

The following requirements are imposed on the material of GTU combustion chambers: high heat resistance, weldability and technological ductility, allowing bending during the manufacturing process, as high as possible thermal conductivity and low thermal expansion coefficient, as well as high levels of yield strength and creep resistance to prevent significant plastic deformation (warping). To meet the requirements for heat resistance, the material should not be embrittled during operation and should not be notch sensitive both in the initial state and after long exposures at operating temperatures. In some cases, the high thermal conductivity of the metal is provided by the use of heat-conducting materials (copper, nickel) with heat-resistant coatings. In other cases, on the contrary, low heat conductive coatings (enamels) are used to prevent possible short-term local overheating of the metal.

1.2.5 Fasteners

Bolts and studs made from high-strength and heat-resistant alloys are used to connect various parts of turbines operating at high temperatures. For example, in some structures, turbine disks are interconnected by coupling bolts, guide vanes are attached to the stator with the help of bolts, etc. During operation, fasteners are subject to variable temperatures and loads. To ensure the operability of fasteners, the material must have:

1. relaxation resistance (to maintain the necessary tension in the compound);
2. structural stability during operation (excluding both material softening and hardening, which are accompanied by a decrease in volume, and in some cases can cause significant increases in tension);
3. long-term strength (to provide the necessary margin of safety);
4. notch insensitivity and high long-term ductility, preventing failures of the thread; ability to resist repeated loads (due to repeated braces); resistance to vibration loads.

1.2.6 Regenerators

The main requirements for materials of regenerators are those of manufacturability, providing the possibility of rolling thin sheets and thin-walled pipes, weldability and punchability. The materials of the regenerator components must have a high corrosion resistance under operating and stand-by conditions in air and combustion products. The material of GTU regenerators operating on sulfur fuels and especially fuels containing vanadium must resist sulfur and vanadium corrosion. The metal of the regenerator and its welded joints must have heat resistance. This requirement arises from the presence of temperature gradients in the regenerator that vary over time (when the turbine starts and stops and changes in the modes of its operation). Since the efficiency of a gas turbine plant is significantly affected by the density of the regenerator, the material of its components during operation must resist the action of various factors causing discontinuity (cracks, ulcers, etc.). Such discontinuities may occur, for example, if the metal is prone to intercrystalline corrosion (ICC) because regenerators operating under marine conditions, or if the metal of welded joints is prone to local damage (in the near-weld zone), or if the metal has a low resistance to vibration loads arising from insufficiently stiff structures.

Certain requirements are also imposed on the technology of welding, brazing and heat treatment of structural elements depending on the material and operating conditions of the regenerator. In the case of soldered structures, the solder joint must resist the effects of vibration.

1.2.7 Blades, Disks and Compressor Housings

The blades of both axial and centrifugal compressors are usually subject to significant vibration loads. In this regard, the main requirements are the high fatigue strength of the material and its ability to damp vibration. Since structural damping plays a comparatively smaller role in compressors compared with aerodynamic and sometimes compared to damping in a material, the choice of material for the blades and the mode of its heat treatment is carried out taking into account the requirement for obtaining the damping factor of the maximum possible value. It should be borne in mind that the logarithmic damping decrement of oscillations of chromium steels widely used for blades increases with increasing temperature, the level of vibration and tensile stresses. However, vibration stresses in the blades sometimes reach 200 MPa. Damage caused by impact by a foreign object or corrosion damage (corrosion cracking) are concentrators that drastically reduce the fatigue strength of the blades. Therefore, all measures are used to increase the fatigue limit, in particular, an appropriate surface treatment is carried out. The requirements of the corrosion resistance of the material and its resistance to corrosion fatigue are particularly important for gas turbine compressors operating in marine environments.

The material of contaminated air compressor blades must resist erosion. Otherwise, erosion resistance should be ensured by the use of special coatings. Under the action of centrifugal forces, tensile stresses arise in the blades. Therefore, the material must also have a certain level of strength properties at operating temperatures. This requirement is particularly significant for high-speed compressors.

In compressors with high degrees of compression, the temperature of the blades can reach a level at which it is necessary to take into account the change in the characteristics of the material with time, in particular the creep resistance.

When turbines operate in cold outdoor air, the material of the first-stage compressor blades should ensure their operability at temperatures below 0°C. This results in a specific requirement for the material with respect to its resistance to brittle fracture, which is associated with the possible action on the blades during operation of shock loads of different origin (for example, from the ingestion of foreign objects of medium size not detained by filters). Therefore, the material of the blades of the first stages of compressors of such GTU should have the lowest possible transition temperature of brittleness.

High demands are placed on the material of the disks and compressor housings only in aircraft structures, where, due to limitations of weight characteristics, high vibration and low-cycle fatigue can occur. In this case, they are subject to the same requirements as for the material of the blades. The parameters of low-cycle fatigue and crack resistance are very important for the material of the disks.

1.3 Materials of Gas Turbine Unite Components

For the manufacturing of gas turbine parts metallic materials of almost all classes are applied. However, the main focus will be given below to heat-resistant materials. They can be divided into the following main classes: pearlitic steels, chromium, ferritic steels, ferritic-martensitic steels, martensitic steels, austenitic-martensitic steels, austenitic steels, titanium alloys and, finally, nickel-based and cobalt-based alloys. As already noted, refractory metals and their alloys can be used for closed-cycle gas turbines. In some cases, GTU blades of very short duration are also made from molybdenum and niobium alloys.

Ceramic materials (cermets), metal oxide alloys of $Al\text{-}Al_2O_3$, $Ni\text{-}Al_2O_3$, etc. and composite materials are currently considered promising for turbine and compressor parts operating at elevated temperatures. These materials consist of a relatively plastic matrix (Al and Ni for compressor parts, Ni and Co for turbine parts), refractory carbides, nitrides or carbonitrides (for cermets), refractory oxides and, finally, filamentous boron fibers and silicon carbides (for composite materials). The use of these materials allows the mass reduction of components (by increasing the specific strength) and to reach operating temperatures of 1100-1300°C, which are too high for heat-resistant materials on iron and nickel bases. The chemical composition of foreign alloys are presented in Table 1.4. Materials, applied in Russia, are given in Table 1.5, in GOST 5632 - 72, 18968 - 73, 23705 - 79, GOST 20072 - 74, and in Khimushin (1969); Kablov (2006); Petrushin et al. (2017). Application temperatures

Table 1.4 Chemical content of foreign heat-resistant nickel-based alloys

Alloy	Cr	Co	Mo	W	Al	Ti	Ta	B	Zr	C	Hf	remaining	ρ, g/cm^3
Single-crystal alloys													
CMSX-10	2	3	0.4	5	5.7	0.2	8	-	-	-	-	6 Re	-
PWA1484	5	10	2	6	5.6	-	9	-	-	-	0.1	3 Re	8.95
PWA1480	10	5	-	4	5	1.5	12	-	-	-	-	-	8.70
PWA1483	12.2	9	1.9	3.8	3.6	4.2	5	0.0001	0.002	0.07	-	-	-
AM1	7	6	2	5.7	5.2	1.1	7.9	-	-	0.003	-	-	-
Rene No. 6	4.2	12.5	1.4	6	5.75	-	7.2	0.004	-	0.05	-	5.4 Re	-
Rene No. 4	9	8	2	6	3.7	4.2	4	-	-	-	-	0.5 Nb	8.56
CMSX-4	6.5	9	0.6	6	5.6	1	6.5	-	-	-	0.1	3 Re	8.7
CMSX-2	8	5	0.6	8	5.6	1	6	-	-	-	-	-	8.56
CMSX-3	8	5	0.6	8	5.6	1	6	-	-	-	0.1	-	8.56
TMS-63	6.9	-	7.5	-	5.8	-	8.4	-	-	-	-	-	8.48
TMS-75	3	12	2	6	6	-	6	-	-	-	0.1	5 Re	-
TMS-138	2.9	5.9	2.9	5.9	5.9	-	5.6	-	-	-	0.1	4.9 Re, 2 Ru	-
TMS-139	2.9	5.8	2.9	5.8	5.8	-	5.5	-	-	-	0.1	4.9 Re, 3 Ir	-
SC16	16	-	3	-	3.5	3.5	3.5	-	-	-	-	-	8.21
SRR99	9	5	-	9.5	5.5	1.8	2.9	-	-	-	-	0.7 Nb	8.5
Oriented-crystallized alloys													
CM186LC	6	9	0.5	8	5.7	0.7	3	0.015	0.05	0.07	1.4	0.3 Re	8.7
CM247LC	8.1	9.2	0.5	9.5	5.6	0.7	3.2	0.015	0.015	0.07	1.4	-	8.54
MM247	8	10	0.6	10	5.5	1	3	0.1	0.01	0.1	1.5	-	8.5
GTD-111	14	9.4	1.5	3.7	3	5	3	0.1	0.01	0.1	0.15	-	8.3
IN792LC	12.5	9	1.85	4.1	3.4	3.8	4.1	0.015	0.02	0.08	1	-	8.21
DS16	16	5	3	-	3.5	3.5	3.5	0.015	0.015	0.06	1	-	8.21
Normal crystalallized alloys													
MM246	9	10	2.5	10	5.5	1.5	1.5	0.015	0.05	0.15	-	-	8.44
IN100	10	15	3	-	5.5	4.7	-	0.014	0.06	0.18	-	1 V	7.7
GTD-111	14	9.5	1.5	3.8	3	4.9	2.8	0.01	-	0.1	-	-	8.3
NFP1916	13.2	4.1	-	1.9	3.7	5	5.1	0.015	0.016	0.07	-	-	8.2
IN738LC	16	8.5	1.7	2.6	3.4	3.4	1.7	0.01	0.05	0.11	-	0.9 Nb	8.1

and physical properties of alloys of different classes are given in Table 1.6[2].

The marked difference in physical properties for materials of different classes (see Table 1.6) significantly affects the stress state and performance of various parts of gas turbines, and especially disks operating under conditions of multiple heat cycles.

[2] In Appendix A are given a list of Russian steel and alloy discussed in the book

Table 1.5 Chemical content of cast blade materials

	Element												
	Cr	Co	W	Mo	Nb	Hf	Ti	Al	C	Y	V	Ta	Re
Alloys for stationary GTU													
ChS70	15-16.2	10.5-11.5	4.7-5.9	1.6-2.3	0.1-0.3	-	4.0-4.8	2.6-3.3	0.06-0.12	-	-	-	-
ChS88	15-16.2	10.5-11.5	4.7-5.9	1.6-2.3	0.1-0.3	0.8-1.4	4.0-4.8	2.6-3.3	0.06-0.12	0.03	-	-	-
SN35	11.4-12.6	7-7.8	6-7	0.7-1.3	0.6-1.0	-	4-4.6	3.9-4.3	0.06-0.12	-	-	-	-
CNK7	14-15.5	8-8.9	6.2-7.5	0.2-0.6	-	-	3.6-4.4	3.4-4.5	0.06-0.1	-	-	-	-
ZMI3	12.5-14	5-6	4-8	0.5-2.0	0.1	-	4.5-5.3	2.6-3.3	0.06-0.1	-	-	-	-
ZMI3U	12.5-14	4-6	6.5-8	0.5-2.0	0.1	-	4-5.5	2.8-4	0.07-0.15	-	-	-	-
ChS104	20-21.8	10.3-12	3-4	0.3-0.9	0.15-0.35	-	3.1-3.9	2.1-2.9	0.07-014	-	-	-	-
EP539LMU (EP958)	15.7-17.1	4-6	3.5-5	1.5-2.4	1.2-1.8	-	2.2-3	3.2-4	0.08-0.14	0.04	-	-	-
Alloys for aviation GTU													
ZhS6K	9.5-12.5	4-5.5	4.5-5.5	3.5-4.5	-	-	2.5-3.2	5-6	0.13-0.18	-	-	-	-
ZhS26	4.3-5.6	8-10	10.9-12.5	0.8-1.4	1.4-1.8	-	0.8-1.2	5.5-6.2	0.13-0.12	-	0.8-1.2	-	-
ZhS32	4.3-5.6	8-10	7.7-9.3	0.8-1.4	1.4-1.8	-	0.8-1.2	5.5-6.2	0.13-0.12	-	-	3.5-4.5	3.5-4.5
ZhS6U	8.5	9.5	10	1.8	1	-	2.5	5.5	0.16	-	-	-	-
VZhL12U	9.5	14.5	1.4	3	0.8	-	4.5	5.5	0.16	-	0.8	-	-
ZhS6F*	5.5	9.4	12	0.9	1.6	0.9	1	5.3	0.11	-	0.9	-	-
ZhS36*	4.5	9	12	1.5	1	-	1	5.9	0.05	-	-	-	2
ZhS40*	6.5	0.5	7	4	0.2	-	-	5.3	0.005	-	-	7	-
ZhS47	2-3	9.5-12	0.1-1.8	1.6-2.4	-	-	-	5.5-6.2	0.002-0.05	-	-	7.8-10	7.8-10
ZhS49	2.4	12	0.8	0.8	-	-	-	6	0.005	-	-	8	12

* Kablov et al. (2002); Petrushin et al. (2017)

Table 1.6 Physical properties of GTU materials

Material	T_M, °C	α, $10^{-6}K^{-1}$, at 20°C - T_M	λ, W/(m·°C) at 20°C	λ, W/(m·°C) at T_M	E_{st}^{20}, 10^5 MPa	E_{dyn}^{20}, 10^5 MPa	ρ, g/cm^3
Perlitic steels	up to 550	13-13.8	33-42	29-42	1.8-2.0	2.1-2.2	7.8-7.9
Modified 12%Cr steels	550 (600)	11.3-12	25	27	1.8-2	2.1-2.2	7.7-7.8
Austenitic steels	600-700	16.5-18	12.5-15	23	1.8-2	2.1-2.2	7.85-8.2
Nickel-based alloys	700-900	15-16	12.5-15	23-29	1.9-2	2.1-2.2	7-8.5
Cobalt-based alloys	700-780	14-15	15	23-25	-	-	8.9
Molybdenum-based alloys	up to 1050	6	130	104	-	3.3	10.2
Niob-based alloys	up to 1000	8.4	42	63	-	11.6	8.65
Titanium-based alloys	up to 600	8.5-9.5	9	15	1.1-1.2	1.2-1.35	4.4-4.6
Ceramics and composites	up to 1300	determined by the composition of the matrix and the amount of the disperse phase					

T_M - limit temperature of metal application, which corresponds to a tensile strength of 1000 h at least 250 MPa, α - thermal expansion coefficient, λ - coefficient of thermal conductivity

The value of the thermal expansion coefficient α of the alloys changes both in the process of polymorphic transformations, and as the second phase precipitates, which occurs with a change in volume. Hence, for materials undergoing such transformations with a diffusion nature in a certain temperature range, the α_T values at these temperatures will depend not only on the temperature T, but also on the heating (cooling) rate, and most importantly on the exposure time at the given temperature. At present, methods for determining α both in fast and slow heating have become widespread, while in the case of long-term operation at specified temperatures, the values of α_T are not known at all. Hence, it seems appropriate to introduce the concepts of the instantaneous value α_{inst} and the values α_t depending on time (heating rate). The relationship between α_{inst} and α_t can be represented by the dependence $\alpha(T,t)$. To determine it, the relationship between the rate of polymorphic transformation (as a function of temperature and its accompanying volume changes) and α must be established. Thus, α is a function of state depending for the given history on the contribution of the corresponding ratio of structural components in the alloy. Changes in the value of α can be expected for high-temperature alloys hardened with the γ' phase (in quantities depending on the time and temperature of operation), as well as for coatings of the Co-Cr-Al-Y system, having polymorphic transformation at temperatures in the range of 800-850°C.

Description of the process of improvement and development of new heat-resistant alloys for parts of aviation gas turbines in the former USSR is given in Kablov et al. (2002); Kablov (2006). For stationary gas turbine materials the corresponding description is given in the works of NPO CKTI and CNIITMASH. Increasing the temperature of the working blades through the use and development of new deformable and cast alloys provided a significant increase in output power and efficiency. Increasing the metal temperature of the working blades through the use and creation of new deformed and cast alloys provided a significant increase in output power and efficiency. Further evolution of blade alloys is associated with the development of directional solidification (NK) alloys, single-crystal alloys, and from 1985 with composite and ceramic materials (Fig. 1.53). Over the years, the rate of improvement of alloys for blades decreases. This is due to the fact that as the temperature rises, the processes of high temperature corrosion become the factors limiting the resource. Therefore, the development and introduction pace of new alloys depends on the effectiveness of the work on the development of protective coatings.

1.3.1 Perlitic Steels

For disks and rotors of gas turbines, pearlitic steels of the grades 20Ch3MWF (EI415) and 26ChN3M2FAA are applied. Alloying provides high structural stability of these steels in conditions of long-term operation. The use of vacuum-arc remelting allows improving the service properties of steel EI415[3]. The requirements regulated by technical specifications for steel 26ChN3M2FAA are given in Table 1.7. Pearlized relaxation-resistant steels 25Ch2M2FA (EI10), 25Cr2Mo1V (EI723), 20Cr1Mo1V1TiB (EP182) have been widely used for fasteners.

1.3.2 Ferritic Steels

The main advantages of ferritic steels compared to austenitic are their low cost due to the low content (or absence) of scarce elements (Ni, Co) and high workability of

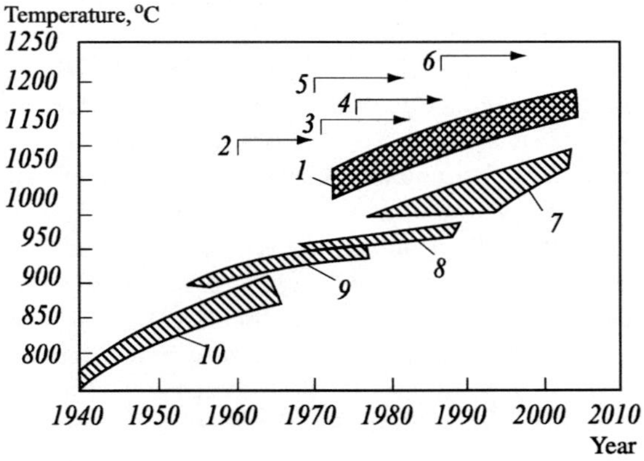

Fig. 1.53 Exploitation temperature of heat-resistant alloys and coatings of high-temperature turbine blades in Rolls-Royce GTU: 1 - heat-protection coating (HPC), 2 - aluminium coating, 3 - modified aluminium coating, 4 - condensed coating, 5 - plasma HPC, 6 - electron beam HPC, 7 - single-crystal alloys, 8 - alloys with oriented crystallization, 9 - cast alloys, 10 - strained materials

[3] The list of Russian steels and alloys, the characteristics of which are given in the book, is presented in Appendix A.

Table 1.7 Mechanical properties of forging pieces made of steel 20ChN3M2FAA

Strength category, MPa	Test temperature, °C	$\sigma_{0.2}$, MPa	δ, % (≥)	ψ, % (≥)	Impact strength on Menage type specimen (ISM), kJ/m² (≥)	Brinell hardness (HB), N/mm² (≤)	T_K^{50}, °C (≤)
KP670	20	670-780	15	48	780	3050	-45
KP780	20	780-900	14	46	580	3240	-40
	250	680	14	46	-	-	-
KP840	20	840-960	13	44	540	3430	-30
	250	730	13	44	-	-	-

KP - strength category, T_K^{50} - critical temperature at which fractured toughness samples have 50% fractured fibers

processing both hot and cold. Systematic studies of the influence of individual alloying elements on the structure, properties, and manufacturability of 12% chromium steels made it possible to determine the optimal C, Mo, W, V, and Nb contents, which provide high heat resistance at optimum free delta ferrite contents. It was found that, on the one hand, free delta ferrite lowers the manufacturability of steels of this class during hot mechanical and heat treatment, leads to a sharp anisotropy of properties after hot machining, causes brittleness and reduces heat resistance. However, on the other hand, delta ferrite prevents hot cracking during welding.

Pearlite and modified 12% chromium steels in the state after heat treatment (quenching or normalization and high tempering), unlike austenitic, are characterized by high yield strength and strength at room temperature, the ability of stress redistributions, a significantly lower thermal expansion and high thermal conductivity. In addition, 12% and 17% chromium heat-resistant steels, as well as heat-resistant pearlitic steels, thermally treated in special modes, are not sensitive to notches during long-term strength tests.

High-chromium steels have increased heat-resistance and digestibility compared with pearlitic steels, but are prone to corrosion in media containing chlorides. The limited heat resistance of ferritic steels allows them to be used for disks of only those GTUs whose design provides sufficiently effective cooling. 15Ch12WNF (EI802) and 20Ch12WNMF (EP428) steels are widely used as such materials.

Steel 20Ch13, 14Ch17N2, 13Ch14N2WFR (EI736) and 13Ch11N2W2MF (EI961), which have high damping ability, are most widely used as materials for compressor blades. In recent years, steels with high nickel content of austenitic-martensitic class, and in particular steel 08Cr17N6T, possessing increased corrosion resistance, has also begun to be used for blades. Steel 30Ch13 and 40Ch13 are used for springs and fasteners. In order to prevent corrosion, when water containing chlorides gets into the compressor, a nickel-cadmium coating is applied to the blades. To eliminate the negative effects of the process of applying this coating to fatigue resistance, blasting of fillets is carried out.

Certain advantages in cold brittleness have steels with a higher nickel content. In addition to the chemical composition, the cold brittleness of ferritic steels is influenced by mechanical properties and heat treatment. It was found that the impact toughness a_H at temperatures in the range from -40°C to -60°C for steel 20Ch13, quenched by temperature 1050-1100°C and having a hardness of HB 207, decreases to 12-35 J/cm^2. For steel 14Ch17N2 after all modes of heat treatment, providing the values of hardness HB 229-363, the impact strength at temperatures up to -70° is not lower than 50 J/cm^2. To obtain high toughness of steel 20Ch13 under test conditions from +20 to -60°C, the hardness should not exceed HB 207, and the quenching temperature should be in the range 950-1000°C.

Steel 08Ch17N6T after heat treatment in the standard mode (quenching from 950°C, holding 3 h, cooling in air, the first two-day tempering at 7500°C, holding 2 h, the second tempering at 510-520°C, holding 8 h) has $R_{p0.2} = 750$MPa and increased brittle fracture resistance compared to steels 20Ch13 and 14Ch17N2. Mechanical properties at the temperature of 400°C of these three steels are shown in Table 1.8. Ferritic steels have different resistance to fatigue in a corrosive environment and have different notch sensitivity (Table 1.9). To reduce the tendency to form slabs in

Table 1.8 Mechanical properties of ferritic steels at temperature 400°C

Properties	20Ch13	14Ch17N2	08Ch17N6T
Strength limit, σ_B, MPa	780	870	900
Long-term strength, σ_{3000}, MPa	390	450	495
Creep resistance, $\sigma_{1\%/1000}$, MPa	390	400	380

Table 1.9 Fatigue strength at temperature 20 °C of stainless steel for compressor blades

Material	σ_B, MPa	σ_{-1}, MPa						
		air ($N = 10^7$) for specimen			sea water ($N = 5\,10^7$) for specimen			sea fog* ($N = 10^8$) for specimen
		smooth	notch r 1 mm	notch r 0.25 mm	smooth	notch r 1 mm	notch r 0.25 mm	smooth
20Ch13	860	360	-	140	180 (270)	-	80	155
EI961	920	570	-	-	305	190	-	-
14Ch17N2	860	420	-	140	240	-	70	-
14Ch17N2	1110	572	-	170	275	-	70	-
Ch16N2M	1100	560	-		410	-	-	-
09Ch16N4B	1100	585	185	-	385	115	320	80
EP718	-	620	190	-	480	130	420	90
08Ch17N6T	1080	550	370	240	300	170	150	-

* no data for notch r = 1 mm

these steels, slag remelting is used.

For the manufacturing of compressor blades of stationary GTU the following steels are applied

20Ch13 Steel 20Ch13 (GOST 5632-72 / GOST 5949-61) is the most widely applicable for the manufacturing of compressor blades due to the high decrement of vibrations and corrosion resistance. The heat treatment of the blades is performed according to the mode: quenching from 1000-1050°C in air, tempering at 660-700°C, air cooling. The fatigue strength of the blades from steel 20Ch13 decreases under the influence of corrosion ulcers formed on the surface and in the presence of contact with salts of sea water. Steel AISI-403 is a foreign analogue of steel 20Ch13.

EP428 (20Ch12WNMF) Steel EP428 (GOST 5632-72 / TU) in comparison with steel 20Ch13 has a higher strength, which allows it to be used for highly loaded compressor blades. Due to the additional alloying, it is possible to use this steel at temperatures higher than for steel 20Ch13. Heat treatment of the blades is performed according to the following conditions: quenching from 1050°C, 1 h, cooling in oil, tempering at 700-720°C, 10-20 h, cooling in air. The structure after heat treatment is sorbitol tempering, oriented according to martensite. In addition, depending on the chemical composition within a given steel grade, delta ferrite may be present in the structure, the amount of which can reach 10-15%. The hardening phase consists of carbides M_{23} C_6 with a size of 100-500 nm, where M is mainly Cr and Fe, as well as fine carbonitrides of type M_2X and MX, where M is Cr and V, and X is C and N. The size of carbonitride phase is 10-50 nm.

EI961 (13Ch11N2W2MF) Steel EI961 (GOST 5632-72 / TU14-1-933-74) has even higher strength properties compared to steels 20Ch13 and EP428. However, its use is possible only in conditions where the environment does not contain chlorides. Otherwise, one should expect intense ulceration and a sharp decrease in the fatigue strength of the blades. Heat treatment of the blades is carried out according to the following conditions: hardening from 1000-1020°C in oil, tempering at 680°C, 1 hour and air cooling.

1.3.3 Austenitic Steels

Austenitic steels used as heat-resistant materials are iron-chromium-nickel or iron-chromium-nickel-manganese alloys with one or another number of other elements. There are a number of classifications of austenitic steels. The most convenient with respect to the materials of components of gas turbines is their division into four groups.

1.3.3.1 Group I

This group includes non-hardened steels, both unstabilized, for example Ch18N9, 20Ch23N18, and stabilized by Ti or Nb (for example, 12Ch18N10T), as well as steels with a hardened solid solution (EI726, EI405). They all have a low yield strength. Nickel-chromium steels with 15-22% Cr and 8-28% Ni with different carbon content (stabilized and non-stabilized) are the main materials for regenerators of various designs and heat exchangers (plate, tubular, rotating)[4], as well as body parts. These steels possess high heat resistance at temperatures up to 600-760°C in the environment of combustion products of the fuel and in air, not containing chlorides. The high manufacturability of nickel-chrome steels makes it possible to obtain from them virtually any type of blanks (including pipes and sheets of various thickness). 12Ch18N10T steel, which, unlike chromium steels (type AISI 430), has a satisfactory corrosion resistance even under sea air conditions, has become widespread for regenerators of gas turbines for various purposes. At present, a good experience has been gained of the long-term operation of regenerators of shipboard gas turbines from this steel. The use of 12Ch18N10T steel as a plate regenerator material for ship GTU in some cases, however, cannot guarantee its performance even when its tendency to intergranular corrosion is eliminated by heat treatment.

The point is that the regenerator zones, in which the metal is heated up during operation (0-150°C), can be subjected to stress corrosion cracking in a medium containing chlorides. The ability for corrosion cracking does not depend much on the content of carbon, titanium and niobium in steel. Eliminating the tendency to corrosion cracking of the metal of the regenerator can be achieved using alloys with a high nickel content (not less than 45%), for example steel Ch20N45B (EP350).

In gas turbine engineering, steel 20Ch23N18 (the material of the flame tubes and other parts of low-temperature turbines), EI726, EI405 (materials of lightly loaded disks of stationary GTU) are used. For the manufacture of stationary gas turbine blades the steel EI726 (09Ch14N19W2B1) GOST 5632-72 / GOST 10500-63 is used. It is highly structure-resistant during long-term operation. Heat treatment of the blades of this steel is performed according to the mode: quenching from 1120-1140°C in air, tempering at 740-760°C for 5 hours and air cooling. The phenomenon of intergranular corrosion (IGC) is observed in steels with C>0.005% and is caused by a lower content of Cr (below 12%) in local areas of carbides, mainly precipitated during operation along the grain boundaries. The IGC warning is provided by the use of sensitizing additives Ti and Nb and heat treatment at a temperature of 850-950°C (after welding).

Welded joints of chromium-nickel steels with titanium and niobium, which are subject to prolonged bending stress at high temperatures (600-700°C), may be prone to local damage in the heat-affected zone, which is one of the factors limiting the performance of housing parts. The nature of local failure is currently not well understood. It is assumed that one of the main reasons contributing to the occurrence of such damage is a sharp decrease in the intergranular strength of the metal of the

[4] For regenerators, chromium steels with a low carbon content are also sometimes used.

heat-affected zone, induced during welding and manifested during operation, and this decrease may be a result of either hardening of the grain interiors or softening of the boundaries. The latter may, for example, be observed in the process of welding niobium-containing steels in connection with the formation of a eutectic consisting of austenite, NbC carbide and the intermetallic compound Fe2Nb. The melting point of this eutectic rises with a relative decrease in the niobium content (relative to carbon). Steels without Ti and Nb are insensitive to local damage. Austenization, which helps to heal the submicroscopic cracks formed during welding along the grain boundaries in the heat-affected zone, leads to an increase in resistance to the propagation of cracks formed and to reduce the likelihood of the occurrence of deep cracks representing the greatest danger in the operation of relatively thick-walled structures. Reducing the tendency of metal to local failure is achieved by technological methods (smelting, hot deformation and heat treatment), leading to an increase in long-term plasticity, as well as by reducing the grain size and the amount of harmful impurities in the base metal (sulfur, phosphorus, lead, gases, etc.), thinning of boundaries, etc. In the cold-worked state, the steels of this group have $\sigma_B \geq 1400$ MPa, which allows them to be used for corrosion-resistant springs.

1.3.3.2 Group II

This group includes steel, dispersion strengthened mainly by intermetallic compounds Ni_3Ti and Ni_3TiAl. Separation of these phases from a solid solution increases the resistance of plastic deformation alloys due to an increase in the resistance of intragranular shear deformation due to the blocking of slip planes and an increase in the resistance to intergranular slip. The representatives of the steels of this group are: 10Ch11N20T3R (EI696), 10Ch12N22T3MR (EI696M, EP33), 10Ch12N20T2R (EI696A), 08Ch15N24W4T, ChN35WT (EI612), EI612K, ChN35WTAlR (EI787), ChN45MWTYuBR (EP718) In these steels, in addition to Cr, Ni, Ti, there are also elements that harden the solid solution and are part of intermetallic compounds. EI612 steel is used for gas turbines for various purposes, including very large ones.

1.3.3.3 Group III

This group includes steel with a carbide dispersion hardening; EI481, EI572 (31Ch19N9MWBT or its foreign analogue 19-9-DL) have received the greatest use among them as materials of disks. Steel EI572 is widely used for impellers of GTU for turbo-charging of diesel engines, disks and blades of stationary turbines and lightly loaded disks of foreign aircraft engines (sometimes in the state of semi-hot working). Steel EI481 is used for disks of aviation and ship GTU. Steel EI572 is used in both deformed and cast condition. Steel EI481 has a pronounced tendency to corrosion cracking. Among many the steels of this group, EI703 and EP126 are used for combustion chambers.

1.3.3.4 Group IV

Steel with mixed hardening is used relatively little. Sinidur steel (0.25% C, 19% Cr, 24% Ni, 2% Mo, 2% Ti, 1% Al), strengthened with a Ni_3TiAl type carbide and intermetallic phase, is used as the material of the working blades. Steels hardened by the carbide phase and the Laves phase (AB_2 type phase), which has a very high hardness (HV 900), are under study.

1.3.4 Nickel, Nickel-Cobalt, and Cobalt Based Alloys

Widely used materials for gas turbine components are nickel-based alloys, strengthened by the dispersed intermetallic γ' phase of Ni_3TiAl, which is precipitated during technological aging. With an additional alloying by cobalt, the strengthening phase is $(NiCo)_3TiAl$. Depending on the amount of the γ' phase (Al+Ti content) and the nature of the alloying of the solid solution, nickel-based alloys have different heat resistance and thermal fatigue resistance (Figs. 1.54 and 1.55). Increasing the heat resistance of Ni-Cr-Ti-Al alloys is achieved by adding molybdenum. Small additives of boron, alkaline-earth and rare-earth elements also have a positive effect on long-term strength. Boron, precipitating during the aging of the alloy in the form of boride phases mainly along the grain boundaries, inhibits diffusion processes, thereby increasing the heat resistance, and in some cases leads to an increase in long-term ductility. The effect of minor additions of alkaline and rare earth elements on long-term strength is determined by their refining action due to chemical activity with respect to harmful impurities (S, Pb, Bi, Sb), as a result of which they are bound to refractory compounds. Crystal chemical studies have established that the

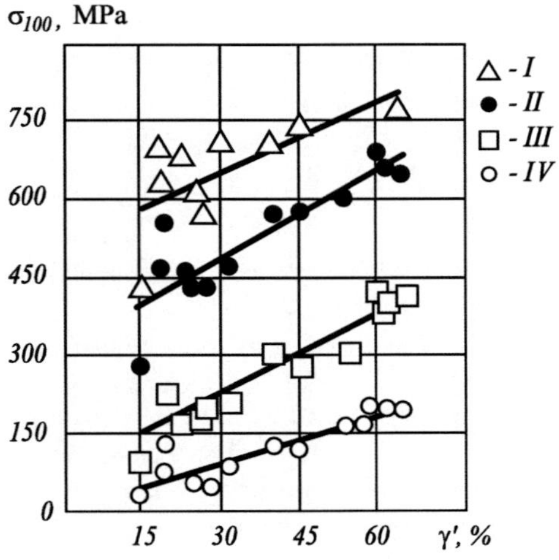

Fig. 1.54 Influence of the content of the γ' phase on the long-term strength of nickel-based alloys at various temperatures: *I* - 700°C, *II* - 760°C, *III* - 870°C, *IV* - 980°C

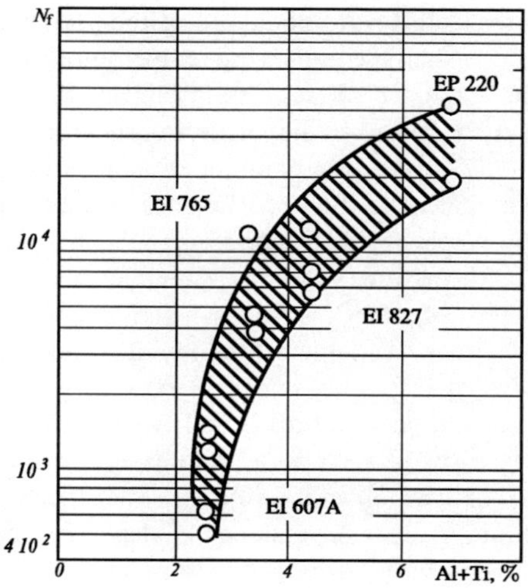

Fig. 1.55 Relation thermal fatigue resistance of nickel based alloys N_f to content of Al+Ti. Testing without holding in the regime 900 ⇆ 700°C, period of the cyclic process 15 s

γ' phase has a lattice parameter very close to that of the solid solution. The smaller the difference of these values, the more intense the decomposition of the γ-solid solution occurs when cooled in air, and the more stable the resulting γ'-phase against temperature effects is. The intensity of the processes of precipitation of the γ'-phase and the size of the particles depend on the temperature, time and degree of alloying of the solid solution.

The kinetics of these processes for the EI437B alloy (Parshin, 1972) can be illustrated by the diagram in Fig. 1.56. In connection with a decrease in the solubility of Ti and Al in a solid solution with a decrease in temperature in the process of technological heat treatment, it is not possible to completely transfer the alloy to a stable

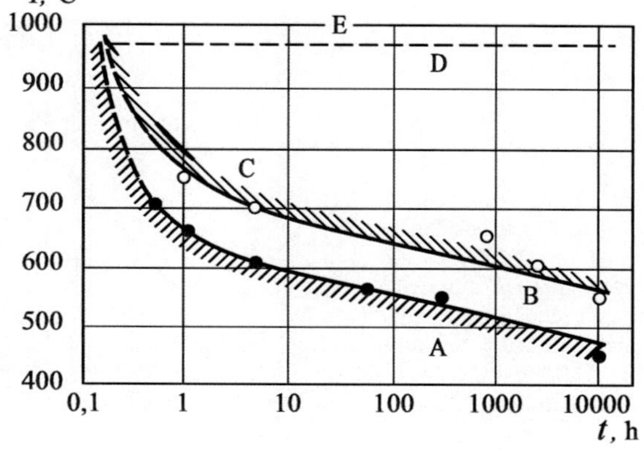

Fig. 1.56 Formation of different structures in alloy EI437B in relation to the temperature T and time t: A - early stage of the "incubation period", B - final stage of the "incubation period" (formation of split-off zones of γ' phase), C - decomposition of split-off zones at grain boundaries with formation of γ' phase, D - isolation and coagulation of stable γ' phase (Ni_3TiAl type), E - dissolution region of the intermetallic phase

state. Prolonged operation at lower temperatures compared to tempering temperatures leads to additional precipitation of the γ'-phase. The operation at temperatures higher than the tempering temperature causes the dissolution of a certain amount of the dispersed phase. The instability of Ni-Cr-Ti-Al alloys at elevated temperatures is also caused by dendritic segregation of Ti and Al. For example, for a wrought EI437A alloy, the titanium content in the inter-axle areas is 2.5 times greater than the content in the dendrite axes (Parshin, 1972). The dispersion and the amount of the γ'-phase after technological heat treatment determine the stability of the alloys and the nature of the time dependence of plasticity. Therefore, to ensure high performance of materials for disks and blades operating in different conditions, the choice of alloy grade and heat treatment mode should be made taking into account the laws of alloy instability under operating conditions.

As the content of Ti and Al increases, the heat resistance of nickel base alloys increases. For example, there is a linear relationship between the limit of long-term strength σ_{100}^{800} and the amount of hardening phase in the range of 8-50% (Khimushin, 1969). Linear dependence is also found between the content of Ti and Al and the temperature at which the alloys have the same long-term strength (Fig. 1.57).

In GOST 23705-79 the application, the characteristics of long-term strength, the values of the coefficients of linear expansion and thermal conductivity of domestic deformed superalloys based on nickel are given. Due to the fact that some alloys do not have structural stability and their properties vary dramatically during operation, according to the specified GOST, they are used for limited service lives. When developing and using deformed and cast nickel-based heat-resistant alloys for blades, the following features of their behavior should be taken into account: relatively low corrosion resistance in contact with the combustion products of sulfur fuels,

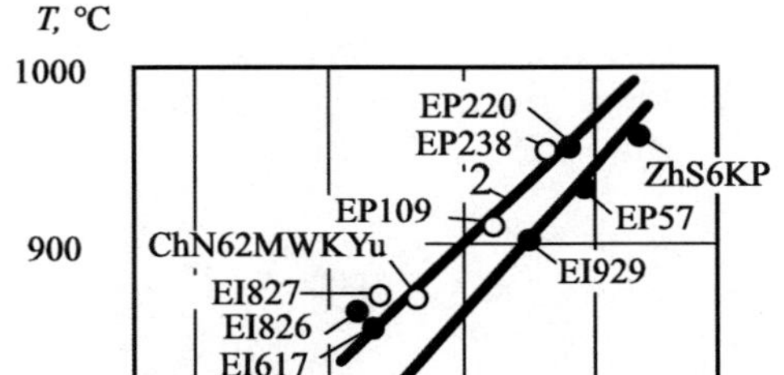

Fig. 1.57 Temperature change T responsible for σ_{100} = 200 MPa for nickel-based alloys as a function of the aluminium and titanium content: 1 - alloy containing Al and Ti (I), 2 - alloy containing aluminium only (II)

possibility of embrittlement during long-term operation, the presence of various metallurgical defects in castings.

A number methods of dealing with these factors, reducing the operational reliability of materials and parts made from them are known as follows:

1. To reduce corrosion, alloys with a high chromium content are developed and protective coatings are applied.
2. The theory of optimal alloying has been developed, which provides an increase in the structural stability of materials. In the alloys developed on the basis of this theory, the formation of topologically close-packed (TCP) phases (such as σ, μ and λ phases) is difficult in connection with long-term operation, and therefore no significant embrittlement of the alloys is observed. In this case, the restricted limits of the content of alloying elements is recommended.
   As for the fine γ'-phase, precipitated at relatively low temperatures, special methods of technological heat treatment are used to suppress these processes. In Pigrova (1980) based on the study of the kinetics of changes in the phase composition of various alloys by X-ray diffraction analysis, a relationship was established between the chemical composition of alloys (in atomic percent) and the tendency to form TCP phases during long-term operation (Fig. 1.58). With the dependence presented in this figure, it is easy to predict the structural stability of the alloys or its absence. Since the formation of TCP phases is one of the main causes of embrittlement and decrease in the performance of alloys during long-term operation, the use of the diagram makes it possible to select only alloys that are not prone to the formation of TCP phases.
3. To increase the ductility alloys with columnar and single-crystal structures obtained by the method of directional crystallization are used. An important role is also played by the mode of technological heat treatment. Special single-crystal alloys (see Table 1.4) contain almost no elements that strengthen the grain boundaries (C, B, Zr, Hf).

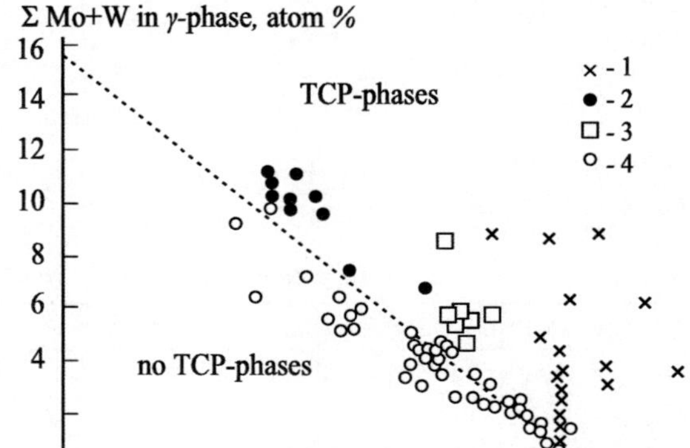

Fig. 1.58 Relation between the type of phase transformation and content of elements of 4th group in the γ phase:
1 - σ phase, 2 - μ phase,
3 - σ phase + μ phase,
4 - no TCP phase

4. To sharply reduce the porosity of cast alloys with an equiaxial structure, gas-static pressing of blanks is used.

Soviet heat-resistant alloys on a nickel basis for turbine blades were developed by teams of various institutions in Russia and Ukraine: VIAM: F.F. Khimushin D.E. Livshits - deformed alloys, S.T. Kishkin, E.N. Kablov - cast alloys; CNIIChermet: G.V. Estullin, A.M. Borzdyka - deformed alloys; CNIITMASH: M.A. Filatova - deformed alloys, E.L. Katz, V.P. Lubenets - cast alloys; CKTI: E.E. Levin, V.V. Rtishchev; Zaporozhye Machine Building Institute: A.G. Andienko, A.D. Koval. As a result, a large number of alloys have been created , see Appendix A. The most widely used alloy for the manufacture of blades of stationary GTU is EI893 (Fig. 1.59).

For the guide blades of stationary gas turbines, the cast alloy EI893(L) is used, which, as established in Getsov (1967), is relatively weakly dependent on the characteristics of the smelting technology. Its ductility, even when melted in air, is at a high level (Table 1.10) and remains at the same level after long-term operation (Table 1.11). Due to the fact that the γ'-phase-phase even in the best single-crystal alloys of the type CMSX-4 above 1100°C dissolves rather quickly, this temperature is currently considered to be the limit for nickel based superalloys.

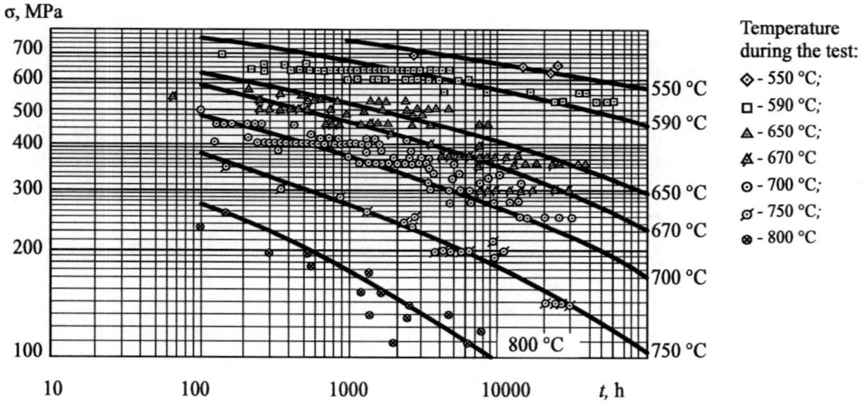

Fig. 1.59 Long-term strength of the metal for rods and blades made of alloy EI893OI, VI after normal heat treatment

Table 1.10 Properties of the alloy EI893(L) at different temperatures

T,°C	$\sigma_{0.2}$, MPa	σ_B, MPa	δ, %	ψ, %	ISM, kJ/m^2	σ_{1000}, MPa	σ_{10000}, MPa
20	490-550	630-750	14-25	16-25	700	-	-
700	-	-	-	-	-	340	280
750	-	440-610	7-19	16-25	1400	220	-
800	360-400	450-500	14-16	28-34	-	165	110

Table 1.11 Mechanical properties of the alloy ZhS47 at temperature 20°C after 5000 h aging

Aging temperature °C	$\sigma_{0.2}$, MPa	σ_B, MPa	δ, %	ψ, %	ISM, kJ/m^2
-	513	734	23.5	22	1020
700	578	817	13.1	18.2	330
750	470	757	16.2	22.7	350-780
800	442	706	16.1	19.1	460

In the 70s-80s of the last century, the most widespread in Russia for working and guide blades were cast alloy grade ZhS6K and its modifications ZhS6U and ZhS6F. In these years, the alloy IN738 was the most widely used abroad (see Table 1.4). Recently, cast alloys ZhS26, ZhS32, ZhS36 and ZhS40, ZhS47 (Kablov et al., 2002; Shalin et al., 1997), were developed for limited service lives, and a number of alloys: CNK7, CNK8, CMI-3, TchS70, EP539LM for long-term operation (Rtishchev, 1992; Rtishchev et al., 1986, see Table 1.5).

Figure 1.60 shows the characteristics of long-term strength of deformable and cast superalloys at 800°C and 900°C, corresponding to 10^2, 10^3 and 10^4 hours. For parts designed to operate at high temperatures, but not for long-term service, alloys with reduced chromium content are used (in alloy ZhS6K 10% Cr, in new alloys ZhS26, ZhS32, ZhS36 and ZhS40 - 4.3. . . 6.5% Cr, and in the recently developed ZhS47 - 2. . . 3%Cr), which provides an increase in strength at high temperature while reducing, unfortunately, the heat resistance. However, this decrease in heat resistance, judging by the results of the tests performed, turned out to be not as striking as expected, and thus their use with protective coatings makes them quite workable for long-term operation. The decrease in the chromium content in these alloys leads to the fact that, like Co, V and C, it decreases the temperature of the γ'-phase dissolution. Therefore, to raise the phase dissolution temperature, such materials are alloyed by Al, Ti, Ta, Hf, W, Mo and Zr. The preference of a particular heat-resistant alloy for turbine blades is determined not only by the characteristics of its long-term strength, but also depends on its density. For example, from three cast alloys used for blades (ZhS6U, ZhS6K and VZhL-12U), having $\gamma = 8{,}4$ g/cm^3, 8.25g/cm^3 and 7.93 g/cm^3, respectively, the preference for operation at a temperature of 800–900°C is VZhL12U alloy, having the maximum ratio σ_{100}/γ, although its long-term strength is not higher than that of ZhS6K and ZhS6U. According to Kablov et al. (2002); Kablov (2006), the long-term strength of the ZhS47 alloy is characterized by the long-term strength values presented in Table 1.12. Abroad and

Table 1.12 Long-term strength (in MPa) of the ZhS47

T,°C	σ_{100}	σ_{500}	σ_{1000}	σ_{5000}
900	535	460	425	350
1000	320	245	215	155
1100	165	115	95	65

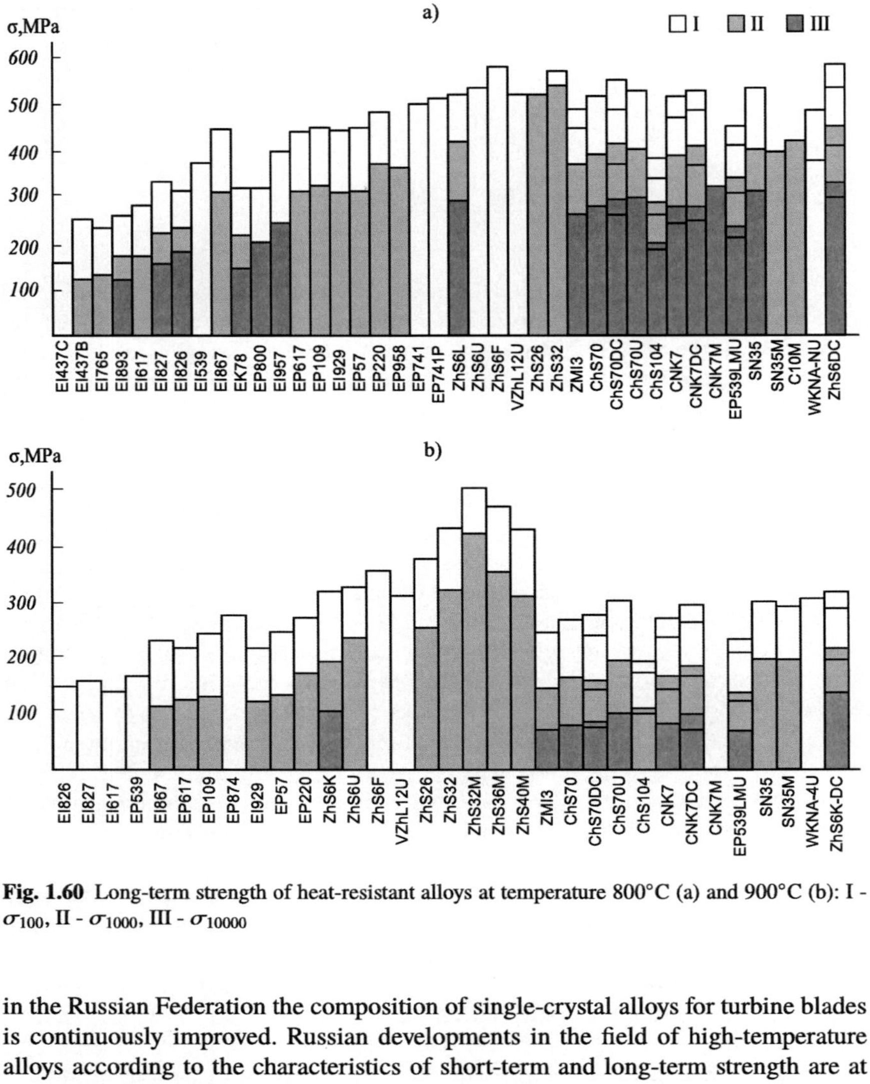

Fig. 1.60 Long-term strength of heat-resistant alloys at temperature 800°C (a) and 900°C (b): I - σ_{100}, II - σ_{1000}, III - σ_{10000}

in the Russian Federation the composition of single-crystal alloys for turbine blades is continuously improved. Russian developments in the field of high-temperature alloys according to the characteristics of short-term and long-term strength are at the level of similar foreign alloys for single-crystal blades and include (Kablov et al., 2002; Kablov, 2006; Shalin et al., 1997)

- The second generation alloys - carbonless rhenium-containing alloy ZhS36 (at the level of alloys CMSX-4, Rene 5, PWA1484 (USA), TMS 71, TMS 82 (Japan).
- The third generation alloys - carbonless rhenium-containing alloys ZhS55, VZhM-1VI (up to 9% Re) (at the level of alloys CMSX-10, Rene 6, TMS 75, Alloy 5A or even exceed their characteristics) (σ_{100}^{1000} = 350MPa). These alloys are superior to those of the second generation ZhS36, CMSX-4 and Rene 5 in long-term durability.

- The fourth generation alloys are carbon-free rhenium-hard alloys such as ZhS52 with 7%Re+2%Ru and ZhS71 with 7%Re+4%Ru (Kablov et al., 2002; Kablov, 2006) at the level of MCNG alloys with 4%Re+4%Ru, France (Snecma, Turbomeca), EPM-102 with 6%Re+3%Ru+0.03%C and its modifications (USA), TMS 138 with 5%Re+2%Ru and its modifications TMS 162 and TMS173 (Japan).

The nickel-based alloys EI607A, EI765, EI437B, EI698, EP742 and EP741 are most common for disks of gas turbines with different service life in Russia. In recent years, the method of hot gas-static pressing (HIP) from powders has become widely used for the manufacturing of disks from materials with high heat resistance (Gessinger, 1984). Disks made from EP741P, EI698P, IN100, Astralloy and other alloys are processed using this method. Promising materials are alloys EK79, EK151, EK152, EP975 developed in recent years for disks. The properties of these materials are listed in Table 1.13 (Shalin, 1994). Studies presented in Fainbron et al. (1993) have shown that the defects in the microstructure of granulated alloy EP741NP during low-cycle fatigue tests are: non-metallic inclusions (81.5%), isolated granules (3%) and micropores up to 200 microns in size (14%).

Nickel-based alloys used for combustion chambers and housing parts contain an increased amount of chromium (EI435, EI868, VZh85, EP99, EI602, EI559, EI652, Nimonic 75, etc.), providing enhanced heat resistance in air atmosphere. These alloys are sufficiently ductile for cold deformation and have relatively low long-term strength and short-term mechanical properties after technological heat treatment (Table 1.14). However, they provide significantly higher performance compared to austenitic steels. Most of these alloys contain only small amounts of elements forming the γ'-phase. The most common for the manufacturing of stator parts in transport and stationary gas turbines has received alloy EI868 (ChN60WT). The flame tubes of this material are successfully operated for 50000 hours or more.

Cobalt heat-resistant alloys are widely used for guide blades of turbines in the USA (alloys X40, X45, MAR-M-509, FSX414, etc.) due to their high heat resistance as a result of the increased chromium content (20-30%). Furthermore at 1100°C cast cobalt alloys are not inferior in heat resistance to cast nickel-based alloys. For working blades, these alloys are usually not applied because of their lower heat resistance compared to nickel alloys at operating temperatures of 800-1000°C. Hardening of

Table 1.13 Properties of new alloys for turbine disks

Alloy	σ_B MPa	$\sigma_{0,2}$ MPa	δ, %	Temperature of tests, °C	Long-term strength		Low-cycle fatigue σ_0, MPa, N = 1000	
	MPA (w.r.t. technical conditions not less)				σ, MPa	t, h	smooth specimen	notched specimen (K_t=3,3)
EK79-IA	1300	900	13	650 (750)	900 (600)	> 100	940 (-)	480
EK151-IA	1450	1050	13	650 (750)	1030 (650)	> 100	1150 (-)	520
EK152-IA	1550	1150	10	650 (750)	1050 (830)	> 100	1200 (-)	500
EP975-IA	1350	1050	14	750 (975)	750 (190)	> 100	1000 (-)	500

Table 1.14 Properties of heat-resistant alloys

Alloy	Mechanical properties at 20°C			Long-term strength					
				σ_{1000}, MPa, at T, °C			σ_{10000}, MPa, at T, °C		
	$\sigma_{0,2}$, MPa	σ_B, MPa	δ, %	700	800	900	700	800	900
VZh85	430	910	48	40	14	15	-	-	-
EI868	540	930	47	140	60	33	85	40	23
EP99	810	1170	49	230	150	65	210	80	38
EI435	175	610	38	-	-	-	-	-	-
EI602	740	36	120	40	-	-	-	-	-
EP914 (VZh131)	650	1000	35	410	140	-	-	-	-

cobalt heat-resistant alloys is carried out mainly by the carbide phases, primarily tungsten carbides. Most industrial cobalt alloys therefore contain from 0.25 to 1% carbon, as well as 10–20% nickel, which makes it possible to increase their heat resistance.

Below a brief description of the alloys used for the manufacture of blades of stationary gas turbines is given.

EI893 vacuum arc (ChN65WMTYuVA) The chemical composition of the alloy is governed by GOST 5632-72. Stamped billet blades are manufactured according to TU 108.02.005-76. Heat treatment of the blades is carried out in one of two modes:

1. heating up to 1160 ± 10°C, for 2 h, air cooling, aging at 900 ± 10°C for 8 h, 820 ± 10°C for 15 h, air cooling,
2. heating to 1160 ± 10°C for 2 h, air cooling, aging at 850 ± 10°C for 12-15 h, air cooling. After heat treatment, the microstructure of the alloy is characterized by the presence of a 10% γ' phase with an average particle size of 0.5 μm, 0.3–1.0% of carbides of the MC, M_6C, $M_{23}C_6$ type and borides of M_3B_2 and M_5B_3. Long-term strength curves of the alloy are shown in Fig. 1.59.

It should be noted that in the state of the alloy after low-temperature quenching and tempering temperatures of 800°C, the structure of the alloy is highly resistant, and therefore it is impossible to prevent its embrittlement at operating temperatures of 580-600°C, which occurs due to the precipitation of a large amount (up to 18-20%) fine γ' phase. In this state, the hardness of the alloy can reach values of 30-40 HR, and the limit of long-term strength at 600°C is reduced to two times.

EI929 vacuum arc (ChN55WMTKYuVA) The chemical composition of the alloy is governed by GOST 5632-72. Stamped billet blades are manufactured according to TU 108.02.003-84. Heat treatment of the blades is carried out in one of two modes:

1. heating up to 1000°C, holding for 2 h, transfer to the furnace heated to 1175 ± 10°C for 3 h, cooling with a furnace to 1000°C, holding 50-60 min, air cooling, aging at 850 ± 10°C, 15-16 h, air cooling,
2. heating to 1000°C, exposure 2 h, heating to 1175 ± 10°C, 3 h, cooling with a furnace to 950°C, exposure 60-90 min, cooling in air, aging at 850 ± 10°C for 15 h, air cooling. After heat treatment, the microstructure of the alloy is characterized by the presence of 38-40% γ' phase with an average particle size of 0.2-0.5 μm and 0.05 μm, 0.75-1.15% of carbides such as MC, M_6C, $M_{23}C_6$ and borides M_3B_2.

1.3.4.1 Cast Heat-resistent Nickel Basis Alloys

EI893(L) The widespread introduction of the deformed alloy EI893 due to its high stability during long periods of operation (Borzdyka, 1990; Levin et al., 1977; Getsov et al., 1965) led the researchers (Borzdyka, 1990; Getsov, 1967) to the idea of the feasibility of studying it in a molten state. It was found that the EI 893L alloy, as well as the deformed EI893 alloy, possesses a set of properties (including technological ones), which made it possible to introduce it for the manufacture of GTU guide blades. Castings of blades of various gas turbines of this alloy are manufactured according to the technical specifications TU 108.1109-82, TU 108.2.01.131-75, TU 108.02.042-82, TU 108.02.046-82, TU 108.02.104-84, TU 108.01.053-85. Heat treatment of the blades is performed according to the following conditions: heating to 1180 ± 10°C for 4 h, air cooling, aging at 850 ± 10°C, 12 h, air cooling. After heat treatment, the microstructure of the alloy is characterized by the presence of a 9-11% γ' phase with a particle size of 0.03-0.05 μm and 0.4-0.8% of carbides of the type MC, M_6C, $M_{23}C_6$ and borides M_3B_2.

IN738 LC The alloy IN738 LC is the most widespread foreign alloy for blades due to its increased heat resistance at a high level of long-term strength. Heat treatment of the blades is performed according to the following conditions: heating to 1120±10°C for 2 h, air cooling, aging at 840±10°C, 24 h, air cooling.

CNK-7 The alloy CNK-7 was developed by NPO CNIITMASH, NPO CKTI, ON LZTL, ON NZL. Blade castings of various gas turbines are manufactured according to TU 108.1086-85, TU 108.01.057-86, TU 481 981. 00010. The blades are heat treated according to the mode: heating up to 1180 ± 10°C, 4h, cooling with a furnace for 30-45 min to 1050 ± 10°C, exposure time 0-2 h, air cooling + heating up to 850±10°C, 24 h, air cooling. After heat treatment, the microstructure of the alloy is characterized by the presence of a 45% γ' phase particle size of 0.02-0.04 μm, as well as MC and $M_{23}C_6$ type carbides in an amount of 0.3-0.4%.

ZMI3U The alloy is designed by Zaporozhye National Technical University, PO LZTL[5] and PO NZL. Heat treatment of the blades (TU 108.1119-82) is performed according to the mode: heating up to 1180±10°C 4 h, air cooling + heating up to

[5] Production Association Leningrad Plant of Turbine Blades

830±10°C for 24 h, air cooling. After heat treatment, the microstructure of the alloy is characterized by the presence of 46-48% γ' phase with particle sizes of 0.3-0.7 μm and 0.03-0.05 μm as well as 0.4-0.5% of MC, $M_{23}C_6$ and borides M_3B_2.

ChS104 (CL-5) The chemical composition of the alloy, given in Table 1.5, indicates that the chromium content is increased to 20% and the lower Al and Ti contents are used for the formation of the γ' phase (compared to the SN-35M alloy). Heat treatment of blades manufactured according to TU 1-809-1040-96 is performed according to the following conditions: heating up to 1140-1170°C for 2-4 h, 1150°C for 4 h, air cooling + heating up to 850 ± 10°C for 16-24 h, cooling on the air.

FSX-414 The manufacturing of guide blades abroad Russia is usually made of heat-resistant cobalt-based alloys due to their good casting properties and high heat resistance. The FSX-414 alloy among them has received the greatest application. Segments of the GTK-25 I guide blades are made of this alloy. Heat treatment of the blades is performed according to the following conditions: heating up to 1150°C for 4 h, cooling with a furnace up to 980°C, holding 4 h, cooling with a furnace + heating up to 540°C, air cooling.

1.3.5 Composite Materials

Composite materials for service at high temperatures are developed both in Russia and abroad (Paton et al., 1987; Fainbron et al., 1993; Portnoi et al., 1979). There are fiber reinforced and layered composites, as well as eutectic and dispersion-hardened materials. The reinforcement of the matrix of nickel-based alloys is carried out practically only with tungsten wire fibers, which can work for a long time in contact with the matrix without losing strength. To obtain composite alloys, hardened with tungsten (40-70% volume fraction), the method of vacuum impregnation, followed by rapid cooling is used. As a matrix, alloys Nimocast 258 and EPD-16 were tested abroad, in the Russian Federation the alloy ZhS6K was applied. The mechanical properties of composites (Table 1.15) are significantly higher than those obtained for the matrix metal. The introduction of these materials for blades is delayed due to their high density, the need to protect against oxidation at high temperatures and the lack of information about the structural strength.

Table 1.15 Mechanical properties of composite alloys

Matrix	Amount of Tungsten fibres, %	$\sigma_B^{20°C}$, MPa	δ, %	$\sigma_{100}^{1000°C}$, MPa	$\sigma_{1000}^{1100°C}$, MPa
ZhS6K	50	580	0.8	-	150
EDP-16	40	670	1.2	260	133
Nimocast 258	40-50	850	0.4	234	130

The feature of composite materials, hardened with tungsten fibers, is their tendency to grow during thermal cycling. The possibility of influencing the tendency to deform by heat treatment has been established. A number of characteristics of these materials (thermal expansion coefficient, thermal conductivity, etc.) obeys the rule of mixtures in cases where the internal stresses do not exceed the yield strength of the matrix. Regarding composites with carbon fibers, they, firstly, have relatively low strength, and secondly, are not applicable at temperatures above 1000°C due to the interaction of fibers with the matrix.

Current practical application has received a different type of composite materials - dispersion-strengthened. There are two such alloys produced by industry: VDU-1 (Ni + 2.5% ThO_2) and VDU-2 (Ni + 2.5% HfO_2). Both alloys have high ductility and long-term strength σ_{1000} at 1100°C (50-100MPa), depending on the type of semi-finished products (sheet, rod), which is 1.5-2 times higher than that of complex alloyed ZhS6K material. At moderate temperatures, the strength properties of these alloys are at a rather low level ($R_{p0.2}$ = 120-200 MPa).

Alloys MA-754, WAZ-D and MA-6000E have been developed abroad Russia. The strength σ_{1000} of the alloy MA-6000E at 1100-1200°C is 105MPa. The alloy resists oxidation, thermal fatigue well and has a ductility of ≈ 3%. In Russia, VKNA-4U alloy was developed, whose long-term strength is characterized by the values shown in Fig. 1.60. One of the most promising in gas turbine construction of heat-resistant composite materials are eutectic alloys, which in fact are naturally-composite materials (Table 1.16). One of the phases of these alloys, usually quite ductile, serves as a matrix (Ni, Co) and the second one - as a hardening phase (TaC, NbC, Ni_3Al, Ni_3Nb, Cr_7C_3).

The potential applications of composite materials are presented in Fig. 1.61. The most promising among eutectic alloys in Russia are the VKLS alloys with the γ/γ'-MeC structure, in which MeC is filament-like niobium or tantalum monocarbide crystals. Alloys of this type are also known in France (COTAS) and the USA (NITAC) (Table 1.17). These alloys provide the requirement of a small value (≈0.2%) of the size mismatch between the periods of the crystal lattices of the γ and γ' phases and

Table 1.16 Composition of eutectic alloys

Alloy	Contents of element, % (remaining Ni)										γ, g/cm^3
	Co	Cr	Al	Nb	Ta	Mo	W	Re	C	V	
VKLS-10	10.0	7.0	5.6	3.8	-	1.0	11.0	-	0.45	1.0	8.53
VKLS-20	9.0	4.5	6.2	4.0	-	1.0	12.5	-	0.40	1.0	8.67
VKLS-20R	9.0	4.0	6.2	4.0	-	-	12.0	4.0	0.40	-	8.00
COTAC-744	10.0	4.0	6.0	3.8	-	2.0	10.0	-	0.45	-	8.51
COTAC-784	10.0	4.0	6.5	4.0	-	4.0	4.0	4.0	0.45	-	-
NITAC-13	3.3	4.4	5.4	-	8.1	-	-	6.2	0.54	5.6	8.69
NITAC-3-116A	1.9	7.0	6.5	-	8.2	-	-	6.3	0.25	4.0	8.60
NITAC-C	3.9	4.0	5.5	-	11.7	3.0	3.0	6.6	0.45	-	-

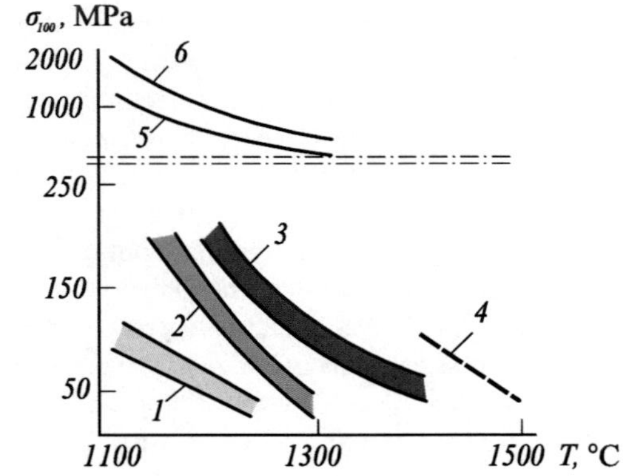

Fig. 1.61 Temperature dependence of long-term strength of composite materials on the base of heat-resistant alloys: 1 - nickel alloys, 2 - NiAl oxyde fibers ($\gamma = 6$ g cm^{-3}), 3 - carbon fibre in Wo ($\gamma = 10$ g cm^{-3}), 4 - NbAl-intermetallic in Wo ($\gamma = 9$ g cm^{-3}), 5 - Wo alloy with Re, Hf, and C, 6 - wire Wo alloy

Table 1.17 Properties of eutectic alloys

t_{expl}, °C	σ_B, MPa	$\sigma_{0,2}$, MPa	σ_{500}, MPa	δ, %	σ_B, MPa	σ_{500}, MPa	δ, %	σ_B, MPa	σ_{500}, MPa	δ, %
	VKLS-20				COTAC-744			matrix		
20	1500	740	-	18	1505	-	13	950	-	20
800	1230	1100	540	8	1170	-	12	-	-	-
900	980	760	340	16	910	310	10	880	305	14
1000	800	690	210	10	570	190	10	690	140	24
1100	480	410	130	8	-	-	130	370	45	21
1150	350	300	85	6	-	-	-	-	-	-

its weak dependence on temperature (Portnoi et al., 1979; Kachanov et al., 1995). An oriented composite microstructure in castings from eutectic alloys is formed in the course of directional solidification under the condition of a flat growth front, i.e. in the absence of concentration supercooling of the melt ahead of the growth front controlling crystallization.

For the formation of a composite microstructure in the process of directional crystallization, the statement is used: the smaller the crystallization temperature interval of the alloy ΔT, the more stable the flat growth front and, therefore, the faster the draw rate by eutectic composites can be obtained. At the same time, as the growth rate increases, the whiskers become thinner, and the distance between them is smaller. The γ/MeC interphase boundaries have a relatively low diffusion permeability, which contributes to increasing the stability of eutectic alloys at a temperature of 1000-1100°C for up to 1000h and more.

To stabilize the microstructure of the γ' phase particles, it is very promising to alloy the material with elements having a high binding energy of atoms in the lattice of the Ni_3Al compound and a low coefficient of self-diffusion of atoms in the γ solid solution, for example, W, Mo and especially Re.

The mechanical properties of eutectic alloys are given in Table 1.17 and Fig. 1.62, indicating their very high strength and ductility at room temperature. Thread-like

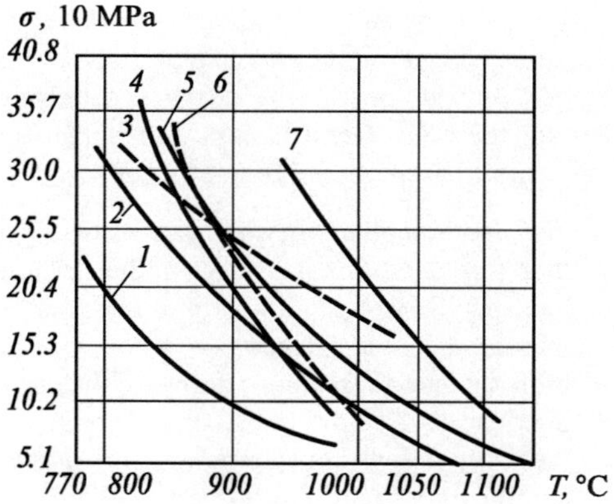

Fig. 1.62 Temperature dependence of long-term strength of eutectic and cast nickel and cobalt heat-resistent alloys:
1 - MAR-M-302,
2 - Co-(Cr, Co)$_7$C$_3$,
3 - (Co, Cr, Ni)-TaC,
4 - MAR-M-200,
5 - Ni-23Nb-4,4Al ($\gamma'/\gamma' - \delta$),
6 - Ni-TaC,
7 - Ni$_3$Al-Ni$_3$Nb-NiCr ($\gamma' - \delta$)

MC crystals prevent the propagation of cracks, and therefore their fatigue strength is high: σ_{-1} = 360 MPa, 320 MPa and 160 MPa at temperatures of 20°C, 1000°C and 1100°C for VKLS-20 alloy based on $2 \cdot 10^7$ loading cycles. The high long-term strength of these alloys (Fig. 1.63) allows them to be used for blades with metal temperatures of ≈1100°C. However, because of the long duration of the process of directed crystallization of castings (15 h) due to the low rate of formation of the composite structure of the product from eutectic alloys, these alloys are very expensive.

1.3.6 Ceramic Materials

Among the various ceramic materials, nitrides and silicon carbides (Si_3N_4 and SiC) are considered promising for components of gas turbines. These materials are significantly different from metallic materials, and therefore their use requires new

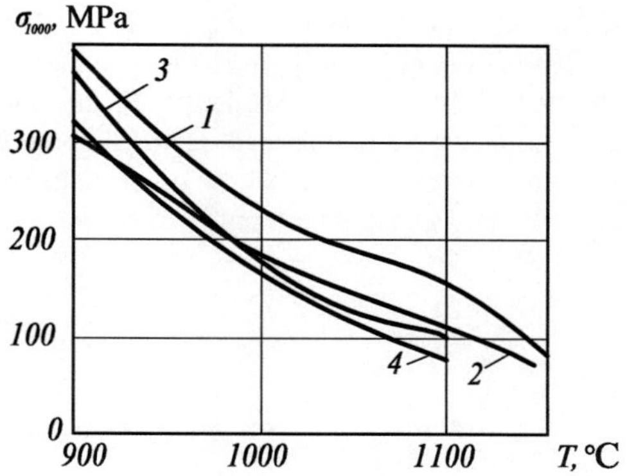

Fig. 1.63 Temperature dependence of long-term strength of heat-resistant nickel alloys with eutectic (1,2), single-crystal [001] (3,4)
1 - VKLS-20R,
2 - VKLS-20,
3 - CMSX-4 (treatment with different terms up to fracture on various temperature and stress levels),
4 - ZhS40 (type SC-83)

design methods and strength analysis of GTU elements. Currently, research works on the introduction of ceramic materials for parts of automotive gas turbines (disks, blades, nozzle apparatuses, combustion chambers, regenerators) are continuing in Russia, the USA, Great Britain, Japan and Germany. The features of the properties of cermets based on Si_3N_4 and SiC are:

- low coefficient of thermal expansion (3-5 times less than that of steel), and therefore the thermal stresses in the ceramic parts of GTU are less than in steel
- low specific weight (2.5-3 times less than that of metals)
- increased heat resistance
- high chemical resistance, allowing the use of low-grade fuels
- brittleness - fracture in the almost complete absence of plastic deformations, and therefore stress concentrators are not allowed in the components
- significant variation in properties associated with the features of the technology of manufacturing parts

The thermal conductivity of silicon nitride is lower than that of steels and nickel-based alloys. The thermal conductivity of silicon carbide is close to that of nickel-based alloys. Components made of silicon nitride (turbine disks) are produced of powder by hot pressing using magnesium oxide as a filler. The second method is isostatic pressing and sintering of parts (blade rims, nozzle apparatuses) in a nitrogen atmosphere. At the same time, lower strength properties are obtained (due to density, which is 20-30% less theoretical) than with hot pressing. However, isostatic pressing can be used to manufacture parts with a complex geometric shape. Due to the absence of filler, creep at high temperatures almost does not occur. Components of silicon carbide (for example, flame tubes) are manufactured either by hot pressing, or by sintering with impregnation of silicon, or slip casting. The disks are made of an alloy of Si_3N_4 + 6% Y_2O_3. The use also received solid solutions Al_2O_3 in silicon carbide.

Advances in the manufacture of ceramic disks can be illustrated by data from the company Daimler-Benz (Germany): for four years (1984-1990) the number of defects decreased from 14-40 to one in one of two disks of silicon nitride, ultimate strength when tested for bending micro-samples increased at a temperature of 20°C from 750 MPa to 980-1030 MPa, and at a temperature of 1200°C from 435 MPa to 480-703 MPa. The Weibull module increased to 11.3-15 at a temperature of 20°C, to 26.3 at 1100°C and to 11.6 at 1200°C.

To seal the flow-through part of relatively low-temperature gas turbines, cermet inserts are used that are installed in the turbine housing. One of the materials of the inserts, successfully operated during long service lives, is UMB-4S, manufactured using a heat-resistant metal base (nichrome) and solid lubricant of boron nitride. The characteristics of this material in the initial state are as follows:

Brinell hardness (HB):

- in the initial state	30-45
- after operation at a temperature of 900°C	90
- porosity,%	10-15
- bending strength, MPa	100
- heat resistance at 900°C for 100 h, mg/cm^2	8-10
- onset temperature of bulk oxidation in air atmosphere, °C	900
- coefficient of dry friction at v = 200m/s and temperature 800-1000°C	0.12
- rate of penetration of the comb of the retaining shelf, mm/km	1.15

It is worth to note the good workability and antifriction properties at temperatures up to 1000°C.

1.3.7 Titanium Alloys

Titanium alloys are used for case parts of disks and compressor blades. Their chemical composition is regulated by OST 1.90013-71. They have increased corrosion resistance, including in sea water (sea fog), and the same level of mechanical properties as stainless steels used for the same purpose. The specific strength of titanium alloys is higher than that of steel. Titanium alloys are used in deformed and cast states for housing parts and various welded assemblies operating depending on the composition and service life at temperatures up to 300-500°C. Marks of these alloys are given in Table 1.18. The mechanical properties are illustrated in Fig. 1.64.

During heat treatment in an air atmosphere and during operation at elevated temperatures, titanium alloys interact with oxygen to form solid solutions. Oxygen diffuses from the surface, forming a fragile, alpha-coated layer. The penetrating oxidation of titanium alloys is the main obstacle against increasing the operating temperature, even in those cases where the characteristics of long-term strength are sufficient to ensure the strength of certain parts. Compared with the deformed titanium alloys, cast titanium alloys have significantly lower fatigue strength and

Table 1.18 Titanium alloy grades used in welded structures

Alloy	T_{max}, °C	σ_B^{20}, MPa	Alloy	T_{max}, °C	σ_B^{20}, MPa
OT4-0	300	450-650	VT6, VT6L	400	900-1100
OT4-1	350	550-800	VT9	500	1050-1250
OT4	350	700-900	VT14, VT14L	400	850-1100
VT4	400	850-1050	VT15	500	900-1350
VT5, VT5L	400	700-950	VT20, VT20L	500	900-1200
VT5-1	450	750-1000	VT21L	500	≥1050
VT6S	450	800-1000	VT23	500	1100-1450

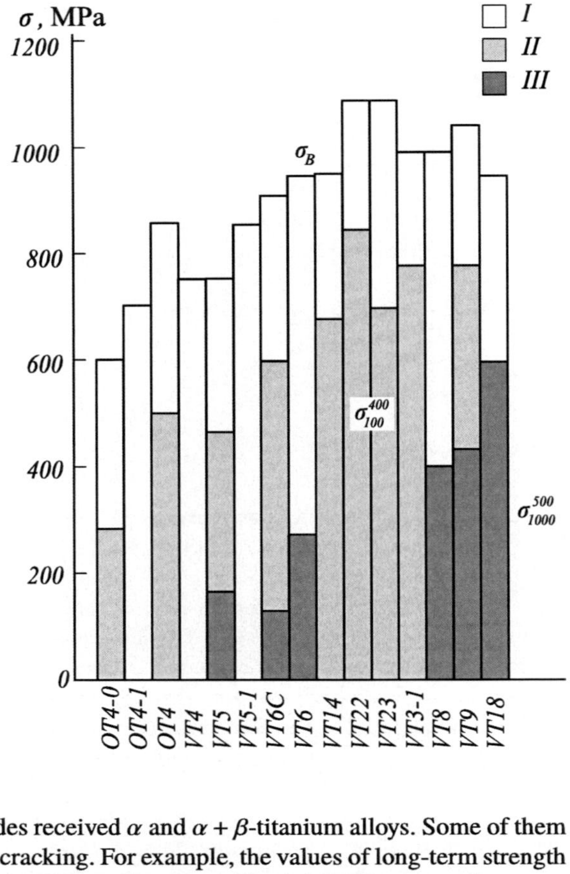

Fig. 1.64 Mechanical properties of titanium alloys (I - σ_{B}^{20}, II - σ_{100}^{400}, II - σ_{1000}^{500})

ductility. Application for blades received α and $\alpha + \beta$-titanium alloys. Some of them have a tendency to corrosion cracking. For example, the values of long-term strength σ_{1000}, obtained on samples from VT3-1 alloy in contact with NaCl, are equal to σ_{1000} in air at temperatures up to 250°C, and at 300°C, 350°C and 400°C less than σ_{1000} in air, by 1.12, 1.33 and 5.5 times, respectively. Due to the negative effect of chlorides on strength, it is usually not recommended to use alloying by Ti at temperatures above 300°C.

A new direction was introduced in the development of high-strength and heat-resistant structural titanium alloys based on a three-phase structure: α and β solid solutions and intermetallic dispersion hardening. It should also be noted that in titanium alloys at low temperatures and stresses lower than the yield point, irreversible deformations can accumulate over time - creep takes place, and under conditions of constant deformation the stress relaxation occurs (Zwicker, 1974; N.N., 1978). At room temperature, the long-term strength is usually 20-30% lower than the tensile strength.

However, the fatigue crack growth rate dl/dN for titanium alloys is significantly higher than for stainless steels. An increase in dl/dN with an increase in the yield strength was noted. Sea water and sea fog have a different effect on alloys with high

(≈ 900MPa) and relatively low (≈ 600MPa) strength levels. The fatigue resistance of high-strength titanium alloys is significantly reduced in sea mist, while for alloys with relatively low strength, the effect of sea salts on σ_{-1} is absent.

The following requirements for the chemical composition of alloys for parts operating in marine conditions are established: Al<5%, O_2<0.1%; Mo≥0.8-1%. Zirconium and nitrogen are undesirable elements, and V and Nb have a positive effect. Burns can be formed on components of titanium alloys with intensive grinding due to low thermal conductivity (see Table 1.6) drastically reducing fatigue strength. Residual tensile stresses in the surface layer also have a negative effect on fatigue strength. In connection with the above, the use of titanium alloys for compressor blades is effective only with a sufficiently high production technology.

At present, titanium alloys of more than 10 grades have been used for blades. The most common in stationary turbine components are high-strength alloys VT5, VT3-1 and TC5. Blades of aviation gas turbine compressors are made of alloys VT6, VT8 and VT9. The chemical composition of titanium alloys used for various parts of gas turbines is given in Table 1.19. A number of new high-strength alloys have been developed for the aircraft engine industry, the properties of which are listed in Table 1.20 (Shalin, 1994).

Table 1.19 Chemical content of titanium alloys

Alloy grade	Al	Sn	V	Zr	Mo	Cr	Si	Fe
VT1-0	-	-	-	-	-	-	<0.1	<0.3
VT5	5.0	-	-	-	-	-	-	<0.3
VT5-1	5.0	2.5	-	-	-	-	-	<0.3
VT-20	6.5	-	1.3	2.0	1.2	-	<0.1	<0.3
TS5	5.0	3.0	2.0	2.0	-	-	-	-
AT6	5.5	-	-	-	-	-	1.5	1.5
VT3-1	6.2	-	-	-	2.5	1.5	0.25	0.5
OT4-1	1.75	-	-	-	-	-	<0.15	<0.3
VT6	6.2	-	5.0	-	-	-	<0.15	<0.3
VT-8	6.5	-	-	-	3.3	-	0.3	<0.3
VT-9	6.5	-	-	1.5	1.5	-	0.3	<0.25

Table 1.20 Properties of high-strength titanium alloys

Alloy	σ_B^{20}, MPa	σ_B^{600}, MPa	$\sigma_{0,2/100}^{600}$, MPa	σ_0, MPa	K_{Ic}, MPa$\sqrt{m}$
VT18U	930	570	120-140	860	58.9
VT36	1050	650	160-180	900	42,16
VT22M	1250	-	-	-	-
834 (USA)	1050	660	180-200	920	40.92

Remark: σ_0 - low cycle fatigue stress with alternating loading and $N = 10^4$ cycles

1.4 Concluding Remarks

In this chapter, in addition to the presentation of issues related to the working conditions of the main parts of GTU for various purposes and with the characteristics of various materials used for their manufacture, the main attention is paid to the consideration of various types of damage in components during operation and laboratory tests. Numerous examples of damage in various components of the GTE are presented hereinafter to illustrate the theoretical and experimental study of the problems of deformation and fracture of materials under various external influences and when developing new computational methods for determining the strength of components in relation to actual operating conditions. The separation of damage caused by static, vibration and thermal cyclic stresses is traditional. Examples of low-cycle fatigue damage are also considered.

It should be noted that the specific design and operation of aircraft engines that had existed for many years now turns out to be implemented to a lesser extent. This is due to the conversion of aircraft engines and their use at compressor stations. On the other hand, in order to ensure competitiveness, the stationary gas-turbine engineering uses in its practice many solutions developed for aircraft engine plants and vice versa.

When examining various materials in this chapter, the issues of manufacturing blanks of parts and features of influencing the structure and properties of materials and their dependence on technology and operating conditions are deliberately excluded. Chapter 4 is devoted to covering these issues.

References

Anurov YM, Fedorchenko DG (2004) Osnovy Obespecheniya Prochnostnoi Nadezhnosti Aviacionnykh Dvigatelej i Silovykh Ustanovok (in Russ., Basics of Providing Strength Reliability of Airplane Engines and Power Plants). SPbGPU publisher, St. Petersburg

Borovkov VM, Getsov LB (eds) (2008) Materialy i prochnost' oborudovaniya TEC (in Russ., Materials and Strength of Thermal Power Station Units). Izd. Politekhnik, St. Petersburg

Borzdyka AM (1990) Yubilei zharoprochnogo splava (K 25-letiyu promyshlennogo primenneniya splava EI893) (in Russ., Anniversary of heat-resistant alloy - On the 25th anniversary of the industrial use of EI893 alloy). Metallovedenie i Termicheskaya Obrabotka Metallov (3):2–10

Dondoshanskiy VK (1978) Dinamika i Prochnost' Sudovykh Gazoturbinnykh Dvigateley (in Russ., Dynamics and Strength of Ship's Gas Turbine Engines). Sudostroenie, Leningrad

Fainbron AS, Pertsovskii NZ, Savin VN (1993) Fracture initiation of EP741NP PM nickel alloy in LCF tests. Metal Science and Heat Treatment 35(6):355–358

Gessinger GH (1984) Powder Metallurgy of Superalloys. Butterworths Monographs in Materials, Butterworth-Heinemann

Getsov LB (1967) Casting heat-resistant alloy for guide vanes of gas turbines. Metal Science and Heat Treatment 9(1):67–69

Getsov LB, Rybnikov AI (1993) Povrezhdeniya lopatochnogo apparata energeticheskikh gtu i ikh klassifikatsiya (in Russ., Damage to the scapular apparatus of energy gas turbines and their classification). Tyazheloe Mashinostroenie (5-6):14–17

Getsov LB, Shovkun VE, Filatova MA (1965) Opyt primeneniya splava EI893 dlya lopatok gazovykh turbin (in Russ., The experience of using EI893 alloy for gas turbine blades). Energomashinostroenie (2):30–32

Getsov LB, Rybnikov AI, Levin AE, Kovalev AN, Osyka AS (1993) The character and reasons of static damages of gas turbine blade (in Russ. Tyazheloe Mashinostroenie (2):30–34

Kablov EN (ed) (2006) Litejnye Zharoprochnye Splavy. Effekt Kishkina (in Russ., Casting Heat-resistant Alloys. Kishkin's Effect). Nauka, Moscow

Kablov EN, Toloraiya VN, Orekhov NG (2002) Single-crystal rhenium-bearing nickel alloys for turbine blades of GTE. Metal Science and Heat Treatment (7-8):274–278

Kachanov EB, Petrushin NV, Svetlov IL (1995) Heat-resistant eutectic alloys with carbide-intermetallic strengthening. Metal Science and Heat Treatment 37:154–159

Khimushin FF (1969) Zharoprochnye Stali i Splavy (in Russ., Heat-resistant Steels and Alloys). Metallurgiya, Moscow

Korsov YG (ed) (1978) Gazoturbinnye ustanovki. Kompressornye mashiny i tyagodut'evoe oborudovanie (in Russ., Gas Turbine Units. Compressor Machines and Draft Equipment), no. 165 in Trudy CKTI, NPO CKTI, Leningrad

Kryukov II, Leont'ev SA, Platonov VS, Kalinin NA (2008) Analiz i sistematizatsiya kharakternykh defektov nagnetateley gpa (in Russ., analysis and systematization of characteristic defects of gas pumping units). Gazoturbinnye Tekhnologii (10):30–33

Kryukov II, Rybnikov AI, Platonov VS, Leont'ev SA, Kovalev AG, Shmelov NG, Andreev PA, Moshnikov AV (2009) Diagnostika treshchin, obrazuyushchikhsya v processe ekspluatacii v elechnikh pazakh diskov gazovykh turbin GTK-10-4 i GT-750-6, metodami nerazrushayushchego kontrolya (in Russ., Diagnostics of cracks formed during exploitation in gas grooves of gas turbine disks GTK-10-4 and GT-750-6 using methods of nondestructiv control). In: Petreniya YK, Getsova LB (eds) Prochnost' materialov i resurs elementov energooborudovaniya, St. Petersburg, no. 296 in Trudy CKTI, p 261–272

Levin EE, Pivnik EM, Generson IG (1977) Opyt primeneniya zharoprochnogo splava EI893 dlya izgotovleniya lopatok gazovykh turbin (in Russ., The Experience of Using Heat-resistant Alloy EI893 for the Manufacture of Gas Turbine Blades). LDNTP, Leningrad

Nikitin VI (1987) (in Russ., Corrosion and Protection of Gas Turbine Blades). Mashinostroenie, Leningrad

NN (ed) (1978) Forging and Properties of Aerospace Materials, Metals Society Series, vol 188, Metals Society, London, Proceedings of an International Conference Organized by Activity Group Committee 2 (Applied Metallurgy and Metals Technology) of the Metals Society, in Association with the Leeds and Bradford Metallurgical Societies, held in the Houldsworth School of Applied Science, University of Leeds, on 5-7 January 1977

Parshin AM (1972) Struktura, Prochnost' i Plastichnost' Nerzhaveyushchikh i Zharoprochnykh Stalei i Splavov, Primenyaemykh v Sudostroenii (in Russ., Structure, Strength and Plasticity of Stainless and Heat-resistant Steels and Alloys used in Shipbuilding). Sudostroenie, Leningrad

Paton BE, Stroganov GB, Kishkin ST, Bokshteyn SZ, Logunov AV, Malashenko IS, Movchan BA, Chumakov VA (1987) Zharoprochnost' liteynykh nikelevykh splavov i zashchita ikh ot okisleniya (in Russ., Heat Resistance of Cast Nickel Alloys and Their Protection Against Oxidation. Naukova Dumka, Kiev

Petrushin NV, Ospennikova OG, Svetlov IL (2017) Monokristallicheskie zharoprochnye nikelevye splavy dlya turbinnykh lopatok perspektivnykh GTD (in Russ., Single crystal heat-resistant nickel alloys for turbine blades of perspective GTU). Aviacionnye Materialy i Tekhnologii (S):72–103

Pigrova GD (1980) O metodakh prognozirovaniya TPU faz v splavakh na nikkelevoi osnove, primenyaemykh dlya gazovykh turbin (in Russ., About methods for predicting topologically close-packed phases in nickel-based alloys used for gas turbines). Trudy CKTI (182):59–65

Portnoi KI, Salibekov SE, Svetlov IL (eds) (1979) Struktura i Svoistva Kompozicionnykh Materialov (in Russ., Structure and Properties of Composite Materials). Mashinostroenie, Moscow

Rtishchev VV (1992) Perspektivnye anizotropnye materialy lopatok stacionarnykh GU so stolbchattoi i monokristallicheskimi strukturami (in Russ., Promising anisotropic materials of blades of stationary GU with columnar and single-crystal structures. Trudy CKTI 270:104–119

Rtishchev VV, Khrabrov PV, Korsov YG (1986) Issledovaniya v oblasti sluzhebnykh kharakteristik materialov - osnova povysheniya nadezhnosti i resursa gazovykh turbin (in Russ., Researches in the field of service characteristics of materials - the basis for increasing the reliability and resource of gas turbines). Energomashinostroenie (6):25–30

Shalin RE (1994) Novye materialy i tekhnologii - perspektiva razvitiya i sozdaniya aviacionnoi tekhniki (in Russ., New materials and technologies - the prospect of development and creation of aviation technology). Problemy Mashinostroeniya i Nadezhnosti Mashin (2):3–12

Shalin RE, Svetlov IL, Kachanov EB, Toloraya VN, Gavrilin OS (1997) Monokristally nikelevykh zharoprochnykh splavov (in Russ., Single Crystals of Nickel Heat-resistant Alloys). Mashinostroenie, Moscow

Starozhuk YP (1978) Kamery Sgoraniya Stacionarnykh Gazoturbinnykh i Parogazovykh Ustanovok (in Russ., Combustion Chamber of Stationary Gas and Steam Turbine Units). Mashinostroenie, Leningrad

Zwicker U (1974) Titan und Titanlegierungen, Reine und angewandte Metallkunde in Einzeldarstellungen, vol 21. Springer, Berlin, Heidelberg

Chapter 2
Deformation and Strength of Heat-resistant Materials under Static and Cyclic Loading

2.1 Mechanical Behavior under Uni-axial Stress State

2.1.1 Elasticity and Plasticity

When discussing the issues of material deformation, macrodeformations (i.e., deformations averaged over the length and section of the sample or part) and microdeformations (varying from grain to grain) are usually distinguished. To assess the strength and plasticity of a material, the most widely used characteristics are determined by uniaxial tensile loading of cylindrical and sometimes flat section samples. For the analysis only macrostrains are taken into account (ultimate strength σ_{B}, yield strength $\sigma_{0.2}$[1], elongation at fracture δ and relative shortening ψ according to GOST 1497-84).

These characteristics, both at room temperature and at elevated temperatures, are usually determined from tests in which the strain rate is constant and the change in load is recorded (tension curves). As a result of such tests, a load-elongation diagram is recorded by the machine. From the diagram, among the generally accepted characteristics listed above, the elastic limit σ_{el}, the proportionality limit σ_{prop} and true separation resistance[2] S_{k} can be determined. To construct a true deformation diagram in the plastic region, the diagram obtained on the machine must be rebuilt using the known value of the Young's modulus[3] E.

For strains that do not lead to necking, the stress values $\sigma^{(j)}$ are calculated based on the current values of the cross-sectional area A_j as follows

[1] In material testing this value is denoted by $R_{\mathrm{p}0.2}$.

[2] The true break strength is the stress in the neck of a tensile specimen, defined as the ratio of the tensile force acting on the specimen immediately before its rupture to the cross-sectional area of the specimen in the neck: $S_{\mathrm{k}} = F_{\mathrm{k}}/A$, where F_{k} is the load at the moment of specimen rupture, A is the true cross-section area of the sample at the moment of rupture .

[3] The angle of slope of the diagram obtained by the testing machine is related to the compliance of the system and has almost no relation to the value of the Young's modulus.

L. B. Getsov, *Materials and Strength of Gas Turbine Parts*, Advanced Structured Materials 150, https://doi.org/10.1007/978-981-16-0534-5_2

$$A_j = A_0(1 - \psi_j) \approx \frac{A_0}{1 + \varepsilon_j}. \tag{2.1}$$

Analysis of stress-strain behavior for a number of materials has shown that for a part of the stress-strain curve before the necking, the values of true stresses σ and total deformations in logarithmic coordinates fit on a straight line. Therefore, the following dependency was proposed

$$\sigma = A\varepsilon^{1/m}, \tag{2.2}$$

where A and m are material parameters. By logarithmizing and differentiating expression (2.2) we obtain

$$m = \frac{\sigma}{\varepsilon}\frac{d\varepsilon}{d\sigma}. \tag{2.3}$$

To determine the parameter m, the graphical method can be applied (Fig. 2.1) such that $m = \sigma_A/(\sigma_A - \sigma_a)$.

With increasing temperature m increases (hardening decreases). However, at relatively small values of ε, the value of m, corresponding to different degrees of deformation, is not constant and a maximum is observed at deformation of 0.8-2%. The value of m, of course, depends on the strain rate. Thus, Eq. (2.2) is unsuitable for the description of stress-strain diagrams of high-temperature alloys. However, knowing the dependence of the parameter m on plastic deformation ε_{pl}, the deformation curve can be represented as consisting of several sections of curves, which correspond to different exponents. The parameter m is widely used in calculating the stress state of components with local stresses exceeding the elastic limit. As was shown in Rabinovich (1966), for $m = 5.5 - 6.0$, the redistribution of stresses in the disks tested for start-up occurs by about 90% with respect to the ideal plastic redistribution. Thus, heat-resistant alloys of grades EI481, EI612, EI787, EI607, having

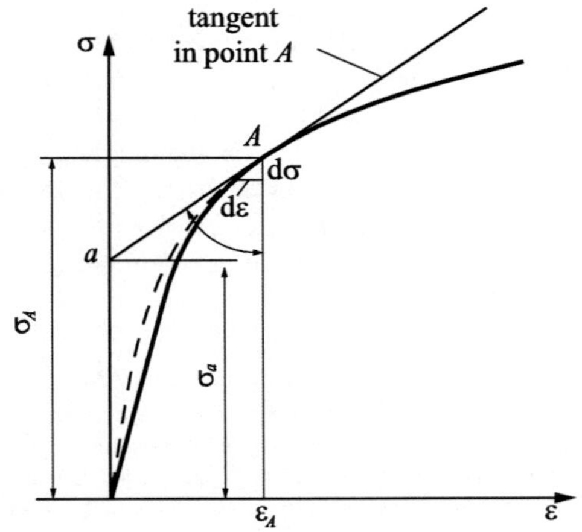

Fig. 2.1 Diagram for the estimation of the material parameter m at the tension curve (solid line - line with elastic modulus incline, dashed line - parabolic approximation)

the values $m = 5.5 \ldots 15$ for temperatures in the range 20 ... 700°C and $\varepsilon_{pl} = 0.2\%$, are able to redistribute the stresses almost completely.

A more general formula for describing stress-strain curves at different temperatures is

$$\varepsilon = \frac{\sigma}{E} + r_0 \left(\frac{\sigma}{\sigma_{el}} \right)^{n_0}, \tag{2.4}$$

where r_0 and n_0 are material parameters depending on the temperature. By increasing the strain rate, starting from a certain value, the creep processes will no longer affect the stress-strain curves and they will merge among themselves. The resulting curve is usually called the instantaneous stress-strain curve. For example, for the alloy EI765 at a temperature of 750°C, the instantaneous stress-strain curve can be obtained in experiments with a rate not exceeding 314%/h, and at 800°C - with a rate not less than about 2300%/h (Fig. 2.2). A further sharp increase in the strain rate may cause an increase in the resistance to plastic deformation.

2.1.2 Young's Modulus

The value of the Young's modulus E and the accuracy of its determination are of great importance when calculating the stresses in the components. Two methods for determining the Young's modulus including static and dynamic are widely used. The relatively high measurement speed and the range of applied stresses virtually exclude the possibility of creep processes for the static method. Nevertheless, between the values of E_{dyn} and E_{st} there is usually a difference reaching 20%, associated with the use of different stress ranges: with the dynamic method $\sigma < 20 \ldots 50$ MPa, while with the static method $\sigma \geq 200 \ldots 400$ MPa (Fig. 2.3). Certain differences in the values of the Young's modulus are obtained when it is measured on samples of different sizes, in different laboratories, on metal of different heats, etc.

It should be noted that in the reference literature, the values of the elastic modulus of materials are given, as a rule, without specifying the method of determination. This, in turn, can lead to erroneous conclusions when comparing stresses (strains)

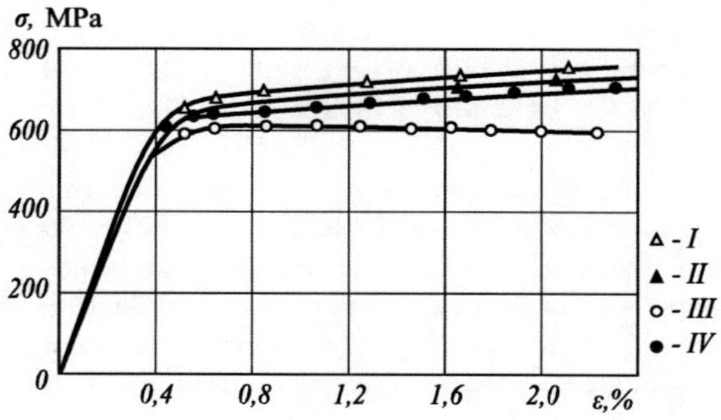

Fig. 2.2 Deformation diagram (tension) for alloy EI765 at temperature 750°C (*I, II*) and 800°C (*III, IV*) and different strain rates $\dot{\varepsilon}$:
I, III - 314%/h,
II, IV - 2400%/h

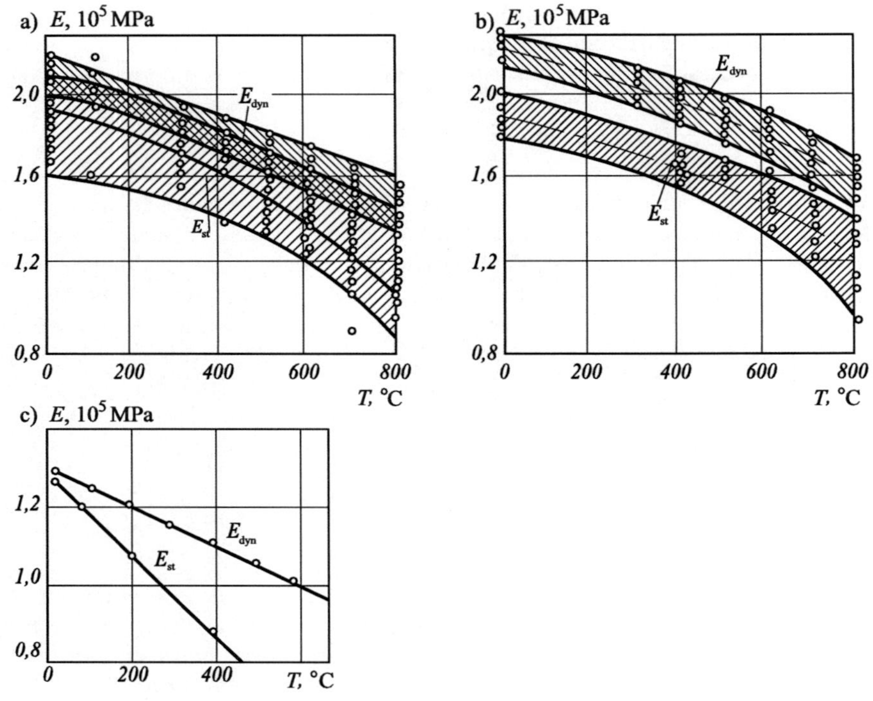

Fig. 2.3 Dependence E_{st} and E_{dyn} from the temperature T: a) austenitic steel, b) nickel-based alloys, c) titanium-based alloys

in parts made of different materials. Relatively small differences in the chemical composition and heat treatment regime have, as a rule, less impact on the values of E compared to the differences introduced by the method of determination.

In some cases, the dependence of the elastic modulus on the angle between the crystallographic orientation of the grain and the direction of load application should be taken into account. So, for example, for a CNK-7DC alloy at a temperature of 20°C $E_{dyn[hkl]} = 126.3 \ldots 298.2$ GPa; for ZhS32 alloy at 20°C $E_{[001]} = 135 \ldots 140$ GPa, and $E_{[111]} = 280 \ldots 300$ GPa. The same picture (Table 2.1) is also observed at other temperatures, as well as for the alloys ZhS32 and Rene 4 (63.17% Ni, 7.53% Co, 9.26% Cr, 1.6% Mo, 5.88% W, 3.77% Al, 4.24% Ti, 3.96% Ta), see Rtishchev (1992).

Changing the orientation leads to the values of the Young's modulus, which differ by up to 2.5 times (see Table 2.1). Similar results were obtained for alloys IN792 and CM247 (Fig. 2.4). In this regard, in the blades in the conditions of fulfilling the hypothesis of flat cross-sections, the stress distribution may differ significantly from the strain distribution. To determine the orientational dependence of the Young's modulus one may use the following relations (Erickson and Harris, 1994)

Table 2.1 Effect of orientation on the Young's modulus

Material	Orientation	E, GPa, at temperature T, °C						
		20	700	750	800	900	1000	1100
CNK7DC	[001]	143,5	117.3	-	114	107.3	97.9	85.8
	[011]	230.6	194.8	-	187.8	177.9	168.5	150.5
	[111]	286.8	243.2	-	236	227.4	213.9	197.3
Rene 4	[001]	-	-	104	-	-	90	-
	[001]	-	-	185	-	-	163	-
	[001]	-	-	253	-	-	226	-
	[001]	-	-	170	-	-	153	-
	[001]	-	-	166	-	-	119	-
	[001]	-	-	181	-	-	166	-
ZhS32	[001]	135-140	-	-	-	-	-	-
	[111]	280-300	-	-	-	-	-	-

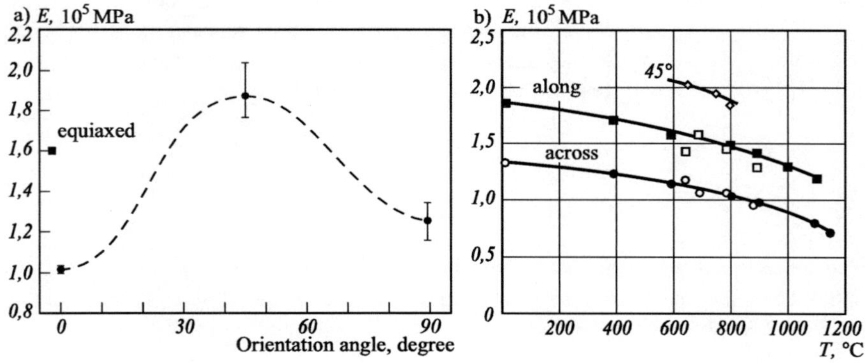

Fig. 2.4 Dependence of the Young's modulus E on the crystal orientation angle in the case of the alloy DSCM247LC at temperature 850°C (a) and the temperature T in the case of the alloy IN 792 DS at $\dot{\varepsilon} = 0.4 \pm 0.1$%/min (b)

$$E_{[\mathrm{hkl}]} = E_{[001]}/f\left(A_{[\mathrm{hkl}]}\right), \tag{2.5}$$

$$A_{[\mathrm{hkl}]} = \frac{\mathrm{h}^2\mathrm{k}^2 + \mathrm{k}^2\mathrm{l}^2 + \mathrm{h}^2\mathrm{l}^2}{(\mathrm{h}^2 + \mathrm{k}^2 + \mathrm{l}^2)^2}, \tag{2.6}$$

$$f(A_{[\mathrm{hkl}]}) = 1 - DA_{[\mathrm{hkl}]}, \tag{2.7}$$

$$D = 2 + 2\nu_{[001]} - \frac{E_{[001]}}{G_{[001]}} \tag{2.8}$$

with ν - the Poisson's ratio and G - the shear modulus. It was found that D weakly depends on the alloy composition and temperature. From (2.6) follows that for orientation [001] $A_{[001]} = 0$ and for [001] $A_{[111]} = 1/3$.

2.1.3 Heterogeneity of Plastic Deformations

The anisotropy of the material properties and the heterogeneity of its structure lead to a non-uniform distribution of plastic deformations in the polycrystalline solid. For example, studies carried out by the polarization-optical method (Kuznecov, 1964) showed that the steel of grade 12Ch18N10T is deformed elastically only for stresses below $\sigma \leq 100$ MPa. For $\sigma = 160$ MPa, about 50% of grains undergo plastic deformation, and at $\sigma = 200$ MPa, plastic deformation is found in all grains. From the very beginning, plastic deformation in grains is very non-uniform: there is a concentration of deformation at double boundaries and grain boundaries. The direction of the principal strains, both elastic and plastic, inside individual grains also varies considerably: the range of changes reaches 33° with the sample deformed by 1.2%, the maximum deviations from the direction of stretching reach 25°. As an illustration, Fig. 2.5 shows the distributions of strains in grains of specimens from 12Ch18N10T steel and titanium alloy (Kuznecov, 1964).

2.1.4 Scale Factor

The mechanical properties of materials are influenced by sample sizes - the so-called scale factor. The degree of this influence is determined by metallurgical and

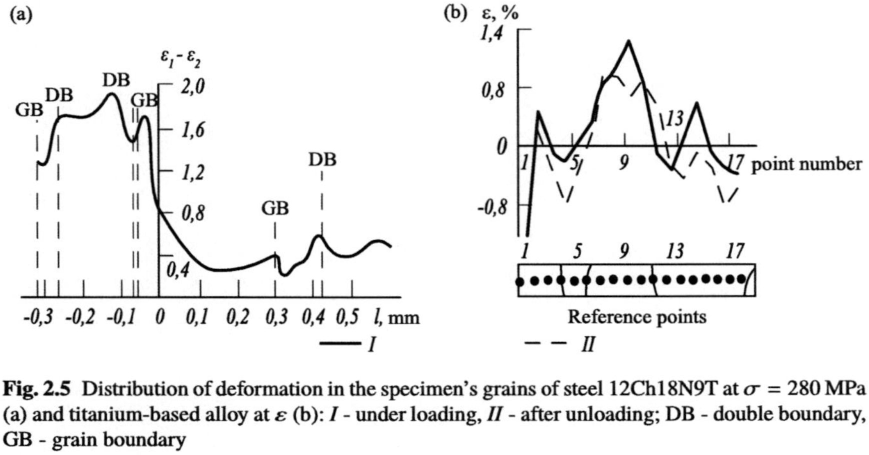

Fig. 2.5 Distribution of deformation in the specimen's grains of steel 12Ch18N9T at σ = 280 MPa (a) and titanium-based alloy at ε (b): *I* - under loading, *II* - after unloading; DB - double boundary, GB - grain boundary

technological reasons and the actual scale factor. It is usually quite difficult to separate the influence of metallurgical reasons (limited hardenability of the metal of large blanks, causing heterogeneity of the structure over the cross section, inclusions of the central zone of the ingot, etc.). In the process of machining in the samples and parts, a surface layer is formed, and residual stresses are created that influence the mechanical properties.

Actually, the scale factor is especially pronounced with brittle fracture of the metal. With ductile failure, an increase in sample sizes leads to a reduction of ductility. For example, when testing the forgings made of steel EI481 at a temperature of 20°C on cylindrical specimen for tension test (d = 5 mm, l = 25 mm), the ductility characteristics are found within the ranges $\delta = 17.5\ldots24\%$ and $\psi = 21\ldots35\%$, while for specimens with d = 8 mm and l = 80 mm, cut from the same blanks, significantly smaller values $\delta = 12.5\ldots22.3\%$ and $\psi = 9.6\ldots31.9\%$ were obtained. The most pronounced non-uniform distribution of strains and the scale factor is observed in samples with stress concentrators.

2.2 Creep and Stress Relaxation

2.2.1 Creep Phenomenon

The effect of the load on most metallic and non-metallic materials causes an increase in strain over time. This property is commonly referred to as creep, regardless of what temperatures and stress levels it takes. The laws governing the creep processes at constant temperatures and stresses may vary depending on the level of temperature (low-temperature creep for $0 < T < 0.25\ T_{\text{melt}}$, creep at medium temperatures, for $0.25\ T_{\text{melt}} < T < 0.5\ T_{\text{melt}}$ and high temperature creep, for $T > 0.5\ T_{\text{melt}}$), stress level ($\sigma > \sigma_{\text{el}}$ and $\sigma < \sigma_{\text{el}}$), the duration of creep tests (short-term creep for $t \ll 1$ h, short duration, for $1\ h < t < 100\ -\ 200$ h, moderate duration for $100\ -\ 200\ h < t < 3000 - 10000$ h and long-term creep 20000 – 300000 h). Here σ_{el} is the elastic limit, T_{melt} is the melting point and t is the time.

The creep curve (Fig. 2.6) usually has the following stages

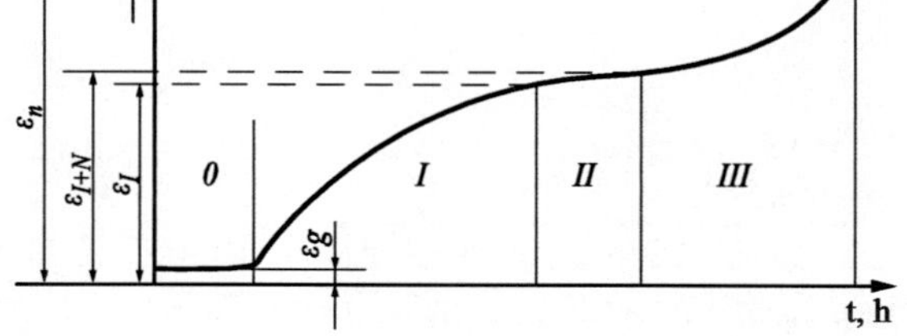

Fig. 2.6 Typical creep curve

- the incubation period during which the creep after the application of the load does not manifest itself, *0* (this stage is usually very short and often not captured even under tests for short-term creep)
- primary stage *I* (creep rate gradually decreases)
- secondary (steady) stage *II* (creep rate is constant)
- tertiary (accelerated) stage *III* (ends with creep fracture)

At very high temperatures under short-term creep conditions, stage *I* may be absent Moroz et al. (1962); Rabotnov and Mileiko (1970): the deformation accumulates at a constant rate. Under some test conditions, and for several materials, secondary creep stage is not observed (Naumenko et al., 2011b,a; Semenov et al., 2014; Eisenträger et al., 2017; Getsov et al., 2020).

For example, at moderately high temperatures and low stresses, the primary creep stage can extend to a very long test period (30000-80000 h). At low temperatures, a continuous decrease in the creep rate during the stage *I* may end after the accumulation of a certain deformation, which does not change with further test duration under load (the creep rate becomes zero). Such a character of creep curves was found in high-strength steels at a temperature of 20°C, α-titanium alloys at $T \approx 350$°C, nickel-based alloys at $20-400$°C. At high stresses, the stage of unsteady creep often passes into the stage of accelerated creep, bypassing the stage of steady-state creep.

Failure in the absence of the accelerated creep stage is also frequently observed. This behavior is common in materials that exhibit relatively low ductility at the test temperature. Cylindrical specimens from such materials fail during the secondary creep stage *II*. Failure is also observed for multi-axial stress state characterized by high stiffness, for compression loading, and when stretching thin-sheet and tubular samples and parts, especially in cases where the number of grains in the cross section of the sample is small. When removing the load, the value of accumulated creep strain decreases slightly. This process, called reverse creep (reverse elastic aftereffect), develops over time and gradually fades. In this connection, creep strain is sometimes divided into recoverable and irrecoverable parts. The recoverable part for metallic materials under non-cyclic loading is usually quite small.

Usually, the creep limit at a given temperature is the stress at which a given value of creep strain accumulates over a specified time (for example $\sigma^T_{0.5/1000}$ - stress at temperature T for 0.5% creep strain after 1000 h). Sometimes the creep limit refers to stress, which corresponds to a given value of the creep rate at the steady-state creep stage (for example $\sigma^{600°C}_{10^{-5}}$ - stress at temperature 600°C for creep rate of 10^{-5}%/h). Creep curves in stages *I* and *II* are well described using the following Soderberg formula (Soderberg, 1936)

$$p = a[1 - \exp(-ct)] + bt \tag{2.9}$$

in which the parameters a, b and c can be represented as power-law dependencies on stress

$$a = A\sigma^k, \quad b = B\sigma^m, \quad c = C\sigma^l$$

Differentiating w.r.t. t expression (2.9) by time, we get

$$\dot{p} = ac\exp(-ct) + b \tag{2.10}$$

The determination of the parameters a, b, c in Eq. (2.9) using the experimental creep curve is carried out in the manner illustrated in Fig. 2.7. The intensity of the creep process at the unsteady stage is very sensitive to the heat treatment mode, the surface work hardening that has arisen during machining (especially for samples and parts whose dimensions are comparable to the thickness of the peeled layer).

The creep resistance characteristics of metals and alloys are non-monotonic functions of temperature (Fig. 2.8). For example, in ARMCO iron, creep was observed at temperatures of -75 and -40°C (Moroz et al., 1962), and in technical titanium and its alloys at room temperature. At higher temperatures (20°C for ARMCO iron, 250 -350°C for titanium), the aging processes sharply inhibit the accumulation of creep deformation at the same σ/σ_B values. High-strength chromium-nickel-molybdenum steels at 20°C are similarly manifested. In nickel-based alloys, at temperatures of 500-550°C, precipitations are formed (Guinier-Preston zones, see Parshin, 1972). As a result the creep strain accumulated at these temperatures is significantly less than at lower temperatures.

The creep rate and the strain accumulated during the primary stage is significantly affected by the preliminary instantaneous plastic deformation (Fig. 2.9). The nature of this influence depends on the type of alloy and the sign of plastic deformation (Rabotnov and Mileiko, 1970). Different effects of plastic deformations of the same sign on the creep behavior of heat-resistant alloys of different classes (see Fig. 2.9a) are associated with different intensity of processes developing under the effect of hardening: an increase in the density of active dislocations, leading to acceleration of creep, on the one hand, and crushing of the substructure slowing down creep on the other. Plastic strains of the opposite sign (compression) cause the formation of a field of residual stresses of the second kind, which, when tensile stresses are applied, are added to them, causing an acceleration of creep in all materials. The creep resistance of cast alloys with an oriented structure depends on the loading

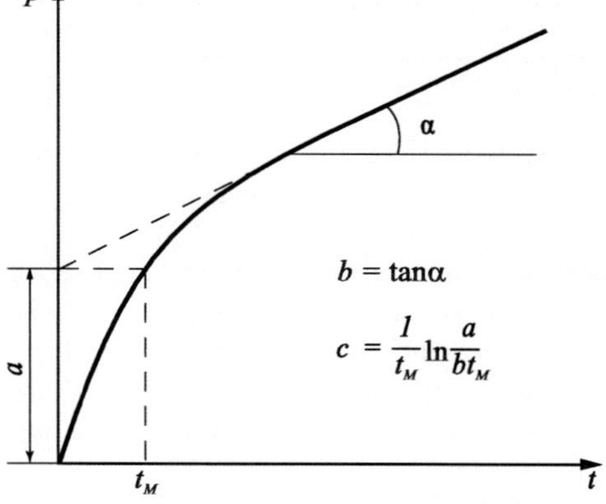

Fig. 2.7 Determination of the parameters in the Soderberg equation (2.9)

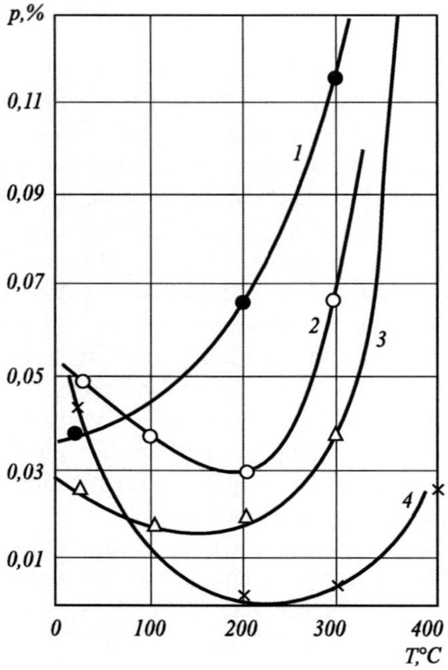

Fig. 2.8 Creep strain p vs. temperature for different materials (1 - 20Ch13, 2 - 35ChGSA, 3 - 25Ch2M2FA, 4 - EI765): t = 300 h, σ = 600 MPa (1-3) and σ = 800 MPa (4)

direction. So, for the IN792 alloy at 650°C and σ = 770 MPa, the creep rate along the [001] crystals is many times smaller than across (Fig. 2.10). At a temperature of 900°C, the difference in the curves at σ = 220 MPa is not so great.

Having postulated the constancy of the creep rate in the secondary stage, the formulas proposed by Ludwik (1909); Norton (1929) and Nadai (1938) are usually applied to describe the creep rate

$$\dot{p} = A\sigma^k \tag{2.11}$$

$$\dot{p} = A\exp(C\sigma) \tag{2.12}$$

$$\dot{p} = C\sinh(m\sigma), \tag{2.13}$$

respectively. Here A, C, k and m are material parameters at a given temperature, determined from experiments. At constant stresses, these formulas correspond to the theory of steady creep rate. The same formulas can be used in cases where the primary stage makes a relatively small contribution to creep deformation or when the average creep rate is used in the analysis. At high temperatures, when the creep rate is practically independent of time, these formulas characterize creep processes throughout the life of the materials. The indicated dependencies, however, do not reflect the turning points observed in the creep curves (Fig. 2.11)[4].

[4] The existence of turning points is observed in some materials and for $\dot{p} < 10^{-5}$%/h (Rozenblyum and Vinogradov, 1973)

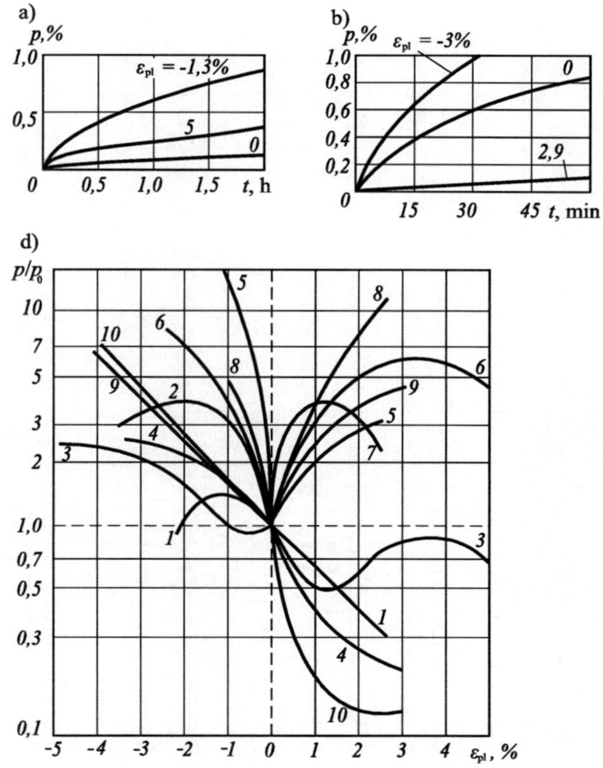

Fig. 2.9 Influence of plastic strain ε_{pl} on creep resistance p of heat-resistant steels and alloys (p_0 - creep deformation in the absence of plastic deformation):
a) EI765 at 750°C, σ = 400 MPa, b) 20Ch23N18 at 750°C, σ = 130 MPa, c) 15Ch12WNMF at 550°C, σ = 320 MPa, d) various alloys (1 - EI723 at 750°C, σ = 550 MPa, t = 1 h, 2 - EI802 at 500°C, σ = 400 MPa, t = 1 h, 3 - EI802 at 550°C, σ = 320 MPa, t = 1 h, 4 - 20Ch23N18 at 750°C, σ = 130 MPa, t = 1 h, 5 - EI765 at 750°C, σ = 400 MPa, t = 1 h, 6 - EI765 at 700°C, σ = 600 MPa, t = 1 h, 7 - EI698 at 700°C, σ = 500 MPa, t = 100 h, 8 - EI698 at 750°C, σ = 420 MPa, t = 1 h, 9 - EI607A at 650°C, σ = 520 MPa, t = 1 h, 10 - 20Ch23N18 at 800°C, σ = 60 MPa, t = 50 h)

To describe the dependence of the creep rate at the steady stage on temperature for constant stresses, the following formula is widely used

$$\dot{p}_{\min} = B \exp\left(-\frac{U}{RT}\right), \tag{2.14}$$

where B is a parameter, U is the creep activation energy and R is the universal gas constant.

If we accept that for the same material the power dependence $\dot{p}_{\min} = A_1 \sigma^{n_c}$ and the formula (2.11) are valid, in which $A_1 = A(T)$, $n_c = n_c(T)$, $B = B(\sigma)$ and

Fig. 2.10 Influence of the orientation on the creep curves of the alloy IN792DS at temperature 650 and 900°C: 1, 4 - T = 650°C, σ = 770 MPa, 2, 3 - T = 900°C, σ = 220 MPa, 2, 4 - fracture in longitudinal direction, 1, 3 - fracture in transverse direction

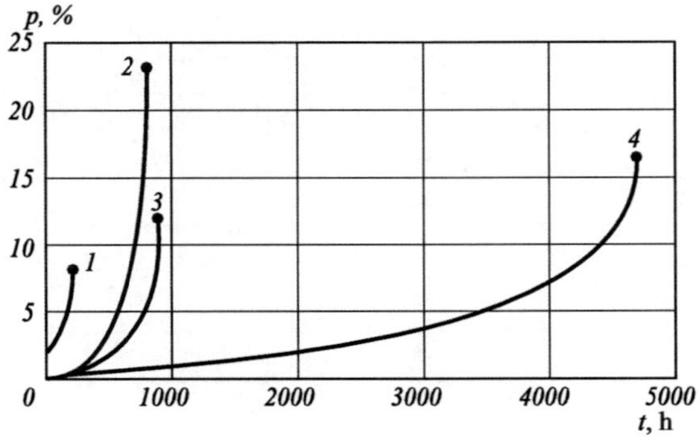

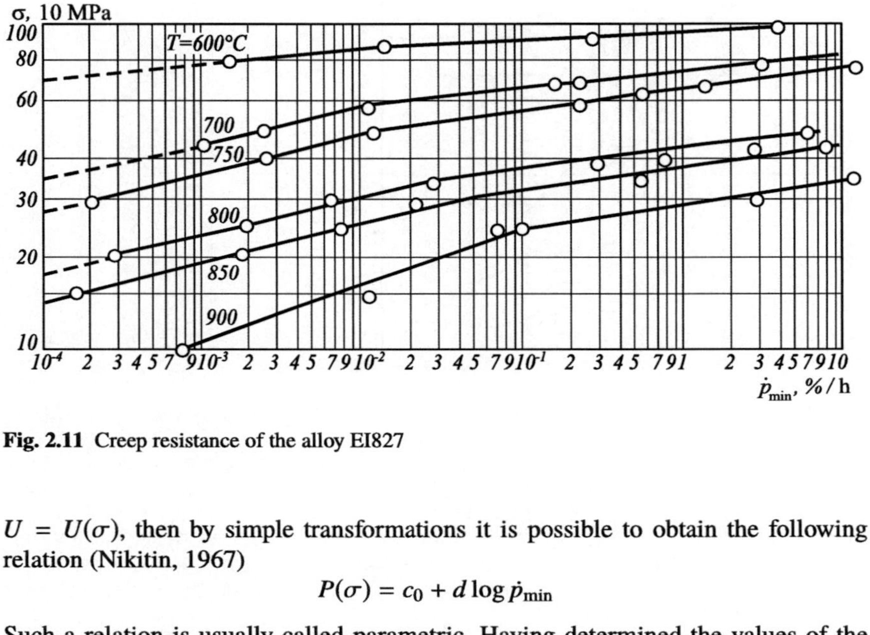

Fig. 2.11 Creep resistance of the alloy EI827

$U = U(\sigma)$, then by simple transformations it is possible to obtain the following relation (Nikitin, 1967)

$$P(\sigma) = c_0 + d \log \dot{p}_{\min}$$

Such a relation is usually called parametric. Having determined the values of the coefficients c_0 and d from the results of tests at two temperatures, for the material one can construct parametric dependencies, schematically presented in Fig. 2.12. Using these dependencies, one can calculate the creep limits at any temperature. Testing the accuracy with the parameter P showed satisfactory agreement between the experimental and calculated values of creep resistance $\dot{p}_{\min}(T)$ of nickel-based alloys EI607A, EI765, EI826, EI827, despite the very low accuracy of determining the slope of creep resistance curves.

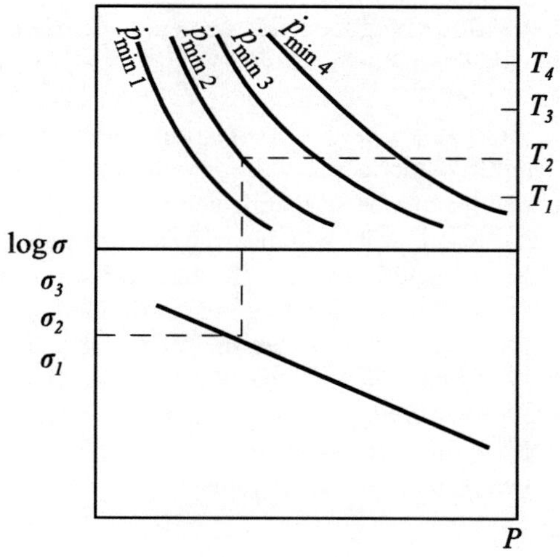

Fig. 2.12 Parametric dependence creep rate vs. temperature

2.2.2 Creep Theories

The considered features of creep behavior of materials were established by tests at constant stresses and temperatures under conditions of uniaxial stress state. Experimental study of creep of metals with a variety of loading conditions encountered in practice is not feasible. Therefore, analytical models are usually applied to solve creep problems.

If the creep constitutive equation were known, i.e. the dependence of the creep rate on stress σ, temperature T, accumulated creep strain p, time t, accumulated instantaneous plastic deformation ε_{pl} and a number of parameters q_i, which can be used to characterize the structural state of the material, i.e. the equation in the form (Rabotnov, 1969)

$$\dot{p} = F(\sigma, T, p, t, \varepsilon_{\text{pl}}, q_1, q_2, \ldots, q_n),$$

then it would be possible in principle to calculate the creep strain for an arbitrary law of stress variation. However, the form of the function $F(\sigma, T, p, t, \varepsilon_{\text{pl}}, q_1, q_2, \ldots, q_n)$ is difficult to establish, and in general case is even impossible. Therefore, various partial hypotheses called creep theories are commonly used. Examples include the aging theory $p = f(\sigma, t, T)$, the flow theory $\dot{p} = f(\sigma, t, T)$ (Kachanov, 1949; Kachanov and Kac, 1949; Kachanov, 1950), strain hardening theory $\dot{p} = f(\sigma, p, T)$, theory of heredity, theory with anisotropic hardening, etc. (Rabotnov, 1969; Getsov, 1973; Malinin, 1975; Borzdyka and Getsov, 1978; Naumenko and Altenbach, 2007, 2016). For the analysis with the theory of hardening, a modified Taira formula can be used (Taira and Otani, 1986)

$$\dot{p} = A\sigma^k p^l + B\sigma^m \tag{2.15}$$

In the Soderberg formula, the dependence $\dot{p} = f(\sigma, p, T)$ is obtained by eliminating the time variable t in Eqs. (2.9) and (2.10). In the Taira formula, and in Eqs. (2.9) and (2.10), $\sigma/(1-\omega)$ can be inserted instead of σ to describe the tertiary creep stage.

Sometimes creep theories are combined to determine the reversible and irreversible components of the creep deformation (combined theories). Furthermore, equations can be supplemented with the introduction of one or more internal state variables (theories with internal state variables), see Rabotnov (1969); Trunin (1976); Borzdyka and Getsov (1978); Shesterikov (1983); Naumenko and Altenbach (2016).

The simplest creep theories are divided into structurally sensitive and structurally insensitive. In the former, the relationship between strains and stresses $p(\sigma)$ is due to structural changes that develop over time and take place regardless of whether the load is acting or not. In the latter, when determining the creep rate, time is not considered as an argument. Trunin (1976) proposed the following temperature-force dependence of the creep rate

$$\dot{p} = AT^{-l}\sigma^m(\varepsilon_{\mathrm{pl}} + p)^{-n} \exp\left(-\frac{U_0 - c\sigma - rp}{RT}\right), \tag{2.16}$$

where l and m are coefficients equal to 1 or 2 and 1.2 or 3, respectively. A, b, c and r are coefficients reflecting the individual characteristics of the material. Due to the fact that creep tests are usually carried out at a constant load and not at a constant stress, the dependencies of $\dot{p}$ on σ are corrected, replacing σ by $\sigma/(1-\omega)$, where ω is the damage caused by the accumulation of creep deformations. In the simplest case

$$\omega = \frac{p}{\varepsilon_{\mathrm{n}}}, \tag{2.17}$$

where ε_{n} is the creep strain accumulated before fracture (deformation ability of the material under creep conditions). However, this approach gives satisfactory agreement with the experiment only in the case of a limited service life, when the mechanisms of damage accumulation during the test are not changed (Getsov, 1964).

In Petrenya (1997), various models for the description of creep damage in various materials (Table 2.2) were considered. Equations given in the second column characterize the rate of damage accumulation as a function of stresses, creep strains and the accumulated damage. Models 3 and 4 consider the damage evolution of various nature, and model 4 assumes that the damage state variable is a vector and the addition occurs according to the law of addition of vectors[5]. Here A, B, C, B, n, r, m, q are material parameters, $\boldsymbol{K}$ is the damage vector, $\boldsymbol{N}$ is the normal to the octahedral plane, σ_{vM} is the equivalent stress according to von Mises[6]. ω_{d} and ω are

[5] Note that in Murakami (2012) the tensorial character of damage is discussed. But damage tensors of the odd-order, including the vector, are not recommended.

[6] In the Russian literature instead of σ_{vM} is often used σ_{i} (stress intensity), the term σ_{eq} is not precise since there are many definitions of equivalent stresses (Altenbach et al., 2014; Kolupaev, 2018).

Table 2.2 Several models for creep damage

Model	Damage evolution equation	Creep constitutive equation
Classical	$\dot{\omega} = B\left(\dfrac{\sigma}{1-\omega}\right)^m$	$\dot{\varepsilon} = A\dfrac{\sigma^k}{(1-\omega)^n}$
Shesterikov	$\dot{\omega} = B\left[\dfrac{(1+\varepsilon)\sigma^m}{(1-\omega)^r}\right]^n$	$\dot{\varepsilon} = A\left[\dfrac{(1+\varepsilon)\sigma}{1-\omega^r}\right]^n$
Othman, Hayhurst and Dyson	$\dot{\omega}_1 = \dfrac{CA(1-\omega_1)\sinh(B\sigma_{\text{vM}})}{(1-\omega_2)^n}$ $\dot{\omega}_2 = \dfrac{DA\left(\dfrac{\sigma_1}{\sigma_{\text{vM}}}\right)^\nu H\sinh(B\sigma_{\text{vM}})}{(1-\omega_1)(1-\omega_2)^n}$ $H = \begin{cases} 1 & \text{for} \quad \sigma_1/\sigma_{\text{vM}} > 0 \\ 0 & \text{for} \quad \sigma_1/\sigma_{\text{vM}} \le 0 \end{cases}$	$\dot{\varepsilon} = \dfrac{A\sinh(B\sigma_{\text{vM}})}{(1-\omega_1)(1-\omega_2)^n}$
Petrenya-Chizhik	ductile damage $\dot{\omega}_\text{d} = \dfrac{A}{\varepsilon_\text{f}}\left(\dfrac{\sigma_{\text{vM}}}{1-\omega_\text{d}}\right)^n$	$\dot{\varepsilon}_{ij} = \dfrac{\dfrac{3As_{ij}}{\sigma_{\text{vM}}}\sinh(B\sigma_{\text{vM}})}{2(1-\omega_d)^n}$
	brittle and quasiplastic damage $\boldsymbol{\omega} = \boldsymbol{\omega}_1 + \boldsymbol{\omega}_2$ $\dot{\boldsymbol{\omega}}_1 = B\left(\dfrac{\sigma_1}{1-\boldsymbol{\omega}\cdot\boldsymbol{K}}\right)^m \boldsymbol{e}_1 \quad \text{if } \sigma_1 > 0$ $\dot{\boldsymbol{\omega}}_1 = \boldsymbol{0} \quad \text{if} \quad \sigma_1 \le 0$ $\dot{\boldsymbol{\omega}}_2 = C\left(\dfrac{\sigma_{\text{vM}}/\sqrt{3}}{1-\boldsymbol{\omega}\cdot\boldsymbol{K}}\right)^l \boldsymbol{N} \quad \sigma_1 \le 0$ $\dot{\boldsymbol{\omega}}_2 = \boldsymbol{0} \quad \text{if} \quad \sigma_1 \le 0$	$\dot{\varepsilon}_{ij} = \dfrac{\dfrac{3As_{ij}}{\sigma_{\text{vM}}}\sinh(B\sigma_{\text{vM}})}{2(1-\omega_1)(1-\omega_2)^n}$

Remark 2.1 The classical model was suggested in the late 50th of the last century by Kachanov (1958); Rabotnov (1959). Kachanov (1958) introduced the scalar variable continuity ($\psi = A/A_0$, A_0 is the initial load carrying area, A is the actual load carrying area with $A \le A_0$) while Rabotnov (1959) has used the scalar variable damage $\omega = (A_0 - A)/A_0$. It is obvious that $\psi = 1 - \omega$. More details are given in Rabotnov (1969); Kachanov (1986).

Remark 2.2 In Shesterikov (1983) the second model was published. It is similar to the classical model, but it takes into account also the strain ε.

Remark 2.3 The third model suggested in (Othman et al., 1993) is a so-called mechanism-based model. In the 90th of the last century several mechanism based models were suggested by Hayhurst and co-workers (see, for example, Kowalewski et al., 1994; Perrin and Hayhurst, 1994). They take several mechanisms based on microstructural observations into account. More information are given in Naumenko and Altenbach (2016).

Remark 2.4 The fourth model was developed in the 70th of the last century (see Petrenya and Chizhik, 1980). This model takes into account different types of damage (for more details see Petrenya, 1997).

damage parameters for ductile and brittle creep damage, respectively and ε_f is the creep strain at fracture.

The Petrenya-Chizhik model was considered in detail in Shesterikov (1983). It suggests the existence of three types of microdamage in creep: transcrystalline, with the formation of wedge-shaped cracks at the junctions of the grain boundaries, caused by maximum normal stresses, and with the formation of micropores on the grain boundaries, caused by maximum tangential stresses. Its use is advisable in relation to the conditions of very long-term operation, when the creep deformation at the primary creep stage can be neglected. The system of equations in this case is

$$\begin{aligned}
\dot{p} &= a\sigma_{\text{vM}}^{m}(1-\omega)^{-q}, \\
\dot{\omega}_1 &= b\sigma_{\text{vM}}^{n}(1-\omega)^{-r} \quad \text{if} \quad \sigma_1 > 0, \\
\dot{\omega}_1 &= 0 \quad \text{if} \quad \sigma_1 \le 0, \\
\dot{\omega}_2 &= c\sigma_{\text{vM}}^{k}(1-\omega_2)^{-s} \quad \text{if} \quad \sigma_1 \le 0, \\
\omega &= A\omega_1\alpha + B\omega_2\beta \quad \text{or} \quad (1-\omega)^q = (1-\omega_1)^{-f} + (1-\omega_2)^{-t},
\end{aligned} \tag{2.18}$$

where a, b, c, m, n, k, q, r, s, f, t are material parameters, and m depends on temperature such that $m = a_1/T + b_1$. The use of this model allows one to obtain a good agreement with experiments with a duration of more than 100000 h, while models with a single damage parameter can adequately describe creep curves up to 50000 h. In particular, creep of materials that exhibit softening and hardening can be well described. Table 2.3 shows the creep models used in the ANSYS finite element package.

An important class of creep constitutive models is based on structural model of the environment (Gokhfel'd and Sadakov, 1984; Novozhilov and Kadashevich, 1990; Ozhoga-Maslovskaja et al., 2015)[7]. The idea of microstructural modeling is that the elementary volume of the material appears to consist of a finite or infinite set of sub-elements that have different characteristics of creep resistance and instant plastic deformation. In this case, the total deformation for all sub-elements is considered to be the same, but due to the difference in the properties of sub-elements, the history of stress changes in each sub-element is different. The elastic deformation in the sub-elements r_k is related to stresses by the Hooke's law $r_k = \sigma_k/E$. The rheological properties of a sub-element are given by the following function

$$\dot{p}_k = \Phi(\sigma/z_k, T), \tag{2.19}$$

which for $\sigma < \sigma_c$ ($\sigma_c$is the creep limit) is zero, for $\sigma_c < \sigma < \sigma_B$ it has the form of the steady-state creep rate vs stress and for $\sigma = \sigma_B$ it is equal to the loading rate. The parameters z_k characterize the scatter of the properties of sub-elements.

The parameters of the volume element r, p, σ are determined by the averaging of the corresponding parameters of the sub-elements. The averaging formulas in the case of an infinite and finite number of sub-elements have the form

[7] The development of structural models of the environment for practical design calculations was most developed at the Department of Materials Resistance of the Chelyabinsk Polytechnic Institute (Gokhfel'd and Sadakov, 1984).

Table 2.3 Creep models in ANSYS package

Number	Creep Model	Equation	Creep stage
1	Strain hardening	$\dot{\varepsilon}_{\text{cr}} = C_1 \sigma_{\text{vM}}^{C_2} \varepsilon^{C_3} \exp(-C_4/T)$	primary
2	Time hardening (flow theory)	$\dot{\varepsilon}_{\text{cr}} = C_1 \sigma_{\text{vM}}^{C_2} t^{C_3} \text{e}^{-C_4/T}$	primary
3	Generalized exponential	$\dot{\varepsilon}_{\text{cr}} = C_1 \sigma_{\text{vM}}^{C_2} r \text{e}^{-rt}, \quad r = C_5 \sigma_{\text{vM}}^{C_3} \text{e}^{-C_4/T}$	primary
4	Generalized Graham law	$\dot{\varepsilon}_{\text{cr}} = C_1 \sigma_{\text{vM}}^{C_2} (t^{C_3} + C_4 t^{C_5} + C_6 t^{C_7}) \text{e}^{-C_8/T}$	primary
5	Blackborn law	$\dot{\varepsilon}_{\text{cr}} = \dot{\varepsilon}_{\text{cr}} = f(1 - \text{e}^{-rt}) + gt, \quad f = C_1 \text{e}^{C_2 \sigma_{\text{vM}}},$ $r = C_3 (\sigma_{\text{vM}}/C_4)^{C_5}, \quad g = C_6 \text{e}^{C_7 \sigma_{\text{vM}}}$	primary
6	Modified flow theory	$\dot{\varepsilon}_{\text{cr}} = C_1 \sigma_{\text{vM}}^{C_2} (t^{C_3+1} \text{e}^{-C_4/T} / (C_3 + 1)$	primary
7	Modified hardening theory	$\dot{\varepsilon}_{\text{cr}} = \left\{ C_1 \sigma_{\text{vM}}^{C_2} [(C_3 + 1)\varepsilon]^{C_3} \right\}^{1/(C_3+1)}$	primary
8	Garofalo law	$\dot{\varepsilon}_{\text{cr}} = C_1 [\sinh(C_2 \sigma_{\text{vM}})]^{C_3} \text{e}^{-C_4/T}$	secondary
9	Exponential	$\dot{\varepsilon}_{\text{cr}} = C_1 \text{e}^{\sigma_{\text{vM}}/C_2} \text{e}^{-C_3/T}$	secondary
10	Norton creep law	$\dot{\varepsilon}_{\text{cr}} = C_1 \sigma_{\text{vM}}^{C_2} \text{e}^{-C_3/T}$	secondary
11	Flow theory	$\dot{\varepsilon}_{\text{cr}} = C_1 \sigma_{\text{vM}}^{C_2} t^{C_3+1} \text{e}^{-C_4/T} / (C_3 + 1) + C_5 \sigma_{\text{vM}}^{C_6} t \text{e}^{-C_7/T}$	primary and secondary
12	Polynomial	$\dot{\varepsilon}_{\text{cr}} = C_1 \dot{\varepsilon}_{\text{c}}, \varepsilon_{\text{c}} = \dfrac{cpt}{1 + pt} + \dot{\varepsilon}_{m^t}, \dot{\varepsilon}_{\text{m}} = C_2 10^{C_3 \sigma_{\text{vM}}} \sigma_{\text{vM}}^{C_4},$ $C = C_7$	primary and secondary
13	Generalized flow theory	$\dot{\varepsilon}_{\text{m}}^{C_8} \sigma_{\text{vM}}^{C_9}, p = C_{10} \dot{\varepsilon}_{\text{m}}^{C_{11}} \sigma_{\text{vM}}^{C_{12}}, \dot{\varepsilon}_{\text{cr}} = f t^r \text{e}^{-C_6/T},$ $f = C_1 \sigma_{\text{vM}} + C_2 \sigma_{\text{vM}}^2 + C_3 \sigma_{\text{vM}}^3, r = C_4 + C_5 \sigma_{\text{vM}}$	primary

$$X = \int_0^\infty x(z) y(z) \mathrm{d}z, \qquad X = \sum_{k=0}^{N} x_k g_k, \tag{2.20}$$

where x is the averaged parameter, $y(z)$ is the density function of the distribution of parameters z, g_k are weights of sub-elements, characterizing the influence of each sub-element in the process of deformation. The following normalization condition is considered

$$\int_0^\infty y(z) \mathrm{d}z = 1, \qquad \sum_{k=0}^{N} g_k = 1. \tag{2.21}$$

To describe the properties of a given material using a microstructural model of a medium, it is necessary to know two functions $\Phi(\sigma/z, T)$ and $y(z)$ (or a set of g_k, and z_k in the case of a finite number of sub-elements). The form of the function $\Phi(\sigma, T)$

corresponds to the dependence of the steady-state creep rate on σ and T. The value of $y(z)$ is determined as the second derivative of the equation of the deformation curve $F(\varepsilon)$. The weights of the sub-elements g_k are determined from the same stress-strain diagram after its approximation in the form of a piecewise-linear curve (Fig. 2.13).

The use of the microstructural model in the analysis of components makes it possible to describe the Bauschinger effect, nonlinear hardening of stress-strain curves, transient creep, creep under isothermal and non-isothermal loading, the mutual influence of instantaneous inelastic deformation and creep processes, and effects during unloading. Solving problems of creep in structures requires knowledge of the creep laws under a complex stress state. Like the theories of plasticity, the theories of creep under a complex stress state are based on some assumptions, only partially confirmed by experience. The difficulty in formulating a general creep theory suitable for a complex stress state is associated with the difficulties of describing the transient creep stage and taking into account anisotropy caused by various factors (initial anisotropy, hardening or softening under the influence of instantaneous plastic deformation and hardening during primary creep stage).

2.2.3 Isochronous Creep Curves

In the numerical analysis of the stress-strain state of structures operating at elevated temperatures, data are usually used, which are represented by various dependencies that describe creep at the primary or secondary stage and use the exponential dependence on temperature (Arrhenius dependence). Such a dependence adequately characterizes creep processes only in a narrow temperature range. In many cases,

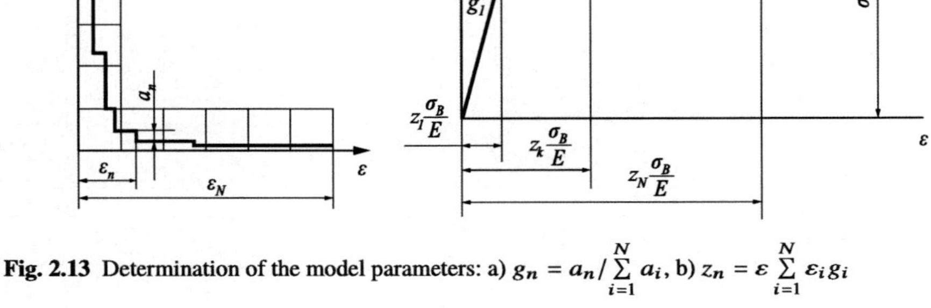

Fig. 2.13 Determination of the model parameters: a) $g_n = a_n / \sum_{i=1}^{N} a_i$, b) $z_n = \varepsilon \sum_{i=1}^{N} \varepsilon_i g_i$

data related to the minimum creep rate are used (Norton's law). In some cases, this way of describing creep leads to large errors, since creep at an unsteady stage can make a significant contribution to the process of strain accumulation. This is especially true for stress relaxation processes (Altenbach et al., 2008b).

Using the experimental data on creep, the values of the elastic modulus and the instantaneous deformation curve, it is possible to construct isochronous creep curves (Fig. 2.14). They provide the most complete and compact representation of the creep characteristics of a material at a constant temperature. Given a certain operating time at a constant stress, from the isochronous curves it is possible to determine all three components of the strain $\varepsilon_{\rm el}$, $\varepsilon_{\rm pl}$ and p. To describe the creep resistance of a material, it is necessary to have isochronous curves at several temperatures. Among the published data, isochronous creep curves of a relatively small number of materials are known (Table 2.4). Among the published data, two approaches are used to represent isochronous creep curves. In the first, the strain curve obtained in simple tensile experiments, the so-called instantaneous curve, is used as the first (zero) curve. To obtain such a curve, the strain rate should be such that the creep processes (stress relaxation) practically do not take place. In the second approach, isochronous curves are constructed starting from some point in time, and the instantaneous tensile curve is not used (this representation is given in norms, NN, 1989).

An analysis of the creep test results of steels and alloys indicates the need to separate this data and the predicted characteristics into (a) two stress ranges (below and above the elasticity limit-yield stress) and (b) the four following temperature ranges:

1. $20\ldots T_{\rm n}$, in which it is believed that creep is absent,
2. $T_{\rm n}\ldots T_{\rm k}$, in which low-temperature creep is observed at stresses above the yield strength,
3. $T_{\rm n}\ldots T_{\rm cr}$, in which creep processes with hardening are realized,
4. $T > T_{\rm cr}$, in which diffusion creep processes prevail and hardening is absent.

Of great practical importance is to interpolate and extrapolate strain values (total and creep) at arbitrary stresses, times and temperatures from isochronous creep curves. In what follows a method to solve this problem is discussed. Various approximation dependencies have been identified. The best results were obtained using expressions of the form

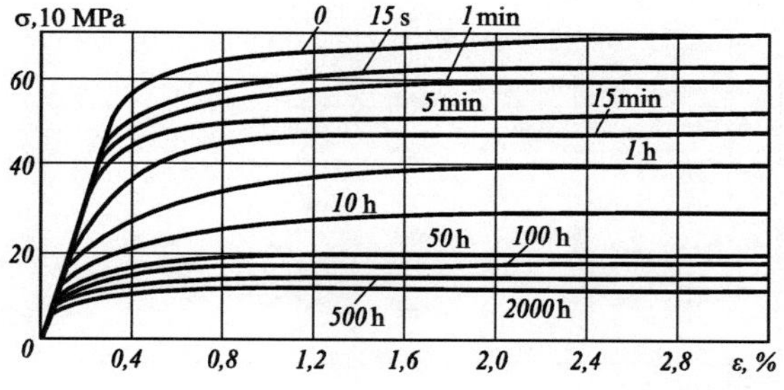

Fig. 2.14 Isochronous creep curves for the EI765 alloy at $T = 800°\text{C}$

Table 2.4 Data on isochronous creep curves (Getsov, 1971; NN, 1981; Pospisil et al., 1987; NN, 1989; Lanin and Balina, 2006; Getsov and Pospisil, 2009)

Nr.	Material	Temperatures	Test duration, h
1	EI481	600, 650	2000
2	EI612	600, 650	50
3	EI607A	600, 650,700, 750, 800	1000-2000
4	EI765	600, 700, 750, 800	2000
5	EI826	600, 700, 750, 800, 850, 900	2000
6	EI827	600, 700, 750, 800, 850, 900	2000
7	EI802	500, 550	25
8	EI415	550	20
9	EI220	850, 900, 950, 1000	2-1000
10	ZhS6K	850, 950, 1000	5000
11	ZhS36	850, 900, 1000	300
12	08Ch17N6T	350, 400	4000
13	14Ch17N2	400	3000
14	20Ch13	400	4000
15	20	400, 450	200 000
16	15ChM,12ChM	500, 550	200 000
17	12Ch2M	450, 500, 550, 565, 600	200000
18	12Ch1MF	500, 550	200000
19	15Ch1M1F	500, 540, 570	200000
20	12Ch18N9	450, 500, 550, 600, 650	200000
21	08Ch15N11M3	450, 500, 550, 600, 650	200000
22	12Ch18N12T	450, 500, 550, 600, 650	200000
23	09G2S	400, 450, 500	200000
24	05Ch12N2M	450, 500, 550	200000
25	16GNMA	400, 450	200000
26	09Ch18N9	450, 500, 550, 600	200000
27	03Ch16N9M2	500, 550, 600, 650	200000
28	R2MA	500, 525, 550	200000
29	EP291	500, 525, 550	200000
30	15Ch1M1FL KP30	475-550	300000
31	20ChMFL KP28	475-550	300000
32	20Ch3MWFL KP60	475-550	300000
33	25Ch1M1F KP50	475-550	300000
34	X22CrMoV12.1 KP65-KP80 data by Mannesmann	550, 600	100000
35	13CrMo, 15ChM	550-600	15000
36	15Mo3	525-575	15000
37	ČSN16236	450-500	10000
38	ČSN15121.5	525-575	600
39	ČSN15128 data by Vitkovice steel	550-600	15000
40	Alloy800H	427-760	300000
41	2.25Cr1Mo	371-650	300000
42	304 SS (08Cr18Ni10)	427-816	300000
43	316 SS (08Cr18Ni12) data by ASME	454-816	300000

$$\sigma = A\exp(B/T)[a + bT + c\log\varepsilon + d\log(t + 10^{-3}) + e\log\varepsilon\log(t + 10^{-3})] \quad (2.22)$$

for the temperature levels within the ranges 2. and 3. and

$$\sigma = [A\exp(B/T)\varepsilon/t]^{1/n} \quad (2.23)$$

for the temperature range 4. A further increase in the accuracy of the calculations is expected to be obtained using equations of the form (2.9), (2.10) and (2.14)

$$\varepsilon = A\sigma^{k_1}\exp[(-B_1 - B_2\sigma)/T][C\sigma^{k_2}\langle 1 - \exp(D\sigma^{k_3}t)\rangle + F\sigma^{k_4}t], \quad (2.24)$$

where

$$k_4 = \frac{k_5}{T} + k_6, \quad F = F_0 + \beta T, \quad C = C_0 + \nu T$$

as well as

$$\varepsilon = A\sigma^k\left[1 + C\sigma^l t\right]\exp\frac{U_1 - H_1\sigma}{T} + B\sigma^m t\exp\frac{U_2 - H_2\sigma}{T}, \quad (2.25)$$

where

$$m = m_0\frac{U_3 - H_4\sigma}{T}$$

Finally, the equations used in ANSYS to describe the creep rate $\dot{\varepsilon}_{cr}$ at the primary and the both stages of creep can be used (Table 2.3). With respect to Help ANSYS the following symbols are used: ε_{cr} - equivalent strain, $\dot{\varepsilon}_{cr}$ equivalent creep rate, σ_{vM} - equivalent stress, T - temperature (in Kelvin), $C_1 - C_{12}$ - constants, t - time, e - base of the natural logarithm. In the case of using the second approach (without a zero curve) in constructing isochronous creep curves, the term 10^{-3} in Eq. (2.22) was excluded. The method was implemented using the special Nonlinear Regression Analysis Program (NLREG) developed by Phillip H. Sherrod.

The method was identified using isochronous creep curves of the EI826 alloy (Getsov, 1971) obtained in tests at six temperatures: 600, 700, 750, 800, 850, and 900°C. The accuracy of the representation of isochronous curves at the same temperature using Eq. (2.22), estimated from the least squares deviation (LSD), LSD = 21.73, 62.2, 21, 13, 28.74, 19.75 and 15.29 MPa for 600, 700, 750, 800, 850, and 900°C, respectively. Further, the calculations were performed for 27 groups of initial data related to two, three, and four temperatures, for which the coefficients of dependences (2.22) and (2.23) were calculated both for the data according to the first and to the second approaches. Interpolation was carried out in relation to temperatures not included in the group used to determine the parameters of the equations. The LSD values given were obtained for deviations along the y axis, while the parameters of the creep curves were calculated from the conditions for obtaining the orthogonal regression curve, i.e. minimum values of least squares deviation over the distances between the experimental points and the calculated curve. In this case, the magnitude of the deviation, as shown by the calculations, will be approximately 2 times less.

The standard deviation of the calculated and experimental stress values corresponding to various strains was determined. When using the entire data array, the value of R^2 was 0.9372, LSD = 60.74 MPa (for data generated by approach 1) and 32.4265 MPa (approach 2). When using only data of one range (700-900°C), the value of R^2 increased to 0.9975, and the standard deviation for data by approach 1 decreased to 34.44 MPa.

It was established that the value of R^2 for the first approach for all temperature groups is 0.90-0.98 and basically groups around 0.95. In case of using data from one temperature range, the standard deviation does not exceed 30-40 MPa. However, the use of these two temperature ranges (600 and 700-900°C) leads to a sharp increase in standard deviation, reaching 174 MPa. It should be noted that for one temperature range the use of the data of the instantaneous deformation curve (approach 1) leads only to a slight decrease in the calculation accuracy (standard deviation increases mainly by 3-4 MPa).

The extrapolated and interpolated values of strains and stresses were compared with experimental isochronous curves at the corresponding temperature. Consideration of the calculation results showed satisfactory interpolation accuracy for temperatures of 800 and 850°C and unsatisfactory results for temperatures of 700 and 750°C. For extrapolation, almost all the results were unacceptable. Therefore, to extrapolate the data to a different temperature range, where data is available at only one temperature, the following method was tested.

According to data at several temperatures within the same temperature range, isochronous curves are calculated at the boundary of the ranges. Further, according to the data at one temperature within another range, for which experimental data are available, together with the calculated data at the border, corrected by the introduction of weights, the regression parameters are calculated and the standard deviation is determined. So, for a temperature of 600°C, together with the calculated values at 650°C, the calculation error was determined. It turned out that in this case LSD = 30.54 MPa, while the standard deviation calculated according to the data without using the proposed standard deviation method was 44 MPa.

The possibility of extrapolating the data obtained up to $\varepsilon = 0.7\%$ was also tested to determine the stress values corresponding to strains $\varepsilon = 2\%$, and the possibility of extrapolation for large durations. The method was also used to determine instantaneous deformation curves from creep tests at constant stresses (approach 2). Further work on the use of the proposed method is carried out in relation to the materials and temperatures given in Table 2.4, as well as for the curves shown in Figs. 2.15-2.18 and Appendix B.

2.2.4 Stress Relaxation

The most common case of accumulation of creep deformation under alternating stresses occurs in a rigidly fixed rod. In such conditions, relaxation (decrease) in stresses[8] is observed. The process of pure relaxation is usually described by the

[8] The metallographic aspects of stress relaxation in various heat-resistant materials, including those used for GTU fasteners, are considered in Borzdyka and Getsov (1978).

Fig. 2.15 Isochronous creep curves for the ZhS6K alloy

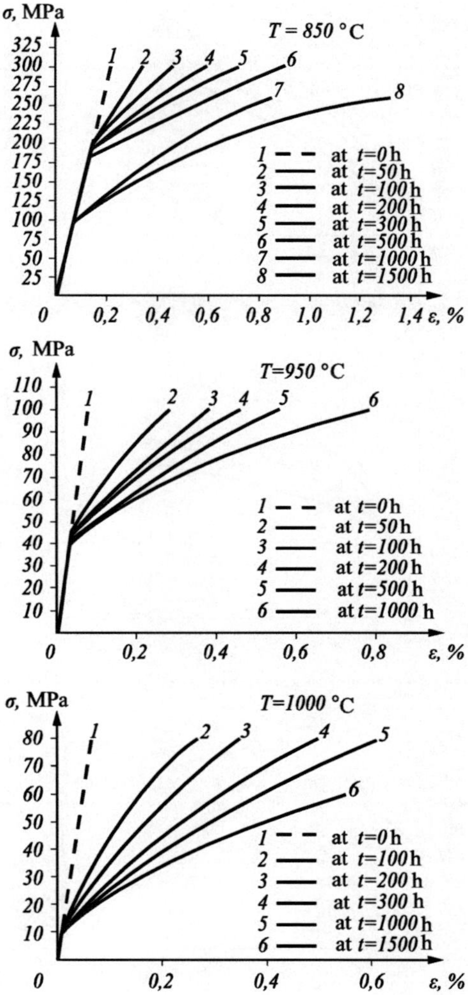

following equation[9]

$$\varepsilon_0 = \varepsilon_{el} + \varepsilon_{pl} + p + \varepsilon_{sd} = \text{const}, \tag{2.26}$$

where ε_{sd} is the structural deformation. From Eq. (2.43) with $\varepsilon_{pl}(0) = \text{const}$, $\varepsilon_{str} = 0$ after differentiation and substitution of the value $\varepsilon_y = \sigma/E$ we obtain

$$\dot{\sigma} = -E\dot{p}$$

[9] Later in the book, the variables will be partly substituted by other variables, for example, $\varepsilon_{el} \doteq r$, $\varepsilon_{pl} \doteq e$.

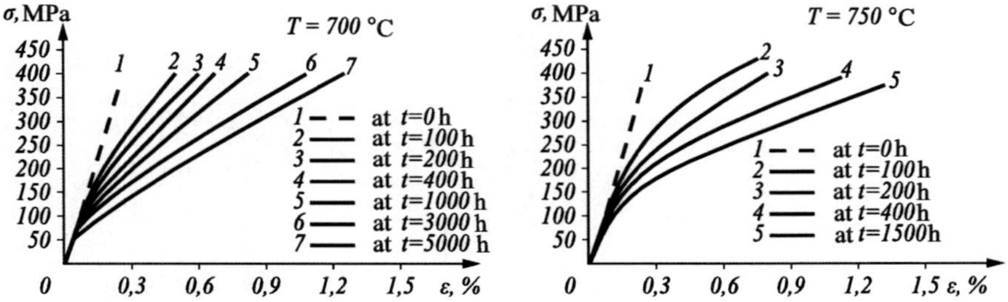

Fig. 2.16 Isochronous creep curves for the ChS104 alloy

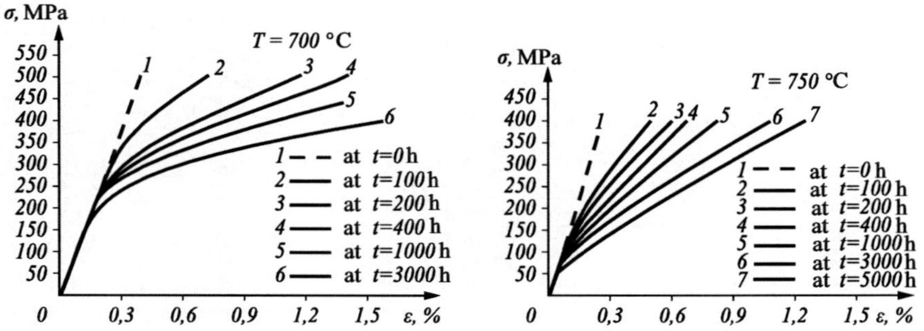

Fig. 2.17 Isochronous creep curves for the ChS70VI alloy

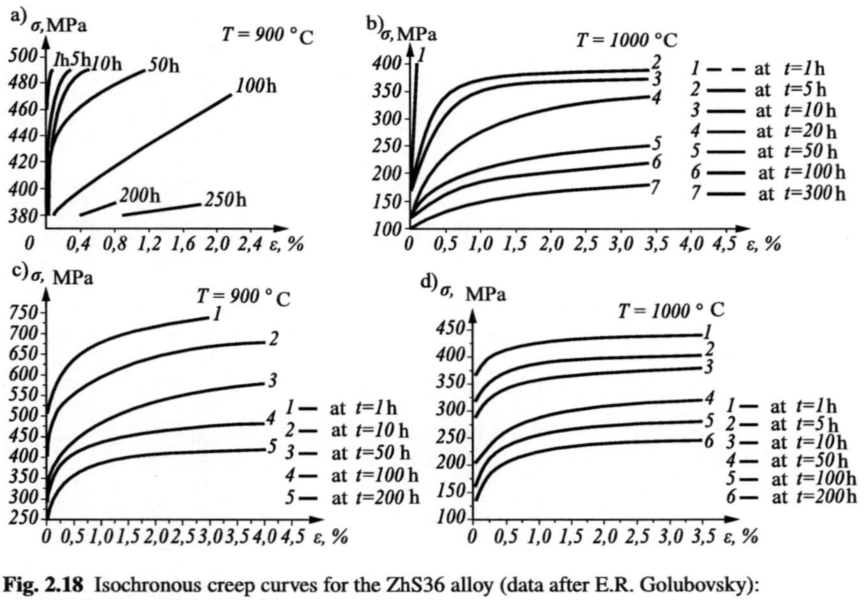

Fig. 2.18 Isochronous creep curves for the ZhS36 alloy (data after E.R. Golubovsky): a), b) - orientation [001], c,) d) - orientation [111]

For a compliant system characterized by a compliance coefficient c we obtain

$$\varepsilon_{\rm el} = \frac{\sigma}{E}(1 + cE). \tag{2.27}$$

In this case

$$\dot{\sigma} = -\frac{E\dot{p}}{1 + cE} \tag{2.28}$$

Integration of Eq. (2.28) can be carried out if the expression for the creep rate $\dot{p}$ is given by a creep constitutive equation[10]. The most simple relaxation estimation is provided by the theory of aging. To this end, it is enough to determine graphically the values of residual stress $\sigma_{\rm res}$ at ε = const from the isochronous deformation curves (for example Fig. 2.14) for a given value of σ_0. The analytical expression for calculating $\sigma_{\rm res}$ has the form

$$\sigma_{\rm res}(t) = \sigma_0 - p_t E, \tag{2.29}$$

where p_t is the accumulated creep strain at time t.

Since relaxation experiments are the simplest tests at variable stresses, they usually judge the applicability of a particular creep theory. Pure relaxation conditions are relatively rare in practice, since absolutely rigid systems at high temperatures are found only in exceptional cases. Therefore, an assessment of the relaxation resistance is usually carried out in relation to the real compliance of the part - machine system (sample - installation).

A comparison of experimental and calculated relaxation curves indicates that the use of strain hardening theory gives the best accuracy in describing the creep of structurally stable materials under conditions of variable stresses not exceeding the yield strength. The accuracy of the estimates obtained using the combined theories is even higher, however, to use it, information on the characteristics of the inverse creep of the material is required, which, as a rule, cannot be determined in ordinary tests. At relatively high temperatures, satisfactory agreement with experimental data can be obtained using the theory of aging. At initial stresses exceeding the yield strength, the experimental relaxation curves are in satisfactory agreement with those calculated according to flow theory with a structural parameter $\varepsilon_{\rm pl}^{(0)}$ (Fig. 2.19). The relaxation curves corresponding to different values of the initial stress intersect each other. Such a behavior of nickel-based alloys under stress relaxation at $\sigma_0 > \sigma_{\rm el}$ was experimentally confirmed in Getsov (1979). In the case of structurally unstable materials the deformation in the direction of the load caused by a change in volume during structural transformations is $\varepsilon_{\rm sd} = f(t) \neq 0$ and the stress rate can be calculated as follows

$$\dot{\sigma} = -E(\dot{p} + \dot{\varepsilon}_{\rm sd}), \tag{2.30}$$

[10] Methods for solving relaxation equations by various theories are described in detail in Rabotnov (1969); Borzdyka and Getsov (1978); Naumenko and Altenbach (2016); Kostenko and Naumenko (2017).

Fig. 2.19 Experimental (*I*) and estimated (*II*) stress relaxation curves of the alloy EI765 at 750°C. Theory: aging (a), yield (b), harding (c) and aging with structural parameters (d), *I* - σ_0 - 750 MPa, *II* - σ_0 - 600 MPa

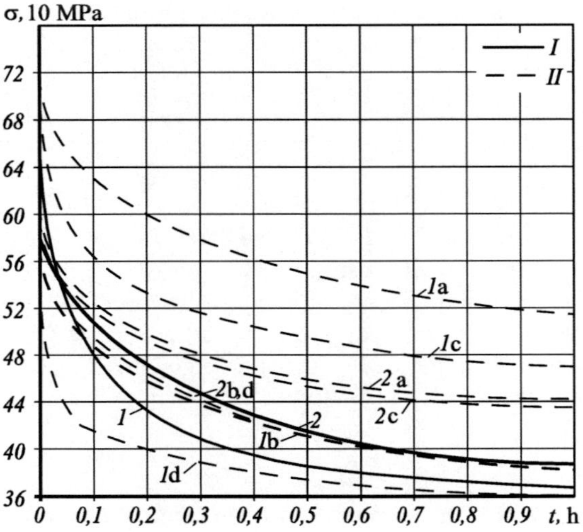

For $\varepsilon_{sd} < 0$ and $\dot{\varepsilon}_{sd} > \dot{p}$ the reverse relaxation is observed, causing an increase in stress with time (Fig. 2.20).

Fig. 2.20 Stress relaxation curves of the steel EI696 at temperature 600°C:
I - long term strength,
II - stress relaxation
(data after B.M. Rakhman and V.I. Maranychin)

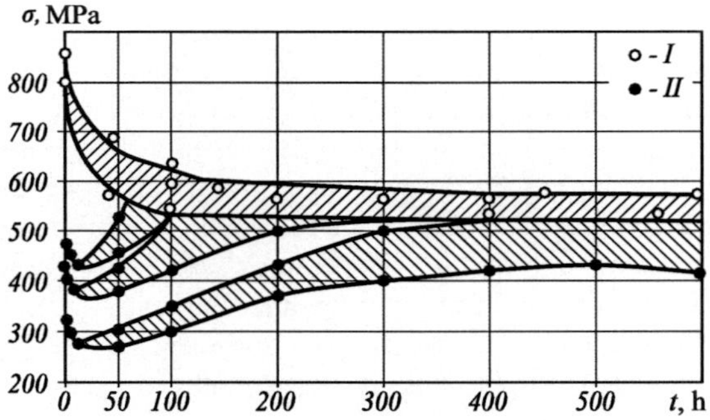

2.3 Multi-axial Strain and Stress States

2.3.1 Elasticity

The principal strains ε_1, ε_2, ε_3 and stresses σ_1, σ_2, and σ_3 in the elastic regime are given by the relations of the theory of elasticity as follows[11]

$$\begin{aligned}\varepsilon_1 &= (\sigma_1 - \nu\sigma_2 - \nu\sigma_3)/E,\\ \varepsilon_2 &= (\sigma_2 - \nu\sigma_1 - \nu\sigma_3)/E,\\ \varepsilon_3 &= (\sigma_3 - \nu\sigma_1 - \nu\sigma_2)/E,\end{aligned} \tag{2.31}$$

where ν is the Poisson's ratio (ratio of the transverse strain and longitudinal strain, in the elastic range $\nu = 0.33$) and E is the Young's modulus. The applicability of Eqs. (2.31) is limited by the yield conditions, i.e. conditions under which the material goes beyond the elastic range. According to the Tresca-St. Venant criterion, such a condition is the equality of the maximum shear stress τ to the shear yield strength. Since there is a relationship between the shear stresses τ_1, τ_2, τ_3 and the normal stresses

$$\begin{aligned}\tau_1 &= 0.5(\sigma_2 - \sigma_3),\\ \tau_2 &= 0.5(\sigma_3 - \sigma_1),\\ \tau_3 &= 0.5(\sigma_1 - \sigma_2),\end{aligned} \tag{2.32}$$

then the Tresca-St. Venant condition means that the tensile yield strength σ_y is equal to the value of the difference $\sigma_1 - \sigma_3$ equal to twice the absolute value of the highest shear stress. However, when determining the plasticity conditions of iron-based structural steels and various heat-resistant steels, the best agreement with the experiment is observed using the von Mises yield condition $\sigma_{vM} = \sigma_y$, where σ_{vM} is the von Mises equivalent stress which can be computed as follows

$$\sigma_{vM} = \sqrt{\frac{1}{2}\left[(\sigma_1 - \sigma_2)^2 + (\sigma_2 - \sigma_3)^2 + (\sigma_3 - \sigma_1)^2\right]}, \tag{2.33}$$

The difference between the two yield conditions can be observed, for example, when considering the case of torsion ($\sigma_2 = 0, \sigma_1 = -\sigma_3 = \tau_{max}$). According to the Tresca-Saint-Venant condition we obtain $\tau_{max} = \sigma_y/\sqrt{2}$, while according to the von Mises condition $\tau_{max} = \sigma_y/\sqrt{3}$ is obtained. Comparison of the yield strengths under torsion and tension (Fridman, 1975a,b) showed that the ratio τ_y/σ_y for various metals can vary from 0.25 (MA-5 magnesium alloy) to 0.74 (U7 steel after quenching and tempering at 200°C). The behavior of the nickel-base alloy EI765 at 700°C, according to the author and A.P. Gusenkov is well described by the von Mises condition.

Any relation between the three principal stresses is a surface in the space of stresses. Therefore, the Tresca-St. Venant and von Mises conditions also determine

[11] The principal stresses are called stresses and strains acting on areas perpendicular to the principal axes, on which there are no shear stresses. To describe the general case of a stress-strain state, the relations between the components of the strain and stress tensors are used (Malinin, 1975).

some surfaces in the stress space. For points inside these surfaces (it is usually called the yield surface), the Hooke's law (2.31) is valid. At stresses corresponding to points on the surface (or even more so points lying outside it), the plastic flow begins. The relation of stresses with strains under these conditions at simple tension can easily be established purely experimentally. However, with a complex stress state, the relationship between $\sigma_1, \sigma_2, \sigma_3$ and the corresponding strains $\varepsilon_1, \varepsilon_2, \varepsilon_3$ the relation in the general form based on tests cannot be established since it requires an infinite amount of tests. Therefore, to establish dependencies between stresses and strain, one has to introduce additional hypotheses, so called theories of plasticity. There are many approaches to the analytical solution of plasticity problems with complex stress states (see Subsect. 2.5). The criterion for their applicability is compliance with the experiment.

2.3.2 Stress Concentrations

One type of complex stress state is a stress state near stress concentrators. The fracture conditions of components with different stress gradients cannot be determined from the fracture conditions of smooth specimens. Therefore, a certain place among the accepted methods of testing materials is occupied by methods for testing specimens with notches designed to create a particular stress concentration. The issues of assessing the stress state of samples and parts with stress concentrators are of significant interest.

Stress concentrators are usually divided into structural (for example, openings, keyways and grooves for mounting blades in disks and rotors) and metallurgical (flakes, accumulations of metallurgical defects, delamination, pores, tears in welds and the weld zones). To determine the degree of increase in stresses in the zone of stress concentrators, Neuber (1961, 2001); Grubin (1970); Eisenträger et al. (2013); Eisenträger et al. (2017) derived formulas that allow one to calculate

- elastic stresses and strains under conditions when the geometric dimensions of the concentrator (radius at the base of the notch, hole diameter) are much larger than grain sizes

$$\sigma_{\mathrm{N}} = \alpha_{\mathrm{N}} \sigma_{\mathrm{nom}}, \qquad \varepsilon_{\mathrm{N}} = \alpha_{\mathrm{N}} \varepsilon_{\mathrm{nom}}, \tag{2.34}$$

  σ_{N} is the notch tension stress, α_{N} denotes the elastic shape number, which depends on the geometrical dimensions and on the stress concentrator type, σ_{nom} is the nominal stress (averaged stress in the minimal cross section of the notch), ε_{N} is the notch strain and $\varepsilon_{\mathrm{nom}}$ is the nominal strain corresponding to the nominal stress σ_{nom}.
- elastic stresses under conditions when the geometrical dimensions of the concentrator are smaller than the grain size (sharp notch, crack)

$$\sigma_{\mathrm{N1}} = \alpha_{N1} \sigma_{\mathrm{nom}}, \tag{2.35}$$

σ_{N1} is the notch tension stress and α_{N1} denotes the elastic shape number depending on the grain size ($\alpha_{N1} \approx \alpha_N$ for the radius in the notch equal to half the grain size).

- stresses and strains in the elastoplastic region

$$\sigma = \alpha_\sigma \sigma_{nom}, \quad \varepsilon = \alpha_\varepsilon \varepsilon_{nom} \tag{2.36}$$

α_σ and α_ε are the stress and strain shape numbers, respectively.

They are related to each other and the elastic concentration coefficient α_N as follows

$$\alpha_\varepsilon \alpha_\sigma = \alpha_N^2 \tag{2.37}$$

Formula (2.37) was derived for a special case of a slice, however, in recent years indirect experimental evidence has been obtained of its applicability for other types of plane stress state. In Makhutov (1981) the refinement of formula (2.37) is proposed by introducing the correction function F_2, which provides better accuracy of the description of experimental data obtained by the moire and dividing nets method. The refined equations can be formulated as follows

$$\alpha_\varepsilon \alpha_\sigma = \alpha_N^2 F_2[\alpha_N \overline{\sigma}_{nom}, f(\overline{\sigma}, \overline{\varepsilon})] \tag{2.38}$$

with

$$F_2 = 1/(\alpha_N \overline{\sigma}_{nom})^{n(1-m)}[1 - (\overline{\sigma}_{nom} - 1/\alpha_N)]$$

Here $\bar{\sigma} = \sigma/\sigma_{prop}$, $\bar{\varepsilon} = \varepsilon/\varepsilon_{prop}$, n is a constant determined from experiments with given α_N and σ_{nom}. $f(\bar{\sigma}, \bar{\varepsilon})$ is the equation of the stress-strain curve in the elastoplastic regime and m is the exponent in power law (2.2).

2.3.3 Simplest Models of Ductility and Creep

The mathematically simplest and therefore the most common are deformation theories of plasticity, which for loading conditions are essentially a generalization of the theory of elasticity to the case of a nonlinear elastic material. However, when using these theories in the case of zigzag and non-proportional loads[12] (for example, for a gas turbine drive operating under conditions when the radial stresses do not change and the tangential ones change sign), as well as in the case of anisotropic hardening of the material, a noticeable discrepancy with experiment is obtained. The deformation theories also do not allow us to correctly describe the experimentally observed differences in the laws of loading and unloading, the effect of preliminary plastic deformation on the nature of deformation under subsequent loading, the Bauschinger

[12] Non-proportional loading, unlike simple (proportional) loading, is called loading in which proportionality between the stress components is not maintained. For a simple loading of an isotropic material, the direction of the principal axes is fixed, while for a non-proportional loading they change.

effect, etc. Thus, deformation theories do not take into account the history of loading and are suitable only for conditions of simple loading at a monotonously varying load. In connection with the above restrictions imposed on the use of deformation theories of plasticity, various modifications have recently been proposed that extend the boundaries of their applicability (Rabotnov, 1969; Malinin, 1975).

The simplest deformation theory is the theory of small elastoplastic deformations. This theory assumes: an elastic change in volume, the similarity of deviators of stress and strain and a unique dependence of the von Mises equivalent stress σ_{vM} on the equivalent strain

$$\varepsilon_{vM} = \frac{\sqrt{2}}{3}\sqrt{(\varepsilon_1 - \varepsilon_2)^2 + (\varepsilon_2 - \varepsilon_3)^2 + (\varepsilon_3 - \varepsilon_1)^2}, \tag{2.39}$$

which does not depend on the kind of the stress state. Under these assumptions, the relationship between stresses and strains can be represented in the form of dependencies containing only one unknown function $\sigma_{vM} = f(\varepsilon_{vM})$

$$\begin{aligned}\sigma_1 - \sigma &= \frac{2}{3}\frac{\sigma_{vM}}{\varepsilon_{vM}}\varepsilon_1 = \frac{2}{3}\frac{f(\varepsilon_{vM})}{\varepsilon_{vM}}\varepsilon_1,\\ \sigma_2 - \sigma &= \frac{2}{3}\frac{\sigma_{vM}}{\varepsilon_{vM}}\varepsilon_2 = \frac{2}{3}\frac{f(\varepsilon_{vM})}{\varepsilon_{vM}}\varepsilon_2,\\ \sigma_1 - \sigma &= \frac{2}{3}\frac{\sigma_{vM}}{\varepsilon_{vM}}\varepsilon_3 = \frac{2}{3}\frac{f(\varepsilon_{vM})}{\varepsilon_{vM}}\varepsilon_3,\end{aligned} \tag{2.40}$$

$$\varepsilon = (\varepsilon_1 + \varepsilon_2 + \varepsilon_3)/3 = (1 - 2\nu)(\sigma_1 + \sigma_2 + \sigma_3)/3E, \tag{2.41}$$

where

$$\sigma = (\sigma_1 + \sigma_2 + \sigma_3)/3 \tag{2.42}$$

Since the form of the function $f(\varepsilon_{vM})$ is assumed to be the same for any stress state, it can be determined, in particular, from experiments on simple tension. Instead of Eqs. (2.40) and (2.41), we can write expressions similar in form to Eqs. (2.40) of the theory of elasticity

$$\begin{aligned}\varepsilon_1 &= (\sigma_1 - \nu\sigma_2 - \nu\sigma_3)/E_{sec},\\ \varepsilon_2 &= (\sigma_2 - \nu\sigma_1 - \nu\sigma_3)/E_{sec},\\ \varepsilon_3 &= (\sigma_3 - \nu\sigma_1 - \nu\sigma_2)/E_{sec},\end{aligned} \tag{2.43}$$

where E_{sec} is the secant modulus, depending on the stress. The change in volume under conditions of plastic deformation is much less than plastic deformation, so it can be neglected and, using the incompressibility condition, the Poisson's ratio can be set $\nu = 0.5$.

The components of the shear strain tensor γ_{xy}, γ_{yz}, γ_{zx} can be calculated using expressions containing the secant shear modulus G_{sec}. For $\nu = 0.5$ we obtain

$$G_{sec} = E_{sec}/[2(1 + \nu)] = E_{sec}/3, \tag{2.44}$$

$$\gamma_{xy} = \tau_{xy}/G_{sec}, \quad \gamma_{yz} = \tau_{yz}/G_{sec}, \quad \gamma_{zx} = \tau_{zx}/G_{sec} \tag{2.45}$$

The use of the deformation theory, the equation of which includes of the current temperature values and the strain values at the moment of rotation of the loading path, leads to unlimited accumulation of one-sided deformation under soft cyclic loading or to unlimited increase in stresses under hard loading. At the same time, the deformation theory makes it possible to fairly accurately calculate the changes in the width of the loop under hard loading.

The simplest theory of creep for multi-axial stress state is the theory of steady creep of an isotropic material[13]. This theory is based on the following assumptions:

1. the volume changes are elastic,
2. the principal directions of the stress tensor and the creep strain rate tensor coincide, and
3. the equivalent creep strain rate is a single-valued function of the equivalent stress.

To experimentally verify the correctness of these assumptions, creep tests were carried out for tubular samples of steel 15Ch1M1F, 15Ch14N14W2M, 12Ch18N10T under tension, torsion, and torsion with tension at various normal and shear stresses σ_x and τ_x from axial force and torque (Trunin, 1965).

The normal $\dot{p}_x$ and shear $\dot{\gamma}_{xy}$ creep rates were also determined. It has been established that the nature of the relationship between the equivalent stress and the equivalent creep rate at the steady creep stage for different types of loading is different, namely $\dot{p}_{\text{min}i}$ for torsion less than for tension and for torsion with tension, the values $\dot{p}_{\text{min}i}$ are between them. This difference is due to some difference in the creep resistance of the metal under compression and tension, as well as the anisotropy of the creep properties. Indeed, the compressive strength of creep materials may differ from the tensile strength of the creep but the difference in strains usually does not exceed 20-30%.

A generalization of experimental data made it possible to propose the following dependence (Trunin, 1965)

$$\dot{p}_{\text{min}i} = A(\sigma_{\text{vM}})^{\bar{n}}(\tau_{\text{max}})^{\bar{m}}, \tag{2.46}$$

The constants A, $\bar{n}$ and $\bar{m}$ can be determined based on the results of tensile and torsion experiments. The formulation of engineering creep theories in the general case of multi-axial stress state can be based on the hypotheses of plasticity theory. Details can be found in Rabotnov (1969); Malinin (1975); Kachanov (1979).

[13] Theories related to the transient creep and anisotropic materials are considered in the monographs Rabotnov (1969); Naumenko and Altenbach (2016).

2.4 Resistance of Materials to Cyclic Deformation

2.4.1 Cyclic Elasto-plastic Deformation

Under cyclic loading (deformation), the properties of materials differ significantly from those under static loading (deformation). If, after the loading beyond the proportional limit, unloading and loading in the opposite direction are performed, the proportional limit will generally decrease. If, after unloading, the metal is loaded in the direction of the initial deformation, then the proportionality limit rises to a value close to the stress from which the unloading was performed. These features of the behavior of materials during cyclic deformation were first discovered by Bauschinger and named in his honor the Bauschinger effect.

In 1924, Masing suggested that the sum of the stress from which the unloading was performed, $\sigma^{(0)}$ and the new proportionality limit $\sigma_y^{(1)}$ under reverse loading at constant temperature, is a constant equal to twice the static proportional limit $\sigma^{(0)} + \sigma_y^{(1)} = 2\sigma_{ps}$ or in relative coordinates

$$\frac{\sigma^{(0)}}{\sigma_{ps}} + \frac{\sigma_y^{(1)}}{\sigma_{ps}} = \bar{S}_y^{(1)} = 2, \tag{2.47}$$

where $\sigma^{(0)} > \sigma_{ps}$.

At the first half-cycle after the initial loading, the dependence of the unloading modulus $G^{(1)}$ on plastic deformation is different for different metals (Fig. 2.21). The possibility of describing strain curves under cyclic loading under S_{max} conditions was established in Shneiderovich (1968); Gusenkov (1979) using generalized diagrams, the parameter of which is the number of loading half-cycles k (Fig. 2.22). S is the stress measured from the moment of loading reversal. The diagrams are constructed in the coordinates $\bar{S}$ and $\bar{\varepsilon}$, where and $\bar{S}$ and $\bar{\varepsilon}$ are stress and strain, referred to stress and strain corresponding to the proportionality limit in the first half-cycle. When constructing deformation curves using a generalized diagram, the beginning of the curve must be combined with the unloading start point in this half-cycle (points A, B, C, N, K, D in Fig. 2.23).

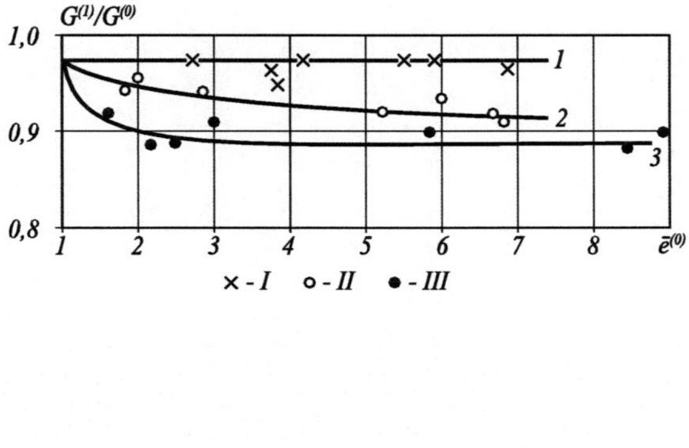

Fig. 2.21 Dependence of the unloading modulus G in relative units on the degree of initial deformation $\bar{\varepsilon}^{(0)}$:
1 - alloy D16T (*I*),
2 - alloy AK-8 (*II*),
3 - steel TC (*III*)

Studies carried out in recent years have shown that for the heat-resistant alloys used to manufacture gas turbine parts, the existence of generalized diagrams – the coincidence of the curves obtained at different strain ranges is not found for all test materials and temperatures (Shevchenko et al., 1998). As the number of half-cycles increases, the geometry of the elasto-plastic hysteresis loop can change. The values of the unloading modulus G and the cyclic proportionality limit ST also undergo some changes.

It should also be noted that the validity of the Masing principle is preserved only until the stresses and strains after unloading exceed the values of $|\sigma^*|$ and $|\varepsilon^*|$ achieved under initial loading. With further loading of the opposite sign, the history of σ^* and ε^* is forgotten and the resistance to deformation, if we neglect the instability of the material, is equal to the initial one. This feature of material behavior is called the generalized Masing principle (Gokhfel'd and Sadakov, 1984). If the width of the hysteresis loop δ increases as the number of semi-cycles grows, this behavior of the material is called cyclic softening, while if δ decreases, this behavior is called cyclic hardening. If the width of the loop remains constant, the material is considered cyclically stable under these test conditions.

Sometimes a more complex dependence of the loop width δ on the semi-cycle number k is also observed: cyclic hardening is replaced by cyclic softening or, conversely, cyclic softening by cyclic hardening. In many cases cyclic hardening is followed by cyclic stability. The nature of the cyclic instability of a material can vary depending on the structural condition and temperature. The following analytical dependence has been proposed to describe cyclic deformation curves (Shneiderovich, 1968; Gusenkov, 1979)

$$\bar{\varepsilon}^{(k)} = \bar{S} + f_1(k)F(\bar{S})f(t) = \bar{S} = \bar{\delta}^{(k)}, \qquad (2.48)$$

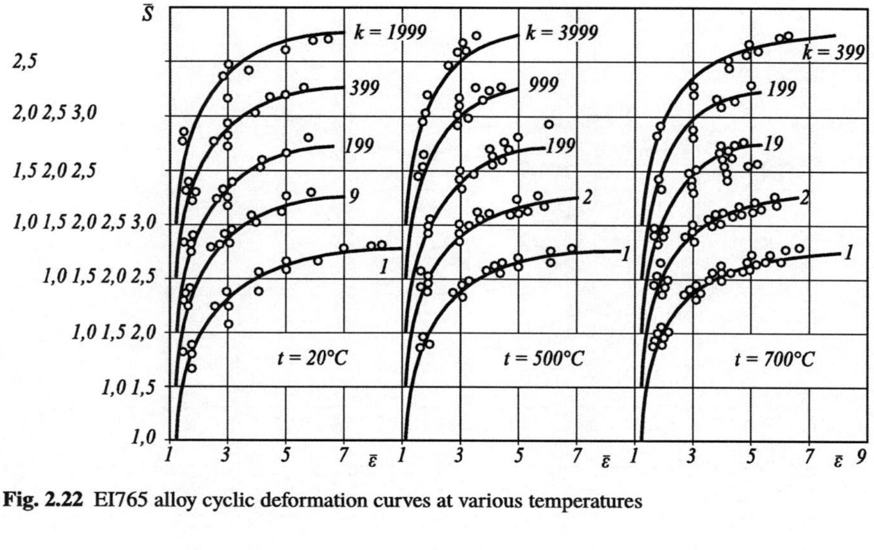

Fig. 2.22 EI765 alloy cyclic deformation curves at various temperatures

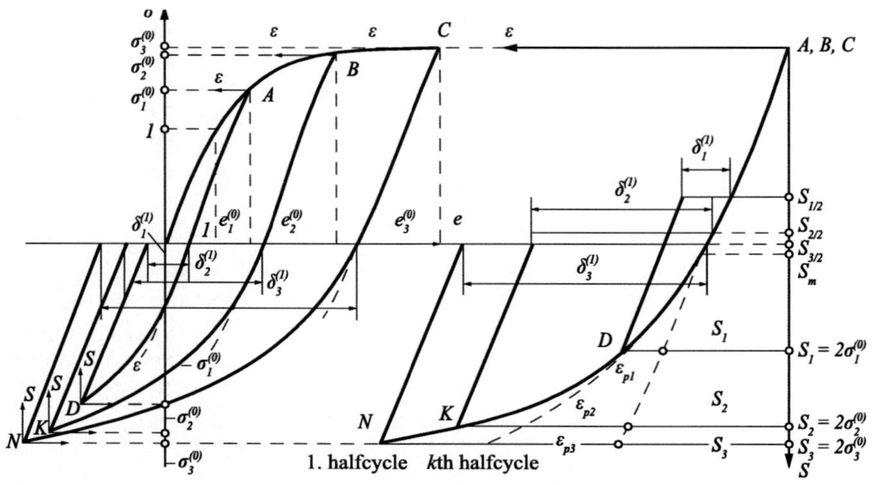

Fig. 2.23 Scheme for constructing a generalized diagram of cyclic elasto-plastic deformation

where $f(t)$ is a function of deformation time, which, like $f_1(k)$, is based on experimental data. The function F is related to the degree of initial deformation by the following dependency

$$F(\bar{S}) = A[f(\bar{S}/2) - \bar{S}_t^{(1)}/2], \tag{2.49}$$

where A is a constant. The value of A for some materials may differ for even and odd halfcycles. Expressions for function $f_1(k)$ depend on the degree of initial deformation and are not necessarily monotonous (Fig. 2.24). In the case of a non-

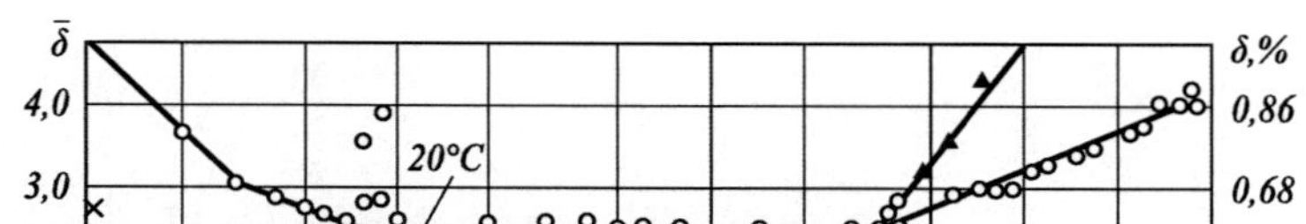

Fig. 2.24 Change in the width of the hysteresis loop $\bar{\delta}$ and δ of the EI765 alloy depending on the number of halfcycles k

stationary (multi-block) cyclic loading as well as in the case of an arbitrary non-cyclic loading program, the use of $f_1(k)$ function requires the introduction of additional assumptions. In these cases, it is suggested to use the functions from the path or energy of cyclic plastic deformation instead of $f_1(k)$ as follows (Getsov, 1978)

$$\lambda_1 = \int |d\varepsilon_{pl}| - \int d\varepsilon_{pl} \mathrm{sign}\left(\int d\varepsilon_{pl}\right), \tag{2.50}$$

$$\lambda_1' = \int \left(|\sigma d\varepsilon_{pl}| - \sigma d\varepsilon_{pl}\right). \tag{2.51}$$

Analytical expressions for $f_1(\lambda_1)$ and $f_1(\lambda_1')$ can be obtained from experiments for stationary cyclic elastic plastic deformation. The currently accumulated experimental data allow, in most cases, to make sufficiently reasonable assumptions about the nature of cyclic instability of a material at room temperature based on knowledge of its structural state and the level of mechanical properties. Structural changes occurring at elevated temperatures in a material, recrystallization processes, and changes in the nature of the relationship between structural state and strength have a very significant impact on the nature of cyclic instability of a material. Different intensities of peening processes at different temperatures can have a very specific impact on the behaviour of a material at variable temperatures. Relative coordinates are often used to reduce the scatter of test results and to improve the reliability of cyclic deformation resistance characteristics when limited experimental data are available.

The solution to the problem on the optimal choice of the main individual characteristics of the specimen material, in relation to which parameters of cyclic tests should be considered, requires in each case to obtain the appropriate statistical analysis. Such a problem, for example, was solved applied to the cyclic curves defined for 10 melts of EI698 alloy tested at 650°C. The cyclic curves $\Delta\sigma = f(\Delta\varepsilon_p)$, $\Delta\sigma/\Delta\sigma_{0,01} = f(\Delta\varepsilon_{pa}/\varepsilon_{0.01})$ and $\Delta\sigma/S_{0,01}^{(1)} = f(\Delta\varepsilon_p/\varepsilon_{0.01})$ constructed in logarithmic coordinates (Fig. 2.25) were compared and the mean square deviations were determined. The dispersions of the values $\Delta\sigma$, $\Delta\sigma_{0.01}$ and $\Delta\sigma_{0.01}^{cycl}$ were found to be $0.078\Delta\sigma$, $0.13\Delta\sigma$ and $0.053\Delta\sigma$ respectively. It should be noted that this dispersion is associated not only with melting features, but also with the accuracy of approximation of cyclic curves by the dependence $\Delta\sigma = k\Delta\varepsilon^m$. Comparison of the obtained values of standard deviations S for the alloy EI698 with the values S for the yield strength shows that they differ little from each other. The obtained results provide the basis to give preference to presentation of results in the form of stress ratio to cyclic limit of elasticity.

2.4.2 Cyclic Elasticity Limit

When studying the features of cyclic deformation of materials, the so-called cyclic proportionality limit value is usually used to describe the results (Shneiderovich,

1968). The accuracy of its determination, as a rule, differs little from the accuracy of static proportional limit determination, i.e., from the accuracy of stress determination, at which there is a 50% increase from the initial tangent of the angle of inclination tangent to the stress axis on the deformation curve[14]. At elevated temperatures, it is advisable to use the elastic limit σ_{el}^{st} and cyclic limit σ_{el}^{cycl} as the ultimate elasticity characteristics under instant loading. Experimental difficulties in determining the value σ_{el}^{st} with high accuracy (with a minimum residual strain tolerance of 0.01% or less) remain in determining. The value σ_{el}^{cycl} also has an independent meaning: for strength calculations of various parts using the theory of adaptability. If the cyclic deformation of the unloading stress is counted from the moment of unloading, then according to the Masing principle the cyclic limits of elasticity and proportionality are $S_{el} = 2\sigma_{el}^{cycl}$, $S_{prop} = 2\sigma_{prop}^{cycl}$ or in relative coordinates we obtain $\bar{S}_{el} = 2\sigma_{el}^{cycl}/\sigma_{el}^{st}, S_{prop} = 2\sigma_{prop}^{cycl}/\sigma_{prop}^{st}$.

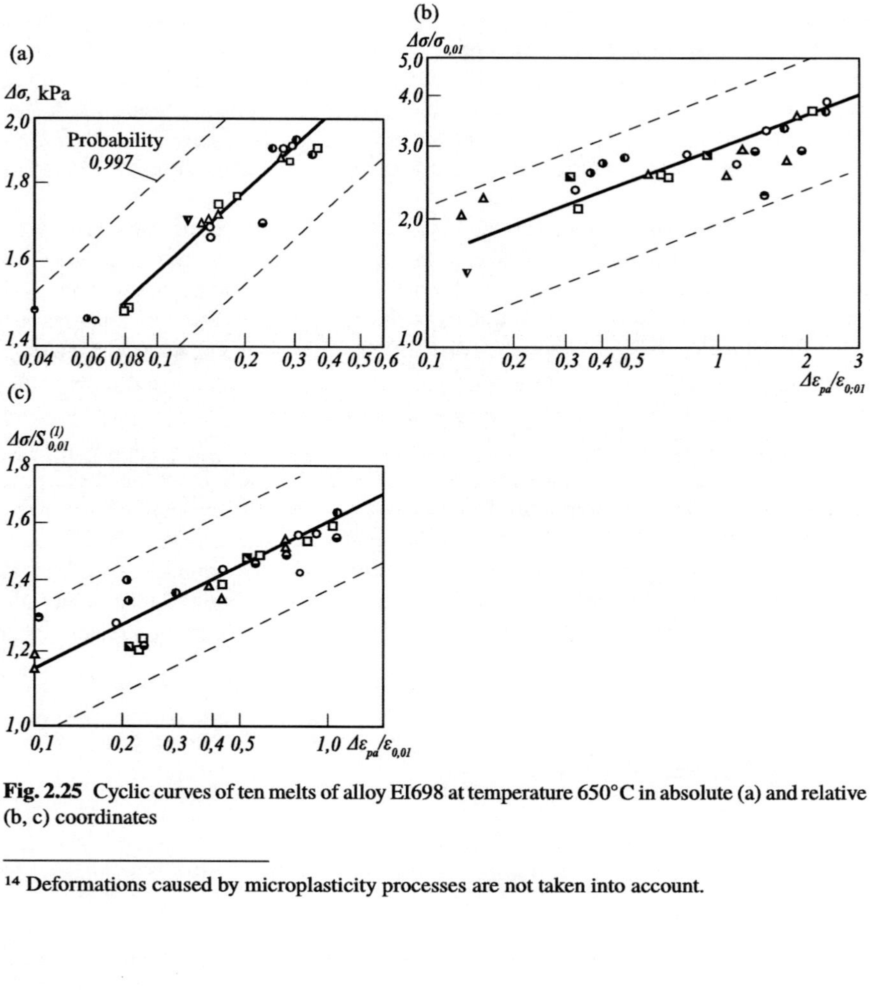

Fig. 2.25 Cyclic curves of ten melts of alloy EI698 at temperature 650°C in absolute (a) and relative (b, c) coordinates

---

[14] Deformations caused by microplasticity processes are not taken into account.

When creep processes are negligible, the parameters $\bar{S}_{el}$ and $\bar{S}_{cycl}$ with decreasing plastic deformation tend to values $\bar{S}_{el} = \bar{S}_{cycl} = 2$. In these conditions, as experiments of Troshchenko (1971) show $\sigma_{prop}^{cycl} = \sigma_{-1}$. Thus, if the amplitude of stress does not exceed σ_{el}^{st}, and $\sigma_{el}^{cycl} = \sigma_{el}^{st}$, respectively, the value of $\lim\limits_{\varepsilon_{pl}\to 0} = \sigma_{el}^{cycl}$ almost equal σ_{el}^{st}. It should be noted that $\lim\limits_{\varepsilon_{pl}\to 0} = \sigma_{el}^{cycl}$ corresponds to the generalization of the Masing principle made by Moskvitin (1965) and is confirmed by experimental data (Fig. 2.26).

For elasto-plastic deformation, the values $\bar{S}_{el}$ for a stable cycle can be either more or less $\bar{S}_{el} = 2$. At moderate temperatures, the values $\bar{S}_{el}$ for strength-strengthening materials are usually less than 2.

In this case, they depend on the test method (at constant or varying amplitude of stresses or strains). If tests are performed first at the amplitudes of stresses $\sigma_a' > \sigma_{el}^{st}$, and then at $\sigma_a'' < \sigma_{el}^{st}$, the plastic deformation will occur at all values $\sigma_a'' > \sigma_{el}^{st}$. If tests are performed at $\sigma_a = \text{const}$ and $\sigma_a < \sigma_{el}^{st}$ there will be no cyclic plastic deformation.

The analysis of elasticity limit characteristics for various steels and alloys at elevated temperatures under quasi-spontaneous cyclic deformation as well as under conditions at which creep processes are realized has been carried out in Getsov et al. (1976). Tests were carried out on materials of different classes under tension-compression and cyclic torsion. It is established that the values $\bar{S}_{el}$ differ significantly for hardening (Fig. 2.27a), softening and stable (Fig. 2.27b) materials. The data obtained indicate that the cyclic limit of elasticity is a function of a number of parameters characterizing the deformation conditions: T, k, δ_ε, $\dot{\varepsilon}(\dot{\gamma})$ ($\dot{\varepsilon}, \dot{\gamma}$ - strain rate under tension-compression and strain rate in experiments on alternating torsion), and loading time in one or both half-cycles (Fig. 2.28). Therefore it can only be accepted as a basic characteristic of the material if certain assumptions are made. Table 2.5 shows the values of the cyclic limit of elasticity of a number of heat resistant steels and alloys. For cyclic loading in conditions when in one half cycle creep has to be considered and in the other only plastic deformation, instead of $\bar{S}_{el}$ the parameter $\bar{S}_{el}^{c}$ is applied. The value of $\bar{S}_{el}^{c}$ is usually much less than $\bar{S}_{c}$ (Getsov et al., 1976). In this case it is reasonable to use the value of cyclic elastic limit $\sigma_{el} = S_{el}^{c} - S_{el}/2$ instead of S_{el} for calculations under creep conditions under alternating loading.

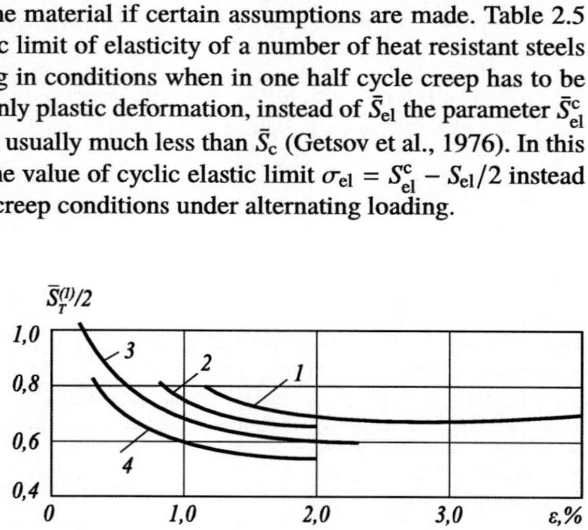

Fig. 2.26 Bauschinger effect under tension-compression:
1 - St3 (Z.P. Kamenceva);
2 - 20Ch (A.E. Chistyakov);
3 - 20 (Z.P. Kamenceva);
4 - 20ChMSA (A.M. Zhukov)

Table 2.5 Values of the cyclic elasticity limit

Material	T,°C	$\bar{S}_{el}$	
		without creep	with creep in both halfcycles
EI802 (15Ch12WNMF)	20	1,57	-
	400	1,63	-
	500	1,47	1,54
	500	1,43-1,53	1,4-1,6
20Ch23N18	600	2,8-3,2	-
	700	2,9-3,2	-
	800	2,2	-
EI868 (ChN60WT)	20	2,4	-
	500	2,1-3,2	-
	600	2,0-2,9	1,8
	700	2,5	1,8
	800	2,5-2,7	-
EI765 (ChN70WMYuT)	600	1,9	-
	700	1,9	1,4-1,7
	800	1,8	-
EI826 (ChN70WMTYuF)	700	-	1,7-1,8
EP220	800	1,9	-
	900	1,9	1,3

2.4.3 Cyclic Creep and Stress relaxation Under Constant Sign Loading

At cyclic loading in the conditions of high temperatures in addition to cyclic elasto-plasticity material accumulates creep deformation. Indeed, under test conditions

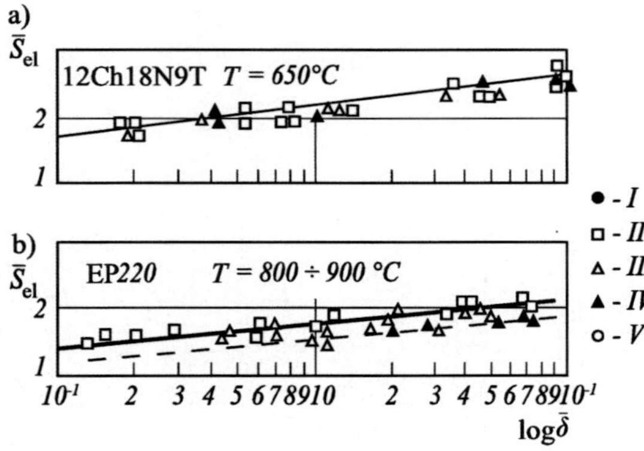

Fig. 2.27 Effect of plastic deformation $\bar{\delta}$ and strain rate $\dot{\gamma}$ on the cyclic elastic limit $\bar{S}_{el}$ in the stabilized state of materials: *I, II, III, IV, V* - $\dot{\gamma}$ equal to 12, 3.2, 1.4, 0.2, 0.15%/min, respectively

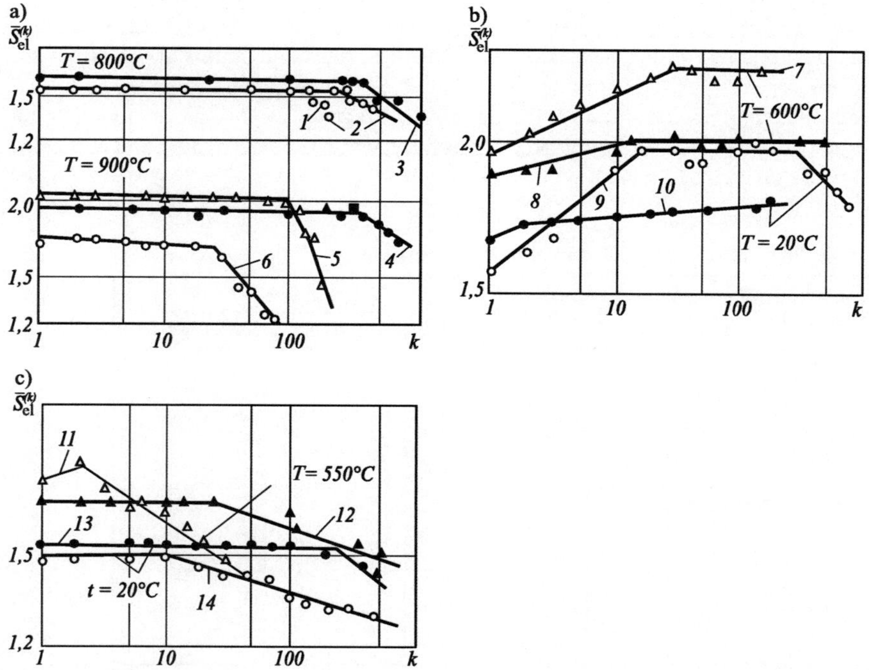

Fig. 2.28 Dependence of the cyclic elastic limit on the half-cycle number for alloys EP220 (a), EI765 (b) and steel EI802 (c): *1* - $\dot{\varepsilon}$ - 0.2%/min, $\Delta\varepsilon = 3\%$; *2* - $\dot{\varepsilon}$ - 0.2%/min, $\Delta\varepsilon = 1.5\%$; *3* - $\dot{\varepsilon}$ - 3.2%/min, $\Delta\varepsilon = 1.5\%$; *4* - $\dot{\varepsilon}$ - 3.2%/min, $\Delta\varepsilon = 3\%$; *5* - $\dot{\varepsilon}$ - 3.2%/min, $\Delta\varepsilon = 3\%$; *6* - $\dot{\varepsilon}$ - 0.2%/min, $\Delta\varepsilon = 3\%$; *7* - $\Delta\varepsilon = 2.8\%$; *8, 10* - $\Delta\varepsilon = 1\%$; *9* - $\Delta\varepsilon = 2.5\%$; *12, 13* - $\Delta\varepsilon = 1\%$; *11, 14* - $\Delta\varepsilon = 2.5\%$

with a given strain amplitude (rigid cyclic loading) depending on the asymmetry of the cycle there is an alternating or repeated relaxation of stresses, and with a given amplitude of loading (soft loading) in each cycle - cyclic constant sign or alternating creep.

Often occurring during operation of GTU components the scheme of cyclic deformation is such, when in one half cycle mainly creep deformations are accumulated, and in another - instant plastic deformations. In contrast to experimental data on creep at constant stresses, there is relatively little data on cyclic creep. Basic features of cyclic creep and relaxation under alternating and constant sign stresses have been studied for different heat-resistant materials in Getsov (1979).

Studies of creep features under repeated cyclic loading ($\sigma_{max} = \sigma$, $\sigma_{min} = 0$) were carried out by testing of cylindrical specimens from heat-resistant alloys under conditions of rectangular cycle with frequencies from 1/48 to 240 cycles per hour. The analysis of primary cyclic creep curves showed that creep curves, as under static loading, have three creep stages: primary, secondary and tertiary (preceding failure). The proportion of deformation of the forward and reverse effect at these three stages is different. The minimum difference between the creep rate of the forward and reverse effect (no more than 30%) is observed at the secondary (steady) stage. At the primary and tertiary stages, the difference in rates and strains at the direct and reverse effect is 50-80% (Table 2.6, Fig. 2.29). Values of the minimum creep rate (for direct and reverse effect) at the established stage (Table 2.6) are attributed to the time under load and without load, respectively, while time of loading and unloading was not taken into account.

Table 2.6 Creep test results under repeated static loading

Material	T, °C	σ_{max}, MPa	Test cycle, h		$\dot{p}_{min}$, %/h		
			loaded	unloaded	average	direct recovery	reverse recovery
EI481	650	320	constant	0	6.2 10^{-4}	-	-
			24	24	3.1 10^{-4}		
			0.5	0.5	3.6 10^{-4}		
			0.04	0.04	0.57 10^{-4}		
EI607A	700	270	constant	0	3 10^{-4}	-	-
			24	24	3.7 10^{-4}		
			2	2	1.3 10^{-4}		
			0.5	0.5	4.4 10^{-4}		
			0.04	0.04	1.6 10^{-4}		
EI826	800	500	constant	0	6	6	-
		500	0.0015	0.0015	2	6.72	4.8
		470	constant	0	3	3	-
		470	0.0015	0.0015	0.16	3.2	3.1
		400	constant	0	0.6	0.6	-
		400	0.0015	0.0015	0.0125	0.77	0.76
EI929	850	550	constant	0	7	7	-
		550	0.0015	0.0015	3.4	6.72	3.5
		500	constant	0	2.8	2.8	-
		500	0.0015	0.0015	1	4.56	3.7
		450	constant	0	1	1	-
		450	0.0015	0.0015	0.08	0.336	0.265
		400	constant	0	0.5	0.5	-
		400	0.0015	0.0015	0.02	0.193	0.175

During the primary and tertiary stages, the ratios of the creep rate $\dot{p}_{rev.recov.}$ of the reverse recovery to the creep rate of the direct recovery $\dot{p}_{dir.recov.}$ for the alloy EI826 were less than for the alloy EI929, which is due to the difference in the resistance to deformation of different elements of the alloys compositions. The specimen creep rate in cyclic tests is several times lower than under static load. This difference

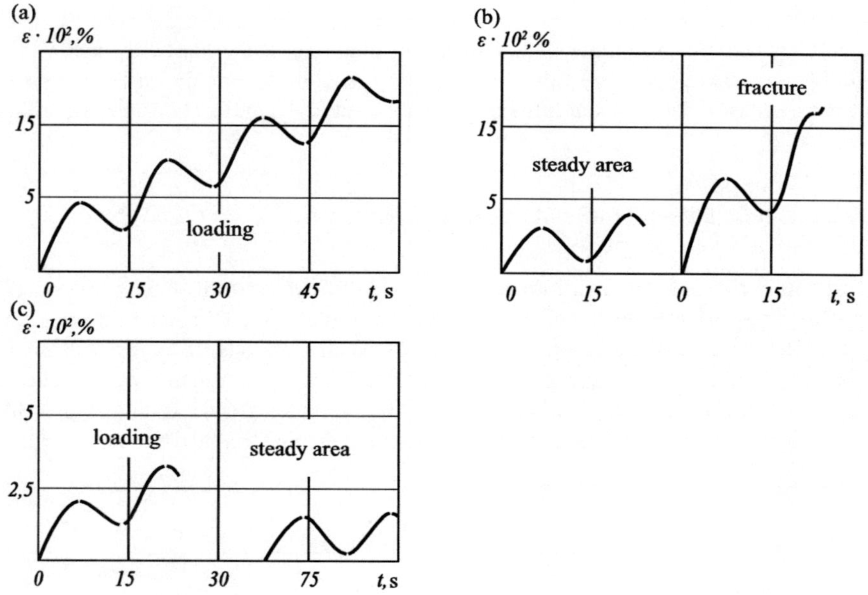

Fig. 2.29 Cyclic creep curves of alloy EI826 at temperature 800°C and $\sigma_{\max} = 500$ MPa (a, b) and 470 MPa (c)

depends on the cycle period, temperature and the specific microstructure of the material. The results of the experiments, accompanied by a thorough measurement of deformation, have shown that deformations of the inverse elastic effect can make noticeable corrections to the behavior of heat resistant alloys during cyclic loading.

It can be assumed that the processes of inverse elastic recovery are determined by the fact that, creep deformations or instantaneous microplastic deformations are observed in microstructure elements under loading. After unloading there are residual stresses σ_{0i} of the opposite sign. Relaxation of these stresses is the reason for the processes of inverse recovery

$$p_t = \frac{1}{A}\sum_i (\sigma_{0i} - \sigma_{t\,i})\frac{\Delta A_i}{E_i}$$

where A is the cross-sectional area of the sample, ΔA_i is the area of the ith element of the structure where the residual stresses occur. E_i is the Young's modulus of the material in microvolume and $\sigma_{t\,i}$ is the residual stress. Depending on the type of the lattice, material and grain size, the nature of hardening, the degree of non-uniformity of deformations along the cross-section and the values of microplastic deformations as well as creep strains at the micro scale change, which causes a difference in the values of the inverse elastic effect for different metals. The analysis of behavior of heat-resistant alloys under conditions of cyclic creep at alternating stresses ($\sigma > 0$)

was carried out based on to the data of repeated relaxation tests at σ_0 = const. The results of tests for a number of heat-resistant materials show that with increasing number of reload cycle the value of remaining tension depending on temperature either increases, decreases or remains constant (Borzdyka and Getsov, 1978).

2.4.4 Cyclic Creep Under Alternating Stress

Tests on cyclic deformation with creep in both half-cycles in the case of identical stresses were performed in Getsov et al. (1978) either with a fixed strain range in the cycle or with a fixed holding time. For all tested materials the creep resistance decreases, i.e. the average creep rate in the unsteady initial section of the creep curve increases from cycle to cycle, then (after 20-30 cycles) becomes constant (Fig. 2.30 and Appendix C). The degree of cyclic instability of the material can be characterized

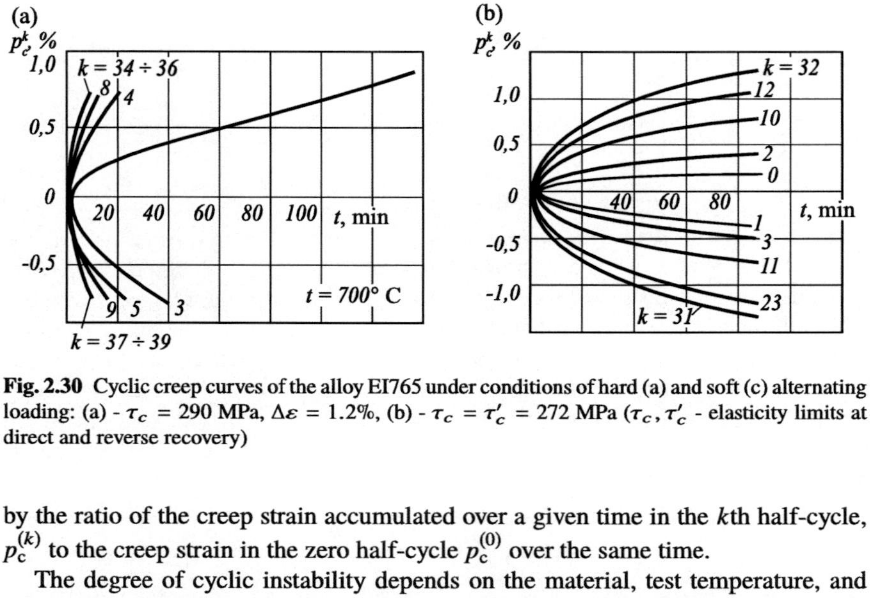

Fig. 2.30 Cyclic creep curves of the alloy EI765 under conditions of hard (a) and soft (c) alternating loading: (a) - τ_c = 290 MPa, $\Delta\varepsilon$ = 1.2%, (b) - $\tau_c = \tau_c'$ = 272 MPa (τ_c, τ_c' - elasticity limits at direct and reverse recovery)

by the ratio of the creep strain accumulated over a given time in the kth half-cycle, $p_c^{(k)}$ to the creep strain in the zero half-cycle $p_c^{(0)}$ over the same time.

The degree of cyclic instability depends on the material, test temperature, and cycle period. The greatest cyclic instability is observed in heat-resistant nickel-based alloys. So, for alloys EI826, EI765 at a temperature of 700°C and EP220 at 900°C, the ratio $p_c^{stab}/p_c^{(0)}$ varies from 4 to 10, for steel 20Ch23N18 at a temperature of 700°C it is 2.3, and for alloy EI868 at a temperature of 600-800°C and steel 15Ch12WNMF at a temperature of 500 and 550°C - about 1.1.

In cycles with creep dwell in one half-cycle, acceleration of creep from cycle to cycle was also observed. In addition, creep had a softening effect on the cyclic deformation curve in the next half-cycle: first, the cyclic elastic limit decreased by $0.5S_{max} - \sigma_c$, where σ_c is the stress at which the exposure was carried out, and secondly, during the exposure, it occurred as there would be some annealing, causing a decrease in cyclic hardening. An analysis of the experimental data shows that the strain curves after creep at different levels are centrally similar with a similarity coefficient α_c with respect to the strain curve for a creep-free cycle (for the same strain range). A typical dependence of α_c on creep stress in relative units $\bar{\tau}_c = \tau_c/\tau_{el}$ is shown in Fig. 2.31.

The deformation resistance under two-frequency loading ($\nu_1 = 1$ cycle/min, $\nu_2 = 1800$ cycle/min) can significantly differ from that for single-frequency (Fig. 2.32) both in unilaterally accumulated strains ε_{pl}, p, and in cyclic strains $\Delta\varepsilon_{pl}$, Δp (Gadenin, 1976).

2.4.5 Cyclic Creep Under Varying Temperature

In 1937, Robinson proposed a number of formulas for assessing the effect of temperature changes on the creep resistance of steels at high temperatures based on the assumption that it is possible to use the equations of the isothermal creep theory (values $\dot{p}$ at t = const) to calculate processes developing under non-isothermal conditions (quasi-stationary approach). Based on this assumption, the values of the average creep rate at a cyclically changing temperature at an established stage can be calculated as follows

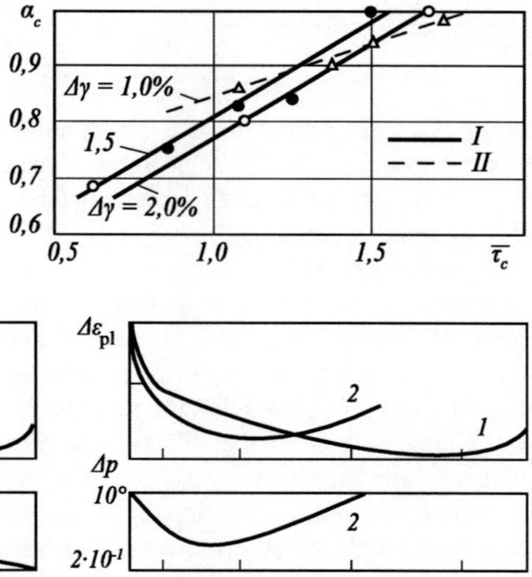

Fig. 2.31 Dependence of the coefficient of central similarity α_c from creep stress τ_c in relative units for alloy EP220 at 900°C (*I*) and steel 12Ch18N10T at 650°C (*II*) and different shear strain $\Delta\gamma$

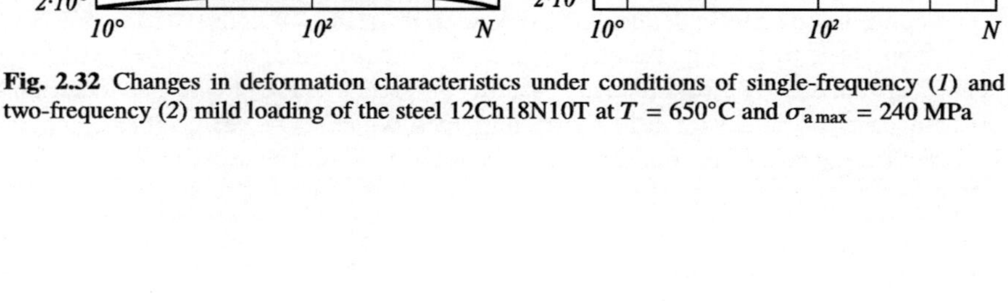

Fig. 2.32 Changes in deformation characteristics under conditions of single-frequency (*1*) and two-frequency (2) mild loading of the steel 12Ch18N10T at T = 650°C and $\sigma_{a\,max}$ = 240 MPa

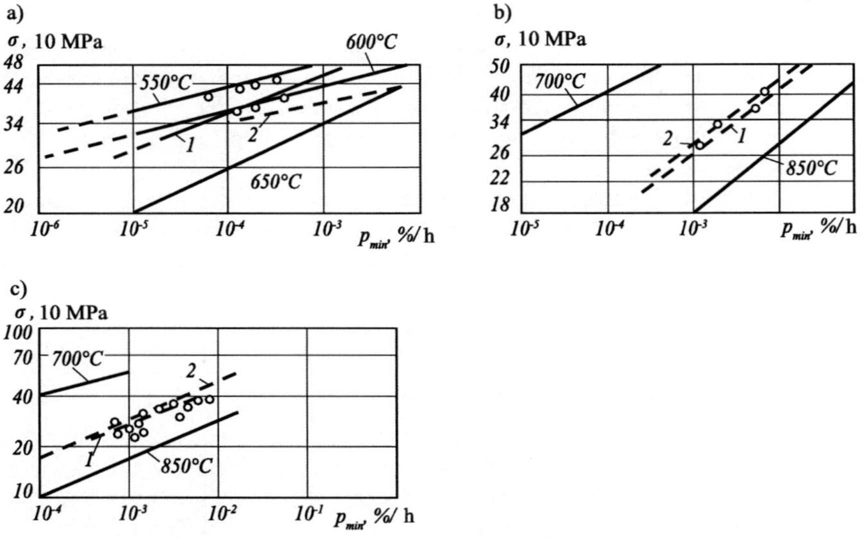

Fig. 2.33 Comparison of experimental (*1*) and calculated (*2*) curves of creep resistance of materials at cyclically changing temperatures with curves (solid lines) at constant temperatures: (a) - EI481, cycle 500°C (8 h) + 550°C (10 h) + 600°C (4 h) + 650°C (1 h), (b) and (c) - EI661 and EI827, cycle 550°C (10 h) + 700°C (12 h) + 850°C (2 h)

$$\dot{p}_{\min e} = \frac{1}{t_{\text{cycl}}} \int_0^{t_{\text{cycl}}} \dot{p}_{\min} dt, \tag{2.52}$$

where t_{cycl} is the period of the cycle. A number of foreign and Russian (I.P. Bulygin, R.N. Sizova, L.B. Getsov, and others) researchers tested the applicability of the quasi-stationary approach to describing the creep of heat-resistant alloys at cyclically varying temperatures.

The experimentally obtained values of creep resistance at cyclically varying temperatures (Fig. 2.33) were compared with the values calculated by the relation (2.52). Temperature step tests were also conducted. Comparison of creep curves during cyclic and temperature-step tests, as well as at the maximum cycle temperature for the same stresses, shows that the same creep strains for structurally stable materials accumulate over a time that differs by no more than 20%. Evaluation of the accuracy of determining the values $\dot{p}_{\min}$ by the formula (2.52) showed that the values $\dot{p}_{\min}^{\exp}$ corresponding to the regression line, the position of which was calculated by the least squares method, are 20% more than the values of $\dot{p}_{\min}^{\text{sim}}$ (Fig. 2.34). This difference corresponds to an error in determining the creep resistance of about 2% for stresses. The effects of temperature aftereffect, the laws of which were studied in Likhachev (1966), are apparently the main reason for this difference: these effects lead to certain differences in the results for different cycle shapes, which, however, can be neglected for practical purposes. It is noted that due to microstructural changes in the material,

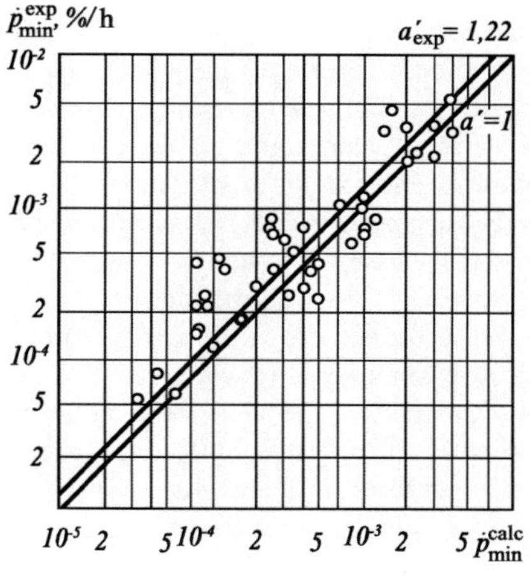

Fig. 2.34 Comparison of experimental data (o) and calculated curves of creep rate at cyclically changing temperatures ($a' = \dot{p}_{min}^{exp}/\dot{p}_{min}^{calc}$)

the accuracy of calculating the minimum creep rate may be quite low in some cases. For the determination of the creep rate at a variable temperature at an unsteady stage, it is necessary to adopt one or another engineering theory of creep (Rabotnov, 1969). The extension of the aging hypothesis (Fig. 2.35b) to the case of variable temperatures contradicts the experimental data, because when the temperature changes abruptly in this case, the value of p also changes abruptly. This contradiction can be

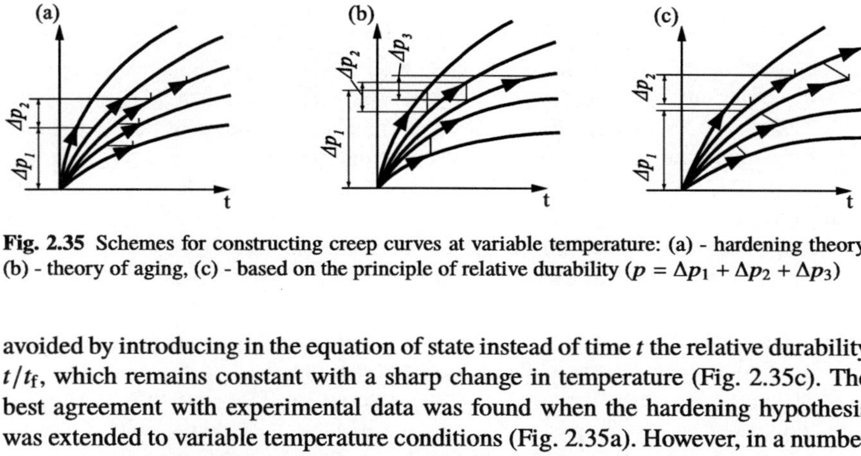

Fig. 2.35 Schemes for constructing creep curves at variable temperature: (a) - hardening theory, (b) - theory of aging, (c) - based on the principle of relative durability ($p = \Delta p_1 + \Delta p_2 + \Delta p_3$)

avoided by introducing in the equation of state instead of time t the relative durability t/t_f, which remains constant with a sharp change in temperature (Fig. 2.35c). The best agreement with experimental data was found when the hardening hypothesis was extended to variable temperature conditions (Fig. 2.35a). However, in a number of cases, the effect of temperature aftereffect should be taken into account, which is not considered in engineering creep theories.

With a short cycle period or a short dwell time at individual cycle temperatures (for example, with the triangular law of temperature change), only an average creep curve can be calculated. For this, using the θ-parameter proposed by Dorn ($\theta = t\exp(-U_0/RT)$, where U_0 is the creep activation energy), the equivalent cycle temperature T_{eq} can be calculated. The validity of the description of creep curves at transient stage using the θ parameter is far from always correct, which reduces the accuracy of the calculations using the formulas proposed for determining T_{eq}.

The variety of structural changes in industrial heat-resistant alloys at elevated temperatures and the different nature of its effect on the behavior of materials at cyclically varying temperatures require special approaches to solving the problem of creep resistance of structurally unstable materials in general. Since structural changes can have a specific effect on creep resistance both at low (T_{min}) and high (T_{max}) temperature steps of the cycle, structural changes occurring at T_{min} and T_{max} should be distinguished (these issues are discussed in more detail in Sect. 4.2.2). Therefore, to analyze the behavior of a material at a cyclically changing temperature, it is necessary to know the nature and degree of influence of microstructural changes at low temperature on the characteristics of the material at high temperature, and vice versa. Microstructural transformations occurring at one or another temperature step of the cycle can be either stable and affect the characteristics of the material at other temperature steps, or unstable (for example, the initial microstructure can be restored during heating or cooling).

When considering the effect of cyclically varying temperatures on the creep characteristics, alloys will be distinguished by the nature of the microstructural transformations that occur upon heating and cooling. Studies have shown that for various alloys, the temperature $T_{structural}$ at which structural changes occur for a long time, negatively affecting the heat resistance, as well as temperatures $T_{restored}$ at which the original microstructure and properties are restored, can differ significantly.

The following approach is proposed for predicting the behavior at a cyclically changing temperature T of materials for which the values of $T_{structural}$ and $T_{restored}$ are known. Microstructural changes affecting the heat resistance at cyclically changing temperatures do not occur if the range of temperature changes during the cycle is such that the conditions $T < \min\{T_{structural}\}$ or $\min\{T_{restored}\} > T, T > \max\{T_{structural}\}$ are met. The indices $\{T\}$ denote the set of temperatures in the range of their variation. When the minimum (or intermediate) cycle temperature is $T_{min} = T_{structural}$, and the maximum $T_{max} = \min T_{restored}$, then the creep resistance at cyclically varying temperatures can be described by formulas of the type (2.52), if the initial heat treatment of the alloy does not contain a tempering operation that causes the precipitation of the hardening phase (for example, heat treatment modes of steels 12Ch18N10T, 20Ch23N18, EI703, EP26, EP99), or if $T_{structural}$ coincides with the tempering temperature $T_{tempering}$. If the maximum cycle temperature is $T_{max} = T_{structural}$, formula (2.52) cannot be used to predict the behavior of the material.

Test modes, in which the ability of the material to regenerate the microstructure and properties (i.e. $T_{restored} = T_{tempering} < \min\{T_{structural}\}$) is manifested, can be characterized by the following conditions: $T_{max} = T_{tempering}$; $T_{min} = T_{restored}$. Calculations made in accordance with the proposed approach to predicting the behavior of materials at variable temperatures are qualitatively confirmed by the results of temperature-step tests.

The established patterns of behavior of structurally unstable materials at varying temperatures are the basis for a quantitative method for taking into account the influence of structural changes of a metal during its long-term operation in the region $T = T_{structural}$ on creep resistance at cyclically varying temperatures. The method can be described as follows: to calculate the creep resistance at variable temperatures, let us accept the additivity of deformations and assume that the characteristics of creep resistance at constant temperature of materials in the microstructural state corresponding to that obtained during aging for a time equal to the full resource are known. We also assume that the rate of microstructural changes depends little on the level of stresses. As experiments with steels EI572 and EI703 have shown, such an assumption for complexly alloyed steels and alloys is quite justified. The explanation for this, apparently, is that the rate of microstructural changes in complex alloys is controlled by the rates of formation of chemical compounds, rather than the rates of diffusion. Taking into account the assumptions made, the following modifications of dependence (2.52) can be used

$$\dot{p}_{\min e} = \frac{1}{t_{cycl}} \int_0^{t_{cycl}} \dot{p}_{\min}[T(t)]\alpha[t_{aging}, T_{aging}, T(t)]dt, \tag{2.53}$$

or

$$\dot{p}_{\min e} = \sum \dot{p}_{\min i}(T)\alpha_i t_i / \sum t_i, \tag{2.54}$$

Coefficients α_i for some materials and aging conditions are given in Table 2.7. The experiments presented by Pavlov and Kurilovich (1978) showed that the creep resistance of heat-resistant alloys under uniaxial and plane stress state, for the same values of equivalent stress can vary significantly. A satisfactory description of transient creep processes can be obtained in the framework of flow and hardening theories if the creep rate is defined as the product of the values $\dot{p}(\sigma)$ and function ψ of the Lode coefficient (Lode, 1926) under creep condition and first loading $\mu_c = (2\sigma_2 - \sigma_1 - \sigma_3)/(\sigma_1 - \sigma_3)_0$. The form of this function is determined from experiments at constant values of σ. The index 0 denotes that the stresses are related to

Table 2.7 Values of coefficients α_i

Material	Low temperature aging		$\alpha_i = \dfrac{\dot{p}_{\min\,aged}}{\dot{p}_{\min\,original}}$
	T, °C	t, h	
20Ch23N18	750	2000	4,5
	750	5000	5,5
	800	5000	2,5
EI481	600	10000	3-4
	650	10000	5
EI827	700	10000	1
	800	10000	5

The designations original and aged refer to the original and aged materials, respectively.

the initial state. This approach is confirmed in experiments with cyclically changing sign-constant and alternating stresses σ and τ (Fig. 2.36).

2.4.6 Cyclic Elasto-Plasticity Under Varying Temperature

The experiments at stepwise increasing and decreasing temperatures, carried out on steel EI481 and alloy EI765 (Kabelevsky, 1972), confirmed that equivalent plastic deformation can be taken as a measure of hardening under non-isothermal loading. On the basis of such experiments, it was suggested that a non-isothermal loading surface exists. Verification of the validity of this assumption for cyclic loading and heating was carried out under the contrasting heating conditions (Fig. 2.37) of 12Ch18N9T steel (Gusenkov, 1979). These experiments showed that the loading surface depends both on the number of loading cycles (intensification of cyclic hardening is observed at variable temperatures), as well as on the frequency and temperature history.

The problems of determining stresses and strains in components working with complex programs of cyclic loading and heating are solved on the basis of theories of plasticity and creep of various types, which can conditionally be divided into two groups: theories, which are modifications of classical deformation theories or theories of plastic flow (Aratyunyan and Vakulenko, 1965; Kabelevsky, 1972; Birger et al., 1975; Kostenko et al., 2013), various creep theories (Naumenko and Kostenko, 2009), theories using the hypothesis of additional stresses (Paley, 1968; Malinin and Khazhinsky, 1969, traditional approach) and theories based on the use of microstructural and physical models (Gokhfel'd and Sadakov, 1984; Novozhilov and Kadashevich, 1990; Naumenko et al., 2011b; Längler et al., 2014; Eisenträger et al., 2018b).

When evaluating the efficiency of the theory in calculating the stress-strain state of solids under complex cyclic loading programs, the following requirements should

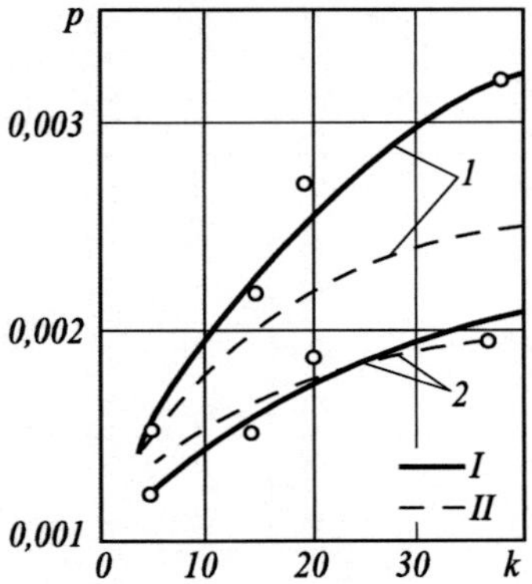

Fig. 2.36 Dependence of the creep strain accumulated over a half-cycle when testing steel EP182 at a temperature of 525°C of the number of cycles for different values of the Lode coefficient: *1* - $\mu_c = -0.39$, *2* - $\mu_c = -0.22$, *I* - experimental data, *II* - simulations based on $p_i = p_1(f_i)\varphi(\mu_c)$ (p_1 - creep strain at tension in the first half-cycle, f_i - function of the cycle number

be satisfied: a) the correct description of the nonisothermal cyclic deformation; b) the need to take into account the cyclic instability of material properties; c) the accurate description of the deformation under complex, in particular cyclic alternating loading; d) the possibility of taking into account the mutual influence of deformations that are time-dependent and time-independent.

2.5 Strain Theories at Complex Loading

The most widespread theories based on fundamentally different approaches than the deformation theory are considered in Sect. 2.3.1 and 2.3.3, namely: various versions of the flow theory; theories reflecting the microstructure of real materials and mechanisms of deformation by using physical and microstructural models, endochronous theories (Kachanov, 1971a; Gokhfel'd and Sadakov, 1984; Novozhilov and Kadashevich, 1990). These theories establish a relationship between increments of strains and stresses. The predictions of flow theory are in good agreement with the experimental data for a complex stress state. For simple (proportional) loading, it provides the same results as those calculated by the deformation plasticity theory. Theories based on the physical and microstructural model of the continuum take into account the fact that, when the type of stress state changes, new microstructural elements pass into the plastic state, and parts of the plastically deformed elements are unloaded, returning into the elastic regime.

The simplest flow theory is the theory of ideal plasticity, which assumes that the position of the yield surface is independent of the loading conditions. According to the theory of isotropic hardening, the radius of the yield surface is determined by some monotonically increasing parameter. To describe the Bauschinger effect

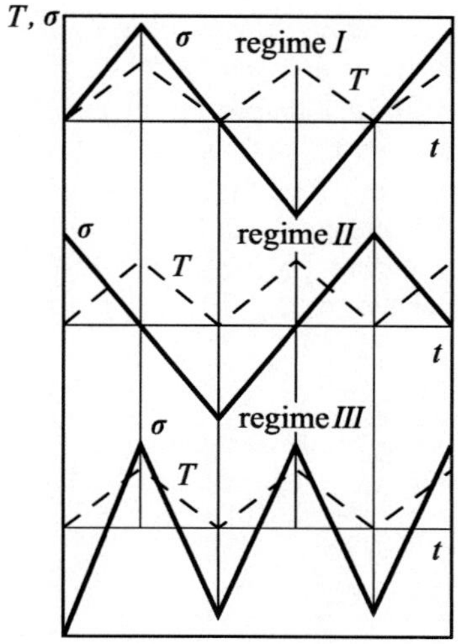

Fig. 2.37 Loading and heating circuit

(changes in deformation resistance when the stress sign changes), various versions of the theory of non-isotropic hardening can be used that describe the translation of the yield surface (Rabotnov, 1969; Kachanov, 1971a; Malinin, 1975; Shesterikov, 2007; Chaboche, 2008; Volkov and Korotkikh, 2008).

The method of experimental determination of yield stress values (the value of the tolerance for plastic deformation) has a significant impact on the accuracy of predicting the conditions of plastic deformation under non-proportional loading. As shown in Kachanov (1971b), the quantity $\chi = (\sigma_{0.2} - \sigma_{0.1})/\sigma_{T_m}$ with $\sigma_{T_m} = (\sigma_{0.2} + \sigma_{0.1})/2$ and T_m as the average temperature at which the stresses are estimated, when moving away from the direction of previous plastic deformation for mild steel, it can reach 0.3, significantly differing from the experimental value of χ. The analysis of the processes of inelastic deformation based on published experimental data allows us to classify the effects of materials behavior.

2.5.1 Effects of the Behavior of Metallic Materials

An analysis of the features of inelastic behavior of metallic materials under uniaxial and multiaxial loading using various programs shows the existence of a large number of effects, as a rule, not described in the framework of many models. A fairly complete list of these effects is schematically presented in Tables 2.8-2.10 for uniaxial stress state, and in Table 2.11 for a complex stress state[15]. It should be noted that in the deformation models included in the packages ABAQUS, ALGOR, ANSYS, COSMOS / M, LS-DYNA, MSC.MARC, MSC.NASTRAN and others, many of the effects presented in Tables 2.8-2.11 are not taken into account.

In Tables 2.8-2.10 the following notations are used: E - Young's modulus; σ is stress; σ_c is the creep stress in the cycle; ε is the strain; ε_{pl} - plastic strain; t is the time; R is the radius of the yield surface; S_T - cyclic yield strength; S_T^c - cyclic yield strength with creep in one half-cycle; δ is the width of the hysteresis loop; p is the creep strain; p_0 is the creep strain of the material in the initial state; ε_p, ε_{cf} - deformation upon failure under conditions of instant deformation and creep; k is the number of half-cycles; λ is the Odqvist parameter during creep; 1,2,... - material number; α is the coefficient of linear thermal expansion; λ_0 is the coefficient of thermal conductivity; t_{aging} is the aging time; T_{aging} - aging temperature.

In the tables $\boldsymbol{\sigma}$ is the stress tensor, $\boldsymbol{s}$ is the stress deviator, $\boldsymbol{\rho}$ is the microstress deviator,

$$\sigma_i = \sqrt{\frac{3}{2}\boldsymbol{s} \cdot\cdot \boldsymbol{s}}$$

is the von Mises equivalent stress, $\boldsymbol{\varepsilon}$ is the strain tensor, $\boldsymbol{\varepsilon}^p$ is the tensor of plastic deformations,

$$\lambda = \int \sqrt{\frac{2}{3}\dot{\boldsymbol{\varepsilon}}^p \cdot\cdot \dot{\boldsymbol{\varepsilon}}^p}\,\mathrm{d}t$$

[15] The material for the Tables 2.8-2.12 was collected by A.S. Semenov and the author.

Table 2.8 Effects of elastoplastic deformation under uniaxial loading

Effect	Scheme	Possible Model
1.1. Hardening (linear, nonlinear, softening, yield point)	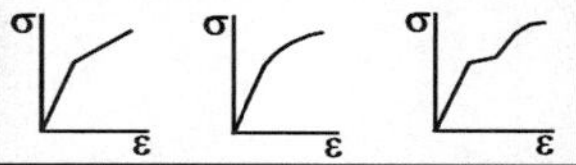	$\varepsilon = \dfrac{\sigma}{E} + r\left(\dfrac{\sigma}{\sigma_{el}}\right)^k$ $\sigma = A\varepsilon^{1/m}$
1.2. Cold hardening (h)	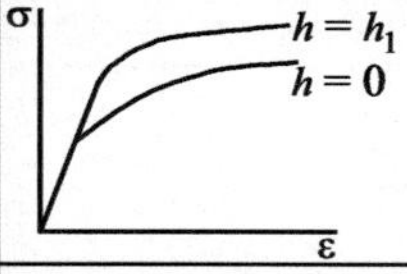	$\sigma_{0,2}^{\text{cold hardening}} = f(\varepsilon_{pl})$
1.3. Yield stress versus loading rate	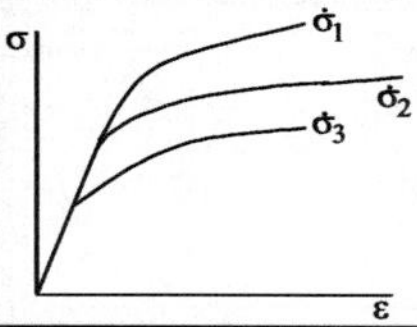	$\sigma_{0,2} = f(\dot{\varepsilon}), \sigma_{0,2} = F(\dot{\sigma})$
1.4. Asymmetry under tension and compression	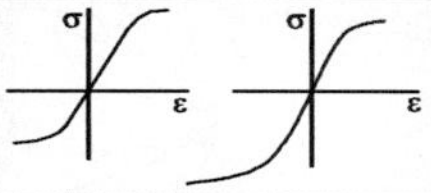	$\sigma = f(\varepsilon)\,\text{sign}\sigma$
1.5. Bauschinger effect	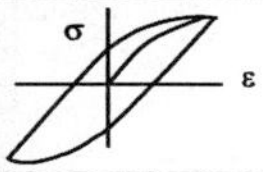	$\sigma_{0,2}^{(-)} = \sigma_T - \sigma_{0,2}^{(+)}$
1.6. Cyclic instability	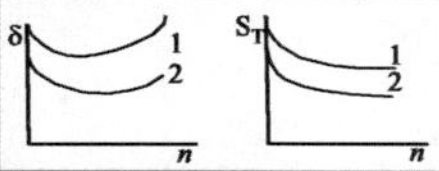	$\delta = f(n), S_T = S_{T0}F(n)$
1.7. Cyclic plastic strain accumulation (ratchetting)	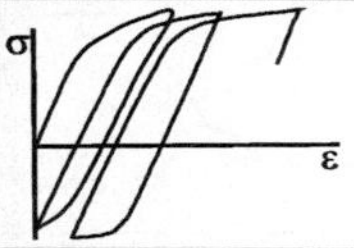	see Gokhfel'd et al. (1978)
1.8. Grain size influence on flow stress	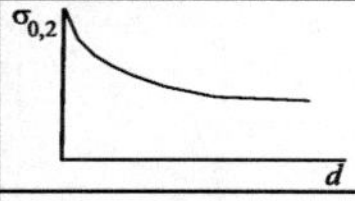	$\sigma_{0,2} = f(1/\sqrt{d})$
1.9. Influence of crystallographic orientation		$\sigma = f(\varepsilon, [\text{hkl}])\,\text{sign}\sigma$ $E = F([\text{hkl}])$
1.10. Thermal dependence material properties	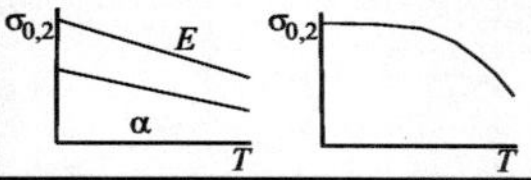	$E = F(T), \alpha = F(T)$ $\lambda_0 = F(T)$; $\sigma_{0,2} = F(T)$
1.11. Aging (structural changes during long exposures at elevated temperatures)	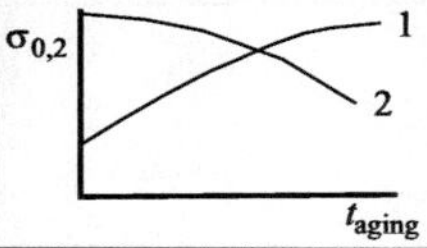	$\sigma_{0,2} = F(t_{aging}, T_{aging})$

Table 2.9 Effects of viscoelastic deformation under uniaxial loading

Effect	Scheme	Possible Model
2.1. Creep stages (primary, secondary, tertiary)	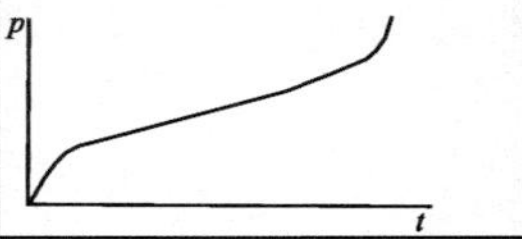	$\dot{p} = a[\sigma/(1-p/\varepsilon_{cf})]^k$ $\exp[-c(\sigma/(1-p/\varepsilon_{cf})^l t]$ $+ b[\sigma/(1-p/\varepsilon_{cf})]^m$
2.2. Stress relaxation	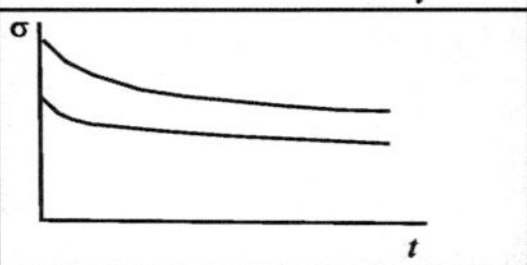	$\dot{\sigma} = -E\dot{p}$
2.3. Bauschinger effect	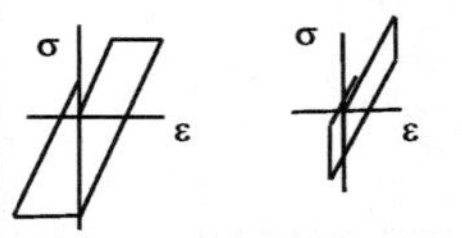	$\dot{p}^{(n)} = \dot{p}_{\min} +$ $\varphi(n, t, p^{(n-1)})(\dot{p}_0 - \dot{p}_{\min})]$ $\psi_1(\varepsilon^+, t)\psi_1(\varepsilon^-, t)$
2.4. Cyclic softening (creep acceleration) under alternating loading	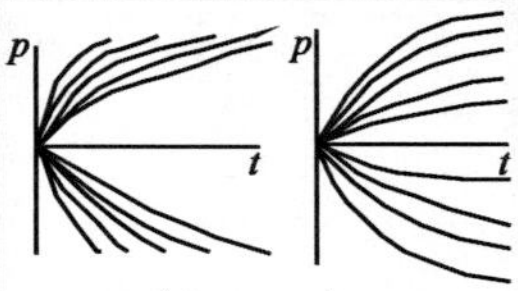 p = const t = const	$\dot{p} = \dot{p}_0 F(\lambda), \quad \lambda = f(k)$
2.5. Creep at variable temperatures between T_1 and T_2	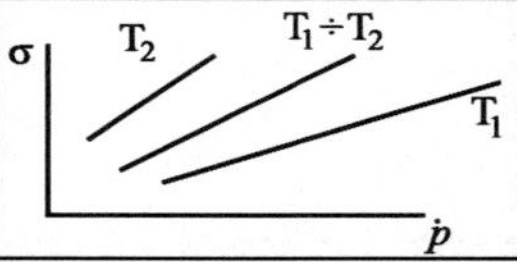	$\dot{p} = F(\Sigma \dot{p}_i t_i / \Sigma t_i)$
2.6. Anomaly in the temperature dependence	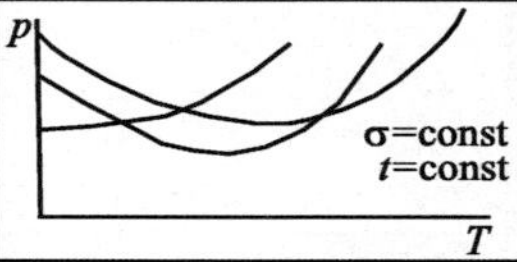	$p = F(T, \sigma), p \neq p_0 \exp(-Q/RT)$
2.7. Microstructural changes during exposure at elevated temperatures (1 - initial state, 2 - after aging)	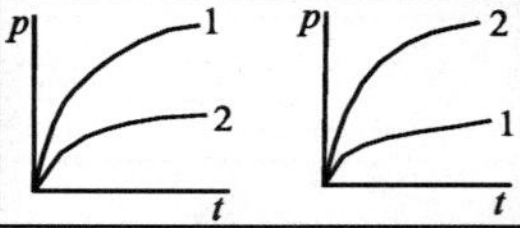	$p = F(\sigma, t_{\text{aging}}, T_{\text{aging}})$
2.8. Cyclic creep anisotropy at alternating loading ($\lvert\sigma\rvert$ = const)	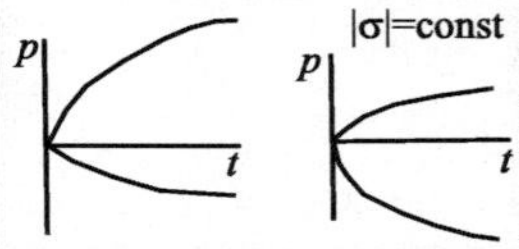	$\dot{p} = F(T, \sigma, p)\,\text{sign}\sigma$
2.9. Influence of crystallographic orientation on creep resistance, including under alternating loading	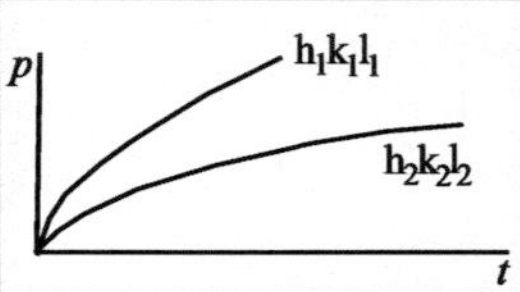	$\dot{p} = F(T, \sigma, p,$ signσ, [hkl])
2.10. Recovery during unloading, including cyclic loading	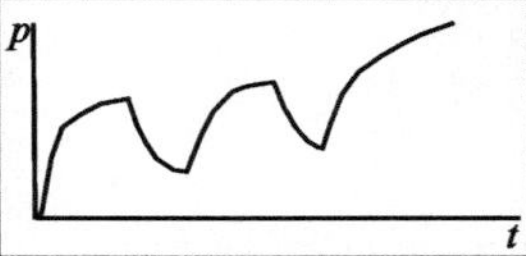	Combined heredity and hardening theories or Maxwell and several Voigt models

Table 2.10 Combined effects of creep and elastoplastic deformation under uniaxial loading

Effect	Scheme	Possible Model
3.1. Viscous effects above and below the yield strength		$\dot{p} = a[\sigma/(1 - p/\varepsilon_{cf})^k]$ $\exp[-c(\sigma/(1 - p/\varepsilon_{cf})^l t]$ $+ b[\sigma/(1 - p/\varepsilon_p)]^m$; $a, b, c, k, l, m = a_1, b_1, c_1, k_1, l_1, m_1$ if $\sigma > \sigma_{0,2}$, and $a_2, b_2, c_2, k_2, l_2, m_2$ if $\sigma < \sigma_{0,2}$
3.2. Intersection of relaxation curves at various initial stresses	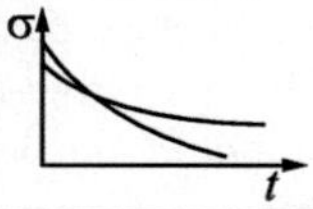	$\dot{p} = f(\sigma, p, \varepsilon_{pl})$
3.3. Dynamic reversibility (restoration of deformation isotropy, elastic modulus and yield strength)		
3.4. Different resistance to instant elastoplastic deformation in the initial state and after creep	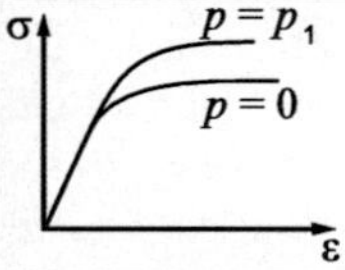	$\sigma = f(\varepsilon + p\varepsilon_f/\varepsilon_{cf})$
3.5. Different creep resistance in the initial state and after instant elastoplastic deformation	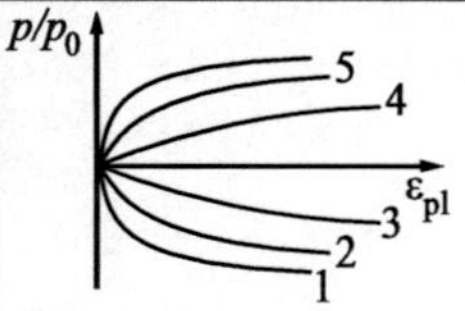	$p = p_0 S(\varepsilon_{pl})$
3.6. Change in cyclic creep strength during alternating elastoplasticity and creep	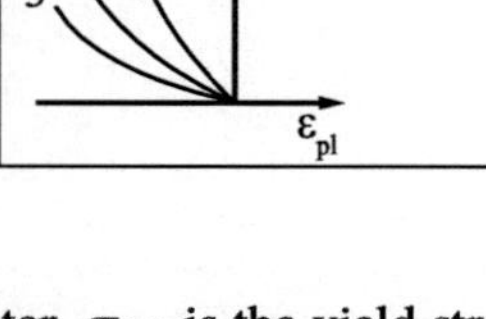	$S_T^c = S_T/2 + \sigma_c; p = f(\sigma_c, t)$
3.7. Creep acceleration after elastoplastic deformation under stress reversals	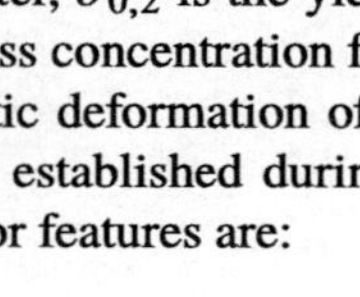	$p = p_0 S(\varepsilon_{pl})$

is the Oqvist parameter, $\sigma_{0,2}$ is the yield strength, f is the current loading surface, α_N is the Neuber stress concentration factor, K_c is the fracture toughness.

Thus, with inelastic deformation of structures, various features of the behavior of materials that are established during the testing under uniaxial stress state are manifested. The major features are:

1. In the case of elastoplastic deformation: Bauschinger effect, cyclic instability of materials (cyclic hardening, cyclic softening, cyclic stability), change in the process of cyclic alternating loading of the cyclic elastic limit (yield strength).
2. In the case of creep (viscoelasticity): the Bauschinger effect during creep, cyclic instability of materials, cyclic softening under creep conditions under alternating loading, non-monotonic temperature dependence of creep resistance, inverse elastic aftereffect during unloading, including cyclic alternating and constant sign loading.
3. With the combined manifestation of creep and elastoplastic deformation (viscoplasticity): the effect on the response to instant elastoplastic deformation of the accumulated creep deformation, the pronounced anisotropy of this effect, the effect on the creep response of instant elastoplastic deformation, the acceleration of creep after elastoplastic deformation under stresses of opposite sign, the change in cyclic elasticity limit under cyclic alternating loading.

Under the conditions of a complex stress state, a number of additional effects are observed, which primarily include: scalar retardation of properties under complex loading, the phenomenon of retardation of vector properties under complex loading, reversible and irreversible isotropic hardening, the effect of preliminary cyclic training on the yield surface, distortion of the surface shape loading. To describe the processes of inelastic non-isothermal deformation, various generalizations of well-known plasticity theories to the case of variable temperatures have been proposed (Birger et al., 1975; Gokhfel'd and Sadakov, 1984; Shevchenko et al., 1992; Längler et al., 2014; Naumenko and Altenbach, 2016, see pp. 115-119).

Table 2.11: Effects of materials response under multi-axial stress state

Effect	Scheme	Possible Model
Delayed scalar properties (as an illustration - a "dive" on the $\sigma_{vM} - \varepsilon_{vM}$ curve, corresponding to a kink in the deformation path)	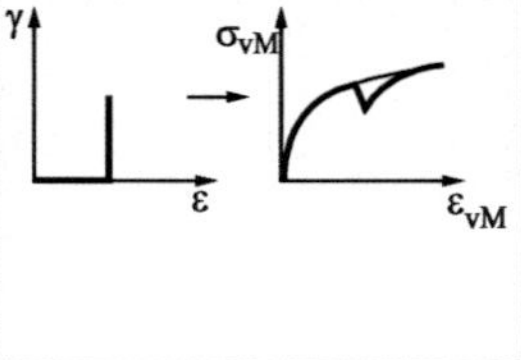	$\sigma = \sigma\left(\varepsilon, \frac{d\varepsilon}{d\lambda}\right)$
Delayed vector properties	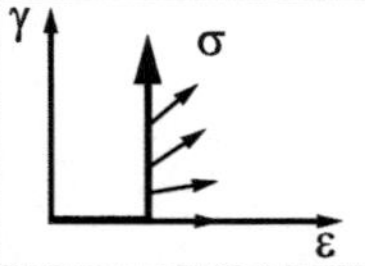	
Additional cyclic hardening (for non-proportional loading cyclic hardening is stronger than for proportional loading)	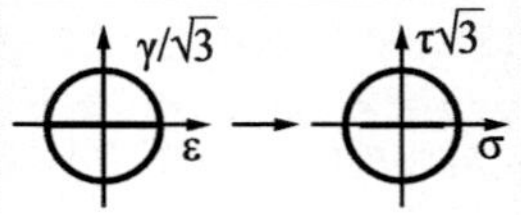	$\dot{\sigma}_T = \gamma[Q(A) - \sigma_T]\dot{\varepsilon}^p_{vM}$ $A = 1-(\dot{\boldsymbol{\varepsilon}}^p \cdot\cdot \dot{\boldsymbol{s}})^2/(\dot{\boldsymbol{\varepsilon}}^p \cdot\cdot \dot{\boldsymbol{\varepsilon}}^p)(\dot{\boldsymbol{s}} \cdot\cdot \dot{\boldsymbol{s}})$

Continuation on next page

Table 2.11: continued

Effect	Scheme	Possible Model
Softening by change to proportional loading Cross cyclic hardening (a change in the direction of proportional cyclic loading is accompanied by hardening followed by softening)	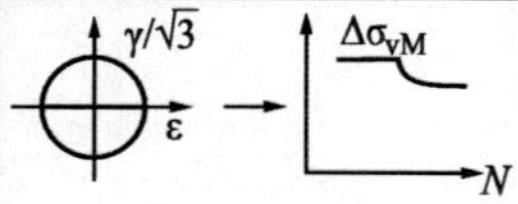 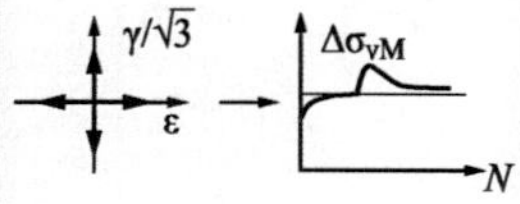	
Effect of strain history on stress (with an equal final deformed state, a dependence of stresses on the shape of the deformation path is observed)	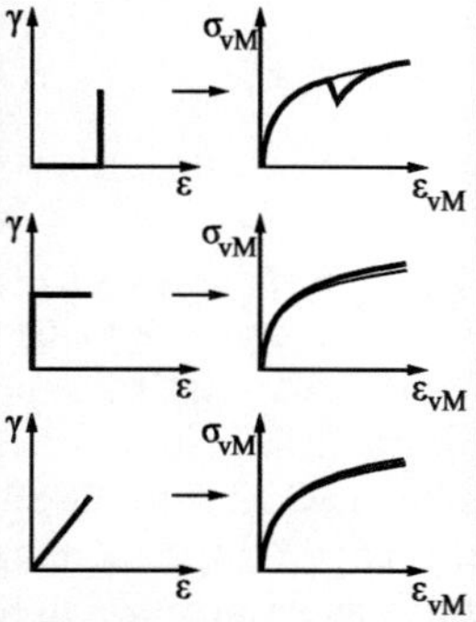	$\sigma = \sigma\left(\varepsilon, \dfrac{\mathrm{d}\varepsilon}{\mathrm{d}\lambda}\right)$
Effect of hydrostatic pressure (high pressure, porosity)	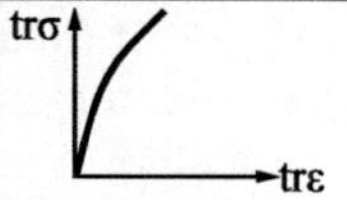	$\mathrm{tr}\boldsymbol{\varepsilon}^{\mathrm{p}} \neq 0$, $\sqrt{\frac{3}{2}(\boldsymbol{s}-\boldsymbol{\rho})\cdot\cdot(\boldsymbol{s}-\boldsymbol{\rho})} - \beta\mathrm{tr}\sigma = \sigma_{\mathrm{T}}$
Non-associated plastic flow (deviation from gradient)	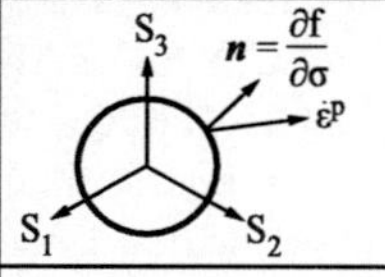	$\dot{\boldsymbol{\varepsilon}}^{\mathrm{p}} = \dfrac{\partial g}{\partial \boldsymbol{\sigma}}; g \neq f$
Irreversible deformation under passive loading (accumulation of plastic deformation along and inside the surface of neutral loading)	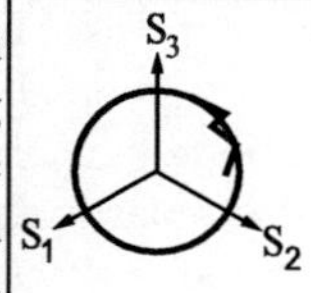	$\dot{\boldsymbol{\varepsilon}}^{\mathrm{p}} = \dot{\lambda}\dfrac{\partial f_k^{\mathrm{active}}}{\partial \boldsymbol{\sigma}}$, $k = 1, 2, \ldots, n$
Sensitivity of yield stress to stress concentration (nonlocal plasticity theories)	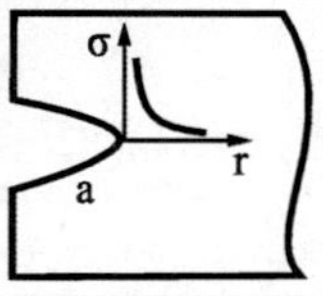	$\sigma_{\mathrm{T}}^* = \dfrac{\sigma_{\mathrm{T}}}{\alpha_{\mathrm{N}}} + \dfrac{K_{\mathrm{c}}\varphi(\alpha_{\mathrm{N}})}{\sqrt{\pi a}}$, $\varphi(1) = 0, \varphi(\infty) = 1$

Continuation on next page

Table 2.11: continued

Effect	Scheme	Possible Model
Strain anisotropy (evolution of the loading surface: expansion, translation, and shape distortion; incomplete ductility and elasticity, ultimate layer)	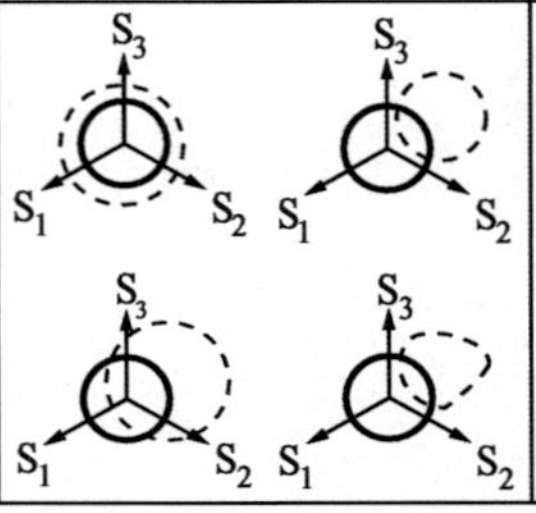	$f[(\boldsymbol{s}-\boldsymbol{\rho})\cdot\cdot\,{}^4\boldsymbol{Q}(\lambda)\cdot\cdot(\boldsymbol{s}-\boldsymbol{\rho})]$ $= \sigma_{\mathrm{T}}(\lambda)$

End of the table

2.5.2 Models of Visco-Elasto-Plasticity

At present, there is no generally accepted theory of plasticity, valid for arbitrary loading paths and a wide class of structural materials. In such a situation, when conducting practical calculations, the most promising approach is that a multimodel analysis of thermoelastic-visco-plastic deformation is used, which allows obtaining reliable results for a specific problem (Chaboche, 1989; Semenov and Melnikov, 1994, 1997a,b; Melnikov and Semenov, 1995a; Krempl, 1999; Melnikov et al., 2001a,b, 2009). It is essential that the material model used should be adequate to the complexity of the processes under study, that is, it should allow to obtain reliable results and at the same time be quite simple. Areas of reliable use of certain models should complement and partially overlap each other.

In modern widely used finite element packages ABAQUS, ANSYS, COMSOL, COSMOS/M, LS-DYNA, MSC.MARC, MSC.NASTRAN, PANTOCRATOR, PERMAS, SAMCEF, ZeBuLoN for inelastic analysis of the stress-strain state of components, a number of models of plasticity and creep is utilized. The selection of models is based on the subjective decision of the user, which, of course, cannot guarantee the adequacy of the analysis. Recommendations to the user on the selection of models that are optimal for solving specific problems are not included in packages.

Most of the models used by commercial finite element analysis systems, unfortunately, do not adequately determine the visco-plastic deformations necessary for calculations within the framework of deformation criteria for the durability of the blades, disks, and rims of guide vanes of gas turbines under thermo-cyclic loading typical for operating conditions.

In Melnikov et al. (2009), a classifier of visco-elasto-plasticity models (theories) was developed, which allows, on the one hand, to see the variety of existing models, and on the other hand, to judge at least the names of the models about their scope. In this classifier, the models of elasto-plasticity, visco-elasticity (creep) and elasto-visco-plasticity (general case) are separately considered, including models of thermoviscoelasticity. In the Table 2.12[16] is a list of models according to the classifier (Melnikov et al., 2009) is presented.

Table 2.12: Classification of models

Approaches (author(s))
ELASTO-PLASTICITY MODELS
1.1. *Deformation theory of plasticity* (Hencky, 1924, 2020; Nádai, 1931; Ilyushin, 2016, etc.) 1.2. *Theories of plastic flow* (Barré de Saint-Venant, 1870; von Mises, 1913; Prandtl, 1925; Reuss, 1930; Melan, 1938; Drucker, 1949, among others) Classification by type of law for plastic deformation: 1.2.1. Associated law of plastic flow 1.2.2. Non-associated law of plastic flow Classification by type of hardening: 1.2.3. Perfectly plastic material (Barré de Saint-Venant, 1870; Levy, 1870, 1871; Prandtl, 1925; Reuss, 1930) 1.2.4. Isotropic hardening (strain hardening, work hardening), see Hill (1950) 1.2.5. Incremental model of thermoplasticity with a piecewise spherical yield surface under proportional cyclic loading (Getsov and Kabelevskii, 1978) 1.2.6. Linear kinematic (translational) hardening (Ishlinsky, 1954; Prager, 1955a; Ziegler, 1958; Barlat et al., 1991; Desmorat and Marull, 2011) 1.2.7. Nonlinear kinematic hardening (Armstrong and Frederick, 1966; Kadashevich, 1967; Backhaus, 1972; Chaboche, 1975; Ohno and Wang, 1991; Frederick and Armstrong, 2007; Altenbach et al., 2019; Naumenko et al., 2020) 1.2.8. Isotropic-kinematic hardening (translational-isotropic hardening) 1.2.9. Anisotropic hardening (Hill, 1950; Zyczkowski, 1981) Classification of non-isothermal effects: 1.2.10. Not counted 1.2.11. Are taken into account Classification by sensitivity to strain rate: 1.2.12. Scleronomic 1.2.13. Rheonomic Classification according to the degree of damage accounting: 1.2.14. Not considered 1.2.15. Are taken into account Classification according to the gradient effect: 1.2.16. Not considered 1.2.17. They are taken into account as the effect of gradient on the yield strength 1.2.18. They are taken into account as the influence of gradient on the form of the constitutive equations Classification by the presence of features in the loading surface: 1.2.19. Regular plasticity (smooth loading surface)

Continuation on next page

[16] This table was prepared by A.S. Semenov.

Table 2.12: Classification of models (Continued)

Approaches (author(s))
1.2.20. Singular plasticity (Prager and Hodge Jr., 1951; Sanders Jr., 1955; Koiter, 1960; Klyushnikov, 1994, among others)
Classification by type of yield condition: For isotropic material
1.2.21. Tresca (1864); Barré de Saint-Venant (1870)
1.2.22. von Mises (1913); Huber (1904)
1.2.23. Ishlinsky (1954); Ivlev and Bykovtsev (1971); Ishlinsky and Ivlev (2003)
1.2.24. Drucker and Prager (1952)
1.2.25. Mohr (1900); Coulomb (1773)
For anisotropic material
1.2.26. Hill (1950)
For single-crystal materials
1.2.7 Schmid and Boas (1935); Qin and Bassani (1992); Cailletaud (1992); Semenov (2014, 2016)
1.3. *Multi-surface Theory of Plasticity*
1.3.1. Mróz (1967)
1.3.2. Chaboche (1975)
1.3.3. Dafalias (1986)
1.3.4. Melnikov and Semenov (1991); Izotov et al. (2000)
1.4. *Theory of Elastoplastic Processes* (Ilyushin and Lensky, 1959, etc.)
1.4.1. Theory of processes of small curvature
1.4.2. Theory of processes of middle curvature (Bik, 1981)
1.4.3. The theory of processes of deformation along two-link trajectories with a kink (Nejescu-Cleja, 1976; Vasin, 1990)
1.5. *Endochronic theory* (Valanis, 1971; Watanabe and Atluri, 1985; Kadashevich and Mosolov, 1989, etc.)
1.6. *Slip Theory* (Batdorf and Budiansky, 1949; Bishop, 1953; Malmeister, 1957; Lin, 1971; Klyushnikov, 1979; Mohel et al., 1983; Chowdhury et al., 2017, 2018, among others)
1.7. *Analytical models* (Klyushnikov, 1994)
1.8. *Rheological models* (Prager, 1955b; Besseling, 1959; Gokhfel'd and Sadakov, 1984; Sobotka, 1984; Novozhilov and Kadashevich, 1990; Shevchenko et al., 1997; Palmov, 1998; Schmicker et al., 2013; Semenov and Melnikov, 2016, among others)
1.9. *Statistical theory of plasticity* (Novozhilov and Kadashevich, 1990, among others)
1.10. *Physical theories of plasticity* (Taylor, 1938; Lin, 1957; Trusov and Shveikin, 2019, among others)
1.11. *Structural and analytical theory* (Likhachev and Malinin, 1993, among others)
1.12. *Multiple Mechanism Model* (Besson et al., 2010, among others)
MODELS OF VISCO-ELASTICITY (CREEP AT STRESSES BELOW YIELD STRENGTH)
2.1. *Classical Creep Theories*

Continuation on next page

Table 2.12: Classification of models (Continued)

Approaches (author(s))
2.1.1. Theory of steady-state creep (Odqvist, 1974; Naumenko et al., 2009; Altenbach et al., 2008a)
2.1.2. Theory of aging (Malinin, 1975; Nordmann et al., 2018)
2.1.3. Flow theory (Kachanov, 1971a)
2.1.4. Hardening theory
2.1.5. Theory of heredity (Rabotnov, 1980)
2.1.6. Mixed (combined) theories (Kheyn, 1966; Chizhik, 1966, among others)
2.2. *Rheological models* (Reiner, 1960)
2.2.1. Maxwell
2.2.2. Voigt
2.2.3. Reiner
2.2.4. Kelvin-Voigt
2.2.5. Combined
2.3. *Creep theory with anisotropic hardening* (Malinin and Khadjinsky, 1972)
2.4. *Creep theories with structural parameters* (Rabotnov, 1969; Lepin, 1976)
2.5. *Creep-damage theories* (Leckie and Hayhurst, 1977; Hayhurst, 1994; Altenbach and Naumenko, 1997; Altenbach and Skrzypek, 1999; Altenbach et al., 2000a,b, 2001; Altenbach and Naumenko, 2002; Volkov and Korotkikh, 2008)
2.6. *Anisotropic creep theories* (Naumenko and Altenbach, 2005; Robinson et al., 2003; Naumenko and Gariboldi, 2014; Gariboldi et al., 2016)
2.7. Creep theories with damage tensors (Murakami and Ohno, 1981; Murakami, 1983; Murakami and Sanomura, 1985; Betten, 2001; Altenbach et al., 2002; Betten, 2008)
ELASTIC VISCO-PLASTIC MODELS
3.1. *Viscoelastic-viscoplastic models*
3.2. *Elastic visco-plastic models*
3.2.1. Norton (1929) model
3.2.2. Lemaitre (1972) model
3.2.3. Perzyna model (Perzyna, 1966; Ugodchikov and Short, 1971; Korotkikh, 1981)
3.2.4. Chaboche and Rousselier (1983) model
3.2.5. Bodner and Partom (1975) model
3.2.6. Miller (1976a,b) model
3.3. *Models of thermo-visco-elasticity* (Shevchenko et al., 1992)
3.3.1. Visco-elasticity model for rectilinear deformation trajectories or for deformation paths slightly deviating from them
3.3.2. Visco-elasticity model for trajectories of small curvature
3.4. *Theory of elasto-plastic processes* (Ilyushin, 1954)
3.4.1. Visco-elasticity model for arbitrary planar trajectories
3.4.2. Thermo-visco-elasticity model for trajectories of medium curvature and small torsion (Shevchenko and Babeshko, 1999)

Continuation on next page

Table 2.12: Classification of models (Continued)

Approaches (author(s))
3.5. *Model of cyclic plasticity and creep under proportional loading* (Getsov and Kabelevskii, 1978)
3.6. *Rheological (structural) models* (Sobotka, 1984; Gokhfel'd and Sadakov, 1984; Novozhilov and Kadashevich, 1990; Palmov, 1998; Semenov and Melnikov, 2016; Eisenträger et al., 2018a, among others)
3.6.1. Models for cyclically stable material
3.6.2. Models for cyclically unstable material
3.6.3. Models for anisotropic material
3.6.4. Models for two-phase material
3.7. *Endochronic theory of a visco-elastic body* (Valanis, 1971; Watanabe and Atluri, 1985, among others)
3.8. *Structural and analytical theory* (Likhachev and Malinin, 1993, among others)
3.9. *Thermodynamic models* (Sidoroff, 1973; Halphen and Nguyen, 1974; de Saxcé, 1992; de Saxcé and Bousshine, 2002, among others)

End of the table

The governing equations, applications, the basic experiments needed to determine the model parameters, questions related to software implementation, and finally, the possibilities of describing and predicting the effects of inelastic deformation under uniaxial and multiaxial loading are covered in special literature and many journal articles by authors and specialists in theories of plasticity and creep. Below, without dwelling on the essence of all the models listed above, we consider some of them that allow us to describe the various effects of the behavior of materials under various complex loading programs. Such models (theories) include the classical theory of plastic flow with linear kinematic hardening (Ishlinsky, 1954; Prager, 1955a), multisurface theory (Melnikov and Semenov, 1991; Izotov et al., 2000), a number models of the Institute of Mechanics of the National Academy of Sciences of Ukraine (Shevchenko and Terekhov, 1982; Shevchenko et al., 1992, 1997, 1998; Shevchenko and Babeshko, 1999), a model of cyclic plasticity and creep under proportional loading (Getsov and Kabelevskii, 1978; Getsov, 1998) and a microstructural model of the medium (Gokhfel'd et al., 1978).

2.5.2.1 Model 1. Theory of Kinematic Hardening

In the model under consideration, the associated law of plastic flow is used, in which the loading function corresponding to the von Mises criterion appears as a plastic potential. The microstress tensor describing the location of the center of the loading surface is proportional to the plastic strain tensor. The proportionality coefficient is determined by the slope of the stress-strain diagram. In the considered version of the one-surface theory of plasticity, the loading surface allows translation in the space of the stress tensor without changing its size and shape. A version of the theory of flow can be formulated as follows

$$\rho = K\varepsilon^{\mathrm{p}}, \tag{2.55}$$

where ρ is the translation tensor of the center of the yield surface and ε^{p} is the tensor of plastic strains.

2.5.2.2 Model 2. Multisurface Theory with one Active Surface

A feature of the theory is that when the load is reversed, an active surface is introduced, with the help of which a simple deformation is described (Melnikov and Semenov, 1991; Izotov et al., 2000). The remaining surfaces are in a passive state. The governing equations of a theory are a variant of incremental theory using the associated or unassociated laws of plastic flow. This theory is oriented towards a refined description of deformation under non-proportional passive loading paths. If used for conditions of active loading, it is a special case of the theory of plastic flow with isotropic-kinematic hardening. The governing equations of the theory satisfy the second law of thermodynamics in the form of a dissipative inequality. The parameters of the model are determined from experiments on loading, unloading and reloading on a different path.

In the works of Yu.N. Shevchenko and co-workers models of two classes were developed. The first class includes models whose defining equations include scalar functions or functionals that are independent of the geometric parameters of the deformation trajectories. The second class includes constitutive laws of plasticity with scalar functions depending on the parameters of the internal geometry of the deformation trajectories. To find these trajectories, the boundary value problem of the theory of plasticity is solved by the method of successive approximations along the paths of deformation or loading. First, the boundary-value problem is solved by loading stages using the governing equations of one of the first-class plasticity theories. If the obtained deformation paths do not correspond to those that can be described by the theory, then the solution is refined using the constitutive equations of the corresponding second-class theory. Among the first-class plasticity models developed by Yu.N. Shevchenko, the following model can be considered (Shevchenko et al., 1992).

2.5.2.3 Model 3. Model of Isotropic Thermo-visco-elasto-plasticity

Model 3.1. Model of Isotropic Thermo-visco-elasto-plasticity Along Rectilinear or Slightly Deviating Deformation Trajectories, Taking into Account the History of Loading and Creep Deformation

The scalar function, which is part of the equations of this theory, is determined experimentally from tensile tests at various constant temperatures, on the basis of which it is possible to construct the so-called instantaneous thermo-mechanical surface and the creep curves. The defining equations are linearized by the method of variable elastic parameters or by the method of elastic solutions and their modifications. The solution of the boundary-value problem is carried out in steps by the method of successive approximations. The constitutive equations are written in the final form (in the deformation form). The law of elastic change in volume and the coaxiality of the stress and strain deviators are used.

Model 3.2. Model of Isotropic Thermo-visco-elastic Deformation Along Trajectories of Small Curvature

The scalar function that is included in the constitutive equations of this theory is determined on the basis of the same experiments as in the previous case (when the solid is deformed along a straight or slightly deviating path from it). The constitutive equations are linearized by the method of additional stresses, they are written in the final form, and the boundary-value problem is solved in steps by the method of successive approximations. When formulating the governing equations, the law of elastic change in volume and the coaxiality of the stress deviators and the increments of inelastic strains are used. Total inelastic strain is defined as the sum of the increments. Both thermo-visco-plasticity theories allow the computing the stress-strain state of a solid with the indicated trajectories of active loading, partial or full unloading and reloading after preliminary inelastic deformation. Moreover, to construct the corresponding thermo-mechanical surfaces, the generalized Masing principle or one of the variants of the micro-structural model of the medium is used.

The second class of plasticity theories, developed by Yu.N. Shevchenko and co-workers, includes the following models.

Model 3.3. Model of Thermo-visco-elasticity Along Arbitrary Flat Trajectories Located in the Five-dimensional Space of Deformations

This model is formulated on the basis of the law of elastic change in volume and expansion of the stress vector corresponding to the stress deviator in the directions of the position vector of the deformation path and tangent to it, or in the expansion of the deformation vector corresponding to the strain deviator in the directions of the position vector of the stress path and tangent to it. The invariance of such a representation of the governing equations is ensured by the Ilyushin's postulate of isotropy.

The scalar functions included in the constitutive equations are defined in the same way as in models 3.1 and 3.2 based on the instantaneous thermo-mechanical surface and associated creep curves, as well as experiments on the deformation of tubular samples along two-link trajectories at various fixed temperatures with a kink angle $\pi/2$, which make it possible to introduce into the equations the delay of vector and scalar properties of the material.

The governing equations linearized by the additional stress method are formulated in the final form. In this case, the physical equations are supplemented by differential equations that determine the angle of non-coaxiality of the stress and strain vectors due to both the delay of the vector properties of the material and the curvature of the deformation path.

Model 3.4. Model of Isotropic Thermo-visco-plasticity Deformation for Arbitrary Trajectories of Deformation (Loading) Located in the same Coordinate Plane of the Corresponding Five-dimensional Ilyushin Space

The model is used to solve boundary value problems of the theory of thin shells of revolution under axisymmetric loading. The relations of this model are obtained from the relations of the previous one by integrating them with a known path of deformation or loading. Moreover, for the angle of non-coaxiality between the stress and strain vectors, a closed-form solution is obtained. The scalar functions included in the defining equations of this theory are specified on the basis of the same experiments as the previous one. The governing equations of this theory also describe the history of loading, the rheological, vector, and scalar properties of the material.

Model 3.5. Model of Thermo-visco-elasto-plasticity for Trajectories of Medium Curvature and Small Torsion

The governing equations of this model are formulated on the basis of the law of elastic change in volume and expansion of the increment of the strain vector in the direction of the vector of the loading path and the tangent to it. The invariance of the governing equations of this model is determined by the isotropy postulate and the coplanarity hypothesis of three vectors: stress, stress differential and strain differential.

2.5.2.4 Model 4. Model of Cyclic Plasticity and Creep Under Proportional Loading

The model describes cyclic plastic deformations and creep by taking into account their mutual influence as well as isotropic and anisotropic hardening (Getsov and Kabelevskii, 1978; Getsov, 1998). Anisotropic hardening is taken into account due to jumps in the yield strength after each unloading. A relatively large number of material parameters and characteristic functions determined from experiments determine the considerable flexibility of the model, which allows a fairly good reflection of the observed effects of the behavior of materials. Identification of the model is quite convenient, since the parameters are well separated. The reflects various manifestations of the hardening properties observed for different materials.

It is assumed that the virgin material in the space of stress deviators is isotropic and that the cyclic loading paths of each point of the solid are enclosed within a cone with a small solid angle a (Fig. 2.38). The region within which the stress-deviator vector should be located during the entire loading cycle is shaded. In the framework of the model under consideration, the additive decomposition of the strain tensor is postulated:

$$\boldsymbol{\varepsilon} = \boldsymbol{r} + \boldsymbol{e} + \boldsymbol{p}, \tag{2.56}$$

where $\boldsymbol{\varepsilon}$ is the total strain, $\boldsymbol{r}$ is the elastic strain, $\boldsymbol{e}$ is the plastic strain and $\boldsymbol{p}$ is the creep strain.

The general expression for the radius of the yield surface is formulated within the framework of the incremental theory of plastic flow as follows

$$f_k = \sqrt{\frac{3}{2}\boldsymbol{s}\cdot\cdot\boldsymbol{s}} - R(e_{\text{eq}}^{(0)}, e_{\text{eq}}^{(1)}, \ldots, e_{\text{eq}}^{(k)}, p_{\text{eq}}^{(0)}, p_{\text{eq}}^{(1)}, \ldots, p_{\text{eq}}^{(k)}, k, T) = 0. \tag{2.57}$$

The half-cycle number increases by one if the following condition is violated

$$\boldsymbol{e}\cdot\cdot\Delta\boldsymbol{e} \geq 0. \tag{2.58}$$

Here

$$\Delta\boldsymbol{e} = \sqrt{\frac{3}{2}}\Delta e_{\text{eq}}\frac{\boldsymbol{s}}{R}, \tag{2.59}$$

where $\Delta\bar{e}$ is the equivalent measure of the increment of plastic deformations. The radius of the loading surface is defined as follows

$$R_k = R_0 + \sum_{i=1}^{k} \Delta R_i. \tag{2.60}$$

The creep strain rate tensor is described within the framework of the modified theory of hardening as follows

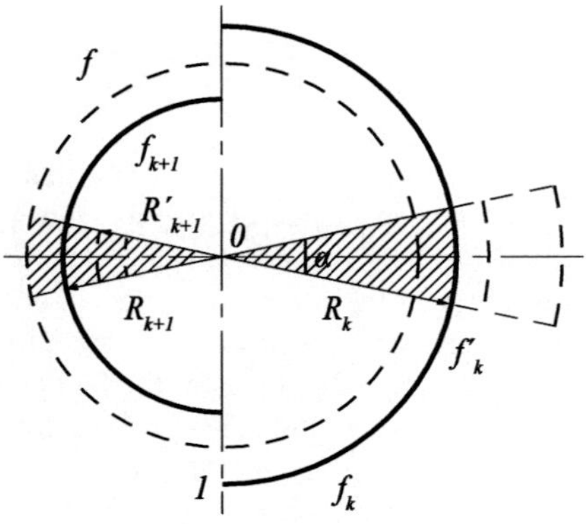

Fig. 2.38 Scheme of the yield surface under cyclic loading: *1* - initial yield surface, f_k, f_{k+1} - surface areas after kth and k + 1st half-cycles, R'_{k+1}, R''_{k+1} - yield surface radii in k + 1st a half-cycle with a corresponding increase in stresses in kth the half-cycle

$$\dot{p}^{(n)} = F(p_{\text{eq}}^{(0)}, p_{\text{eq}}^{(1)}, \ldots, p_{\text{eq}}^{(n)}, e_{\text{eq}}^{(0)}, e_{\text{eq}}^{(1)}, \ldots, e_{\text{eq}}^{(n)}, \sigma_{\text{vM}}, n, T)\frac{\boldsymbol{s}}{\sigma_{\text{vM}}}, \tag{2.61}$$

where $\boldsymbol{s}$ is the stress deviator, s_{ij} are the stress deviator components, $\bar{\sigma}$ is the von Mises equivalent stress, $p_{\text{eq}}^{(n)}$ is the equivalent creep strain and $\varepsilon_{\text{eq}}^{(n)}$ is the equivalent plastic strain, determined as follows

$$p_{\text{eq}}^{(n)} = \int\limits_{t_{n-1}}^{t_n} \sqrt{\frac{2}{3}\dot{p}_{ij}\dot{p}_{ij}}\text{d}t, \qquad e_{\text{eq}}^{(n)} = \int\limits_{t_{n-1}}^{t_n} \sqrt{\frac{2}{3}\dot{e}_{ij}\dot{e}_{ij}}\text{d}t, \tag{2.62}$$

where n is the half-cycle number, changing by one if the following condition is violated

$$p_{ij}\Delta p_{ij} \geq 0. \tag{2.63}$$

When changing cyclic loading conditions, instead of the number of half-cycles, the paths of cyclic deformation λ_1 and λ_2 are accepted as parameters characterizing the change in the properties of the material due to its cyclic instability as follows

$$\begin{aligned}
\lambda_1 &= \left| \int \text{d}\varepsilon - \int \text{d}\varepsilon \,\text{sign}(\varepsilon \text{d}\varepsilon) \right|, \\
\text{d}\varepsilon &= \sqrt{\frac{3}{2}\text{d}\varepsilon_{ij}\text{d}\varepsilon_{ij}} \\
\Delta\lambda_1 &= \lambda_1^{(k)} - \lambda_1^{(k-1)} > 0, \quad \frac{\partial \lambda_1^{(k)}}{\partial t} > 0 \quad \text{if} \quad t < t_k, \\
\frac{\partial \lambda_1^{(k)}}{\partial t} &= 0 \quad \text{if} \quad t > t_k, \\
\lambda_2 &= \left| \int \sqrt{\frac{3}{2}\text{d}p_{ij}\text{d}p_{ij}} - \sqrt{\frac{3}{2}\text{d}p_{ij}\text{d}p_{ij}}\,\text{sign}(p\text{d}p) \right|.
\end{aligned} \tag{2.64}$$

The method for the practical implementation of this model is described in more detail below.

2.5.2.5 Model 5. Microstructural Model

The microstructural model of the medium (see Sect. 2.2.1), generalized to the multiaxial stress state, is described by the following equations (Shneiderovich, 1968)

$$\varepsilon_{ij} = \varepsilon_{ij}^{(k)}, \quad e_{ij} = r_{ij} + p_{ij} = r_{ij}^{(k)} + p_{ij}^{(k)} = e_{ij}^{(k)}, \tag{2.65}$$

$$\varepsilon_{ij} = \sum_{k-1}^{N} \int \sigma_{ij}^{(k)} g_k =< \sigma_{ij}^{(k)} >, \quad r_{ij} =< r_{ij}^{(k)} >, \quad p_{ij} =< p_{ij}^{(k)} >, \tag{2.66}$$

$$r_{ij}^{(k)} = S_{ij}^{(k)}/2G, \quad \varepsilon_0^{(k)} = \sigma_0^{(k)}/3K + \varepsilon_T, \tag{2.67}$$

$$\frac{\mathrm{d}c_{ij}^{(k)}}{\mathrm{d}t} = \dot{c}_{ij}^{(k)} = \Phi(S_\mathrm{i}^{(k)}/z_k, T) S_{ij}^{(k)}/S_\mathrm{i}^{(k)}, \tag{2.68}$$

where ε_0 and σ_0 are the mean strain and stress, ε_{ij} is the strain deviator, ε_T is the thermal strain tensor, r_{ij}, $p_{ij} = b_{ij} + c_{ij}$ are deviators of elastic and inelastic deformations, b_{ij} and c_{ij} are time-independent and time-dependent parts of p_{ij}. G and K are elastic constants for subelements, $k = 1, ..., N$, S_i is the von Mises equivalent stress. The values of b_{ij} are determined from the ideal flow condition according to the law associated with the Mises surface.

When solving the problem of inelastic deformation of components using the microstructural model of the medium, the calculation is carried out in time steps Δt. At each step, the increments of inelastic deformation and stress are calculated for all microstructural elements and sub-elements of each element. The increase in rheonomic (time-dependent) deformation $\Delta c^{(k)}$ is computed assuming that c remains constant over time Δt. The increments of scleronomic (time-independent) deformation in subelements are determined by successive approximations using the additional deformation method (Birger, 1964). The use of the microstructural model of the medium makes it possible to describe the effects of vector and scalar retardation under non-proportional loading, the processes of strain accumulation under cyclic isothermal and non-isothermal loading and the generalized Masing principle under isothermal and non-isothermal loading (Gokhfel'd et al., 1978).

In order to describe the effects of cyclic instability in recent years, model extensions have been proposed in Gokhfel'd et al. (1978) with the introduction of additional hardening and recovery parameters and some functions determining the cyclic instability of the medium into Eq. (2.68).

Table 2.13 presents the classification of materials, while Tables 2.14 and 2.15 give the assessment of the models 1-4 as they describe the inelastic behavior of various materials under different loading paths. The following notation is applied: 1 - a description by theory is fundamentally possible, but testing has not been conducted or convincing testing is unknown; 2 - is described by theory and the comparison with the experiment shows a satisfactory agreement; 3 - in a known implementation of the model, the behavior of the material is not described by theory; 4 - is described by theory, but there is an unsatisfactory agreement with experimental data. An analysis of the data given in the tables shows that the considered models in many cases provide opportunities for solving boundary value problems for various materials and various loading programs. On the other hand, classical models, as was shown in Borzdyka and Getsov (1978); Lamba and Sidebottom (1978); Mel'nikov et al. (1990); Semenov and Melnikov (1994); Melnikov and Semenov (1995a,b); Getsov

Table 2.13 Classification of materials

Material type number	Type of classification	Property
1.1	By degree of homogeneity	Homogeneous
1.2		Heterogeneous
1.2.1		With hardening phases
1.2.2		Composite
2.1	By chemical composition	Pearlitic low carbon steels
2.2		Pearlitic medium carbon steels
2.3		Pearlitic high carbon steels
2.4		Cast iron
2.5		Martensitic steels
2.6		Austenitic steels
2.7		Nickel based alloys
2.8		Titanium alloys
2.9		Aluminum alloys
2.10		Copper alloys
2.11		Magnesium alloys
3.1	By strength (σ_B) and ductility (δ, ψ) at room temperature	High ductility
3.2		Medium strength
3.3		High strengt
3.4		Brittle
4.1	Elasto-plastic behavior under cyclic loading	Cyclically hardened (at a given temperature)
4.2		Cyclically softened (at a given temperature)
4.3		Cyclically stable (at a given temperature)
4.4		With variable deformation behavior
5.1	Orientation dependence in the initial state	Isotropic
5.2		Anisotropic
5.2.1		Anisotropic polycrystals
5.2.2		Direction solidified
5.2.3		Single crystals
6.1	Orientation dependence in the process of deformation	Deformation isotropy
6.2		Deformation anisotropy
7.1	Microstructural stability (during operation)	Stable
7.2		Unstable

Table 2.14 Applicability of models for different materials

Material properties	Version	Subversion	Model			
			1	2	3.1-3.5	4
Degree of homogeneity	Homogeneous	–	1	2	2	2
	Heterogeneous	With hardening phase	1	1	3	2
		Composite	4	1	3	
Strength and ductility	High ductility		1	2	2	2
	Moderate strength		1	1	2	2
	High strength		1	1	2	2
	Brittle		-	1	2	2
Features under cyclic loading	Cyclic hardening		4	2	2	2
	Cyclic softening		4	1	2	2
	Cyclic stability		1	1	2	2
	Changing behavior during deformation		3	1	1	2
Orientation dependence in the initial state	Isotropic		2	2	2	2
	Anisotropic	Polycrystals	2	1	3	1
		Directionally solidified	4	1	3	1
		Single crystals	4	1	3	1
Orientation dependence in the process of deformation	Deformation isotropy		1	2	2	2
	Deformation anisotropy		3	2	3	3
Microstructural stability	Stable		1	2	2	2
	Unstable		3	1	1	2

(2004b), are suitable only for relatively simple loading programs. Therefore, in the case of relatively complex loading regimes, in order to obtain adequate calculation results, it is necessary to use rather complex models (for example, the models 2-4), which require a large number of basic experiments to determine the parameters of the corresponding equations. In the case of simple loading programs, simple classical models should be used.

Table 2.15: Applicability of models for various loading programs

Loading conditions	Stress state	Version	Subversion	Model			
				1	2	3	4
Loading direction	Uni-axial	Constant sign of constant loading		1	2	2	2
		Constant sign of changing loading		1	2	2	2
		Varying sign		3	2	2	2
Thermal loading	Uni-axial	Isothermal		1	1	1	2
		Non-isothermal		3	3	2	2
Loading type	Uni-axial	Hard		1	2	2	2
		Soft		1	2	2	2
		Varying stiffness		3	1	1	2
Deformation type	Uni-axial	Elastic-plastic, no creep		1	1	2	2
		Elastic-plastic with creep		3	3	2	2
Loading regime	Uni-axial	Stationary		2	2	2	2
		Nonstationary	Non-cyclic	2	2	3	2
			Cyclic symmetric	2	2	3	2
			Cyclic non-symmetric	1	2	3	2
			Cyclic with varying parameters within cycle	3	1	3	2
Thermal regime	Multi-axial	Isothermal		1	1	2	2
		Non-isothermal		3	3	2	2
Type of stress state	Multi-axial	Plane		1	2	2	1
		Triaxial		1	2	3	1
Deformation type	Multi-axial	Elastic		3	1	2	1

Continuation on next page

Table 2.15: Applicability of models for various loading programs (Continued)

Loading conditions	Stress state	Version	Subversion	Model			
				1	2	3	4
		Creep		3	3	1	1
		Elastic-plastic		1	1	2	1
		Elastic-plastic with creep		3	3	2	1
Loading (cyclic) regime	Multi-axial	Stationary		1	2	1	1
		Nonstationary		1	2	3	1
Loading type	Multi-axial	Hard with spherical part		3	3	3	1
		Hard with deviatoric part		1	2		1
		Soft with spherical part		3	3		1
		Soft with deviatoric part		2	2	2	1
		With variable stiffness		3	1	3	1
Stress concentrators	Multi-axial	Without stress (strain) concentrations		1	2	2	1
		With stress (strain) concentrations		3	2	3	1
Deformation trajectory type	Multi-axial	Straight		2	2	2	1
		Curvilinear	remark 2.5	1-5	2	2	1

Continuation on next page

Table 2.15: Applicability of models for various loading programs (Continued)

Loading conditions	Stress state	Version	Subversion	Model			
				1	2	3	4
					1,2,5	1-4	
				1 6.1-6.4	2 6.1-6.4	2 6.1,6.2	1
Strain levels	Multi-axial	Small strains		1	2	2	2
		Finite strains		1	3	3	1

End of the table

Remark 2.5 For loading with a curved deformation path, the following options are used: with very small curvature (1), with small curvature (2), with medium curvature (3), with instant curvature (at break points) (4), complex passive loading (5), multilink spatial trajectories (6). In addition, the following notation is used: 6.1 - a two-link broken line path, 6.2 - a three-link broken line path, 6.3 - a closed curved path, 6.4 - a multi-link broken line.

Note that, nevertheless, the set of models 2-4 provides the possibility of a quantitative description of most of the listed effects of the materials behavior. On the other hand, unfortunately, many of the above effects are not taken into account in the deformation models included in the commercial packages ANSYS, MARC, NASTRAN, ABAQUS, LUSAS, LS-DYNA, COSMOS, ALGOR, etc.

The limiting conditions for the use of certain models depend on the conditions of failure of materials. In turn, these conditions are different for materials with different ductility. For example, for brittle materials, elasto-plastic deformation is possible only under conditions of comprehensive compression, high-strength materials can become brittle under certain biaxial tension conditions, and for brittle materials, creep strains can accumulate, etc.

Therefore, when certifying models of elastic-viscoplasticity, in addition to loading conditions and types of materials, one should distinguish these materials by the level of plasticity under uniaxial tension and the ratio of yield strength to tensile strength. The differences between the experimentally determined inelastic deformations and stresses from those calculated without taking into account the listed effects are illustrated in some examples using the values given in Table 2.16 (according to the data of Getsov, 1996), where the experimental data are compared with the calculations performed by the theory of creep flow (without taking into account the effect of cyclic loading).

Table 2.16 Difference between the experimental data and simulations of viscoelastic deformations and stresses under uniaxial stress state

Effect (Table 2.9)	Material	T,°C	p/p_{calc}	$\sigma/\sigma_{\text{calc}}$
1.6	Various	20-900		0,5-2,6
1.11	Various	500-900		0,36-1,5
2.1	EI765	700	5-7	
2.5	EI481	550-650	1-5	
	EI827	700-850		
2.6	Various	20-400	up to 25	
2.7	EI481	600	150	
2.8	IN792	650, 900	up to 50	
	ZhS6F	900		up to 3
	ZhS32	1000		
2.9	EI826	800	0,5	
3.2	EI765	750		0,75
3.4	EI723	550		1,07
3.5	Various	Various	0,12-10	
3.6	Various	Various		up to 2
3.7	Various	Various	up to 15	

2.5.2.6 Methods for Implementing a Multi-model Approach

The practical implementation of the multi-model approach is based on the use of a number of classifiers developed by the authors of publications as control elements (Getsov et al., 2000; Melnikov et al., 2001a,b, 2009) as follows

- classification of models of plasticity, visco-elasticity (creep) and elastic-visco-plasticity (see Table 2.12)
- materials classification
- classification of loading conditions
- classification of effects during viscoelastic. elasto-plastic and elasto-visco-plastic deformation under uniaxial and complex stress conditions (see Table 2.8-2.11).

In developing the classification of models, the necessity of separating irreversible deformations into time-dependent and time-independent was taken into account (Getsov, 2004b). The next step in implementing a multi-model approach is the certification of models.

To use the multimodel approach to solving boundary value problems, a scheme for the automated selection of models has been developed. Within the framework of this scheme, the selection of the constitutive equation is made by analyzing the stress-strain state of the structure using material models based on a hierarchically ordered sequence (in order of complexity starting from elastic) and by comparing with the contents of three tables (filled out on the basis of the questionnaire of the models).

Further development and implementation in modern finite element packages of the multi-model method, i.e. the creation of a "universal" theory of deformation requires the solution of a large number of problems. These include

- creating a library of these models built into finite element packages,
- determination of the system of criteria for selecting models in connection with the characteristics of the material, the geometry of the structure in question and, most importantly, the conditions of its loading,
- development of algorithms and codes for multivariate calculations that determine the ranges of adequate application of different models.

Of course, a modern model library should primarily use those models that are included in commercial FE packages. However, with complex loading paths, it is not enough to adequately describe the ongoing processes of these models, as was shown before. In particular, in relation to deformation processes occurring under conditions of prolonged residence at high temperatures, it is advisable to modify existing models by including in the governing equations the structural parameters of the material associated with the features of their microstructure and depending on the conditions of prolonged residence at high temperatures. On the other hand, it is advisable to include in the model library a number of models developed in recent years that allow predicting the features of processes occurring under neutral loading paths under cyclic loading and describing various effects of the behavior of materials in both uniaxial and complex stress states Getsov and Kabelevskii (1978); Shevchenko and Terekhov (1982); Melnikov and Semenov (1991); Getsov (1998); Izotov et al. (2000); Naumenko and Altenbach (2016).

The proposed strategy for multimodel solution of boundary value problems is partially implemented in the framework of the PANTOCRATOR finite element software package Semenov (2003), oriented to solving nonlinear problems in the solid mechanics. Multimodel analysis allows automatic selection of a material model in the process of solving a boundary value problem. The presence in the package of an extensive and easily expandable by means of object-oriented programming library of inelastic models and a wide range of methods for solving non-linear problems opens the possibility of correct and effective modeling of the behavior of materials and structures with complex physical and mechanical properties.

It should be noted that the algorithm for selecting adequate models within the framework of a multi-model approach includes a multi-stage solution of a number of problems. When starting to solve a boundary-value problem with given geometry of the structure, material properties, boundary conditions and the nature of external influences, first of all, it is necessary to determine by what paths of loading or deformation the process is carried out in each structural element. To do this, at the first stage, elastic calculations should be carried out in relation to the entire program of loading the object. At the stages following the elastic calculation, inelastic deformation models are selected that make it possible to fairly simply and with the necessary accuracy describe the effects that occur during deformation. Since the shape of the deformation path is often slightly dependent on the choice of governing equations, the shape of the resulting trajectories can be used as the first iteration the corresponding governing equations adequately describing the studied process of loading (deformation) of the structural element. In this case, the selection of an adequate theory can be carried out on the basis of a comparison of the radius of curvature of the deformation trajectory and the value of the delay trace. Therefore, when processing the data of the corresponding basic experiments, it is rational to determine the delay parameters of the vector (and scalar) properties of this material.

Classification of the shape of the strain path in this case can be made according to the deviation of the strain path from the straight line passing through the origin in the space of strains and the point of exit of the strain path beyond the elastic work of the material. If this deviation turns out to be less than the value of the delay trace for a given material, and the radius of curvature of the trajectory is much larger than the delay trace, then, as experiments show, the relations of the theory of small elasto-plastic deformations (the theory of simple loading processes), taking into account the history of loading, satisfactorily describe the processes under study. In the case when the radius of curvature of the deformation path is larger than the delay trace, and the deformation path substantially deviates from the straight line indicated above, one should use the theory of plasticity based on the coaxiality of deviators of stress tensor and rates of plastic deformations. This is a theory of plastic flow or a theory of deformation processes along trajectories of small curvature. In these theories, the scalar functionals of the constitutive equations are independent of the parameters that determine the shape of the deformation paths. In all other cases, one should use the relations of plasticity theories in which the functionals depend on the parameters determining the shape of the deformation trajectories. For many practically feasible loading trajectories after such an analysis with a numerical solution, it is advisable to linearize the governing equations of theories for which there are experimental justifications for the corresponding paths of deformation.

2.5.2.7 Database of Material Properties

To use different models, it is necessary to have the results of a rather large number of basic experiments on determining the characteristics of materials (Getsov and Zhitkov, 2000; Getsov, 2001, 2004a). For this purpose, the viscoelastic models used in the implementation of the multimodel approach are expediently divided into two groups. The parameters of the models of the first group are determined on the basis of experiments under conditions of proportional loading under uniaxial stress state. To determine the parameters of the models of the second group, test results are also used for non-proportional loading mainly under conditions of a complex stress state. The number of basic experiments for the models of the second group includes experiments with changes in the loading trajectory during the tests (for example, tensile and torsion sequentially).

Of course, the models of the first group are more preferable from the point of view of basic experiments, but in some cases they do not adequately describe a number of effects observed with complex loading programs under complex stress conditions. Certain possibilities for describing these processes are achieved using models of the second group. Among them, for example, is the multisurface theory considered above with one active surface can be applied. As an example, we consider the lists of basic experiments when applying the ANSYS finite element package in calculations (Table 2.17). It should be noted that the number of model parameters and, therefore, the volume of basic experiments depend on the characteristics of the material and, first of all, on whether it is cyclically hardened, softened, or stable.

A sufficiently detailed analysis of the situation associated with the basic experiments was carried out by Vasin (1990). An example of the Benallal-Marquis model (Benallal and Marquis, 1987) is given in which the equations of the theory of elastic-visco-plasticity with internal variables for non-proportional cyclic loading are formulated so that they contain only material parameters in the form of constants. For their full determination, one test is carried out according to a complex program: uniaxial cyclic loading, deformation along a circular path, proportional loading, cyclic out-of-phase deformation and stress relaxation test. Thus, the authors of the models either rely on the available experimental data, or develop special procedures for the experimental determination of the parameters of their governing equations.

Table 2.17 Basic experiments used for various models

Creep and plasticity models	Basic tests
FLOW THEORY	
linear and multilinear isotropic hardening	stress-strain curve
kinematic hardening	stress-strain curve
isotropic-kinematic hardening	hysteresis loop in the first cycle or stabilized cyclic stress-strain curves
anisotropic hardening	stress-strain curves in different directions
non-linear kinematic hardening (Chaboche)	families of loading/unloading stress-strain curves
VISCO-ELASTICITY THEORY	
Maxwell model	creep curves or isochronous curves
CREEP THEORY	
ageing theory	creep curves, stress relaxation curves or isochronous curves
flow theory	creep curves, stress relaxation curves or isochronous curves
isotropic hardening theory	creep curves, stress relaxation curves or isochronous curves
kinematic hardening theory	creep curves, stress relaxation curves or isochronous curves
VISCO-PLASTICITY (RATE DEPENDENT PLASTICITY) THEORY	
Anand model	stress-strain curves for different strain rates
isotropic hardening	stress-strain curves for different strain rates

In the latter case, the convergence and stability of these procedures will not always be ensured.

However, users can apply the models of both the first and second groups only if it is possible to obtain very specific information on this material, and in the absence of it, they are forced to refuse to carry out calculations according to the model that gives even the most adequate results under the conditions of the considered loading path. Of course, in the process of testing the theory, in addition to standard basic experiments, it is advisable to conduct special experiments under conditions close to the conditions corresponding to the operational model of the part, the calculations of which are supposed to be carried out. Therefore, one of the most important problems of the practical application in the calculations of the multimodel approach is the accumulation of experimental data - basic experiments on materials used to manufacture parts in which inelastic deformations occur. On the other hand, a number of models do not allow these features to be taken into account, and therefore, significant inconsistencies with the real picture can be obtained in the calculations. Let us consider the list of tests, the results of which, according to the authors, are basic, allowing calculations using a multi-model approach. These include those obtained from experiments under uniaxial stress state as follows

- Tensile curves under instant loading,
- Isochronous strain curves constructed from creep tests at constant stresses,
- Limit values of ductility $\varepsilon_{\mathrm{fpl}}$ under instant tensile loading,
- Limit values of creep ductility $\varepsilon_{\mathrm{fc}}$ as a function of time to fracture t_{f},
- Characteristics of the effect of preliminary plastic deformation on creep rate,
- Resistance to alternating elasto-plastic deformation,
- Resistance to cyclic creep under alternating loading.

Such data for 72 grades of materials are given in Gokhfeld et al. (1996). Among the basic experiments that must be carried out in order to use the multimodel approach to solving structural deformation problems with complex non-proportional loading programs, as noted above, also include experiments on determining the deformation resistance for a two-link deformation path including loading - unloading and loading with a different ratio of stress tensor components.

It should also be noted that in order to carry out calculations on different models, including those used in modern finite element packages, it is necessary to have a large number of material characteristics, which in many cases are not available in the reference literature. In this regard, it can be considered an element of a multimodel approach that is being developed at SPbSPU an electronic database on the properties of metallic materials, focused on interaction with finite element packages (Getsov and Zhitkov, 2000; Getsov, 2004a).

2.5.3 Implementation of Cyclic Plasticity and Creep Models under Proportional Loading

In Getsov and Kabelevskii (1978); Getsov (1979, 1998), a method for the practical implementation of the model 4 is described. In what follows let us explain this approach in detail. The dependence $\dot{p} = f(\sigma, p, T, \varepsilon_{\mathrm{pl}}, \lambda)$ is obtained by eliminating t from equations (2.9) and (2.10). However, in the general form with the accepted dependence (2.61) it cannot be done in a closed analytical form. Therefore, it is proposed that the values of the function in solving any specific problems be found numerically by eliminating t from equations (2.9) and (2.10) at each step of time integration. The specific form of the dependence for describing the creep curves at the first and second stages under conditions of constant loading and the absence of instant plastic deformations is adopted similarly to Eq. (2.9) as follows

$$p = A\sigma^k \left[1 - \exp(-C\sigma^l t)\right] + B\sigma^m t, \tag{2.69}$$

where A, B, C, k, l, m are parameters which are constant for a given test temperature. The creep curve at plastic strains can be presented by a modified Eq. (2.69)

$$p = A\sigma^k \left\{1 - \exp\left[-C\sigma^l (t - t_{\mathrm{a}})\right]\right\} S(\varepsilon_{\mathrm{pl}}) + B\sigma^m (t - t_{\mathrm{a}}), \tag{2.70}$$

with t_{a} as time up to last plastic strain. This refers to creep under discrete instant plastic strain and not continuous plastic strain. Here p is the strain accumulated from the moment of acting plastic strain $\varepsilon_{\mathrm{pl}}$. To describe the tertiary creep stage, instead of stress, the ratio $\sigma/(1 - p/\varepsilon_{\mathrm{f}})$, where ε_{f} is the strain at fracture, is substituted into Eq. (2.70).

In the case when the plastic strain $\varepsilon_{\mathrm{pl}}$ accumulates at stresses under the conditions of creep tests, the influence of the value $\varepsilon_{\mathrm{pl}}$ is automatically taken into account by the values of the coefficients A, B, C, k, l, m. The values of the function S of the effect of plastic deformation on creep depend on the sign of $\varepsilon_{\mathrm{pl1}}$. The form of the function $S(\varepsilon_{\mathrm{pl}})$ of the influence of plastic deformation on the creep rate, depending on the sign of $\varepsilon_{\mathrm{pl}}$, is determined from a series of experiments at different values of $\varepsilon_{\mathrm{pl}}$ and is specified either pointwise or using the following approximating function

$$S(\varepsilon_{\mathrm{pl}}) = 1 + h\varepsilon_{\mathrm{pl}}^q. \tag{2.71}$$

It was found that the values of the material parameters h and q weakly depend on the stress level. The dependence on the stress sign is taken into account as follows

$$\dot{p} = f(|\sigma|)\mathrm{sign}\sigma. \tag{2.72}$$

The effect of cyclic loading on the creep rate is taken into account by including in the Eq. (2.70) as a factor the function $f_2(\lambda)$ by the following expression

$$p = A\sigma^k \left\{1 - \exp\left[-C\sigma^l (t - t_{\mathrm{a}})\right]\right\} S(\varepsilon_{\mathrm{pl}}) f_2(\lambda) + B\sigma^m (t - t_{\mathrm{a}}), \tag{2.73}$$

where ta is the time until the last plastic deformation or until the last change of the stress sign. The function $f_2(\lambda)$ is determined from experiments under cyclic loading and constant parameters of the cycle, under these conditions $f_2(\lambda) = f_3(k)$, where k is the number of the cycle. For $k = 1$, $f_3(k) = f_2(\lambda) = 1$. Thus, at the selected test temperatures $T_i (i = 1, \ldots, N$, we have the following expressions for p and $\dot{p}$

$$\dot{p} = \left\{ A_i C_i \left(\frac{|\sigma|}{1 - \frac{|p|}{\varepsilon_{pi}}} \right)^{k_i + l_i} \exp\left[-C_i \left(\frac{|\sigma|}{1 - \frac{|p|}{\varepsilon_{pi}}} \right)^{l_i} (t - t_a) \right] S[\varepsilon_{pl}\mathrm{sign}(\sigma_{pl}, \sigma)] f_2(\lambda) \right.$$
$$\left. + B_i \left(\frac{|\sigma|}{1 - \frac{|p|}{\varepsilon_{pi}}} \right)^{m_i} \right\} \mathrm{sign}\sigma, \tag{2.74}$$

$$p \cong \left\{ A_i \left(\frac{|\sigma|}{1 - \frac{|p|}{\varepsilon_{pi}}} \right)^{k_i} \left\{ 1 - \exp\left[-C_i \left(\frac{|\sigma|}{1 - \frac{|p|}{\varepsilon_{pi}}} \right)^{l_i} (t - t_a) \right] \right\} S[\varepsilon_{pl}\mathrm{sign}(\sigma_{pl}, \sigma)] \times \right.$$
$$\left. \times S[\varepsilon_{pl}\mathrm{sign}(\sigma_{pl}, \sigma)] f_2(\lambda) + B_i \left(\frac{|\sigma|}{1 - \frac{|p|}{\varepsilon_{pi}}} \right)^{m_i} (t - t_a) \right\} \mathrm{sign}\sigma, \tag{2.75}$$

In order to find $\dot{p}$ according to Eq. (2.74) with the value of p known at each step of integration, it is necessary to exclude the parameter $t - t_a$ from Eqs. (2.74).

Determination of $\dot{p}$ at an arbitrary temperature T is carried out by interpolation of the values $\dot{p}$ found by the method described above for several nearest values of T_i. As another material parameter, the limiting temperature T_0 is taken such that for $T < T_0$, $p = 0$.

In the case of calculations with a changing sign of stresses, whenever the sign of σ changes (i.e., when the continuous function σ passes through zero), the value of p changes as follows

$$\dot{p} = f(\sigma, p - \alpha, T, \varepsilon_{pl}, \lambda_2), \tag{2.76}$$

where α is the creep strain accumulated to the time point as the stress sign changes.

For complex stress conditions, an additional assumption should be made that the cyclic loading and preliminary plastic deformation affect only the scalar properties of materials. Then instead of Eq. (2.74) we can use again Eq. (2.61) with (2.62).

2.5.3.1 Methodology for Determining the Material Parameters

The system of equations (2.74), (2.75) contains the following material parameters

$$\begin{aligned} & A_1, B_1, C_1, k_1, l_1, m_1, \varepsilon_{p1}(t), h_1, q_1; \\ & A_2, B_2, C_2, k_2, l_2, m_2, \varepsilon_{p2}(t), h_2, q_2; \\ & A_N, B_N, C_N, k_N, l_N, m_N, \varepsilon_{pN}(t), h_N, q_N; \\ & T_0 \quad \text{and the function} \quad S[\varepsilon_{pl}\mathrm{sign}(\sigma_{pl}, \sigma)] f_2(\lambda). \end{aligned} \tag{2.77}$$

The coefficients $A_i, B_i, C_i, k_i, l_i, m_i$ are determined by processing the data on creep under uniaxial stress state in the primary and secondary stages at N constant temperatures T and several (at least six) stress levels. For each experimental creep curve, we find three parameters a, b, c defining the primary and secondary creep stages (Fig. 2.7). According to values of a, b, c for different values of σ and T = const, we find the values of A, B, C, k, l, m using the least squares method, based on the following power dependencies

$$a = A\sigma^k, \quad b = B\sigma^m, \quad c = C\sigma^l. \tag{2.78}$$

The parameters H_i and q_i of the function $S(\varepsilon_{\text{pl}})$ in system (2.77) are determined by processing the results of creep tests at n temperatures T_i, one stress value for each T_i and several values of preliminary plastic deformation ε_{pl} ranging from $-3\ldots5$ to $+3\ldots5\%$ (at least three values $\varepsilon_{\text{pl}} < 0$ of three values $\varepsilon_{\text{pl}} > 0$ and $\varepsilon_{\text{pl}} = 0$). For materials for which the function $S(\varepsilon_{\text{pl}})$ depends on the level of stresses, it is advisable to use the results of determining $S(\varepsilon_{\text{pl}})$ using the point-like setting of the values of $S(\varepsilon_{\text{pl}}, \sigma$. The function $f_2(\lambda)$ is found by processing the results of the creep tests under alternating loading with the profile shown in Fig. 2.39. For this purpose, the dependence of the strain accumulated at the primary creep stage in each half-cycle on the cycle number (creep strain accumulated during the half-cycle) is determined, Fig. 2.40. The values of the coefficients in Eqs. (2.74), (2.75) for the alloy EI826 are given in Table 2.18.

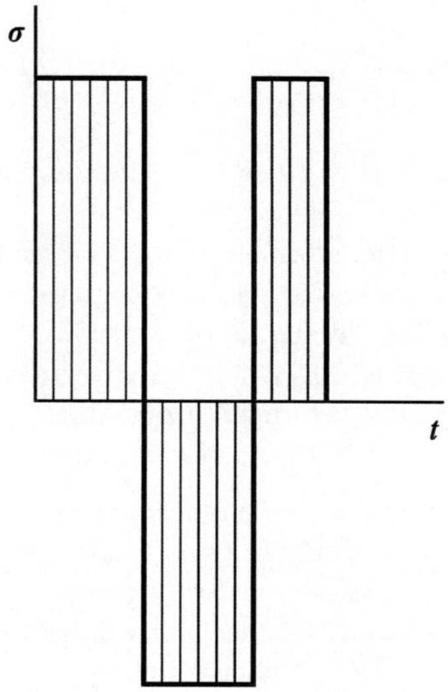

Fig. 2.39 Sample loading scheme

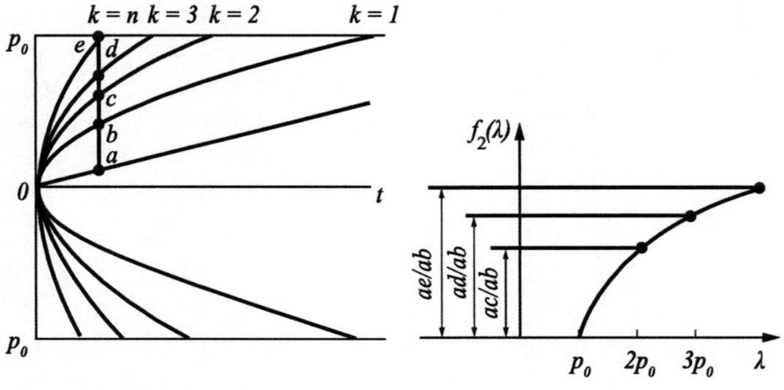

Fig. 2.40 Sample loading scheme

Table 2.18 Material parameters for creep of alloy EI826

T,°C	ε_p, %	h	q	σ_y, MPa	stress level	$-\lg A$	$-\lg B$	$-\lg C$	k	l	m
600	2	-	-	580	$\sigma < \sigma_a$	21,33	26,25	8,68	9,8	4,94	12,35
700	4-6	10^6	3,54	500	$\sigma < \sigma_y$	36,5	7,8	16,65	18,2	9,3	13,3
					$\sigma > \sigma_y$	36,5	27,1	16,65	18,2	9,3	13,3
750	6-12	-	-	500	$\sigma < \sigma_B$	13,2	23,34	18,4	6	10,75	12,05
800	6-12	-	-	400	$\sigma < \sigma_y$	14,75	14,54	7,65	7,4	5,3	7,6
					$\sigma > \sigma_y$	14,75	18,4	7,65	7,4	5,3	10
850	6-14	6,7	0,6	245	$\sigma < \sigma_y$	8,1	8,03	2,35	3,71	2,34	3,03
					$\sigma > \sigma_y$	8,1	20,62	12,35	3,71	2,34	12,05

2.5.3.2 Calculation Methodology

Creep deformation is determined for an arbitrary law of variation in stress, temperature, and instantaneous plastic deformation by numerically integrating equations (2.76) by the fourth-order Runge-Kutta method using a computer-implemented algorithm. In a simplified version of changes in stresses $\sigma(t)$, temperature $T(t)$ and plastic deformation $\varepsilon_{pl}(t)$, the algorithm can be implemented without a computer.

As examples of calculation, the following options were considered.

- *Creep* - in the primary to tertiary stages (σ = const, T = const); at a changing temperature (σ = const); at alternating stress (T = const); under cyclic plastic deformation (σ = const, T = const); under cyclic plastic deformation of the opposite sign; under conditions of plastic deformation and varying stress (T = const); with alternating stress and changing temperature.
- *Stress relaxation* - at $\sigma_0 < \sigma_y$; at $\sigma_0 > \sigma_y$ and additional preliminary plastic deformation; with loading and cyclic plastic deformation.

An analysis of the correspondence of the calculated and experimental data for various steels and alloys of 20 grades showed the acceptability of the proposed creep model and the methodology for determining its parameters (Fig. 2.41). The considered version of the theory, however, is unsuitable for cases where the deviator vector changes by angles close to straight, realized, for example, when the loading conditions change from $\sigma_1 = \sigma_2 = A$ to $\sigma_1 = A, \sigma_2 = 0$.

2.5.3.3 Comparison of Calculated and Experimental Data

Differences between experimentally determined inelastic deformations and stresses from those calculated without taking into account the listed effects can be illustrated in some examples using the parameters given in Table 2.18. Here, the experimental data are compared with calculations performed by the theory of flow. The values of p/p_{calc} correspond to the ratio of the experimental and calculated values of the accumulated creep strain, $\varepsilon_{pl}/\varepsilon_{pl\,calc}$ is the ratio of the experimental and calculated values of the accumulated plastic strain, and σ/σ_{calc} is the ratio of the experimental stresses to those obtained in viscoelastic calculations.

2.5.3.4 Criteria for Choosing a Material Model

The definition of criteria for choosing an adequate theory for solving specific problems is associated with the classification of the proposed models and consideration of ranges of their reliable application. Criteria are introduced to distinguish from a whole set of realized material models a certain subset suitable for solving a specific problem. Next, the simplest model is selected from the formed subset, and iterative search for a material model that is adequate for the complexity of the deformation process in the problem under consideration begins. Conventionally, all the criteria used can be broken into four groups corresponding to the four classifiers introduced above.

Classical examples of the criteria under consideration are: the magnitude of plastic deformation compared with elastic deformation (cf. the Prandtl-Reuss and

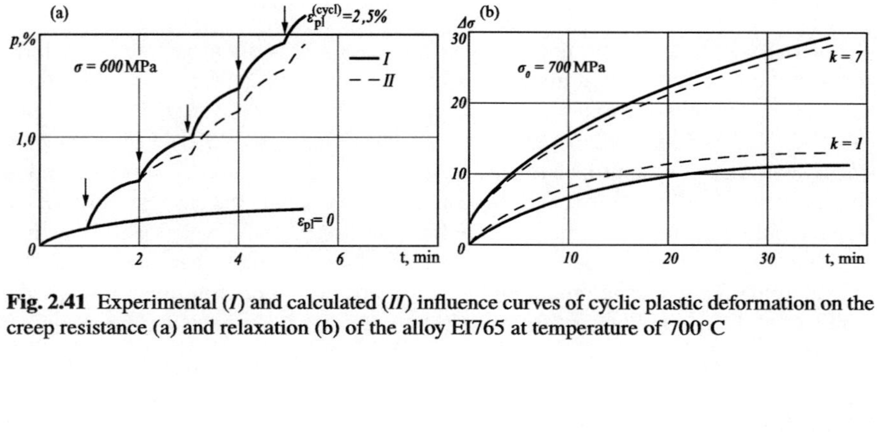

Fig. 2.41 Experimental (*I*) and calculated (*II*) influence curves of cyclic plastic deformation on the creep resistance (a) and relaxation (b) of the alloy EI765 at temperature of 700°C

Levy-von Mises theory) and the proportionality of the loading path (cf. flow theory and deformation theory).

The next criterion is related to the consideration of curvatures and twists of the deformation path. So, in the case of proportional unidirectional active loading, the deformation theory can be successfully applied, in the case of partial unloading, it is most often possible to apply the theory of plastic flow with isotropic hardening, and for unloading accompanied by reverse plastic deformation, it is necessary to use the theory of flow with anisotropic hardening, which allows one to take into account the Bauschinger effect. For problems where trajectories of deformation of small curvature are realized, a reliable solution can be found using the theory of plastic flow. The presence of sharp kinks in the deformation path, loading with frequent changes in the loading direction require the use of more complex models, such as multi-surface theories and rheological models. However, the use of generalized versions of the theory of plastic flow with a nonlinear law of kinematic hardening and the introduction of additional variables characterizing the non-proportionality of loading allow us to describe the effects of cyclic deformation under complex non-proportional loading (see Tables 2.14 and 2.15).

The central criterion is to select from a library of material model materials a variant sufficient to describe the effects of plastic deformation realized in the class of processes under study. The problem of choosing the most adequate material model can be solved on the basis of the computer aided complex loading method (Ilyushin, 1971; Zubchaninov, 2000), which requires a series of experiments in the process of solving the boundary-value problem, which allow us to establish empirically the parameters of the approximations of the functionals characterizing the reaction of the material to the loading program implemented at the considered representative points. The experiments are carried out on uniformly deformable samples. This approach can be used to automate the process of selecting a material model and increase the reliability of calculations, however, it is associated with the need to conduct numerous experiments with complex non-proportional loading, which is not always available when performing calculations.

2.5.3.5 Example of the Structural Analysis

A rational way to analyze complex structures with a non-trivial geometry, including stress concentrators and allowing a heterogeneous structure, is a multi-model approach. A number of boundary value problems were solved on the basis of the PANTOCRATOR finite element software package (Semenov, 2003). In Fig. 2.42 an example of using this approach to the analysis of structures - models of blades under thermo-cyclic loading - is given. Blade models were tested at the gas-dynamic bench in the laboratory of the Institute for Problems of Strength of the National Academy of Sciences of Ukraine under thermo-cyclic loading. It can be seen that the use of calculations by the elastic model and by the theory of a constant creep rate yields results that differ significantly from cycle-by-cycle calculations using the modified hardening theory.

2.6 Concluding Remarks

A review of various theories of plasticity and creep, which allow predicting the behavior of solids under cyclic non-isothermal loading, makes it possible to formulate their following features. On the one hand, it would be tempting to widely use microstructural or physical models of the medium, which allow us to describe a wide range of features of the behavior of the material under complex loading programs using a relatively small number of material parameters determined experimentally. However, such models, unfortunately, do not provide sufficient confidence in the quantitative reliability of the calculations, since they assume monotonic changes in the properties of the material with distance from the conditions of basic experiments. On the other hand, it is obvious that the modification of traditional models of plastic flow and various creep theories as applied to the indicated loading conditions, ensuring high accuracy of calculations, requires an extremely large number of basic experiments to obtain an adequate description of the behavior of materials. Discussions on the appropriateness or inappropriateness of separating inelastic deformations into time-dependent and time-independent (creep and plastic deformations) are still ongoing. Below is the position of Getsov (2004b) on this issue, based on the analysis of a large number of experimental data.

Obviously, time-independent micro-plastic strains and micro-creep strains can occur during deformation with stresses both large and lower than the elastic limit, including the relaxation of internal stresses of the second kind. In Rovinsky and Lutzau (1957) a relationship between micro- and macro-creep processes is established: the nature of the relaxation process of oriented micro-stresses is similar to the nature of macro-stresses relaxation. Which deformations, micro-plastic or micro-creep, are large in magnitude under stresses less than the elastic limit σ_{el}, including

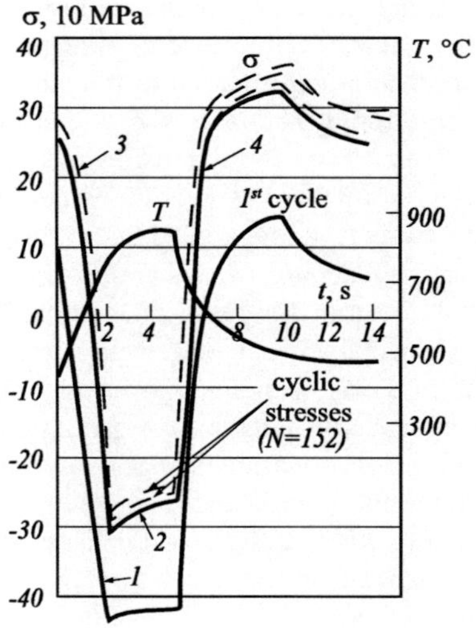

Fig. 2.42 Stress distribution in the blade edge under thermal cyclic loading: — elastic calculation and calculation according to the theory of constant rate, - - - calculation by modified hardening theory, 1, 4 - $\sigma_{st} = 0$ MPa, 2, 3 - $\sigma_{st} = 180$ MPa

cyclic ones, is a question requiring experimental study. As the analysis of the data of micro-deformations of different materials under single loading shows (McMahon, Jr., 1968), at low temperatures the former are significantly larger than the latter. However, during cyclic deformation, the process of development of micro-plastic deformations in many materials is damped with increasing number of cycles, while the rate of accumulation of micro-creep deformations under cyclic alternating loading increases with increasing cycle number (see Fig. 2.30). At high temperatures, micro-creep strains are substantially larger in magnitude than micro-plastic strains. This is confirmed by the results of experiments using an electron microscope to study the strain relief on specimens of heat-resistant alloys with dispersion hardening after they are instantly stretched in vacuum at elevated temperatures to stresses noticeably below the elastic limit (a) after holding in this state for some short time (b) and after the creep test (c). So, in the state after unloading, the deformation relief of the EI826 alloy (ChN70WMTYuF) was studied at 850°C after instant and slow loading to 220 MPa ($0.6\sigma_y$). After slow loading and creep testing for several hours (1-4 h), accompanied by the accumulation of macro creep deformation of 0.1%, an intensive development of fine slip processes was detected in the overwhelming number of grains. Under conditions of instant loading (at a rate of 180-2400%/h), after loading and unloading, individual thin slip bands were observed in only about 30% of grains.

It should be noted that particles of the second phase in heat-resistant alloys are effective barriers that prevent the development of micro-plastic deformations, which is possible only at high stresses by the mechanisms of transverse sliding, cutting, and bending. During prolonged exposure to stress at stresses $\sigma < \sigma_{eq}$, the accumulation of creep deformations is controlled by the creep of dislocations at the barriers. In addition, in contrast to instant plastic deformation, in which sliding along grain boundaries (due to a highly misoriented lattice) are impossible, under conditions of creep, intergranular slip processes develop, causing damage accumulation.

Deformations that accumulate at the time of fracture of materials under instant loading and under creep conditions, as a rule, differ significantly. Table 2.19 shows the values of the residual elongation of two groups of steels and alloys during short-term deformation to fracture and in tests for long-term strength (under creep conditions). The table shows that the materials of group A have similar plasticity values for short-term tension and in tests for long-term strength, while for materials of group B these values differ significantly. Hence, using the deformation criterion of fracture, it can be argued that for materials of group B, the identical residual strains accumulated during instant loading (ε) and creep conditions (p) can correspond to significantly different values of the degree of damage ($\varepsilon/\varepsilon_f$ and p/ε_n).

Special experiments were carried out to study the effect of preliminary long-term aging without load on the microstructure of alloys and their resistance to deformation under instant loading and under creep conditions. It has been established that microstructural changes observed after prolonged aging affect the instantaneous deformation resistance and creep resistance of materials to varying degrees, and the deformation ability p and q. Thus, the microstructure of the material is a parameter of its mechanical state - a microstructural parameter. Metallographic studies were carried out in the process of deformation of samples of alloy EP787 (KhN35VTAl)

Table 2.19 Deformation ability of heat-resistant alloys

Material	Group	T_{test}, °C	Residual elongation, % for	
			Long term fracture (100-1000h)	Instantaneous fracture
22K	A	500	21-23	24
15ChM		520	15-31	21
12Ch1MF		500	34-41	21-22
10Ch18N10T		650	9-45	27-37
EI723	B	550	1,0	14
EI481		600	1-3	12
EI612		600	1-3	15-25
EI765		700	10-13	20
EI826		800	4-7	14
EI827		800	2-6	18
EP220		900	2-3	10

after aging for 50000 h at 650°C (state 2) in comparison with the initial state (state 1). Both states of the material corresponded to approximately the same values of yield strength (880 and 900 MPa) in tensile tests (under instant loading). The values of ductility and time to fracture of the alloy at a temperature of 650°C and a stress of 350 MPa turned out to be significantly different.

Thus, the adequacy of the results of calculating the stress-strain state of components with experimental data can be obtained only for parts from materials of group A. For materials of group B, which include all austenitic hardened steels and heat-resistant alloys based on nickel, as well as pearlite steels with a high yield strength, in order to obtain adequate calculation results, it is necessary to separate irreversible strains into dependent and time-independent. Among the latest published works of domestic authors on the issues discussed in this chapter we refer to Aratyunyan (2004); Semenov (2008); Semenov et al. (2008); Kadashevich and Pomytkin (2009); Besson et al. (2010); Khazhinsky (2011); Brovko and Georgievsky (2016); Lokoshchenko (2016).

On the success achieved in recent years by scientists and organizations, mentioned in the foreword, on the deduction of perfect models of solid mechanics, and the accumulation of experimental data allowing their verification, in particular is reported in Zubchaninov (2000); Krivtsov (2007); Bondar and Danshin (2008); Semenov (2008); Volkov and Korotkikh (2008); Besson et al. (2010); Zhilin (2012); Palmov (2014); Trusov and Shveikin (2019). The creation of powerful computers allow us to hope for significant progress in the field of providing industry with reliable computing facilities for an adequate determination of the resource of various energetic systems operating under complex operating conditions in the next decade.

Remark 2.6 After the publication of the 4th Russian edition of Getsov (2010), studies were carried out with the participation of the author, concerning the creep processes in the bends of steam lines (Katanaha and Getsov, 2011; Popova et al., 2011; Katanakha et al., 2013a,b) as well as works on creep of single-crystal alloys (Grishchenko et al., 2015; Semenov et al., 2016a,b; Getsov et al., 2020).

References

Altenbach H, Naumenko K (1997) Creep bending of thin-walled shells and plates by consideration of finite deflections. Computational Mechanics 19(6):490–495

Altenbach H, Naumenko K (2002) Shear correction factors in creep-damage analysis of beams, plates and shells. The Japan Society of Mechanical Engineers - International Journal Series A, Solid Mechanics and Material Engineering 45:77–83

Altenbach H, Skrzypek JJ (eds) (1999) Creep and Damage in Materials and Structures, CISM International Centre for Mechanical Sciences, vol 399. Springer

Altenbach H, Breslavsky D, Morachkovsky O, Naumenko K (2000a) Cyclic creep damage in thin-walled structures. The Journal of Strain Analysis for Engineering Design 35(1):1–11

Altenbach H, Kolarow G, Morachkovsky O, Naumenko K (2000b) On the accuracy of creep-damage predictions in thinwalled structures using the finite element method. Computational Mechanics 25:87–98

Altenbach H, Kushnevsky V, Naumenko K (2001) On the use of solid- and shell-type finite elements in creep-damage predictions of thinwalled structures. Archive of Applied Mechanics 71:164–181

Altenbach H, Huang C, Naumenko K (2002) Creep-damage predictions in thin-walled structures by use of isotropic and anisotropic damage models. The Journal of Strain Analysis for Engineering Design 37(3):265–275

Altenbach H, Gorash Y, Naumenko K (2008a) Steady-state creep of a pressurized thick cylinder in both the linear and the power law ranges. Acta Mechanica 195(1-4):263–274

Altenbach H, Naumenko K, Gorash Y (2008b) Creep analysis for a wide stress range based on stress relaxation experiments. International Journal of Modern Physics B 22(31-32):5413–5418

Altenbach H, Bolchoun A, Kolupaev VA (2014) Phenomenological yield and failure criteria. In: Altenbach H, Öchsner A (eds) Plasticity of Pressure-sensitive Materials, Springer, Berlin, Heidelberg, pp 49–152

Altenbach H, Breslavsky D, Naumenko K, Tatarinova O (2019) Two-time-scales and time-averaging approaches for the analysis of cyclic creep based on armstrong–frederick type constitutive model. Proceedings of the Institution of Mechanical Engineers, Part C: Journal of Mechanical Engineering Science 233(5):1690–1700

Aratyunyan RA (ed) (2004) Problema deformacionnogo stareniya i dlitel'nogo razrusheniya v mekhanike materialov (in Russ., Problem of strain aging and long-term fracture in material mechanics). Publishing house SPbSU, St. Petersburg

Aratyunyan RA, Vakulenko AA (1965) O mnogokratnom nagruzhenii uprugo-plasticheskoy sredy (in Russ., On multiple loading of an elastic-plastic medium). Izvestiya AN SSSR - Mekhanika Tverdogo Tela (4):93–103

Armstrong PJ, Frederick CO (1966) A mathematical representation of the multiaxial bauschinger effect. Report RD/BfN 731, Central Electricity Generating Board and Berkeley Nuclear Laboratories

Backhaus G (1972) Zur Analytischen Erfassung des Allgemeinen Bauschingerefekts. Acta Mechanica 14(1):31–42

Barlat F, Lege DJ, Brem JC (1991) A six-component yield function for anisotropic materials. International Journal of Plasticity 7(7):693–712

Barré de Saint-Venant AJC (1870) Sur l'établissement des équations des mouvements intérieurs opérés dans les corps solides au delà des limites où l'élasticité pourrait les ramener à leur premier état. Comptes rendus de l'Académie des sciences, Paris 70:473–480

Batdorf SB, Budiansky B (1949) A mathematical theory of plasticity based on concept of slip. Technical Note 1871, NACA, Washington

Benallal A, Marquis D (1987) Constitutive equations for nonproportional cyclic elasto-viscoplasticity. Journal of Engineering Materials and Technology 109(4):326–336

Besseling JR (1959) A theory of elastic, plastic and creep deformations of an initially isotropic material showing anisotropic strain-hardening, creep recovery, and secondary creep. Transaction of the American Society of Mechanical Engineering - Journal of Applied Mechanics 25(4):529–534

Besson J, Cailletaud G, Chaboche JL, Forest S, Blétry M (2010) Non-Linear Mechanics of Materials, Solid Mechanics and its Applications, vol 167. Springer, Dordrecht, Heidelberg, London & New York

Betten J (2001) Mathematical modelling of materials behavior under creep conditions. Applied Mechanics Review 54(2):107 – 132

Betten J (2008) Creep Mechanics, 3rd edn. Springer, Berlin

Bik DZ (1981) Modifikatsiya sootnosheniy uprugoplasticheskikh protsessov sredney krivizny (in Russ., Modification of the relations of elastoplastic processes of middle curvature). Vestnik MGU Mathematicians and Mechanics (5):103–106

Birger IA (1964) Metody uprugikh resheniy v teorii plasticheskogo techeniya (in Russ., Elastic solution methods in the theory of plastic flow). Izvestiya AN SSSR - Mekhanika i Mashinostroyeniye (2):73–78

Birger IA, Shorr BF, Demyanushko IV, Dul'nev RA, Sizova RN (eds) (1975) Termoprochnost' detaley mashin (in Russ., Heat Resistance of Machine Parts). Mashinostroenie, Moscow

Bishop JFW (1953) VI. A theoretical examination of the plastic deformation of crystals by glide. The London, Edinburgh, and Dublin Philosophical Magazine and Journal of Science 44(348):51–64

Bodner SR, Partom Y (1975) Constitutive equation, for elastic-viscoplastic strain hardening materials. Transaction of the American Society of Mechanical Engineering - Journal of Applied Mechanics 42:385–389

Bondar VS, Danshin VV (2008) Plastichnost'. Proportsonal'nyye i neproportsional'nyye nagruzheniya (in Russ., Plasticity. Proportional and Non-proportional Loads). FIZMATLIT, Moscow

Borzdyka AM, Getsov LB (1978) Relaksaciya napryazhenii v metallakh i splavakh (in Russ., Stress Relaxation in Metals and Alloys), 2nd edn. Metallurgiya, Moscow

Brovko GL, Georgievsky DV (eds) (2016) Uprugost' i Neuprugost'. Materialy Mezhdunarodnogo nauchnogo simpoziuma po problemam mekhaniki deformiruyemykh tel, posvyashchennogo 105-letiyu so dnya rozhdeniya A.A. Il'yushina (Moskva, 20-21 yanvarya 2016 goda) (in Russ., Elasticity and Inelasticity. Materials of the International Scientific Symposium on Problems of Mechanics of Deformable Bodies, dedicated to the 105th anniversary of the birth of A.A. Ilyushin (Moscow, January 20-21, 2016)). Moscow University Press, Moscow

Cailletaud G (1992) A micromechanical approach to inelastic behaviour of metals. International Journal of Plasticity 8(1):55–73

Chaboche JL (1975) Viscoplastic constitutive equations for the description of cyclic and anisotropic behavior of metals. Bulletin de l'Académie Polonaise des Sciences - Série des sciences techniques 25:33–41, 42–48

Chaboche JL (1989) Constitutive equations for cyclic plasticity and cyclic viscoplasticity. International Journal of Plasticity 5:247 – 302

Chaboche JL (2008) A review of some plasticity and viscoplasticity constitutive equations. International Journal of Plasticity 24:1642 – 1693

Chaboche JL, Rousselier G (1983) On the plastic and viscoplastic constitutive equations — Part I: Rules developed with internal variable concept. Journal of Pressure Vessel Technology 105(2):153–158

Chizhik AA (1966) Vzaimosvyaz' relaksacii napryazheniy i yavleniy polzuchesti v metallakh (in Russ., Relationship between stress relaxation and creep phenomena in metals). In: Svoystva i primeneniye zharoprochnykh splavov, Moscow, pp 69–76

Chowdhury H, Naumenko K, Altenbach H, Krüger M (2017) Rate dependent tension-compression-asymmetry of ti-61.8 at% al alloy with long period superstructures at 1050° c. Materials Science and Engineering: A 700:503–511

Chowdhury H, Naumenko K, Altenbach H (2018) Aspects of power law flow rules in crystal plasticity with glide-climb driven hardening and recovery. International Journal of Mechanical Sciences 146-147:486 – 496

Coulomb C (1773) Sur une application des règles de maximis et minimis à quelques problèmes de statique relatifs à l'architecture. Mémoires de la Mathematique et de Physique 7:343–382

Dafalias YF (1986) Bounding surface plasticity. I: Mathematical foundation and hypoplasticity. Journal of Engineering Mechanics 112(9):966–987

de Saxcé G (1992) Une généralisation de l'inégalitêê de fenchel et ses applications aux lois constitutives. Comptes rendus de l'Académie des sciences, Paris - Série II 314:125–129

de Saxcé G, Bousshine L (2002) Implicit standard materials. In: Weichert D, Maier G (eds) Inelastic Behaviour of Structures under Variable Repeated Loads, Springer, Vienna, International Centre for Mechanical Sciences (Courses and Lectures), vol 432, pp 59–76

Desmorat R, Marull R (2011) Non-quadratic kelvin modes based plasticity criteria for anisotropic materials. International Journal of Plasticity 27(3):328–351

Drucker DC (1949) The significance of the criterion for additional plastic deformation of metals. Journal of Colloid Science 4:299–311

Drucker DC, Prager W (1952) Soil mechanics and plastic analysis for limit design. Quarterly of Applied Mathematics 10(2):157–165

Eisenträger J, Naumenko K, Altenbach H, Lenz W (2013) Approximative methods to calculate notch stresses and strains for plasticity and creep. Forschung im Ingenieurwesen 77(3):71–80

Eisenträger J, Naumenko K, Altenbach H (2017) Robust approaches for the assesment of stress concentration in the creep range. In: Tonti A, di Gianfrancesco A (eds) Creep & Fracture in High Temperature Components - Design & Life Assessment Issues, 3rd ECCC Creep Conference, pp 493–501

Eisenträger J, Naumenko K, Altenbach H, Gariboldi E (2017) Analysis of temperature and strain rate dependencies of softening regime for tempered martensitic steel. The Journal of Strain Analysis for Engineering Design 52(4):226–238

Eisenträger J, Naumenko K, Altenbach H (2018a) Calibration of a phase mixture model for hardening and softening regimes in tempered martensitic steel over wide stress and temperature ranges. The Journal of Strain Analysis for Engineering Design 53(3):156–177

Eisenträger J, Naumenko K, Altenbach H (2018b) Numerical implementation of a phase mixture model for rate-dependent inelasticity of tempered martensitic steels. Acta Mechanica 229(7):3051–3068

Erickson GL, Harris K (1994) Ds and sx superalloys for industrial gas turbines. In: Coutsouradis D, Davidson JH, Ewald J, Greenfield P, Khan T, Malik M, Meadowcroft DB, Regis V, Scarlin RB, Schubert F, Thornton DV (eds) Materials for Advanced Power Engineering 1994 - Proceedings of a Conference held in Liège, Belgium, 3–6 October 1994, Springer, Netherlands, vol II, pp 1055–1074

Frederick CO, Armstrong PJ (2007) A mathematical representation of the multiaxial bauschinger effect. Materials at High Temperatures 24(1):1–26

Fridman YB (1975a) Mekhanicheskie svoistva metallov (in Russ., Mechanical Properties of Metals), vol 1. Mashinostroenie, Moscow

Fridman YB (1975b) Mekhanicheskie svoistva metallov (in Russ., Mechanical Properties of Metals), vol 2. Mashinostroenie, Moscow

Gadenin MM (1976) Osobennosti razvitiya deformatsiy i nakopleniya povrezhdeniy pri dvukhchastotnom malotsiklovom nagruzhenii pri povyshennykh temperaturakh (in Russ., Features of the development of deformations and damage accumulation during two-frequency low-cycle loading at elevated temperatures). Mashinovedenie (1):69–77

Gariboldi E, Naumenko K, Ozhoga-Maslovskaja O, Zappa E (2016) Analysis of anisotropic damage in forged Al-Cu-Mg-Si alloy based on creep tests, micrographs of fractured specimen and digital image correlations. Materials Science and Engineering: A 652:175 – 185

Getsov LB (1964) Methods of evaluating deterioration in heat-resisting materials (a survey). Industrial Laboratory 30(1):1707–1716

Getsov LB (1971) Soprotivlenie polzuchesti nekotorykh zharoprochnykh stalei (in Russ., Creep resistance of some heat resistant steels). Raschety na Prochnost' (15):426–438

Getsov LB (1973) Materialy i prochnost' detalei gazovykh turbin (in Russ., Materials and Strength of Gas Turbine Parts). Mashinostroenie, Leningrad

Getsov LB (1978) Strain resistance of heat-resisting alloys under complex loading program conditions. Strength of Materials 10(6):638–643

Getsov LB (1979) Razrabotka kriteriyev razrusheniya materialov dlya obosnovaniya nadezhnosti detalei gazovykh turbin, rabotayushchikh pri nestacionarnykh rezhimakh (in Russ., Development of criteria for the destruction of materials to justify the reliability of parts of gas turbines operating under unsteady conditions). Abstract doctoral thesis, Institute for Strength Problems, Academy of Sciences of the Ukrainian SSR, Kiev

Getsov LB (1996) Materialy i prochnost' detalei gazovykh turbin (in Russ., Materials and Strength of Gas Turbine Parts), 3rd edn. Nedra, Moscow

Getsov LB (1998) Model' deformirovaniya dlya slozhnykh programm nagruzheniya (in Russ., Deformation model for complex loading programs). In: Nauchnyye trudy 2go mezhdunarodnogo seminara *Sovremennyye problemy prochnosti* im. V.I. Likhacheva, Staraya Russa, vol 2, pp 35–43

Getsov LB (2001) Svoystva metallicheskikh materialov i elektronnaya baza dannykh (in Russ., properties of metallic materials and electronic database). In: Trudy 4-oy mezhdunarodnoy konferencii *Nauchno-tekhnicheskiye problemy prognozirovaniya nadezhnosti i dolgovechnosti konstrukciy i metody ikh resheniya*, St. Petersburg, pp 75–80

Getsov LB (2004a) Baza dannykh po svoystvam materialov dlya provedeniya raschetov napryazhenno-deformirovannogo sostoyaniya konstruktsiy metodom konechnykh elementov (in Russ., Database on the properties of materials for calculating the stress-strain state of structures by the finite element method). In: Trudy SPbGPU, St. Petersburg, Mechanics of materials and structural strength, vol 489, pp 128–136

Getsov LB (2004b) O razlichnoy prirode zavisyashchikh i nezavisyashchikh ot vremeni neobratimykh deformatsiyakh (in Russ., On the different nature of time-dependent and time-independent irreversible deformations). In: Trudy SPbGPU, St. Petersburg, Mechanics of materials and structural strength, vol 489, pp 94–101

Getsov LB (2010) Materialy i Prochnost' Detalei Gazovykh Turbin (in Russ., Materials and Strength of Gas Turbine Parts), vol I. Gazoturbinnye Tekhnologii, Rybinsk

Getsov LB, Kabelevskii MG (1978) The question of theories of plasticity and creep in cyclic nonisothermal loading. Strength of Materials 10(6):660–666

Getsov LB, Pospisil B (2009) Izokhronnyye diagrammy polzuchesti i metody ikh obrabotki (in Russ., Isochronous creep diagrams and methods of their processing). In: Petreniya YK, Getsova LB (eds) Prochnost' materialov i resurs elementov energooborudovaniya, St. Petersburg, no. 296 in Trudy CKTI, p 140–153

Getsov LB, Zhitkov DB (2000) K voprosu sozdaniya universal'noy elektronnoy bazy dannykh po sostavu i svoystvam metallov i splavov (in Russ., On the issue of creating a universal electronic database on the composition and properties of metals and alloys). Tyazheloye mashinostroyeniye (12):13–15

Getsov LB, Kononov KM, Rebyakov YN (1976) K opredeleniyu ciklicheskogo predela uprugosti (in Russ., Determination of the cyclic elastic limit). Zavodskaya Laboratoriya (11):1400–1403

Getsov LB, Gorskii SV, Kononov KM, Rebyakov YN (1978) Special features of cyclic deformation of heat-resistant materials at elevated temperatures. Strength of Materials 10(7):781–785

Getsov LB, Melnikov BE, Semenov AS (2000) K voprosu vybora modeley dlya raschetov napryazhenno-deformirovannogo sostoyaniya elementov mashinostroitel'nykh konstruktcii (in Russ., On the issue of choosing models for calculating the stress-strain state of elements of engineering structures). In: Physics of the processes of deformation and fracture and prediction of the mechanical behavior of materials. Proceedings of the XXXVI International Seminar *Actual Problems of Strength*, Vitebsk, vol 1, pp 40–47

Getsov LB, Semenov AS, Golubovsky ER, Grishchenko AI, Semenov SG (2020) Osobennosti i yedinoye opisaniye I, II i III stadiy polzuchesti monokristallicheskikh zharoprochnykh splavov (in Russ., Features and uniform description of I, II, and III stages of the creep in single-crystal superalloys). Industrial laboratory Diagnostics of materials 86(3):44–54

Gokhfel'd DA, Sadakov OS (1984) Plastichnost' i polzuchest'v elementov konstrukcii pri povtornykh nagruzheniyakh (in Russ., Plasticity and Creep in Structural Elements During Repeated Loading). Mashinostroenie, Moscow

Gokhfel'd DA, Kononov KM, Sadakov OS, Chernyaskiy OF (1978) Problemy prochnosti termonapryazhennykh konstrukcii (in Russ., Problems of the Strength of Thermally Stressed Structures), Mekhanika deformiruyemogo tela, vol 12. VINITI AN SSSR, Moscow

Gokhfeld DA, Getsov LB, Kononov KM (1996) Mekhanicheskiye svoystva staley i splavov pri nestatsionarnom nagruzhenii (in Russ., Mechanical Properties of Steels and Alloys under Nonstationary Loading). Ural Branch of RAS, Yekaterinburg

Grishchenko AI, Semenov AS, Getsov LB (2015) Modeling inelastic deformation of single crystal superalloys with account of γ/γ' phases evolution. Materials Physics and Mechanics 24(4):325–330

Grubin AN (1970) Raschet na prochnost' yelochnogo zamka lopatok turbin (in Russ., Calculation of the Strength of the Herringbone Lock of Turbine Blades). Mashinostroenie, Leningrad

Gusenkov AP (1979) Prochnost' pri izotermicheskikh i neizotermicheskikh malociklovom nagruzhenii (in Russ., Strength Under Isothermal and Non-isothermal Low-cycle Loading). Nauka, Moscow

Halphen B, Nguyen QS (1974) Plastic and visco-plastic materials with generalized potential. Mechanics Research Communications 1(1):43–47

Hayhurst DR (1994) The use of continuum damage mechanics in creep analysis for design. The Journal of Strain Analysis for Engineering Design 25(3):233–241

Hencky H (1924) Zur theorie plastischer deformationen und der hierdurch im material hervorgerufenen nachspannungen. ZAMM - Journal of Applied Mathematics and Mechanics / Zeitschrift für Angewandte Mathematik und Mechanik 4(4):323–334

Hencky H (2020) On the theory of plastic deformations and the residual stresses caused by them in the material. ZAMM - Journal of Applied Mathematics and Mechanics / Zeitschrift für Angewandte Mathematik und Mechanik 100(3):e202002019, translated from the German original: Zur Theorie plastischer Deformationen und der hierdurch im Material hervorgerufenen Nachspannungen, Z Angew Math Mech 4(1924):323-334 by O. Bruhns with kind support of B. Brechan and B. Nijboer. Certain misprints of Hencky's and a small mistake of minor importance in Eq. (4) and the respective consequences in subsequent relationships have been tacitly corrected in our translation

Hill R (1950) The Mathematical Theory of Plasticity. Clarendon Press, Oxford

Huber MT (1904) Właściwa praca odkształcenia jako miara wytężenia materyału. przyczynek do podstaw teorii wytrzymałości. Czasopismo Techniczne XXII(3-6):80–81, 49–50, 61–62, 80–81, translated from the original in Polish and published in Arch Mech 56, 3, pp. 173–190, Warszawa 2004

Ilyushin AA (1954) O svyazi mezhdu napryazheniyami i malymi deformatsiyami v mekhanike sploshnykh sred (in Russ. On the relationship between stresses and small deformations in continuum mechanics). Prikladnaya Matematika i Mekhanika 18(6):641–666

Ilyushin AA (1971) Resheniye zadach mekhaniki sploshnoy sredy s odnovremennym ispol'zovaniyem EVM i SN (in Russ., Solving problems of continuum mechanics with the simultaneous use of computers and SN). In: Uprugost' i neuprugost' (in Russ., Elasticity and Inelasticity), Moscow State University Publishing House, Moscow

Ilyushin AA (2016) Plastichnost': osnovy obshchey matematicheskoy teorii (in Russ., Plasticity: Foundations of General Mathematical Theory), 2nd edn. URSS-LENAND, Moscow

Ilyushin AA, Lensky VS (1959) Soprotivleniye materialov (in Russ., Resistance of Materials). Fizmatgiz, Moscow

Ishlinsky AY (1954) Obshchaya teoriya plastichnosti s lineynym uprochneniyem (in Russ., General theory of plasticity with linear hardening). Ukrainskiy matematicheskiy zhurnal 6(3):314–325

Ishlinsky AY, Ivlev DD (2003) Matematicheskaya teoriya plastichnosti (in Russ., Mathematical Theory of Plasticity). FIZMATLIT, Moscow

Ivlev DD, Bykovtsev GI (1971) Teoriya uprochnyayushchegosya plasticheskogo tela (in Russ., The Theory of a Hardening Plastic Body). Nauka, Moscow

Izotov IN, Kuznetsov NP, Melnikov BE, Mityukov AG, Musienko AY (2000) New variants of the multisurface theory of plasticity: comparison with the experimental data. In: Melker AI (ed) Third International Workshop on Nondestructive Testing and Computer Simulations in Science and Engineering, International Society for Optics and Photonics, SPIE, vol 4064, pp 362–367

Kabelevsky MG (1972) Ciklicheskoye nestatsionarnoye nagruzheniye turbinnogo diska (in Russ., cyclic unsteady loading of a turbine disk). Izvestiya AN SSSR - Mekhanika Tverdogo Tela (1):169–174

Kachanov LM (1949) Hekotorye voprosy teorii polzuchesti (in Russ., Some Questions of Creep Theory). Gostekhizdat, Moscow

Kachanov LM (1950) K teorii neustanovivsheisya polzuchesti (in Russ., On the theory of transient creep). Prikladnaya Matematika i Mekhanika XIII(4):381–390

Kachanov LM (1958) O vremeni razrusheniya v usloviyakh polzuchesti (in Russ., On the time to rupture under creep conditions). Izvestiya AN SSSR Otdelenie Tekhnicheskikh Nauk (8):26–31

Kachanov LM (1971a) Foundations of the Theory of Plasticity, North-Holland Series in Applied Mathematics and Mechanics, vol 12. North-Holland, Doordrecht

Kachanov LM (1971b) K voprosu ob eksperimental'nom postroyenii poverkhnostey tekuchesti (in Russ., On the question of the experimental construction of yield surfaces). Izvestiya AN SSSR - Mekhanika Tverdogo Tela (4):177–179

Kachanov LM (1979) Teoriya polzuchesti (in Russ., Creep Theory). Fizmatgiz, Moscow

Kachanov LM (1986) Introduction to Continuum Damage Mechanics. Martinus Nijhoff, Dorddrecht

Kachanov LM, Kac SN (1949) O teoriyakh polzuchesti (in Russ., On the creep theories). Kotlostroenie (1)

Kadashevich Y, Pomytkin S (2009) Endokhronnaya teoriya neuprugosti dlya razuprochnyayushchikhsya materialov s uchetom bol'shikh deformatsiy (in Russ., Endochronic theory of inelasticity for softening materials taking into account large deformationsy). In: Sovremennyye problemy resursa materialov i konstrukciy. Trudy 3ey shkoly-seminara MAMI, Moscow, pp 158–165

Kadashevich YI (1967) Obobshchennaya teoriya plasticheskogo techeniya (in Russ., Generalized theory of plastic flow). In: Issledovaniya po uprugosti i plastichnosti, vol 6, Izdate'stvo Leningradskogo Gosudarstvennogo universiteta, Leningrad, pp 25–38

Kadashevich YI, Mosolov AB (1989) Endokhronnyye teorii plastichnosti: osnovnyye polozheniya, perespektivy razvitiya (in Russ., Endochronic theories of plasticity: basic provisions, development prospects). Izvestiya AN SSSR - Mekhanika Tverdogo Tela (1):161–168

Katanaha NA, Getsov LB (2011) Characteristics of creep in conditions of long operation. Materiali in tehnologije / Materials and technology 45(6):523–527

Katanakha NA, Semenov AS, Getsov LB (2013a) Modifikatsiya modeli polzuchesti povyshennoy tochnosti prognoza pri bol'shoy dlitel'nosti nagruzheniya i identifikatsiya yeye parametrov (in Russ., Modification of creep model of higher forecast accuracy at long loading time and identification of its parameters). Deformaciya i Razrushenie Materialov (10):16–23

Katanakha NA, Semenov AS, Getsov LB (2013b) Unified model of steady-state and transient creep and identification of its parameter. Strength of Materials 45(4):495–505

Khazhinsky GM (2011) Modeli deformirovaniya i razrusheniya metallov (in Russ., Deformation and Fracture Models of Metals). Nauchnyi Mir, Moscow

Kheyn YA (1966) Ob opisanii polzuchesti metallov s uchetom uprugogo posledeystviya (in Russ., On the description of the creep of metals taking into account the elastic aftereffect). Inzhenernyi Zhurnal - Mekhanika Tverdogo Tela (2):103–105

Klyushnikov VD (1979) Matematicheskaya teoriya plastichnosti (in Russ., Mathematical Theory of Plasticity). Izdatelstvo Moskovskogo Gosudarstvennogo universiteta, Moscow

Klyushnikov VD (1994) Fiziko-matematicheskiye osnovy prochnosti i plastichnosti (in Russ., Physical and Mathematical Foundations of Strength and Plasticity). Izdatel'stvo Moskovskogo Gosudarstvennovogo universiteta, Moscow

Koiter WT (1960) General theorems for elastic-plastic solids. In: Sneddon I, Hill R (eds) Progress in Solid Mechanics, North Holland, Dordrecht, vol VI, p 165–221

Kolupaev VA (2018) Equivalent Stress Concept for Limit State Analysis, Advanced Structured Materials series, vol 86. Springer, Cham

Korotkikh YG (1981) Uravneniya teorii termovyazkoplastichnosti s kombinirovannym uprochneniyem (in Russ., Equations of the theory of thermoviscoplasticity with combined hardening). In: Makhutov NA, Gadenin MM, Gokhfeld DA (eds) Uravneniya sostoyaniya pri malociklovom nagruzhenii, Nauka, Moscow, pp 129–167

Kostenko Y, Naumenko K (2017) Prediction of stress relaxation in power plant components based on a constitutive model. In: Turbo Expo: Power for Land, Sea, and Air, American Society of Mechanical Engineers, vol 50954, p V008T29A017

Kostenko Y, Almstedt H, Naumenko K, Linn S, Scholz A (2013) Robust methods for creep fatigue analysis of power plant components under cyclic transient thermal loading. In: ASME Turbo Expo 2013: Turbine Technical Conference and Exposition, The American Society of Mechanical Engineers, p V05BT25A040

Kowalewski ZL, Hayhurst DR, Dyson BF (1994) Mechanisms-based creep constitutive equations for an aluminium alloy. The Journal of Strain Analysis for Engineering Design 29(4):309–316

Krempl E (1999) Creep-plasticity interaction. In: Altenbach H, Skrzypek J (eds) Creep and Damage in Materials and Structures, CISM Lecture Notes, vol 399, Springer, Wien, New York, pp 285–348

Krivtsov AM (2007) Deformirovaniye i razrusheniye tverdykh tel s mikrostrukturoy (in Russ., Deformation and Fracture of Microstructured Solids). FIZMATLIT, moscow

Kuznecov BA (1964) Izuchenie nachal'noi stadii plasticheskoi deformacii metallicheskikh polikristallov polyarizacionno-opticheskim metodom (in Russ., Study of the initial stage of plastic deformation of metal polycrystals by the polarization-optical method. Metallovedenie (8):35–42

Lamba HS, Sidebottom OM (1978) Cyclic plasticity for nonproportional paths: Part 1 — Cyclic hardening, erasure of memory, and subsequent strain hardening experiments. Journal of Engineering Materials and Technology 100(1):96–103

Längler F, Naumenko K, Altenbach H, Ievdokymov M (2014) A constitutive model for inelastic behavior of casting materials under thermo-mechanical loading. The Journal of Strain Analysis for Engineering Design 49(6):421–428

Lanin AA, Balina VS (2006) Zharoprochnye metally i splavy: Spravochnye materialy (in Russ., Heat Resistant Metals and Alloys: Reference Materials). Energotekh, St. Petersburg

Leckie FA, Hayhurst DR (1977) Constitutive equations for creep rupture. Acta Metallurgica 25:1059–1070

Lemaitre J (1972) Viscoplastic models. In: Proceedings of the Conference of the Polish Academy of science, Warsaw, pp 1089–1098

Lepin GF (1976) Polzuchest' metallov i kriterii zharoprochnosti (in Russ., Creep of Metals and Criteria for Heat Resistance). Metallurgiya, Moscow

Levy M (1870) Mémoire sur les équations générales des mouvements intérieurs des corps solides ductiles au delà des limites où l'élasticité pourrait les ramener à leur premier état. Comptes rendus de l'Académie des sciences, Paris 70:1323–1325

Levy M (1871) Extrait du mémoire sur les équations générales des mouvements intérieurs des corps solides ductiles au delà des limites où l'élasticité pourrait les ramener à leur premier état; présenté le 20 juin 1870. Journal de Mathématiques Pures et Appliquées 2e série 16:369–371

Likhachev VA (1966) Issledovaniya polzuchesti pri peremennykh temperaturakh (in Russ., Creep studies at variable temperatures). Zavodskaya Laboratoriya (1):70–86

Likhachev VA, Malinin VG (1993) Strukturno-analiticheskaya teoriya prochnosti (in Russ., Structural-analytical Theory of Strength). Nauka, St. Petersburg

Lin TH (1957) Analysis of elastic and plastic strains of a face-centred cubic crystal. Journal of the Mechanics and Physics of Solids 5(2):143–149

Lin TH (1971) Physical theory of plasticity. In: Yih CS (ed) Advances in Applied Mechanics, vol 11, Elsevier, pp 255–311

Lode W (1926) Versuche über den Einfluß der mittleren Hauptspannung auf das Fließen der Metalle Eisen, Kupfer und Nickel. Zeitschrift für Physik 36(11-12):913–939

Lokoshchenko AM (ed) (2016) Polzuchest' i dlitel'naya prochnost' metallov (in Russ., Creep and Long-term Strength of Metals). FIZMATLIT, Moscow

Ludwik P (1909) Elemente der Technologischen Mechanik. Springer, Berlin/Heidelberg

Makhutov NA (1981) Deformacionnye kriterii razrusheniya i raschet elementov konstrukcii na prochnost' (in Russ., Deformation Criteria of Fracture and Structural Strength Analysis). Mashinostroenie, Moscow

Malinin NN (1975) Prikladnaya teoriya plastichnosti i polzuchesti (in Russ., Applied Theory of Plasticity and Creep). Mashinostroenie, Moscow

Malinin NN, Khadjinsky GM (1972) Theory of creep with anisotropic hardening. International Journal of Mechanical Sciences 14(4):235–246

Malinin NN, Khazhinsky GK (1969) K postroyeniyu teorii polzuchesti s anizotropnym uprochneniyem (in Russ., On the construction of creep theory with anisotropic hardening). Izvestiya AN SSSR - Mekhanika Tverdogo Tela (3):148–152

Malmeister AK (1957) Uprugost' i neuprugost' betona (in Russ., Elasticity and Inelasticity of Concrete). Izdatelstvo AN Latviskoi SSR, Riga

McMahon, Jr CJ (ed) (1968) Microplasticity. Interscience, New York

Melan E (1938) Zur Plastizität des räumlichen Kontinuums. Ingenieur-Archiv 9(2):116–126

Melnikov BE, Semenov AS (1991) Mnogopoverkhnostnaya teoriya plastichnosti s odnoy aktivnoy poverkhnost'yu plasticheskoy podatlivosti (in Russ., The multisurface theory of plasticity with one active surface of plastic compliance). In: Proceedings of LSTU *Strength of Metals and Structures*, Leningrad, 441, pp 26–31

Melnikov BE, Semenov AS (1995a) Multimodel analysis as strategy of reliable description of elastic-plastic deformation processes. In: Computational Plasticity: Fundamentals and Applications. Proceedings of the Fourth International Congress, Barcelona, pp 133–139

Melnikov BE, Semenov AS (1995b) Strategy of multimodel analysis of elastic-plastic stress-strain state. In: Proceedings of 6th International Conference on Computing in Civil and Building Engineering., Berlin, pp 1073–1079

Melnikov BE, Getsov LB, Semenov AS (2001a) Metod avtomatizirovannogo vybora modeley termovyazkouprugoplastichnosti, obespechivayushchikh adekvatnoye opredeleniye napryazhenno-deformirovannogo sostoyaniya konstrukcii pri proizvol'nykh programmakh nagruzheniya (in Russ., A method for the automated selection of thermoviscoelasticity models providing an adequate determination of the stress-strain state of a structure with arbitrary loading programs). In: Trudy 4oy mezhdunarodnoy konferencii *Nauchno-tekhnicheskiye problemy prognozirovaniya nadezhnosti i dolgovechnosti konstrukciy i metody ikh resheniya*, St. Petersburg, pp 213–215

Melnikov BE, Getsov LB, Semenov AS (2001b) Metod avtomatizirovannogo vybora modeley termovyazkouprugoplastichnosti, obespechivayushchikh adekvatnoye opredeleniye napryazhenno-deformirovannogo sostoyaniya konstruktsii pri proizvol'nykh programmakh nagruzheniya. soobshcheniya 1-4 (in Russ., A method for the automated selection of thermoviscoelasticity models providing an adequate determination of the stress-strain state of a structure with arbitrary loading programs. Part 1-4). In: Nauchnyye trudy 5go mezhdunarodnogo seminara *Sovremennyye problemy prochnosti* im. V.I. Likhacheva, Velikiy Novgorod, pp 266–287

Melnikov BE, Semenov AS, Getsov LB (2009) Mnogomodel'noe modelirovanie termouprugovyazkoplasticheskogo deformirovaniya (in Russ., Multimodel modeling of thermoelastic visco-plastic deformation). In: Petreniya YK, Getsova LB (eds) Prochnost' materialov i resurs elementov energooborudovaniya, St. Petersburg, no. 296 in Trudy CKTI, p 307–354

Mel'nikov OB, Izotov II, Kuznetsov NP (1990) Calculation and experimental examination of complex modes of elastoplastic deformation. Strength of Materials 22(8):1122–1127

Miller A (1976a) An inelastic constitutive model for monotonic, cyclic, and creep deformation: Part I — Equations development and analytical procedures. Journal of Engineering Materials and Technology 98(2):97–105

Miller A (1976b) An inelastic constitutive model for monotonic, cyclic, and creep deformation: Part II — Application to type 304 stainless steel. Journal of Engineering Materials and Technology 98(2):106–112

Mohel AN, Salganik RL, Khristianovich SA (1983) O plasticheskom deformirovanii uprochnyayushchikhsya metallov i splavov. Opredelyayushchiye uravneniya i raschety po nim (in Russ., On plastic deformation of hardening metals and alloys. Governing equations and calculations for them). Izvestiya AN SSSR - Mekhanika Tverdogo Tela (4):119–141

Mohr CO (1900) Welche Umstände bedingen die Elastizitätsgrenze und den Bruch eines Materials? Zeitschrift des Vereins Deutscher Ingenieure 24:1524–1530, 1572–1577

Moroz LS, Khesin YD, Marinec TK (1962) Issledovanie polzuchesti i dlitel'noi prochnosti zheleza pri nizkikh temperaturakh (in Russ., Study of creep and long-term strength of iron at low temperatures). Metallovedenie (6):145–153

Moskvitin VV (1965) Plastichnost' pri peremennykh nagruzheniyakh (in Russ., Plasticity Under Variable Loads). Moskovskii Gosudarsvennyi universitet, Moscow

Mróz Z (1967) On the description of anisotropic workhardening. Journal of the Mechanics and Physics of Solids 15(3):163–175

Murakami S (1983) Notion of continuum damage mechanics and its application to anisotropic creep damage theory. Journal of Engineering Materials and Technology 105:99 – 105

Murakami S (2012) Continuum Damage Mechanics, Solid Mechanics and its Applications, vol 185. Springer, Dordrecht et al.

Murakami S, Ohno N (1981) A continuum theory of creep and creep damage. In: Ponter ARS, Hayhurst DR (eds) Creep in Structures, Springer, Berlin, pp 422 – 444

Murakami S, Sanomura Y (1985) Creep and creep damage of copper under multiaxial states of stress. In: Sawczuk A, Bianchi B (eds) Plasticity Today - Modelling, Methods and Applications, Elsevier, London, New York, pp 535–551

Nádai A (1931) Plasticity. Mc Graw–Hill, New York–London

Nadai A (1938) The influence of time upon creep. the hyperbolic sine creep law. In: Contributions to the Mechanics of Solids dedicated to Stephen Timoshenko by his friends on the occasion of his sixties birthday anniversary, MacMillian, New York, pp 155–170

Naumenko K, Altenbach H (2005) A phenomenological model for anisotropic creep in a multi-pass weld metal. Archive of Applied Mechanics 74:808 – 819

Naumenko K, Altenbach H (2007) Modeling of creep for structural analysis. Springer Science & Business Media

Naumenko K, Altenbach H (2016) Modeling High Temperature Materials Behavior for Structural Analysis: Part I: Continuum Mechanics Foundations and Constitutive Models, Advanced Structured Materials, vol 28. Springer

Naumenko K, Gariboldi E (2014) A phase mixture model for anisotropic creep of forged Al-Cu-Mg-Si alloy. Materials Science and Engineering: A 618:368 – 376

Naumenko K, Kostenko Y (2009) Structural analysis of a power plant component using a stress-range-dependent creep-damage constitutive model. Materials Science and Engineering: A 510:169–174

Naumenko K, Altenbach H, Gorash Y (2009) Creep analysis with a stress range dependent constitutive model. Archive of Applied Mechanics 79(6-7):619–630

Naumenko K, Altenbach H, Kutschke A (2011a) A combined model for hardening, softening, and damage processes in advanced heat resistant steels at elevated temperature. International Journal of Damage Mechanics 20(4):578–597

Naumenko K, Kutschke A, Kostenko Y, Rudolf T (2011b) Multi-axial thermo-mechanical analysis of power plant components from 9-12%Cr steels at high temperature. Engineering Fracture Mechanics 78:1657–1668

Naumenko K, Gariboldi E, Nizinkovskyi R (2020) Stress-regime-dependence of inelastic anisotropy in forged age-hardening aluminium alloys at elevated temperature: Constitutive modeling, identification and validation. Mechanics of Materials 141:103262

Nejescu-Cleja S (1976) Issledovaniye krayevykh zadach teorii uprugo-plasticheskikh protsessov s tochkoy izloma (in Russ., Research of boundary value problems of the theory of elastic-plastic processes with a break point). Avtoref. diss. cand. phys.-math. sciences, MGU. Mekhaniko-matematicheskii fakul'tet, Moscow

Neuber H (1961) Theory of stress concentration for shear-strained prismatical bodies with arbitrary nonlinear stress-strain law. Transaction of the American Society of Mechanical Engineering -Journal of Applied Mechanics 28(4):544–550

Neuber H (2001) Kerbspannungslehre, 4th edn. Springer, Berlin, Heidelberg

Nikitin VI (1967) Parametricheskii metod opredeleniya kharakteristik polzuchesti metallov (in. Russ., Parametric method for determining the creep characteristics of metals). Zavodskaya Laboratoriya (11):1428–1420

NN (1981) Rukovodyashchie ukazaniya CKTI, vyp. 43. Sluzhebnye svoistva kotel'nykh materialov (in Russ., CKTI Guidelines, vol. 43. Service properties of boiler materials). CKTI, Leningrad

NN (1989) Pravila i normy v atomnoi energetike. Normy rascheta na prochnost' oborudovaniya i truboprovodov atomnykh energeticheskikh ustanovok. PNAE G-7-002-86 (in Russ., Rules and Regulations in Nuclear Energy. Standards for Calculating the Strength of Equipment and Pipes of Nuclear Power Plants). Energoatomizdat, Moscow

Nordmann J, Thiem P, Cinca N, Naumenko K, Krüger M (2018) Analysis of iron aluminide coated beams under creep conditions in high-temperature four-point bending tests. The Journal of Strain Analysis for Engineering Design 53(4):255–265

Norton FH (1929) Creep of Steel at high Temperatures. McGraw-Hill, New York

Novozhilov VV, Kadashevich YI (1990) Mikronapryazheniya v konstrukcionnykh materialakh (in Russ., Microstresses in Structural Materials). Mashinostroenie, Leningrad

Odqvist FKG (1974) Mathematical Theory of Creep and Creep Rupture. Oxford University Press, Oxford

Ohno N, Wang JD (1991) Transformation of a nonlinear kinematic hardening rule to a multisurface form under isothermal and nonisothermal conditions. International Journal of Plasticity 7(8):879–891

Othman AM, Hayhurst DR, Dyson BF (1993) Skeletal point stresses in circumferentially notched tension bars undergoing tertiary creep modelled with physically based constitutive equations. Proceedings of the Royal Society of London Series A: Mathematical and Physical Sciences 441(1912):343–358

Ozhoga-Maslovskaja O, Naumenko K, Altenbach H, Prygorniev O (2015) Micromechanical simulation of grain boundary cavitation in copper considering non-proportional loading. Computational Materials Science 96:178–184

Paley IZ (1968) K postroyeniyu neatermicheskoy teorii tsiklicheskikh nagruzhenii (in Russ., On the construction of a non-thermal theory of cyclic loading). Izvestiya AN SSSR - Mekhanika Tverdogo Tela (1):130–134

Palmov VA (1998) Vibrations of Elasto-plastic Bodies. Foundations of Engineering Mechanics, Springer, Berlin

Palmov VA (2014) Nelineynaya mekhanika deformiruyemykh tel (in Russ., Nonlinear Mechanics of Deformable Bodies). Izdatelstvo SPbPU, St. Petersburg

Parshin AM (1972) Struktura, prochnost' i plastichnost' nerzhaveyushchikh i zharoprochnykh stalei i splavov, primenyaemykh v sudostroenii (in Russ., Structure, Strength and Plasticity of Stainless and Heat-resistant Steels and Alloys used in Shipbuilding). Sudostroenie, Leningrad

Pavlov PA, Kurilovich NN (1978) Predel'noye soprotivleniye konstruktsionnykh staley v usloviyakh vysokotemperaturnoy polzuchesti (in Russ., Ultimate resistance of structural steels under conditions of high-temperature creep). In: Prochnost'i ustoychivost' inzhenernykh konstruktsiy, Izdatel'stvo Altayskogo politekhnicheskogo instituta, Barnaul, pp 39–49

Perrin IJ, Hayhurst DR (1994) Creep constitutive equations for a 0.5Cr-0.5Mo-0.25V ferritic steel in the temperature range 600-675°C. The Journal of Strain Analysis for Engineering Design 31(4):299–314

Perzyna P (1966) Fundamental problems in viscoplasticity. In: Chernyi GG, Dryden HL, Germain P, Howarth L, Olszak W, Prager W, Probstein RF, Ziegler H (eds) Advances in Applied Mechanics, vol 9, Elsevier, pp 243–377

Petrenya YK (1997) Fiziko-mekhanicheskie osnovy kontinual'noi mekhaniki povrezhdaemosti (in Russ., Physico-mechanical Foundations of the Continuum Damage Mechanics). NPO CKTI, St. Petersburg

Petrenya YK, Chizhik AA (1980) Defects and long-term strength of materials with low deformability. Metal Science and Heat Treatmen 22(12):877–879

Popova IP, Oryschenko AC, Getsov LB (2011) A computational method to determine creep characteristics of first and second deformation stage using limited number of isochronal creep curves. Inorganic Materials: Applied Research 2(6):640–650

Pospisil B, Kvitka AL, Tretyachenko GN, Kravchuk LV, Kuriat RI, Semenov GR, Voroshko PP (1987) Prochnost' i dolgovechnost' elementov energeticheskogo oborudovaniya (in Russ., Strength and Durability of Elements of Power Equipment). Naukova Dumka, Kiev

Prager W (1955a) Probleme der Plastizitätstheorie. Birkhäuser, Basel

Prager W (1955b) The theory of plasticity: A survey of recent achievements. Proceedings of the Institution of Mechanical Engineers 169(1):41–57

Prager W, Hodge Jr PG (1951) Theory of Perfectly Plastic Solids. John Wiley & Sons, New York

Prandtl L (1925) Spannungsverteilung in plastischen Körpern. In: Biezeno CB, Burgers JM (eds) Proceedings of the 1st International Congress for Applied Mechanics, J. Waltman, Jr., Delft, pp 43–46

Qin Q, Bassani JL (1992) Non-schmid yield behavior in single crystals. Journal of the Mechanics and Physics of Solids 40(4):813–833

Rabinovich VP (1966) Prochnost' turbinnykh diskov (in Russ., Turbine Disk Strength). Mashinostroenie, Moscow

Rabotnov YN (1959) O mechanizme dlitel'nogo razrusheniya (in Russ., A mechanism of the long term fracture). In: Voprosy Prochnosti Materialov i Konstruktsii, AN SSSR, pp 5–7

Rabotnov YN (1969) Creep Problems in Structural Members. North-Holland, Amsterdam

Rabotnov YN (1980) Elements of Hereditary Solid Mechanics. Mir Publisher, Moscow

Rabotnov YN, Mileiko ST (1970) Kratkovremennaya polzuchest' (in Russ., Short-term Creep). Nauka, Moscow

Reiner M (1960) Deformation, Strain and Flow: an Elementary Introduction to Rheology. H. K. Lewis, London

Reuss A (1930) Berücksichtigung der elastischen Formänderung in der Plastizitätstheorie. ZAMM - Journal of Applied Mathematics and Mechanics / Zeitschrift für Angewandte Mathematik und Mechanik 10(3):266–274

Robinson DN, Binienda WK, Ruggles MB (2003) Creep of polymer matrix composites. I: Norton/Bailey Creep Law for transverse isotropy. Journal of Engineering Mechanics 129(3):310 – 317

Rovinsky BI, Lutzau VG (1957) Relaksaciya oriyentrovannykh mikronapryazheniy (in Russ., Relaxation of oriented microstresses). Zhurnal Tekhnicheskoi Fiziki 27(2):345–351

Rozenblyum VI, Vinogradov NN (1973) Calculation of the creep of structures with low stress levels. Strength of Materials 5(12):1464–1465

Rtishchev VV (1992) Perspektivnye anizotropnye materialy lopatok stacionarnykh gu so stolbchattoi i monokristallicheskimi strukturami (in Russ., Promising anisotropic materials of blades of stationary gi with columnar and single-crystal structures). Trudy CKTI 270:104–119

Sanders Jr JL (1955) Plastic stress–strain relations based on linear loading functions. In: Naghdi PM (ed) Proceedings of Second U.S. National Congress Applied Mechanics, American Society of Mechanical Engineers, New York, pp 455–460

Schmicker D, Naumenko K, Strackeljan J (2013) A robust simulation of direct drive friction welding with a modified carreau fluid constitutive model. Computer Methods in Applied Mechanics and Engineering 265:186–194

Schmid E, Boas W (1935) Kristallplastizität: Mit Besonderer Berücksichtigung der Metalle, Struktur und Eigenschaften der Materie in Einzeldarstellungen, vol 17. Springer, Berlin, Heidelberg

Semenov A, Melnikov B (2016) Interactive rheological modeling in elasto-visco-plastic finite element analysis. Procedia Engineering 165:1748–1756, 15th International scientific conference "Underground Urbanisation as a Prerequisite for Sustainable Development" 12-15 September 2016, St. Petersburg, Russia

Semenov AS (2003) PANTOCRATOR – konechno-elementnyy programmnyy kompleks, oriyentirovannyy na resheniye nelineynykh zadach mekhaniki (in Russ., PANTOCRATOR - finite element software package focused on solving non-linear problems of mechanics). In: Trudy V Mezhdunarodnoy konferentsii *Nauchno-tekhnicheskiye problemy prognozirovaniya nadezhnosti i dolgovechnosti konstruktsiy i metody ikh resheniya*, Publishing house SPbSPU, St. Petersburg, pp 466–480

Semenov AS (ed) (2008) Vychislitel'nyye metody v teorii plastichnosti: Uchebnoye posobiye (in Russ., Computational Methods in the Theory of Plasticity: A Training Manual). Publishing house of Polytechnic University, St. Petersburg

Semenov AS (2014) Identifikatsiya parametrov anizotropii fenomenologicheskogo kriteriya plastichnosti monokristallov na osnove mikromekhanicheskoy modeli (in Russ., Identification of the anisotropy parameters of the phenomenological criterion of plasticity of single crystals based on the micromechanical model). Nauchno-tekhnicheskiye vedomosti SPbGPU - Fiziko-matematicheskiye nauki 194(2):15–29

Semenov AS (2016) Dvukhinvariantnyy stepennoy kriteriy plastichnosti dlya monokristallov kubicheskoy simmetrii (in Russ., Two-invariant power-law plasticity criterion for single crystals of cubic symmetry). In: Uprugost' i Neuprugost'. Materialy Mezhdunarodnogo nauchnogo simpoziuma po problemam mekhaniki deformiruyemykh tel, posvyashchennogo 105-letiyu so dnya rozhdeniya A.A. Il'yushina (Moskva, 20-21 yanvarya 2016 goda) (in Russ., Elasticity and Inelasticity. Materials of the International Scientific Symposium on Problems of Mechanics of Deformable Bodies, dedicated to the 105th anniversary of the birth of A.A. Ilyushin (Moscow, January 20-21, 2016)), Moscow University Press, Moscow, pp 250–256

Semenov AS, Melnikov BE (1994) Multimodel numerical analysis of the elasto-plastic deformation of constructions. In: Internationales Kolloquium über Anwendungen der Informatik und der Mathematik in Architectur und Bauwesen, Weimar, pp 336–341

Semenov AS, Melnikov BE (1997a) Multimodel analysis of the elasto-plastic and elasto-visco-plastic deformation processes in material and structures. In: Rie KT, Portella PD (eds) Proceedings of 4th International Conference Low Cycle Fatigue and Elasto-Plastic Behavior of Materials, Elsevier, pp 659–664

Semenov AS, Melnikov BE (1997b) Multimodel numerical analysis of the elasto-visco-plastic deformation of materials and constructions. In: Internationales Kolloquium über Anwendungen der Informatik und der Mathematik in Architectur und Bauwesen, Weimar, p 6

Semenov AS, Getsov LB, Semenov SG, Grishchenko AI (2014) Deformation and fracture strength of single-crystal superalloys under static and cyclic loading. Vestnik Samarskogo gosudarstvennogo aerokosmicheskogo universiteta (5, part 3):66–75

Semenov SG, Semenov AS, Melnikov BE (2008) Modelirovaniye protsessov neuprugogo deformirovaniya na osnove endokhronnoy teorii (in russ., modeling of inelastic deformation processes based on endochronous theory). In: Nauchno-tekhnicheskiye problemy prognozirovaniya nadezhnosti i dolgovechnosti konstruktsiy i metody ikh resheniya. RELMAS'2008, Publishing house of Polytechnic University, St. Petersburg, vol 2, pp 309–317

Semenov SG, Getsov LB, Semenov AS, Petrushin NV, Ospennikova OG, Zhivushkin AA (2016a) Increasing the lifetime of gas-turbine engine nozzle blades using a new monocrystalline alloy. Journal of Machinery Manufacture and Reliability 45(4):316–323

Semenov SG, Getsov LB, Tikhomirova EA, Semenov AS (2016b) Special features of creep and long-term strength of single-crystal refractory nickel-base alloys. Metal Science and Heat Treatment 57(11):731–738

Shesterikov SA (ed) (1983) Zakonomernosti polzuchesti i dlitel'noi prochnosti (in Russ., Laws of Creep and Long-term Strength). Mashinostroenie, Moscow

Shesterikov SA (2007) Izbrannyye trudy (in Russ., Selected Works). Moscow State University, Moscow

Shevchenko YN, Babeshko ME (1999) Method of approximately describing the lag of scalar properties of isotropic materials. International Applied Mechanics 35(7):671–677

Shevchenko YN, Terekhov RG (1982) Fizicheskiye uravneniya termovyazkoplastichnosti (in Russ., Physical Equations of Thermoviscoplasticity). Naukova Dumka, Kiev

Shevchenko YN, Babeshko ME, Terekhov RG (1992) Termovyazkouprugoplasticheskiye processy slozhnogo slozhnogo deformirovaniya elementov konstruktsii (in Russ., Thermo-viscoelastic-plastic Processes of Complex Deformation of Structural Elements. Naukova Dumka, Kiev

Shevchenko YN, Terekhov RG, Braikovskaya NS, Tormakhov NN (1997) Constitutive equations of thermoelastovisco-plastic processes of material deformation along three-dimensional load paths with orthogonal segments. International Applied Mechanics 33(12):966–974

Shevchenko YN, Terekhov RG, Braikovskaya NS, Tormakhov NN (1998) Complex loading beyond the elastic limit on three-dimensional trajectories. International Applied Mechanics 34(12):1573–8582

Shneiderovich RM (1968) Prochnost' pri staticheskom i povtorno-staticheskom nagruzheniyakh (in Russ., Strength Under Static and Repeated-static Loading). Mashinostroenie, Moscow

Sidoroff F (1973) The geometrical concept of intermediate configuration and elastic-plastic finite strain. Archives of Mechanics 25(2):299–308

Sobotka Z (1984) Rheology of Materials and Engineering Structures, Rheology series, vol 2. Elsevier

Soderberg CR (1936) The interpolation of creep tests for machine design. Transaction of the American Society of Mechanical Engineering 58:733–743

Taira S, Otani R (1986) Teoriya vysokotemperaturnoi prochnosti materialov (in Russ., Theory of High Temperature Strength of Materials). Metallurgiya, Moscow

Taylor G (1938) Plastic strains in metals. Journal of the Institute of Metals 62:307–324

Tresca HE (1864) Mémoire sur l'écoulement des corps solides soumis à de fortes pressions. Comptes rendus de l'Académie des sciences 59:754–758

Troshchenko VT (1971) Ustalost' i neuprugost' metallov (in Russ., Fatigue and Inelasticity of Metals). Naukova Dumka, Kiev

Trunin II (1965) Kriterii prochnosti v usloviyakh polzuchesti pri slozhnom napryazhennom sostoyanii (in Russ., Creep strength criteria for complex stress conditions). Prikladnaya Mekhanika 1(7):77–83

Trunin II (1976) A mechanical equation of the condition of metal materials and prediction of high-temperature strength characteristics. Strength of Materials 8(9):1007–1012

Trusov PV, Shveikin AI (2019) Mnogourovnevyye modeli mono- i polikristallicheskikh materialov: teoriya, algoritmy, primery primeneniya (in Russ., Multilevel Models of Mono- and Polycrystalline Materials: Theory, Algorithms, Application Examples). Izdatelstvo SO RAN, Novosibirsk

Ugodchikov AG, Short YG (1971) Nekotoryye metody resheniya na ECVM fizicheski nelineynykh zadach teorii obolochek i plastin (in Russ., Some methods for solving physically nonlinear problems in the theory of shells and plates on a digital computer). Naukova Dumka, Kiev

Valanis KC (1971) A theory of viscoplasticity without a yield surface. Archiwum Mechaniki Stosowanej 23(4):517–551

Vasin RA (1990) Opredelyayushchiye sootnosheniya teorii plastichnosti (in russ., constitutive relations in the theory of plasticity. In: Itogi Nauki i Tekhniki, VINITI AN SSSR, Moscow, Mekhanika Deformiruyemogo Tverdogo Tela, vol 21, pp 3–75

Volkov IA, Korotkikh YG (2008) Uravneniya sostoyaniya vyazkouprugoplasticheskikh sred s povrezhdeniyami (in Russ., Equations of state for damaged viscoelastoplastic media). FIZMATLIT, Moscow

von Mises R (1913) Mechanik der festen Körper im plastisch deformablen Zustand. Nachrichten Königlichen Gesellschaft der Wississenschaften, Mathematisch-Physikalische Klasse, Göttingen pp 582–592

Watanabe O, Atluri SN (1985) A New Endochronic Approach to Computational Elastoplasticity: Example of a Cyclically Loaded Cracked Plate. Transaction of the American Society of Mechanical Engineering - Journal of Applied Mechanics 52(4):857–864

Zhilin PA (2012) Racional'naya mekhanika sploshnykh sred (in Russ., Rational Mechanics of Continuous Nedia). Izdatelstvo SPbPU, St. Petersburg

Ziegler H (1958) An attempt to generalize Onsager's principle, and its significance rheological problems. Zeitschrift für angewandte Mathematik und Physik 9(5-6):748–763

Zubchaninov VG (ed) (2000) Mekhanika sploshnykh deformiruyemykh sred (in Russ., Mechanics of Continuous Deformable Media), 2nd edn. ChuDo, Tver

Zyczkowski M (1981) Combined Loadings in the Theory of Plasticity. PWN-Polish Scientific Publishers, Warszawa

Chapter 3
Deformation Response of Heat Resistant Materials at Static and Cyclic Loading

3.1 Ductile and Brittle Fracture Criteria

3.1.1 Fracture Criteria

To assess the conditions for the fracture of materials in the general case of a complex stress state according to the data of experiments in a uniaxial stress state and the simplest experiments in a biaxial stress state, the strength theory[1] is used. Fracture criteria within the framework of various theories include: the criteria of von Mises, Tresca, Sdobyrev, Mohr, Nadai, Pisarenko-Lebedev, Trunin, maximum principal stresses (the maximum principal stress must be positive) and maximum principal strains (the maximum principal strain must be positive). These and other criteria are discussed in Birger (1977); Altenbach et al. (1995); Lebedev et al. (2003); Altenbach et al. (2014); Kolupaev (2018) among others.

To determine the conditions of macroscopic fracture in a complex stress state, the critical (limit) stress $\sigma_{\text{limit}}(T)$ must be compared with the equivalent stress σ_{eq}. The further presentation of the problem of fracture criteria for a complex stress state will be carried out based on the ideas developed in Birger (1977), where the value of σ_{eq} is defined as a function of principal stresses $\sigma_i, i = 1,2,3$, material parameters $\lambda_j, j = 0,1,2,3$ and internal state variables $\chi_k, k = 1,2,\ldots$, reflecting the loading history and other effects (Volkov, 1959; Altenbach et al., 2014). Rejecting the consideration of internal state variables and accepting the linear dependence $\sigma_{\text{eq}} = f(\sigma_1, \sigma_2, \sigma_3)$ we have a four-parameter fracture criterion[2]

$$\sigma_{\text{eq}} = \lambda_0 \sigma_{\text{vM}} + \lambda_1 \sigma_1 + \lambda_2 \sigma_2 + \lambda_3 \sigma_3 = \sigma_{\text{limit}} \tag{3.1}$$

[1] Strength theories are explained in the textbooks on the strength of materials. The term "strength theory" is preferred in Russian references, in other references the term "strength criteria" is used.

[2] In a similar way a six-parameter criterium was introduced in Altenbach et al. (1993).

L. B. Getsov, *Materials and Strength of Gas Turbine Parts*, Advanced Structured Materials 150, https://doi.org/10.1007/978-981-16-0534-5_3

To define λ_0, λ_1, λ_2, λ_3 four basic experiments[3] are required. Using experiments on tension, compression, torsion and testing of thin-walled tube under inner pressure as such experiments, we obtain

$$\begin{aligned}
\lambda_0 &= \frac{1}{2-\sqrt{3}}\left(1+\frac{\sigma_{\text{tension}}}{\sigma_{\text{compression}}}-\frac{\sigma_{\text{tension}}}{\tau_{\text{torsion}}}\right),\\
\lambda_1 &= 1-\lambda_0,\\
\lambda_2 &= 2\frac{\sigma_{\text{tension}}}{\sigma_{\text{inner pressure}}}-1+\frac{\sigma_{\text{tension}}}{\sigma_{\text{compression}}}-\frac{\sigma_{\text{tension}}}{\tau_{\text{torsion}}},\\
\lambda_3 &= \frac{1}{2-\sqrt{3}}\left[1+(\sqrt{3}+1)\frac{\sigma_{\text{tension}}}{\sigma_{\text{compression}}}-\frac{\sigma_{\text{tension}}}{\tau_{\text{torsion}}}\right],
\end{aligned} \tag{3.2}$$

where σ_{tension}, $\sigma_{\text{compression}}$, τ_{torsion} and $\sigma_{\text{inner pressure}}$ are ultimate strengths (maximum of the stress-strain diagrams) for the listed types of loading. If we accept the values of the relations $\sigma_{\text{tension}}/\sigma_{\text{compression}}$, $\sigma_{\text{tension}}/\tau_{\text{torsion}}$ and $\sigma_{\text{tension}}/\sigma_{\text{inner pressure}}$ constant, then instead of (3.2) we have one-parameter criteria[4].

Using as basic two experiments (for example, tension and compression or tension and torsion), from which we determine the ratios $\lambda = \sigma_{\text{tension}}/\sigma_{\text{compression}}$ or $\lambda = \sigma_{\text{tension}}/\tau_{\text{torsion}}$, we obtain two-parameter criteria. The most popular criteria are one- and two-parameter criteria. They are obtained from the ratio (3.1) at various values of the ratio of strength characteristics, Table 3.1. In addition, in Paul (1968); Birger (1977); Altenbach and Zolochevsky (1996); Altenbach et al. (2014); Kolupaev (2018), three- and higher-parameter criteria were proposed, which make it possible to obtain greater accuracy when verifying them and comparing with experimental data.

For materials that fail under a given stress state without significant plastic deformations, the best agreement with the experimental results is obtained by the criterion of the maximum positive principal stress. If the fracture of the component occurs as a result of shear and is accompanied by significant plastic deformations, the theory of the maximum shear stress (Tresca criterion) gives the best agreement with the experimental data. For biaxial tension conditions, when $\sigma_3 = 0$, the failure criteria for both theories coincide. According to Mohr's theory, failure occurs when the most unfavorable combination of normal and shear stresses occurs on a certain plane.

Using only one assumption, for example, about the value of $\sigma_{\text{tension}}/\tau_{\text{torsion}}$, we have three-parameter criteria. One of the variants of this criterion is one for which $\lambda_2 = 0$, i.e. the influence of σ_2 on σ_{tension} is neglected. In this case, for

$$\frac{\sigma_{\text{tension}}}{\tau_{\text{torsion}}} = (\sqrt{3}+1)/2 = 1.36,$$

[3] The term "basic experiment" is under discussion. In solid mechanics basic experiments are tests, where homogeneous stress states are realized. Considering this and statements in material testing, such tests are (Blumenauer, 1996): tension, compression, torsion and hydrostatic compression.

[4] Other possible sets of relations are widely discussed in Altenbach et al. (1995, 2014); Kolupaev (2018). The question of the number of parameters for the fracture criteria is debatable and is determined by the uniqueness of the used material characteristics, which have sometimes a great scattering.

Table 3.1 Fracture criteria (Birger, 1977)

Criterion	Expression	$\dfrac{\sigma_{\text{tension}}}{\sigma_{\text{compression}}}$	$\dfrac{\sigma_{\text{tension}}}{\tau_{\text{torsion}}}$	$\dfrac{\sigma_{\text{tension}}}{\sigma_{\text{inner pressure}}}$
1	$\sigma_{\text{tension}} = \sigma_{\text{vM}}$	1	$\sqrt{3}$	$\dfrac{\sqrt{3}}{2}$
2	$\sigma_{\text{tension}} = \sigma_1 - \sigma_3 = 2\tau_{\text{torsion}}$	1	2	1
3	$\sigma_{\text{tension}} = \sigma_1$	0	1	1
4	$\sigma_{\text{tension}} = \sigma_1 - \dfrac{1}{2}(\sigma_2 + \sigma_3)$	$\dfrac{1}{2}$	$\dfrac{3}{2}$	$\dfrac{3}{4}$
5	$\sigma_{\text{tension}} = \dfrac{1}{2}(\sigma_{\text{vM}} + \sigma_1)$	$\dfrac{1}{2}$	$\dfrac{1}{2}\left(\sqrt{3}+1\right)$	$\dfrac{1}{4}\left(\sqrt{3}+2\right)$
6	$\sigma_{\text{tension}} = \sigma_1 - \chi\sigma_3$	χ	$1+\chi$	1
7	$\sigma_{\text{tension}} = \dfrac{1}{2}(1+\chi)\sigma_{\text{vM}} + \dfrac{1}{2}(1-\chi)(\sigma_1 + \sigma_2 + \sigma_3)$	χ	$\dfrac{1}{2}\sqrt{3}(1+\chi)$	$\dfrac{1}{4}\left[\sqrt{3}(1+\chi) + 3(1-\chi)\right]$
8	$\sigma_{\text{tension}} = \chi\sigma_{\text{vM}} + (1-\chi)\sigma_1$	χ	$1 + (\sqrt{3}-1)\chi$	$1 - \left(1 - \sqrt{\dfrac{3}{2}}\right)\chi$

1 - Huber, von Mises, Hencky; 2 - Coulomb, Tresca, St. Venant, 3 - maxiumum principal stress (the stress must be positive), Galilei, Leibniz; 4 - maximum principal strain (special case if $\nu = 0.5$, Rankine, Lamé, Clebsch; 5 - Sdobyrev; 6 - Mohr; 7 - Nadai, Botkin-Mirolyubov, Schleicher, Prager-Drucker; 8 - Pisarenko-Lebedev

the criterion coincides with the condition of Sdobyrev and Pisarenko-Lebedev while for

$$\frac{\sigma_{\text{tension}}}{\tau_{\text{torsion}}} = 1 + \frac{\sigma_{\text{tension}}}{\sigma_{\text{compression}}} = 1.5,$$

it coincides with Mohr's criterion. The use of two-, three-, and even more so four-parameter criteria makes it possible to refine the conditions of failure in comparison with the one-parameter von Mises and maximum shear stresses criteria, which is essential for critical components[5]

The following criterion proposed by Trunin et al. (1979) somewhat falls out of the approach proposed by Birger

$$\sigma_{\text{eq}} = \frac{1}{2}(\sigma_{\text{vM}} + \sigma_1)\alpha^{1-2\eta} = \sigma_{\text{tension}}, \tag{3.3}$$

where α is a material parameter, independent of the kind of the stress state, and

$$\eta = (\sigma_1 + \sigma_2 + \sigma_3)/(\sigma_{\text{vM}} + \sigma_1) = 3\sigma_{\text{m}}/(\sigma_{\text{vM}} + \sigma_1). \tag{3.4}$$

[5] However, as noted in Bugay et al. (1994), the four-parameter criterion has a drawback associated with the fact that at triaxial tension ($\sigma_1 = \sigma_2 = \sigma_3 > 0$) the strength of the material coincides with the compressive strength ($\sigma_1 = \sigma_2 = \sigma_3 < 0$), which contradicts the experimental observations. A similar statement concerning the behavior under hydrostatic tension and compression is given in Kachanov (1974).

It was found that when criterion (3.3) is used for various metallic materials and stress states, in most cases the accuracy turns out to be at least 15 %.

The type of fracture of materials (separation or shear), and, consequently, the decision on the choice of a particular theory of strength depends on a number of factors, which include: the rate of deformation, test temperature, characteristics of material plasticity under uniaxial tension, geometric dimensions of the component, the stiffness of the stress state.

The nature of the processes preceding fracture can be judged by the value of the accumulated plastic deformation. As an example, consider the case of a symmetric triaxial stress state $\sigma_2 = \sigma_3 = A\sigma_1$ for which most of the particular theories of plasticity give similar results. From expressions of the theory of elasticity, for example, for ε_1 we obtain

$$\varepsilon_1 = \frac{1}{E}[\sigma_1 - \nu(\sigma_2 + \sigma_3)]. \tag{3.5}$$

For the considered case we have

$$\varepsilon_1 = \frac{\sigma_1}{E_{eq}}, \tag{3.6}$$

where

$$E_{eq} = \frac{E}{1 - 2\nu A}. \tag{3.7}$$

From Eq. (3.6) it can be seen that the conditions $A > 0$ and $\sigma_1 > 0$ correspond to a lower elastic strain ε_1 than in a uniaxial stress state. Relations (3.5) and (3.6) also apply to plasticity conditions, if in them, instead of the elastic modulus, the secant modulus is included.

Consider the situation with the possibility of brittle fractures on the basis of the approaches proposed in Davidenkov (1936); Fridman (1946, 1974); Shanley (1961), see Fig. 3.1. In this approach, strains are determined according to the stress-strain curve under uniaxial stress state, and the fracture conditions are determined by the first theory of strength (maximum positive principal stress criterion): $\sigma_1 = \sigma_{lim}$.

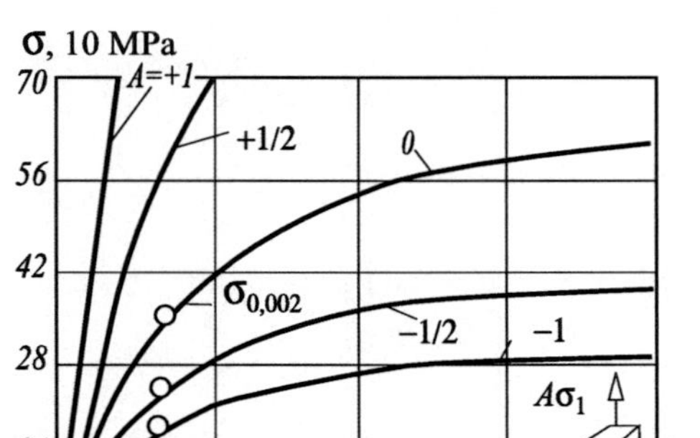

Fig. 3.1 Influence of the symmetric complex stress state on the stress-strain diagram ($A = 0$ - curve of deformation at simple tension)

Here $\sigma_{\lim}$ is the resistance to separation or, with a certain degree of conservatism, ultimate strength.

Equating the expressions for the von Mises equivalent stress in simple tension to the case under consideration $\sigma_2 = \sigma_3 = A\sigma_1$, we obtain the relation

$$\sigma_1 = \sigma_1^{\text{axial tension}}/(1 - A) \tag{3.8}$$

from which it follows that the same equivalent plastic strain under tension and triaxial stress state will occur at different stresses that differ by a factor of $(1 - A)$. Here, the subscript ut refers to uniaxial tension conditions. On the other hand, fracture caused by the action of the maximum positive principal stress $\sigma_1 = \sigma_{\text{tension}}$ will be accompanied by plastic deformation. From the known stress-strain curve for simple tension using relations (3.6) and (3.8), it is possible to construct stress-strain curves corresponding to the given values of A (Fig. 3.1). For the case under consideration $\gamma = 1/(1-A)$ (the ratio of the maximum normal stress σ_1 to the von Mises equivalent stress σ_{vM}), the curves for $A > 0$ are located above the curve corresponding to pure tension. Under these conditions, fracture can occur brittle at deformations lower than those corresponding to the conditional ultimate strength in uniaxial tension σ_{tension}. Conditional yield stress also turns out to be related to loading stiffness as follows

$$\sigma_{0.2} = \sigma_{0.2}^{\text{axial tension}}\gamma \tag{3.9}$$

If $\sigma_{\text{tension}}^{\text{axial tension}}/\sigma_{0.2}^{\text{axial tension}} = \gamma$ then from Eq. (3.9) we obtain $\sigma_{0.2} = \sigma_{\text{tension}}$ and the fracture occurs brittle.

For biaxial tension, characterized by the conditions $\sigma_2 = 0.5\sigma_1$ and $\sigma_3 = 0$, we obtain $\gamma = 1.155$. If the ratio $\sigma_{0.2}^{\text{axial tension}}/\sigma_{\text{tension}}^{\text{axial tension}}$ found from the results of uniaxial tensile tests is greater than 0.866, then the component in the case under consideration should fail with very small plastic deformation (Rabinovich, 1970). The uniform triaxial tension $\sigma_1 = \sigma_2 = \sigma_3$ corresponds to $\gamma = \infty$, i.e. any material must fail brittle. It should be noted that the brittle fracture conditions described above apply only to simple loading.

3.1.2 Conditions for Ductile and Brittle Fracture

In practice, strength calculations are usually carried out using simplified fracture criteria focused on methods of elastic analysis using a limited set of experimental data (mainly the yield stress). For certain classes of materials, this approach is justified theoretically and tested experimentally (Trunin et al., 1979). In this regard, it is advisable to distinguish the following groups of materials used in power engineering, which differ in the nature of fracture under static loading, including creep conditions

1. Pearlitic steels (with a bcc-lattice) exhibit depending on the temperature and loading rate both intragranular and intergranular (for example, wheel disks of centrifugal compressors) fracture.

2. High-chromium steels (with 12-17% Cr, also with a bcc lattice) show intragranular fracture in a wide temperature range (disks and blades of turbines and compressors).
3. Austenitic steels (with fcc lattice) depending on temperature and loading rate exhibit both intragranular and predominantly intergranular fracture (disks and blades of gas turbines, regenerators, pipelines).
4. Heat-resistant nickel-based alloys (with fcc lattice) show intergranular fracture under static loading (blades and disks of gas turbines).
5. Titanium alloys with a single-phase hexagonal α-structure (VT5, VT5-1, OT4-1, VT18, VT20), two-phase ($\alpha + \beta$) alloys (VT6, VT3-1, VT8, VT9, VT14, VT16, VT22) and single-phase alloys with a cubic β-structure (VT15, VT30, TS6) fail both along slip bands and by separation (disks and blades of compressors and steam turbines, Chechulin and Khesin, 1987). Depending on the size of the grains in the alloy and the corrosive environment, the nature of the fracture may differ, either transcrystalline or intergranular. It is not uncommon for the process of damage by cleavage, as in the case of monocrystalline nickel heat-resistant alloys, the fracture mode can be changed to separation.

Materials that exhibiting in a uniaxial stress state intragranular fracture, under certain conditions of the stress state, fail by separation and manifest themselves as brittle. Materials behave similarly under creep conditions in cases of intergranular fracture. At the same time, materials that show intergranular fracture under a uniaxial stress state in the absence of creep fail by shear accompanied by significant plastic deformations.

Usually, metallic materials, according to their behavior in structures operating under static loading, are conventionally divided into the following three groups

1. materials with $\sigma_{0.2}^{\text{axial tension}}/\sigma_{\text{tension}}^{\text{axial tension}}$ not more than 0.3-0.5 (such as, for example, austenitic steel 12Ch18N10T),
2. materials with $\sigma_{0.2}^{\text{axial tension}}/\sigma_{\text{tension}}^{\text{axial tension}}$ in the range of 0.6-0.8,
3. materials with $\sigma_{0.2}^{\text{axial tension}}/\sigma_{\text{tension}}^{\text{axial tension}} > 0.87$ (such as high-strength steels).

For each of these groups, different fracture criteria are preferable, namely: for group 1, when calculating local static stresses, it is advisable to take the separation resistance as the limiting stress, for groups 2 and 3, strain or strain-force criteria are optimal, and in the presence of macrocracks - criteria known from fracture mechanics (K_{Ic}, J_{Ic}). Of course, the proposed boundaries between different materials depend on the loading conditions of the structure, its stress state.

Let us assume that $\sigma_2 = A\sigma_1$, $\sigma_2 = A\sigma_1$, where the constants A and B can vary from $-\infty$ to 1. According to this approach, depending on the relationships between σ_1, σ_2 and σ_3, there is a brittle fracture region that depends on the $\sigma_{\text{lim}}/\sigma_{0.2}$ ratio, in which it is advisable to take the separation strength as the ultimate strength characteristic for local stresses. These ratios can also be written in a slightly different form. Consider the case when $\sigma_1/\sigma_{\text{vM}} > 1$. Let us take an approximation of the stress-strain curve in the form

$$\sigma_{\text{vM}} = \sigma_{0.2} + A\left(\varepsilon_{\text{vM}}^{\text{p}}\right)^m \tag{3.10}$$

and the fracture condition, according to the first theory of strength, is $\sigma_1 = \sigma_{\text{lim}}$.

Let $q = \sigma_1/\sigma_{\text{vM}}$, then we have $q[\sigma_{0.2} + A(\varepsilon^{\text{p}}_{\text{vM}})^m] = \sigma_{\text{lim}}$, and the plastic strain can be calculated as follows

$$\varepsilon^{\text{p}}_{\text{vM}} = \left(\frac{\frac{\sigma_{\text{lim}}}{q} - \sigma_{0,2}}{A}\right)^{1/m} \tag{3.11}$$

Let $\sigma_{\text{lim}}/\sigma_{0.2} = k$ then

$$\varepsilon^{\text{p}}_{\text{vM}} = \left(\frac{\sigma_{0.2}\frac{k}{q} - \sigma_{0,2}}{A}\right)^{1/m} = \varepsilon_1 = \left[\frac{\sigma_{0.2}\left(\frac{k}{q} - 1\right)}{A}\right]^{1/m}$$

where $k/q > 1$. For $k/q = 1$ the brittle fracture can be expected.

To identify the conditions of brittle fracture within the framework of the considered approach, a program was developed and calculations were performed (Getsov and Turgenev, 2006), some of the results of which are shown in Fig. 3.2.

Despite the familiarity of the term brittleness, the issue of brittle fracture requires more careful consideration, which is advisable to carry out on the basis of the ideas presented in Karzov et al. (1993). The brittle fracture is usually defined to be the fracture mode which is not accompanied by the macroplastic deformations. This definition is not entirely correct, since brittle fracture of a material with high ductility is possible. After the formation of a neck, the failure of the sample can occur by a brittle mechanism, as evidenced by the consideration of its fracture. That is, it is not a matter of plasticity before fracture, but of the speed of propagation of the resulting discontinuities in the material. Hence, it is clear that S_{f} (resistance to fracture) and S_{sep} (resistance to separation) can have different values. To determine S_{sep}, its independence from temperature is used. The value can be found from tensile experiments in the region of brittle fractures, when $S_{\text{sep}} = S_{\text{k}}$. By varying the test temperature and the stiffness of the stress state, it is possible to determine the dependence of the separation resistance on the magnitude of plastic deformation, which is widely used in models (Karzov et al., 1993).

On the other hand, brittle fracture is a fracture mode in which sharp microcracks nucleate (bluntness is equal to the lattice parameter). They initiate and cleavage microcracks propagate through various effective barriers (microstresses or boundaries of the deformation substructure of the material). If during the initiation of a sharp microcrack the Griffith fracture condition is not satisfied, then its further growth, like the growth of pores, can only be stable, due to plastic deformations at its tip. In the case of viscous intragranular fracture, the mechanisms of formation, growth, and coalescence of pores are usually valid. In this case, the criterion for fracture is plastic deformation at the moment of rupture, and the rate of pore growth is determined by the rate of plastic deformation and the stiffness of the stress state.

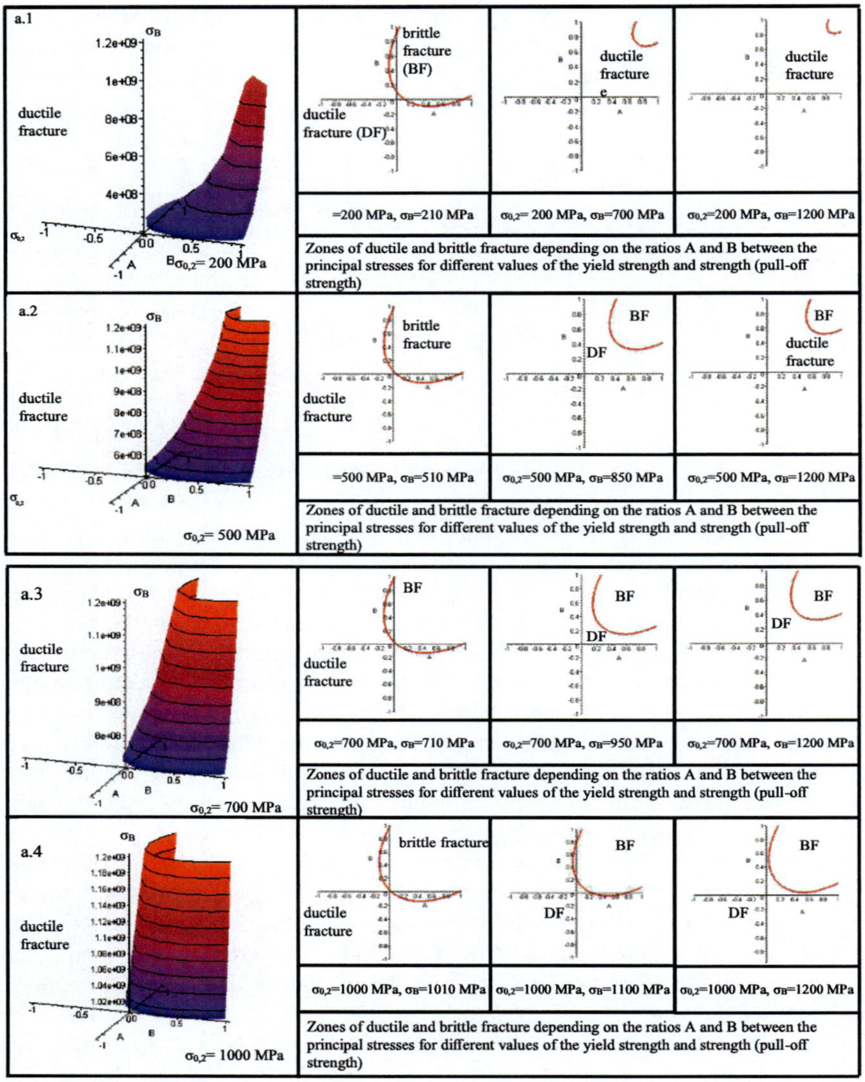

Fig. 3.2 Zones of ductile (DF) and brittle (BF) fracture depending on the ratios A and B between the main stresses for different values of the yield strength σ_B and pull-off strength S_{po}. a - spatial diagrams for $\sigma_{0.2}$ = 200, 500, 700 and 1000 MPa and various values of pull-off resistance

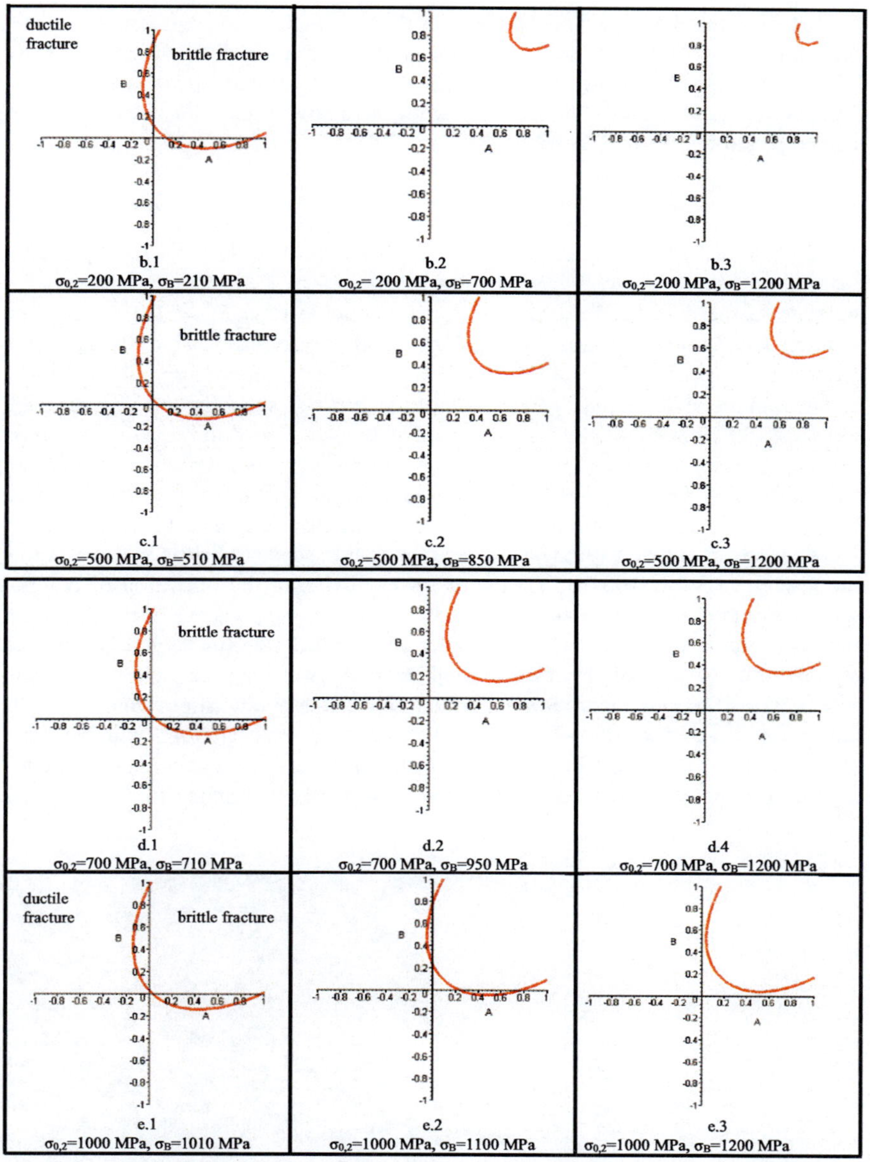

Fig. 3.2 (continued) Zones of ductile (DF) and brittle (BF) fracture depending on the ratios *A* and *B* between the main stresses for different values of the yield strength σ_B and pull-off strength S_{po}. b - diagrams for $\sigma_{0.2} = 200$ MPa, $\sigma_B(S_{po}) = 210, 700$ and 1200 MPa, c - diagrams for $\sigma_{0.2} = 500$ MPa, $\sigma_B(S_{po}) = 510, 850$ and 1200 MPa, d - diagrams for $\sigma_{0.2} = 700$ MPa, $\sigma_B(S_{po}) = 710, 950$ and 1200 MPa, e - diagrams for $\sigma_{0.2} = 1000$ MPa, $\sigma_B(S_{po}) = 1010, 1100$ and 1200 MPa

This explains the experimental data on ductile fracture of high-strength steels (with $\sigma_{0.2}/\sigma_{B} > 0.87$) under plane deformation conditions. The value of ductility for a complex stress state, with which it is necessary to compare the deformations ε_{p} obtained in elastic-plastic calculations, the following two expressions can be applied (Hancock and Mackenzie, 1976; Makhutov, 1981)

$$\varepsilon^{*} = 1.7\varepsilon_{\lim} \exp(-1.5\sigma_{m}/\sigma_{vM}), \qquad \varepsilon^{*} = K_{e}\varepsilon_{\lim}\sigma_{vM}^{2}/(3\sigma_{1}\sigma_{m}) \tag{3.12}$$

leading to conservative estimations. Here $\varepsilon_{\lim}$ is the ultimate strain (strain capacity), determined from experiments with short-term tension, K_{e} is the characteristic of the state of the material (brittle state $K_{e} = 1$, ductile state $K_{e} = 1.2$). The value of ε_{f} is determined within the framework of an elastic-plastic analysis using the corresponding plasticity model and the lower envelope of the deformation curves.

Thus, in the elasto-plastic calculation of the stress-strain state of a structure, the assessment of its performance depends both on the stiffness of the stress state and on the properties of the material. Unfortunately, in the reference literature, most often there is no data on the values of the separation strength of materials, and therefore the assessment is carried out according to the ultimate strength.

The tendency of pearlitic, ferritic and martensitic steels to brittle fracture at low temperatures (below 100-200°C) manifests itself during the operation of components, especially welded joints, as a rule, in the case of the presence of metallurgical defects in them, which, growing to cracks of a critical size (unstable cracks), are capable of causing brittle fracture at rated stresses lower than the yield point (up to $\sigma \geq 0.25\sigma_{0.2}$). As an illustration of brittle fracture in a biaxial stress state, Fig. 3.3a shows a pump casing collapsed during the hydraulic test made of 15Ch5MF steel with $\sigma_{0.2}^{ut}/\sigma_{b}^{ut} = 0.9$. Experimental studies and the experience accumulated in power engineering indicate the need to take into account the conditions of brittle fracture

Fig. 3.3 Failure of pump housings (a) from steel 15Ch5MF and (b) from steel 20Ch13(L)

considered above when choosing a material operating under a complex stress state, for example disks with stress concentrators, parts with various defects and cracks obtained during the manufacture of blanks or during welding, etc.

The above considerations about the tendency of materials to brittle fracture do not allow predicting the nature of the fracture of a material in which a crack has already formed. Perhaps this is due to the fact that for most materials, with an increase in the deformation rate, the yield stress sharply increases. Microcracks in the material can form in the zones of deformation localization. Thus, knowing only the characteristics of macroplasticity (stress-strain curves) during tension of smooth specimens, it is impossible to reliably estimate in the general case the tendency of the material to brittle fracture. An example of the destruction of a steel part with a ratio of $\sigma_{0.2}/\sigma_B < 0.87$ was the brittle fracture of a pump casing operating under a complex stress state, for which the value of $A = \sigma_2 < \sigma_1 = 0.4$ at the place of crack formation. The body was made of cast steel 20Ch13(L), which had a rough structure and the following mechanical properties $\sigma_{0.2} = 293$ MPa, $\sigma_b = 451$ MPa, $\delta = 10\%$, $\psi = 9.8\%$ (Fig. 3.3b). The failure of the casing was caused by an emergency overpressure. Samples of the Menage type for impact bending tests with a notch radius of 1mm were made from the metal of the fractured component. The value of impact strength (specific work of fracture) was found to be $70 - 100$ kJ/m^2.

The above example, as well as many others, indicate that the values of impact strength, previously used only to assess the characteristics of resistance to the action of dynamic loads and control compliance with the heat treatment regime, can also be used to assess the tendency to brittle fracture under static loading conditions. However, the answer to the question of what value of impact toughness is sufficient to exclude the possibility of brittle fracture of the component depends on the material, the nature of the stress state, and the scale factor.

To assess the tendency of materials to brittle fracture, there are now about 70 test methods. The most common are groups of methods can be classified by

- temperature criteria, in particular, the transition temperature of brittleness (temperature corresponding to a decrease in α_H to 300 kJ/m^2 or a decrease in the fiber content in a fracture to 50%,
- critical values of stress intensity factors under plane strain conditions for specimens with cracks developing by opening K_{Icr} as well as transverse K_{IIcr} and longitudinal K_{IIIcr} shears,
- the critical value of the J-integral, i.e. the specific energy until the maximum load is reached for the sample with a crack, as well as the critical value δ_{cr} (CTOD) of the displacement of the crack edges until the crack starts to propagate,
- the strength coefficient of notched disks K_{σ_B} under acceleration test conditions, equal to the ratio of the average tangential stress in the minimum section of the disk (minus the notches and holes) to the ultimate strength of the material.

Temperature criteria for brittle fracture are used for pearlitic, ferritic and martensitic steels due to their property of dramatically changing the resistance to crack propagation in a relatively narrow temperature range. With increasing temperature, the work of fracture increases and its character changes. This made it possible to introduce the concept of the critical (transitional) temperature T_{cr} of the material. The safety of the structure is ensured if the condition $T_{work} \gg T_{cr} + \Delta T$ is met, where T_{work} is the operating temperature and ΔT is the temperature margin of brittleness, the value of which depends on the yield point of the material and the size of the possible defect.

Fracture toughness criteria are much more universal. According to Irwin (1957), crack propagation can occur without stopping if the stress intensity factor K_{I} reaches a critical value equal to the fracture toughness K_{Ic}, which is a material property, but depends on the specimen geometry, stress state at the crack edge, and loading conditions. Experimental determination of the characteristics of crack resistance (fracture toughness) K_{cr}, K_{Icr}, δ_{cr}, J_{cr}, J_{Ic} under static loading is carried out in accordance with GOST 25.506-85. The values of the stress intensity factors K_{I}, K_{II} and K_{III} of a particular part under operating loads in relation to one or another size and shape of defects are calculated by methods of the theory of elasticity (Parton and Morozov, 1974; Cherepanov, 1979).

The applicability of strength calculations according to Irwin's criterion ($K_{\text{I}} = K_{\text{Icr}}$) within the framework of linear fracture mechanics is limited to small values of fracture stresses $\sigma \ll \sigma_{0.2}$, when asymptotic estimates of the stress state in the vicinity of the crack are valid. In the general case of a complex stress state, there is a certain limiting surface in the space of stress intensity factors, which cuts off on the coordinate axes the segments equal to K_{Icr}, K_{IIcr} and K_{IIIcr}. The intersection of the vector with the projections K_{I}, K_{II} and K_{III} with this surface determines the beginning of the failure of the component and the mechanism of this failure (Chizhik, 1970). To determine the values of K_{Icr} from the test results of samples, the relations of the theory of elasticity are applied in the following form

$$K_{\text{Icr}} = M\sigma\sqrt{a_{cr}}, \tag{3.13}$$

where σ is the critical nominal stress under conditions of a crack of length a_{cr} and M is a parameter depending on the geometry of the specimen, the shape of the crack and the type of stress state. Methods for determining K_{Icr}, K_{IIcr} and K_{IIIcr} are described in NN (1964); Chizhik (1970).

The fracture toughness characteristics determine the maximum allowable crack size (metallurgical defect) that the material in question can withstand without fracture if loaded to a level close to that at which fracture occurs due to excessive plastic deformation. Since it is practically impossible to hope for the absence of any cracks, stress concentrators, etc. in the workpiece of the part, which have arisen during smelting, welding, forging and heat treatment, then in order to develop effective measures for the control of workpieces, it is necessary to have information on the size of permissible defects in this material.

With an increase in the level of fracture stresses, the zone of plastic deformations surrounding the crack tip increases, which can distort the asymptotic assessment of the stress state and the concept of the stress intensity factor. The transition in this case to a quasi-brittle state requires methods of nonlinear fracture mechanics for its description. In this case, the concept of the strain intensity factor K_{Ie} is used (Makhutov, 1981). Furthermore, fracture criteria based on the use of the values of the critical crack opening δ_c and the J-integral should be applied. The J-integral is a promising fracture criterion in nonlinear mechanics, since it exhibits invariance to the sample size under conditions of plastic flow at the crack tip. The value of the specific energy of a specimen with a crack (J-integral) by the time of the first subcritical breakaway of the crack (reaching the maximum load) for the mode I is specified by J_{Icr}. After recalculating into K_{Icr} according to the formula

$$K_{\mathrm{Ic}} = \sqrt{\frac{E J_{\mathrm{Ic}}}{1 - \nu^2}}$$

it is used for estimates of the conditions of failure in structures.

The method for estimating K_{σ_B} from the results of testing flat models of notched discs became widespread with Rabinovich (1966). As an example, in Table 3.2 shows the values of K_{σ_B} of two heat-resistant pearlitic steels used for gas turbine disks in comparison with other criteria of brittle fracture.

In some cases, however, the connections between different criteria characterizing the tendency of a material to brittle fracture are ambiguous. So, for example, Table 3.3 shows the results of testing samples with an initiated fatigue crack at a constant rate of deflection, carried out on the plant with feedback developed by Chizhik (1970). The table shows that the relationship between δ_{cr}, K_{Icr} and J_{Icr} is ambiguous.

The data obtained during acceleration tests at a temperature of 20°C for disks with a diameter of 230-610 mm (solid, with an inner hole with a diameter of 80 mm and with holes provided with two notches of different depths) from two pearlitic chromium-molybdenum-vanadium steels of types I and II (Khinsky, 1968) indicate about the inadequacy of the values of K_{σ_B} and T_{cr}, Table 3.4.

When calculating the stress-strain state by the finite element method of power plant components, including main gas pipelines operating at the compressor station, in some cases local stresses exceed the yield point. At the same time, taking into account the redistribution of stresses in the calculations according to the theories of plasticity and creep, which ensure the adequacy of the results obtained, does not change the situation of uncertainty in solving the question of the admissibility of this under static loading conditions. Such cases include stresses in disks and blades of

Table 3.2 Test results on disk models

Steel	$\sigma_{0.2}$, MPa	K_{Icr}, MPa m$^{1/2}$	T_{cr}, °C	K_{σ_B}
EI415	600	41	140-160	0.18
25Ch1M1F (P2M)	520	52	120-140	0.22
	560	91	90-100	0.36

1. The temperature below which a sharp drop in toughness occurs is called the critical brittleness temperature T_{cr}.
2. Until this version, K_{Icr} could be evaluated using this test method as well as by using Test Method E399. To avoid duplication, the evaluation of K_{Icr} has been removed from this test method and the user is referred to Test Method E399.

Table 3.3 Fracture toughness of several heat-resistant alloys

Material	T	δ_{cr}	K_{Icr}	$K_{I\delta}$	a_{pl}	a_y	J_{Icr}
	°C	mm	MPa m$^{1/2}$	MPa m$^{1/2}$	KJ m^2 10^{-2}	KJ m^2 10^{-2}	KJ m^2 10^{-2}
EI612	600	0.54	194	52	13.4	6.7	20.1
	650	0.18	110	47	6.4	4.9	11.3
EI827	700	0.19	132	51	1.5	3.9	6.4
EI929	700	0.10	110	62	0	2.3	2.3
ZhS6K	700	0.12	145	72	0	8.2	8.2
EI607A	500	1.50	-	-	105	-	-
	650	0.40	-	-	7	-	-

In Table 3.3 the index cr means critical, $K_{I\delta}$ is the conditional stress intensity factor during crack propagation by separation determined by the formulas of elasticity theory without correction for the plastic zone at the crack tip.

Table 3.4 Influence of concentrations on brittle strength of disks from steels I (T_{cr} = 4°C) and II (T_{cr} = 135°C)

Disk	K_{σ_B}, steel I	K_{σ_B}, steel II	$K^{I}_{\sigma_B}/K^{II}_{\sigma_B}$
full	1	1	1
with a hole	1	0.72	1.39
notched	0.57	0.29	1.97

gas turbines, impellers of centrifugal compressors. The question arises when such stresses can be tolerated depending on the material and the stiffness of the stress state. The widespread methods of calculating the safety margins in terms of average stresses and bearing capacity give, as a rule, a positive answer to this question, since usually the zones in the details where the yield strength is exceeded are small (Demyanushko and Birger, 1978). Nevertheless, in relation to problems of very long service life, this issue requires additional consideration.

3.1.3 Bearing Capacity of Structures

The question of methods for determining the bearing capacity of structures under plastic flow conditions deserves special consideration. There are known methods for calculating the number of revolutions to failure of turbine disks, based on a complete redistribution of stresses with an increase in the number of revolutions (Demyanushko and Birger, 1978), using the results of calculations of elastic stresses. However, as shown by experimental data (Vasilchenko, 1962; Kutyrev, 2001), disks made of low-plastic materials and disks of complex microstructures fail at noticeably lower rotational speeds. This is due to their inability to completely redistribute stresses.

Carrying out elastic-plastic calculations with the FEM makes it possible to noticeably reduce the indicated difference between the calculated and experimental values of the failure speed. For example, in Gorokhov et al. (2004), to assess the critical rotation frequencies, various failure criteria were used

- for small strains

1. maximum principal stress criterion

$$\sigma_{1\,\max} = \sigma_{\rm B} \tag{3.14}$$

2. maximum principal residual strain criterion

$$\varepsilon^{\rm p}_{1\,\max} = \delta \tag{3.15}$$

- for large strains

3. maximum principal stress criterion

$$\sigma_{1\,\max} = \frac{\sigma_{\rm B}}{1 - \psi} \tag{3.16}$$

4. maximum principal strain criterion

$$\varepsilon^{\rm p}_{1\,\max} = \ln\left[1 + \delta + \frac{\sigma_{\rm B}}{E(1 - \psi)}\right] \tag{3.17}$$

Calculations of the critical rotational speeds of disks with holes made of plastic material using various material models and fracture criteria showed that the highest accuracy (up to 3%) is provided by calculations based on the theory of elastic-plastic flow with isotropic hardening for large deformations. For the disks under consideration, the use of the formulas for the limiting equilibrium and the criterion for maximum stress intensities leads to significant errors in determining the rotation frequency (up to 30%).

A similar picture is observed for gas turbine blades in the absence of creep. FEM calculations in a number of cases give elastic stresses at some points of the blade, which significantly exceed the yield point. Elastic-plastic calculations show that these stresses are redistributed, remaining above the yield point, but below the ultimate strength (separation strength). Therefore, the old calculation methods based on the rod design of a blade retain their usefulness for blades made of plastic materials. The results of calculation by the finite element method have to be averaged over the section and related to the maximum temperature in the section.

Thus, the widespread use of the finite element method in the practice of calculating the stress-strain state of gas turbine parts makes it possible to increase their reliability during long-term operation. In this case, it is necessary to take into account basic features of the materials used and their characteristics such as separation resistance, shear resistance, isochronous creep diagrams and limiting characteristics of long-term plasticity. The use of the given dependencies of the nature of fracture on the stiffness of the stress state makes it possible to select the appropriate fracture criteria for assessing the strength of structures operating under conditions of a long service life.

3.2 Long-term Strength at Constant and Variable Temperatures and Stresses

The ability to resist long-term stress of a material is called long-term strength. Long-term strength decreases with increasing temperature and time. Therefore, the accepted criterion for the long-term failure of the material - the long-term strength $\sigma_{\mathrm{lt\,s}}$ - is a function of time and temperature. The long-term strength of a material is the maximum stress value that a material can withstand without failure for a given time at a given temperature. In this case, the material should be understood to mean the material of different heats and cross-sectional sizes, heat-treated in production conditions according to the regime regulated by the technical conditions, and having the grain sizes and mechanical properties corresponding to the technical conditions. This is due to the fact that the characteristics of the long-term strength of materials are greatly influenced by their microstructural state, for example, the presence or absence of work-hardening, grain size, the amount and size of the hardening phase, depending on the modes of technological heat treatment or long-term aging during operation, etc. These issues will be discussed separately in Chapt. 4. The difference in the characteristics of the long-term strength of the metal of different heats, as a rule, exceeds the difference in these characteristics for the metal of blanks of different sizes, therefore, the long-term strength is usually referred to the material regardless of the type and size of blanks (rolled sections, forging, stamping). This, of course, does not apply to the ultimate strength of forged and cast metal.

When determining the long term strength, a number of problems arise associated with the need to predict the values of $\sigma_{\mathrm{lt\,s}}$: for the time to fracture greater than that experimentally obtained when testing a given material at a given temperature and for temperatures at which tests were not carried out at all. Let us consider some aspects of solving these problems.

3.2.1 Time Dependency

A number of empirical formulas have been proposed to describe the dependence of σ_{lt} on time until fracture t_{f} of smooth samples at constant temperature

- power law

$$t_{\mathrm{f}} = A\sigma^{m_{\mathrm{c}}}, \quad (3.18)$$

- exponential

$$t_{\mathrm{f}} = A_1 \exp(-B\sigma), \quad (3.19)$$

- mixed type

$$t_f = A_2\sigma^{m_0}\exp(-\gamma\sigma), \tag{3.20}$$

- Krisch's formula (Krisch, 1962)

$$t_f = \exp\sqrt{C - D_1 \lg\sigma} \tag{3.21}$$

Here $A, m_c, A_1, B, A_2, m_0, \gamma, D_1, C$ are parameters. The indicated dependencies make it possible to describe with a sufficient degree of reliability the characteristics of the long-term strength of industrial alloys on the basis of a limited amount of experimental data (tests at several stress levels), which make it possible to determine the constants included in them.

The use of interpolation formulas (3.18) and (3.19) to determine the extrapolated values of long-term strength can in some cases lead to significant errors. The fact is that the long-term strength curves have changes in slope (Fig. 3.4), which for steels of the pearlite class are associated with a change in the nature of fracture (intergranular - with long test durations and intragranular - at small). In addition, the use of semilogarithmic coordinates (exponential dependence) usually gives overestimated values of σ_{lt}. In Chizhik and Petrenya (1978); Petrenya (1997); Naumenko and Kostenko (2009); Naumenko et al. (2011a) it is shown that mechanisms of microdamage, but not the nature of failure are responsible for the change in the slope of the curves.

To assess the ultimate long-term strength of a material, it is first of all necessary to determine the maximum possible scatter of test results for a given material and to establish a relationship between the time to fracture and stress suitable for extrapolation. The spread of data in some cases can reach ±40% in voltage. Methods of statistical processing of results are used to solve these problems. Studies carried out by various authors have established a number of patterns.

Processing of the results of tests for long-term strength of 250 specimens of one heat showed that the distribution of the logarithm of the time to failure at σ = const corresponds to the normal law in the range 1 . . . 99% and is limited by two sensitivity thresholds (Borodin and Stepanov, 1969). Therefore, to determine the characteristics of long-term strength at low and high fracture probabilities, it is proposed to use the

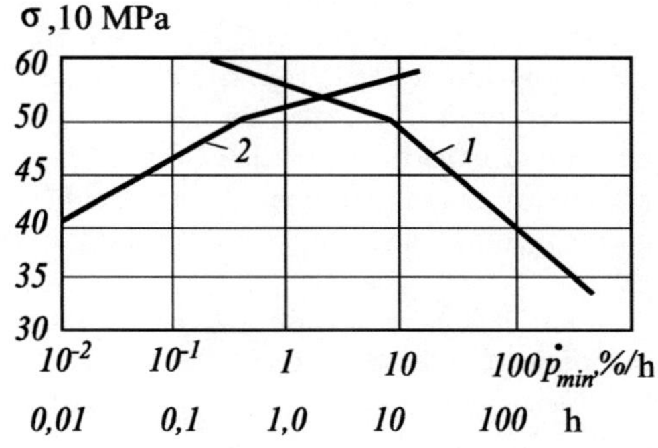

Fig. 3.4 Long-term strength (*1*) and creep resistance (*2*) of EI415 steel at T = 550°C

distribution function of the logarithm $\chi = (t_f - t_{fl})/(t_{fh} - t_f)$, where t_{fl} and t_{fh} are the lower and upper thresholds in time to failure, respectively.

The study of long-term strength, carried out on 20 heats of two materials, showed that the distribution of the logarithm of durability obeys the normal law not only for the metal of one heat, but also for the grade of material (Bulygin et al., 1977). With a test duration of at least 5000 h, the expansion of the confidence interval in the area extrapolated to 100000 h can be neglected. In the case of using samples of different sizes, the scatter band of test data for samples of several heats does not expand, but is located inside it.

A number of authors consider it expedient to exclude from consideration the results of tests with a duration of less than 100 h, regardless of the stress level. This approach, in our opinion, is incorrect, since if we exclude from the group of test results at relatively high stresses some of the results corresponding to $t_{f\,mean} < 100$ h, this will lead to an unjustified increase in $t_{f\,mean}$ and, consequently, to a change in the slope of the long-term strength curve as a result of which underestimated values of extrapolated values of ultimate strength will be obtained.

As an example, Fig. 3.5 shows the calculated mean lines of the orthogonal regression S_{orth} and the corresponding values of the least square deviation $S_{orth} = S_x S_y / \sqrt{S_x^2 + S_y^2}$ for the material of forging of discs of two austenitic steels at different temperatures. Here S_x and S_y are deviations, determined by the least squares method in relation to different positions of the axes, Fig. 3.6 and $r = \sqrt{b_x b_y}$ is correlation coefficient. The values of a_x, b_x, a_y, b_y are determined by methods of mathematical statistics.

It is seen that the level of long-term strength of both steels is approximately the same. The experimental points corresponding to long-term tests of metal forgings turned out to be close to the midline calculated from the results of tests of limited duration, which testifies in favor of the method of extrapolating the long-term strength curves (after fracture) by one order of magnitude in double logarithmic coordinates. With long-range extrapolation, the best agreement between the calculated and experimental values of $\sigma_{lt\,s}$ is observed in the case of using the Krisch dependence (3.21).

The use of the lower values of the confidence interval $\sigma_{lt\,s} - 3S_d$ as characteristics of long-term strength makes it possible to more reasonably approach the choice of the required safety factor for components operating at high temperatures. In a number of cases, this allows one to significantly reduce the value of the margin in comparison with the norms established for the case of using data averaging a relatively small number of experimental values. As an example, illustrating the influence of the probability of non-failure on the values of long-term strength, Table 3.5 shows the corresponding values for the cast alloy ZhS6U according to the data of Bulygin et al. (1977).

A special attention is placed by the question of the durability of materials under compressive load conditions. Many authors believe that damage does not accumulate under compression. However, a number of experimental data contradict this assumption.

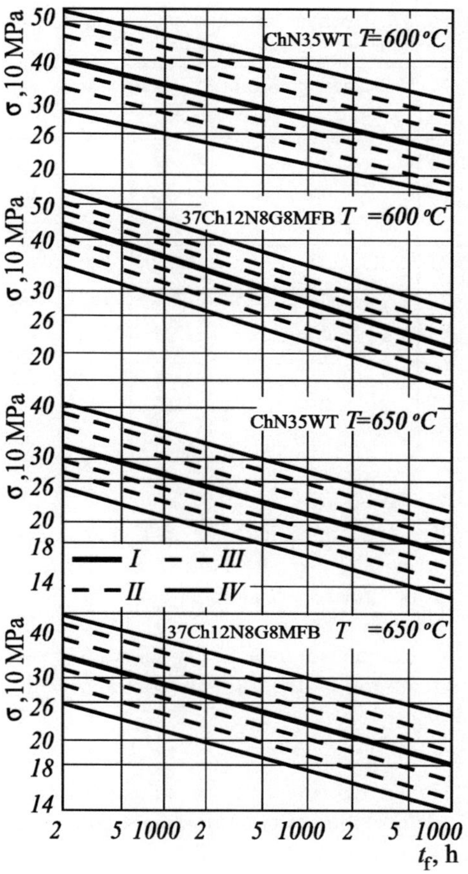

Fig. 3.5 Medium lines and scatter bands characteristics of long term strength of EI612 (ChN35WT) and EI481 (37Ch12N8G8MFW) steels at different temperatures:
I - medium lines;
II - lines corresponding to S_d;
III - lines corresponding to $\pm 2S_d$;
IV - lines corresponding to $\pm 3S_d$

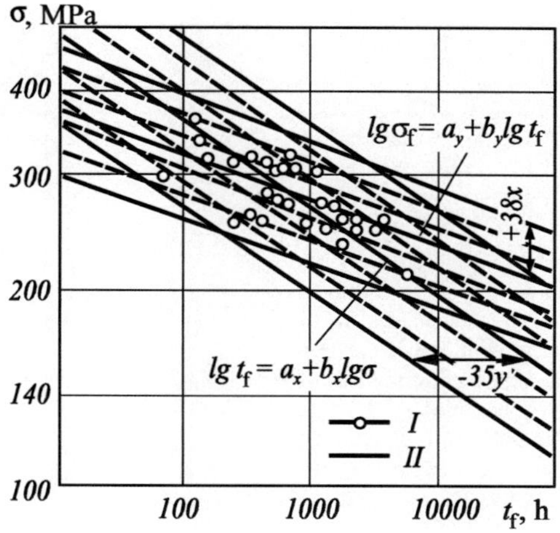

Fig. 3.6 Experimental (*I*) and calculated (*II*) values of long-term durability of smooth samples from forging EI612 steel disks at $T = 650°C$ ($r = 0.744$)

Table 3.5 Experimental values of the long-term strength of the ZhS6U alloy depending on the given probability of non-destruction

T,°C	Non-failure probability	Long-term strength, MPa			
		σ_{10}	σ_{100}	σ_{500}	σ_{1000}
900	0.5	470	370	280	240
	0.9	440	350	260	210
	0.99	425	330	230	190
1000	0.5	250	180	130	115
	0.9	225	160	120	105
	0.99	215	150	110	95
1050	0.5	180	120	75	58
	0.9	160	95	55	45
	0.99	140	85	45	36

1. By R.N. Sizova (Birger et al., 1975) it is found that holding in compression for from 0.83 to 2.3t_f at $|\sigma| = \sigma_0$ leads to a decrease in the time to fracture during subsequent tension under stress s_0 (Fig. 3.7). Here, t_f is the fracture time under tension. The authors consider the loading duration $t_{f\,c}$ that corresponds to the condition when the last application of tensile stress leads to almost instantaneous failure to be the ultimate compressive life. It was found that at the same stress $|\sigma_0|$
$$t_{f\,c} = (2.5\ldots3)t_f.$$
2. In the experiments of the author, according to the influence of preliminary plastic deformation of compression on long-term tensile strength, a noticeable decrease in durability was established (Table 3.6).

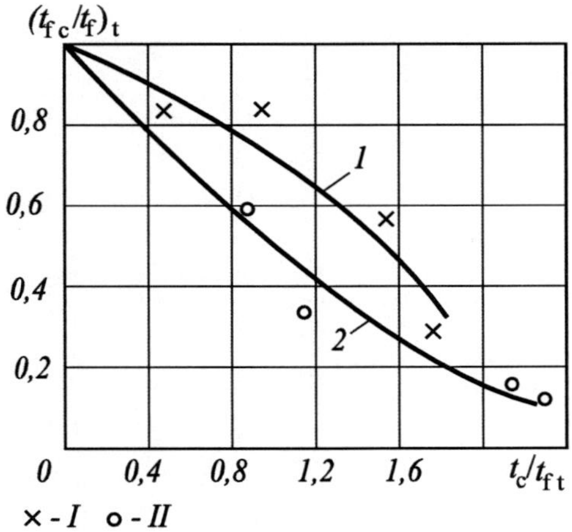

Fig. 3.7 Relative reduction of time to failure at tension depending on the relative duration of the preliminary compression action (T = 900°C): *1* - alloy ZhS6K (*I*), σ = 300 MPa, σ = 400 MPa; 2 - alloy EP109 (*II*), σ = 270 MPa, σ = 400 MPa

Table 3.6 Influence of compression prestrain on fracture time

Material	T,°C	ε_p, %	σ, MPa	t_f, h	$t_f/t_{f\,\mathrm{initial}}$
EI802	550	0	400	793	1
		-1.5		413	0.52
	550	0	320	32	1
		-2.4		24	0.75
		-3.0		18	0,56
		-5.0		22	0.69
EI765	700	0	600	19	1
		-1.05		7	0.37
		-2.5		6^{10}	0,32
	750	0	400	55^{45}	1
		-0.9		16	0.29
		-1.3		15^{20}	0.28
20Ch23N18	700	0	150	89	1
		-0.5		46	0.52
		-1.5		34	0.38
	750	0	130	44^{50}	1
		-3		11^{5}	0.25
	800	0	60	453	1
		-3.6		144	0.32
EI723	550	0	550	55	1
		-1.42		39^{20}	0.71

3. The features of damage accumulation during tension and compression are manifested (according to the data of Lepin and Garpinich) in different orientations of the formed cracks. The authors proposed the following expression

$$t_{f\,c} = t_f(0,7 - 1,7/\chi)^{-(\lambda+1)} \tag{3.22}$$

where $\chi = 1/\varepsilon_{\mathrm{limit}}$ and λ is the creep exponent in the creep constitutive equation

$$\dot{p} = K_m p^{-\lambda} \exp(\sigma/A) \tag{3.23}$$

Of great practical importance is the effect of the dimensions of components (samples) on the long-term strength. With a sample thickness comparable with the grain size, intergranular fractures can be observed at lower stresses than in thick samples (Nikitin and Taubina, 1965; Naumenko and Altenbach, 2016). For nickel-based alloys with the ratio of the grain size to the wall thickness of tubular specimens $d_3/\Delta_0 = 0.08\ldots0.5$, a decrease in durability according to the data of statistical processing of the results shown in Fig. 3.8, was only 30%. With a significant increase in the size of the samples, the total number of defects increases, which increases the degree of scatter in the values of time to fracture.

With very long service lives, the period of unsteady creep for damage accumulation can be neglected. The creep exponent under such conditions changes over time. In Chizhik and Petrenya (1978) the following criterion of failure is proposed

$$\{\varepsilon, \omega_1, \omega_2\} = [\varepsilon_{\lim} \cup 1] \tag{3.24}$$

where $\varepsilon, \omega_1, \omega_2$ are determined from Eqs. (2.18), $\varepsilon_{\text{limit}}$ is the ultimate plastic strain during creep under conditions of ductile fracture and *1* means one damage variable ω_1 or ω_2.

As an example in Shesterikov (1983) the case was considered when the micromechanism of the formation of wedge-shaped cracks predominates during prolonged fracture, and $\mu = B/A$ is taken as a small parameter. After integrating (2.18) and expanding them in a series in μ, the authors, using the criterion $\omega = A\omega_1^{\alpha} + B\omega_2^{\beta} = 1$ at $\alpha = \beta = 1$, obtained the following relation for the brittle fracture time with the interaction of creep micromechanisms

$$t_f = \frac{1}{A(n+1)b\sigma_1^n}\left(1 - \beta\frac{n+1}{2n-s+1}\frac{c}{b}\sigma_0^{s-n}\right). \tag{3.25}$$

Under such conditions, the dependence of long-term strength on time cannot be approximated by a straight line in logarithmic coordinates.

3.2.2 Influence of Temperature

The effect of temperature on the resistance of a material to long-term rupture has two effects: both the level of long-term strength and the slope of the curves change.

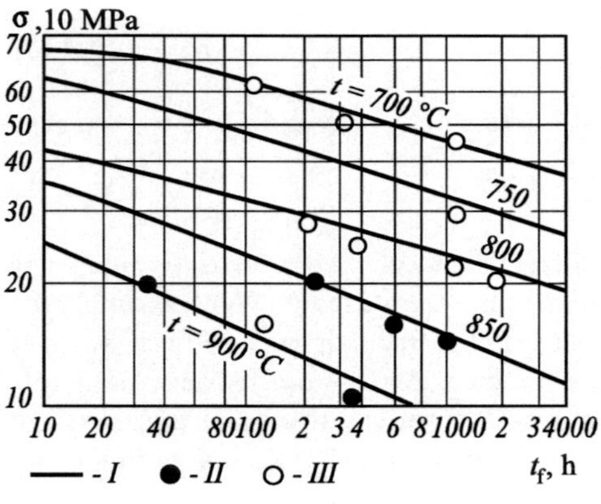

Fig. 3.8 Long-term strength of alloy EI826: *I* - cylindrical samples; II - tubular specimens with a wall thickness of $\Delta_0 = 1$ mm; *III* - the same, with a wall thickness $\Delta_0 = 0.5$ mm

As the test temperature increases, the slope of the curves and the exponent m_c in expression (3.18), as a rule, decreases. The temperature dependence of m_c is rather complex its nature is determined by the corresponding dependencies of the exponents in the expressions for the creep rate and plasticity. It was found that for a number of nickel-based alloys it has the form

$$m_c = a + \frac{b}{t} \tag{3.26}$$

Table 3.7 shows the values of the coefficients a and b in the temperature range for which the formula (3.26) is valid. To assess the long-term strength for time t_f at a certain temperature T A.B. Stanyukovich proposed the following empirical formula

$$\sigma_{lt\,s} = D \exp \frac{C}{T}, \tag{3.27}$$

where D and C are constants determined from two values of $\sigma_{lt\,s}$ at temperatures T_1 and T_2 for t_f = const. The analysis of experimental data on long-term strength tests of various materials at constant stresses (and not at constant load, as is usually accepted), allowed S.N. Zhurkov to propose the following formula connecting the time to fracture t_f, temperature T and stress σ

$$t_f = t_0 \exp\left(\frac{U_0 - \gamma_0 \sigma}{kT}\right), \tag{3.28}$$

where k is the Boltzmann constant, t_0, U_0, and γ_0 are material constants in a given microstructural state. t_0 is a value close to the period of thermal vibrations of atoms, U_0 is the heat of sublimation of atoms in the crystal lattice, γ_0 is the degree of decrease in the potential barrier U_0 by applied external stresses. The use of formula (3.28) for practical assessments of the long-term strength (related to tests under constant load) in some cases leads to significant errors both due to instability and heterogeneity of materials, on the one hand, and due to an actual increase in stress during testing of plastic materials in constant load conditions, on the other. In addition, the use of expression (3.28) at relatively low values of stresses gives underestimated results like the use of (3.19), and for $\sigma = 0$ gives $t_f \neq \infty$.

Long-term strength curves can be represented in parametric form

$$P = F(\sigma_{lt\,s}) = \Phi(T, t_f).$$

Table 3.7 Values of coefficients a and b in Eq. (3.26)

Alloy	Temperature range, °C	a	$b, 10^{-4}$
EI765	600-800	-43.6	5.1
EI826	700-900	-11.4	2.0
EI827	700-800	-13.8	2.1

Different authors propose different parameters P, which, with varying accuracy, as experiments have shown, for a given material are only a function of the stress σ_{lt}. The following parametric dependencies are widely used, connecting the initial stress σ, time to fracture and temperature

1. Larson-Miller parameter (Fig. 3.9a, Larson and Miller, 1952)

$$P_{\mathrm{L-M}} = F(\sigma) = T(C + \lg t_{\mathrm{f}}) \tag{3.29}$$

2. Orr-Sherby-Dorn parameter (Orr et al., 1954)

$$P_{\mathrm{D}} = F(\sigma) = \lg t_{\mathrm{f}} - B/T \tag{3.30}$$

3. Manson-Succop-Clauss parameter[6] (see, for example Manson and Succop, 1956; Clauss, 1960)

$$P_{\mathrm{M-S-C}} = F(\sigma) = \lg t_{\mathrm{r}} + AT \tag{3.31}$$

4. Manson-Haferd parameter (Fig. 3.9b, Manson and Haferd, 1953)

$$P_{\mathrm{M-H}} = F(\sigma) = \frac{T - T_a}{\lg(t_a - t_{\mathrm{f}})} \tag{3.32}$$

where A, B, C, T_α, and t_α are material constants. Often, the value of C is taken to be in a first approximation equal to 20. However, the value of C can be refined based on the results of processing experimental data. Similar to the Manson-Haferd parameter the following parameter was proposed in Murry (1963)

$$P_{\mathrm{M}} = F(\sigma) = \frac{\lg t_{\mathrm{f}} - p^*\sigma}{1/T - n^*}, \tag{3.33}$$

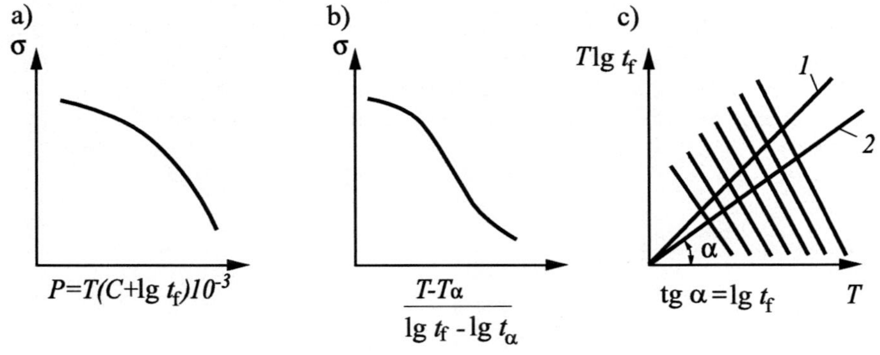

Fig. 3.9 Parametric dependencies of Larson - Miller (a), Manson - Haferd (b) and Nikitin (c): *1, 2* - lines equal to the time of fracture

[6] A parameter similar to (3.31) was proposed by Chateau and Duval.

where p^* and n^* are material constants.

For different stresses, dependencies (3.29)-(3.32) can be represented as straight lines in the form $\lg t_f = f(T)$ or $\lg t_f = f(1/T)$ shown in Fig. 3.10. The difference between the Manson-Haferd dependence and the Larson-Miller dependence can be seen from a comparison of Figs. 3.10a and 3.10d: the straight lines corresponding to σ = const, according to the Larson-Miller dependence, intersect at one point on the ordinate axis, namely at the point $1/T = 0$, and according to the Manson-Haferd dependence, at a point not lying on the axis.

The expediency of using one or another parametric dependence for a specific material can be established by comparing the curves obtained from experimental data with the curves shown in Fig. 3.10. It should be noted that, with the exception of the Manson-Haferd dependence, all parametric equations can be derived from the unified theoretical Arrhenius equation for the rate of a physical process

$$\nu = \nu_0 \exp\left(\frac{-U_0}{RT}\right), \tag{3.34}$$

where U_0 is the activation energy. For creep processes $\nu = \dot{p}$, and U_0 is the creep activation energy. If we assume that the creep deformation accumulates at a constant rate, and the elongation at rupture does not depend on time, stress, and temperature, then from Eq. (3.34) we obtain an exponential dependence of Zhurkov and Sanfirova (1955)

$$t_f = t_0 \exp\left(\frac{U_0}{RT}\right), \quad \lg t_f = \frac{U_0}{2.3RT} - C \tag{3.35}$$

Setting in Eq. (3.35) $U_0 = f(\sigma)$, and C = const, we obtain the Larson-Miller parameter. In the case U_0 = const, $C = f(\sigma)$, the Dorn parameter can be obtained.

Expressing temperature in degrees Celsius and decomposing

$$\frac{U_0}{RT} = \frac{U}{R273(1 + T/273)}$$

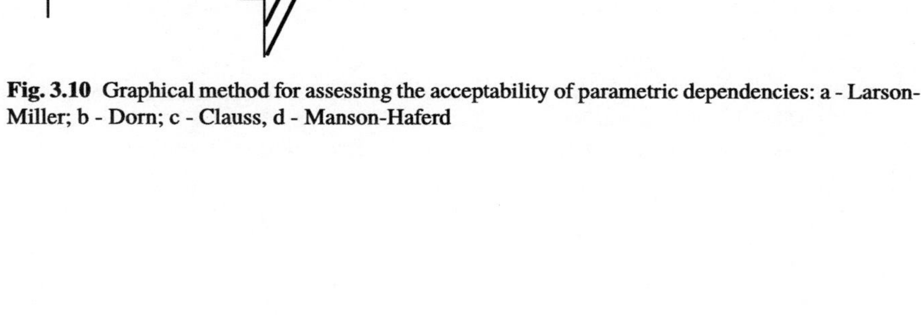

Fig. 3.10 Graphical method for assessing the acceptability of parametric dependencies: a - Larson-Miller; b - Dorn; c - Clauss, d - Manson-Haferd

in a series in $T/273$, we obtain, by keeping the first term, the following relation

$$\frac{U_0}{2,3\,R\,273} = \lg t_\mathrm{f} + C + T\frac{C + \lg t_\mathrm{f}}{273}$$

If we accept that U_0 = const, $(C + \lg t_\mathrm{f})/273$ = const = A, then, transferring C to the left-hand side, we obtain the parametric dependence (3.31). A large number of works have been devoted to checking the validity of parametric dependencies[7], some used statistical methods. A number of authors indicate a change in the slope of the parametric Larson-Miller curve with a change in the nature of fracture and a change in the value of C in the case of data processing for work-hardened metal.

Based on the processing of a large number of experimental data, Nikitin (1959) proposed the following refinement of the Larson-Miller parameter. He assumed that C is not constant, but depends on the material and, in some cases, on the stress

$$C = \varphi' - \gamma' \ln \sigma,$$

where φ' and γ' are constants for a given material. In this case, the use of the dependence of $\lg t_\mathrm{r}$ on T (Fig. 3.9c) for test results at σ = const makes it possible to find the values of C as the tg of the slope of the corresponding straight lines, $C = \mathrm{tg}\beta$ (Fig. 3.11).

As an example, Fig. 3.12b shows the parametric curve of long-term strength of the IN738LC alloy, which is widespread for blades of foreign gas turbine units. Rtishchev et al. (1986) determined the dependencies of the parameter P (according to OST 108.901.102-78) on stress and temperature for a number of domestic heat-resistant alloys. So, for alloys CNK-7 and ZMI3U, the parametric dependence of the long-term strength is approximated by the following expressions:

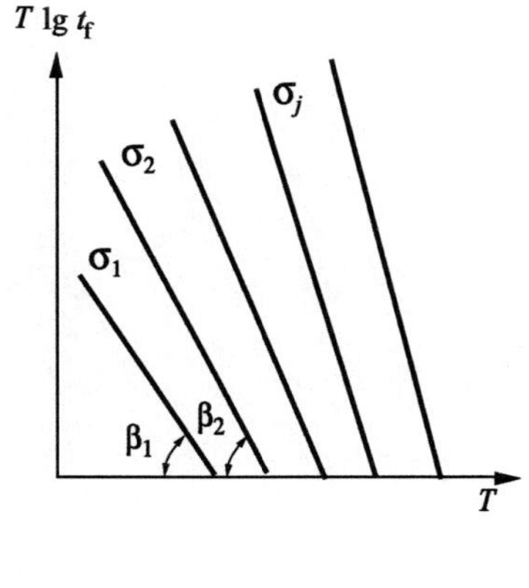

Fig. 3.11 Graphical method for assessing the acceptability of Nikitin's parametric dependence. The straight lines correspond to σ_j = const, $\mathrm{tg}\beta_j = C_j$

[7] It should be noted that all the specific data available in the literature on parametric dependencies for various materials are given for the engineering stress $\sigma = F/A_0$ (F is the load and A_0 is the initial cross-sectional area), i.e. do not take into account changes in the true stress due to a decrease in the section during creep.

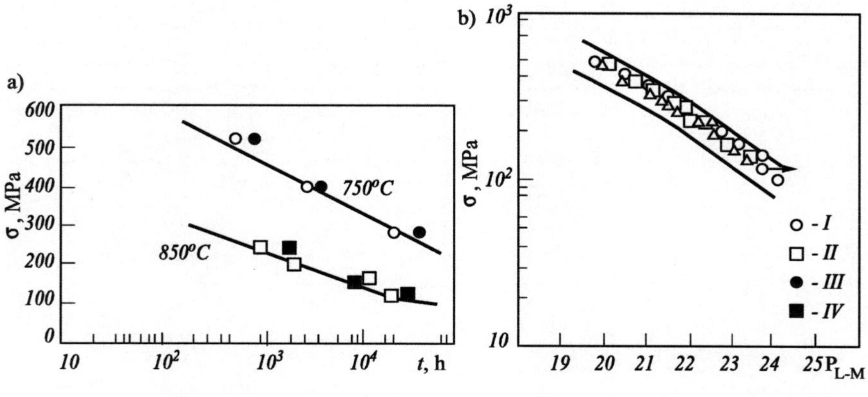

Fig. 3.12 Long-term strength curves (a) and parametric dependence ($C = 20$), (b) IN738LC alloy: *I, II* - uncoated; *III, IV* - coated

- for alloy CNK-7

$$P = T10^{-3}(\lg t - 2\lg T + 25.938) = 30.679 - 2.4\lg\sigma - 0.060669\sigma,$$
$$P = [\lg t - 211100/T + 22.71]^{1.5} = (\lg 72 - \lg\sigma)/0.05598,$$

- and for alloy ZMI3U

$$P = T10^{-3}(\lg t - 2\lg T + 26.69) = 31.43 - 2.4\lg\sigma - 0.06321\sigma.$$

From the incorrectness inherent in Eq. (3.28) for $\sigma = 0$, Trunin (1976) proposed the formula

$$t_f = A_2 T^l \sigma^{-\beta} \exp(U_0 - c_1\sigma), \tag{3.36}$$

which is a generalization of formulae of S.N. Zhurkov (3.28) and B.Ya. Pines[8]. In the case of determining the coefficients of formula (3.36) by methods of statistical processing of test results, it allows a high degree of accuracy to describe the long-term strength of the material in a wide range of temperatures and times (up to 100000 h). So, for example, from Fig. 3.13 it can be seen that Eq. (3.36) allows us to reliably predict the values of durability for a resource up to 10^4 h according to data up to 10^3 h (black dots), as well as outside (at 800 and 1100°C) used to determine the coefficients of the temperature range 900-1050°C.

In OST 108.901.102-78, to determine the extrapolated values of the creep rupture strength σ_{lt} according to the data of arbitrary tests at two temperatures, it is recommended to use the formula

$$\lg\sigma_{\text{lt s}}^{(T)} = a_1^{(T_2)} + b_1^{(T_2)}[\lg(t/\gamma_{\text{lt s}})]^{3/2}. \tag{3.37}$$

[8] Formula (3.36) is used in OST 108.901.102-78 to determine the conventional limits of long-term strength according to the data of processing the results of tests. In this case, the coefficient β is taken equal 2.

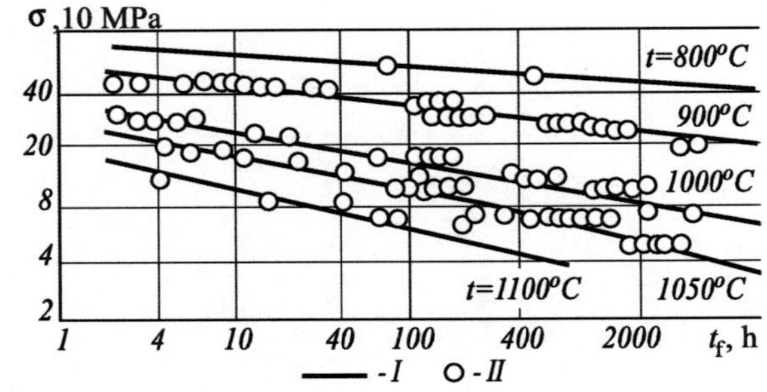

Fig. 3.13 Long-term strength curves calculated according to Eq. (3.36) (*I*) and results of tests (*II*) for ZhS6U alloy

where $\lg \gamma_{lt} = \lg t_{max}^{(T_1)} - [(\lg \sigma_{min}^{(T_1)} - a_1^{(T_2)})/b_1^{(T_2)}]^{3/2}$ is the coefficient of extrapolation. Here $\lg t_{max}^{(T_1)}$ is the maximum time to rupture in tests at temperature T and $\sigma_{min}^{(T_1)}$ is the minimum test stress at temperature T_1.

In the case when noticeable microstructural changes occur at a particular temperature (for example, dissolution of the hardening phase), the use of parametric dependencies leads to errors in determining σ_{lts}, reaching 30-40%. Certain advantages in this case are possessed by the method of generalized diagrams proposed by V.I. Kovpak (Shesterikov, 1983). The method is based on the assumption that the long-term strength diagrams are geometrically similar in several (depending on the nature of one or another micromechanism of creep failure) time intervals. The author confirmed the validity of this assumption by studying the structure and properties of materials under conditions of equivalent temperature-force loading. Similarity conditions for each interval are determined by the relations

$$[\sigma'(T_n) - \sigma_0]/[\sigma'(T_m) - \sigma_0] = [\sigma''(T_n) - \sigma_0]/[\sigma''(T_m) - \sigma_0] = \text{const}, \quad (3.38)$$

where σ_0 is the ordinate of the pole point for a given time interval (Fig. 3.14). To determine the position of the poles, numerical methods are used to process the results of tests of limited duration. Using the generalized diagram method, it is possible to reliably determine the extrapolated values of creep rupture using data of limited duration at higher temperatures. Extrapolated creep rupture strength values determined using different methods can differ significantly from each other. This, in particular, is evidenced by the calculation results for $\sigma_{5 \cdot 10^4}$ obtained using the results

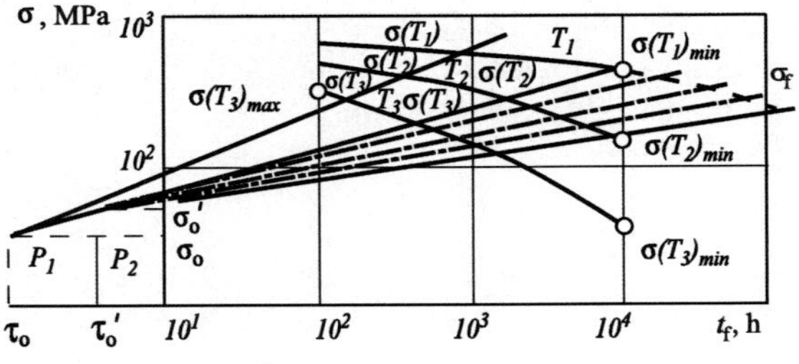

Fig. 3.14 Scheme for predicting long-term strength by the method of generalized diagrams at two poles

of tests of the EI765 and EI826 alloys at several temperatures with a maximum duration of 2500 h, carried out on a metal of increased homogeneity, Table 3.8. The reason for this is not only different methods, but also different input information used in each of them. In Shalin et al. (1981); Kablov and Golubovsky (1998), it is recommended that the minimum characteristics of the time to fracture of heat-resistant alloys at arbitrary temperatures can be determined by formula (3.36) for the given confidence probability β (usually $\beta = 0.9$) and the probability of non-fracture l (usually $l = 0.8$). Calculations are carried out according to the formula

$$\ln t_{\mathrm{f\,min}} = \ln t_{\mathrm{f\,m}} - S_{\mathrm{f}}(\ln t_{\mathrm{f}}) \left[\frac{t_{\beta,n-1}}{\sqrt{n}} + 0.8798 \left(1 + \frac{\sqrt{n-1}}{\chi^2} \right) \right], \qquad (3.39)$$

where $S_{\mathrm{f}}(\ln t_{\mathrm{f}})$ is the variance of the logarithm of the time to failure, $t_{\beta,n-1}$ is the tabular value of the Student's criterion, χ^2 is the tabular value of Pearson's criterion and n is the number of tests.

The above methods of temperature-time correspondence of long-term strength values are advisable to use when assigning control test modes for materials regulated by technical conditions. The duration of such tests is 50... 500 h. They are carried out at temperatures exceeding the operating ones, but not causing new processes of microstructural changes in time. In addition to temperature and time, long-term strength is influenced by the deformation capacity of the material, the nature of the stress state, the size of components (samples) and the presence of stress concentrators in them.

3.2.3 Damage Mechanisms and Deformation Capacity of Materials During Creep

Creep processes in metals may not develop infinitely. When considering models of long-term fracture, a distinction is made between brittle fracture, which develops under conditions of small deformations, non-brittle fracture, which occurs due to the vanishing of the continuity, and, finally, ductile fracture, realized according to

Table 3.8 Results of calculating average values of σ_{lt}

Material	T, °C	$\sigma_{5\cdot10^4}$, 10 MPa, computed by the methods of			
		linear extrapolation	I.I. Trunin OST 108.907.102-78	V.K. Adamovich OST 108.901.102-78	V.I. Kovpak
EI765	600	49	53	50.1	45
	700	22	20.5	21.7	24
EI826	600	65	56	53.1	50
	700	29.5	29.4	26.7	28

the Hoff's scheme (Hoff, 1953), when the cross-sectional area becomes zero due to creep before the continuity is exhausted. The ultimate deformation that a material can withstand without fracture under these test conditions is called its deformation capacity or plastic resource. The characteristics of the deformation capacity are the so-called values of long-term plasticity: relative elongation $\delta_f = \delta_{t\,pl} - \delta_0$, where $\delta_{t\,pl}$ is the total irreversible plastic strain and δ_0 is the plastic strain during loading, and relative contraction ψ, accumulated due to creep at the time of fracture. These characteristics are very sensitive to the temperature and duration of test[9]. For a material tested at constant temperature, as the time to fracture increases, the ductility at rupture, as a rule, decreases, Fig. 3.15. In Stanyukovich (1967) the following relationship between the average creep rate $\dot{p}_{avr} = (\delta_p - \delta_0)/t$ and stress

$$\dot{p}_{avr} = B_1 \sigma^{n_1} \tag{3.40}$$

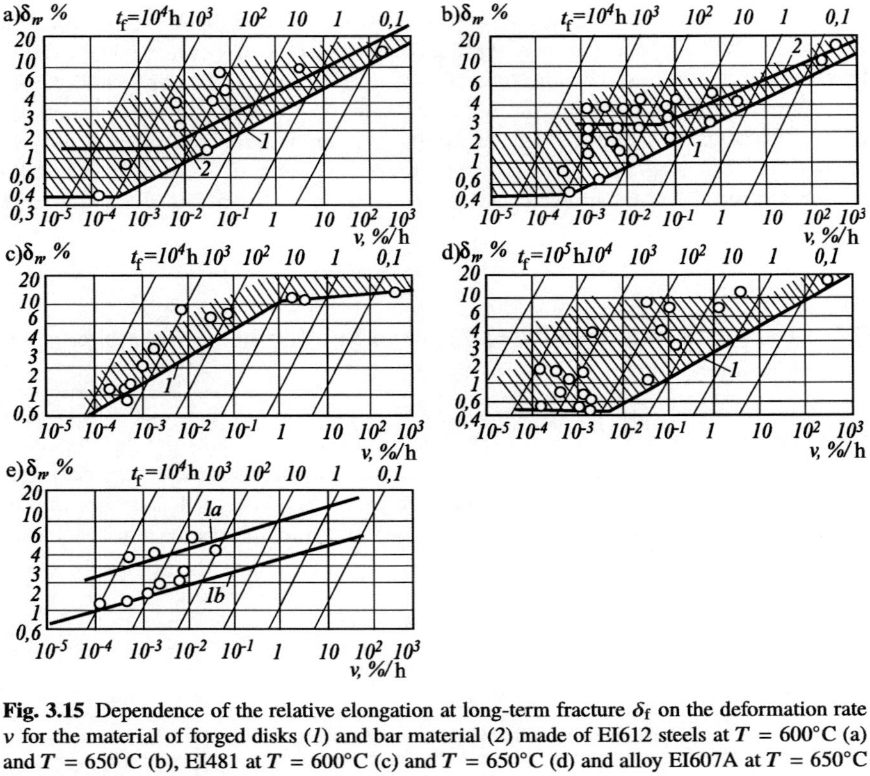

Fig. 3.15 Dependence of the relative elongation at long-term fracture δ_f on the deformation rate v for the material of forged disks (*1*) and bar material (2) made of EI612 steels at $T = 600°C$ (a) and $T = 650°C$ (b), EI481 at $T = 600°C$ (c) and $T = 650°C$ (d) and alloy EI607A at $T = 650°C$ (straight line *1a*) and 700°C (straight line *1b*) (e)

[9] Detailed studies of the laws governing the plasticity of heat-resistant materials during creep were carried out by Stanyukovich (1967).

which is similar to the minimum creep rate vs. stress dependence

$$\dot{p}_{\min} = A_1 \sigma^{n_c}, \tag{3.41}$$

where $n_1 \approx n_c$.

At low creep rates and stresses, long-term ductility either does not change, or in some cases even increases, which is caused, as a rule, by microstructural changes occurring in the metal during testing. This increase in ductility in long-term experiments is especially clearly observed for pearlitic steels. To predict the characteristics of the deformation capacity (Trunin, 1976) proposed the following equation

$$\delta_f = A_3 \sigma^{-k_0} \exp[(U_0 - \gamma\sigma)/(RT)], \tag{3.42}$$

So, for example, for the ZhS6U alloy, Eq. (3.42), according to test data with a duration of up to 1000 h, has the form

$$\delta_f = \exp(-12.01)\sigma \exp[(13848 - 7.3\sigma)/T]/10,$$

where δ_f is in %.

Since the dependence of the plasticity of materials on the average creep rate manifests itself quite clearly, in connection with the difficulties encountered in some cases in separating plastic strains and creep strains, a method for assessing the deformation capacity of materials using tests at constant strain rates is used (Stanyukovich, 1967). This method, possessing high efficiency, made it possible to determine the plasticity characteristics of many heat-resistant materials, their dependence on temperature, heat treatment mode, etc. It was found that the temperature dependence of the δ_f value has a minimum, and the $\delta_{f\,\min}$ values change with a change in the strain rate (Fig. 3.16). It should be noted that the plasticity characteristics determined in experiments on long-term strength and in experiments with constant deformation rates are the same. The application of the test method with constant strain rates is especially advisable in a comparative study of the influence of the technology for manufacturing samples by varying the melting technology, forging, heat treatment mode, grain size, etc. It has been established that when the values of long-term and short-term plasticity change for the same material, the temperature which corresponds $\delta_{p\,\min}$ does not change.

When analyzing the process of long-term fracture of metals, the periods of initiation and development of defects (pores, cracks) are usually distinguished. However, now there is no definite point of view regarding the critical dimensions of cracks and the patterns of damage accumulation in materials of various classes under conditions of prolonged static loading. For heat-resistant steels from Eq. (2.18), dependencies for the values of ε_f on stress were obtained: ε_f = const, $\varepsilon_f = A\sigma^{m-n}$ and $\varepsilon_f = A\sigma^{m-k}$, the first of which refers to transcrystalline fracture ($m = n$), the second - to fracture due to damage along the grain boundaries ($m > n$) and the third - to fracture due to pore formation. The ratio between m and k depends on the steel grade (Balina and Myadekshas, 1994).

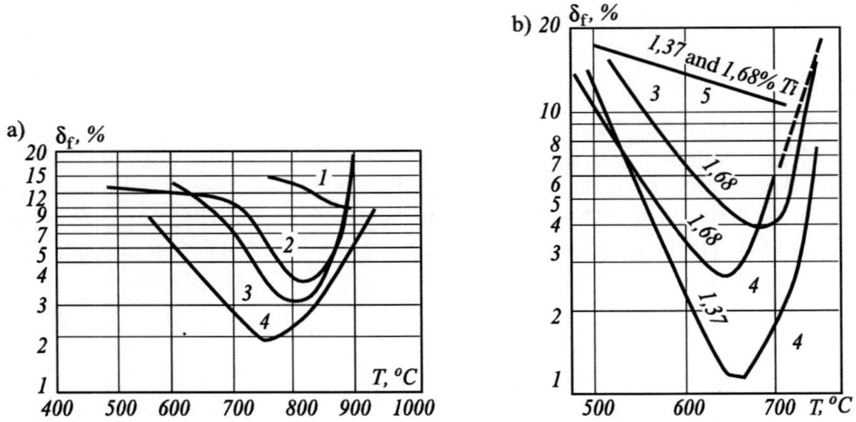

Fig. 3.16 Temperature dependence of the relative elongation at fracture δ_f in tests with constant strain rates of EI481 (a) and EI612 steels with different titanium content (b) for different tensile rates v, %/h: *1* - 314; *2* - 3.6; *3* - 0.8; *4* - 0.08; *5* - 180

To describe the effect on the deformation capacity under creep conditions of the material state and the degree of volumetric stress state of the part, it is advisable to use formulas similar to (3.12),

$$p^* = 1.7\varepsilon_f \exp(-1.5\sigma_m/\sigma_{vM}), \qquad p^* = \varepsilon_f K_e \sigma_{vM}^2/(3\sigma_1\sigma_m) \tag{3.43}$$

where ε_f is the ultimate creep strain under uniaxial loading, p^* is the ultimate equivalent creep strain (deformation capacity) in a complex stress state, K_e is a characteristic of the state of the material (in the brittle state $K_e = 1$ and in the ductile state $K_e = 1.2$). These formulas, like (3.12), give a conservative estimate of the ultimate deformation.

The issue of local stresses of structures operating under creep conditions deserves a separate consideration. If we carry out calculations of the redistribution of stresses in the structure due to creep, then we obtain stress relaxation curves, for comparison of which with the ultimate strength at the operating temperature T, it is necessary to calculate the equivalent stress

$$\sigma_{eq} = \left(\frac{1}{t}\int \sigma^m \mathrm{d}t\right)^{1/m} \tag{3.44}$$

where m is the slope of the long-term strength curve.

In this case, we usually obtain the maximum values of σ_{eq} greater than the average stress in the structure, obtained on the basis of the assumption of a complete redistribution of stresses. However, the value of σ_{eq} highly depends on which creep model is used for the calculation. So, assuming a constant creep rate at constant stress and Norton's law, we obtain noticeably larger values of σ_{eq} (smaller values of safety margins) compared to the results of calculations using the theory of flow (or hardening) taking into account the primary stage of creep. Therefore, performing calculations of stress redistribution due to creep using adequate models makes it possible to increase the reliability of assessing the strength of a structure and to reduce the safety factors adopted on the basis of a complete redistribution of stresses (for example, in calculating stresses in a turbine blade according to a model of rod). Unfortunately, in the domestic reference literature for the majority of industrially used materials, only stress dependences of the minimum creep rates are given (isochronous creep curves are known only for a limited number of materials and a small number of temperatures).

3.2.4 Fracture Under Stress Relaxation Conditions

It is usually assumed that the redistribution of stresses in structures due to creep (stress relaxation) cannot lead to their failure. The theoretical possibility of crack formation under conditions of stress relaxation for materials with low deformation capacity under certain conditions was considered in Borzdyka and Getsov (1978). As an example, Fig. 3.17 shows the results of tests for relaxation of the EI481 austenitic steel in a low-plastic state. There are also known cases of failure of fasteners under conditions of repeated pulling, leading to the exhaustion of long-term plasticity under conditions of stress relaxation. In this case, one should also bear in mind the decrease in the deformation capacity of the material under conditions of a complex stress state (Getsov et al., 2009a). And finally, the third case is the failure of fasteners made of a material that undergoes structural changes during operation, accompanied by an increase in volume. Under these conditions, the initial stress may not only not decrease, but, on the contrary, increase, leading to failure. There are also several examples of the failure of power plant components under conditions of stress relaxation.

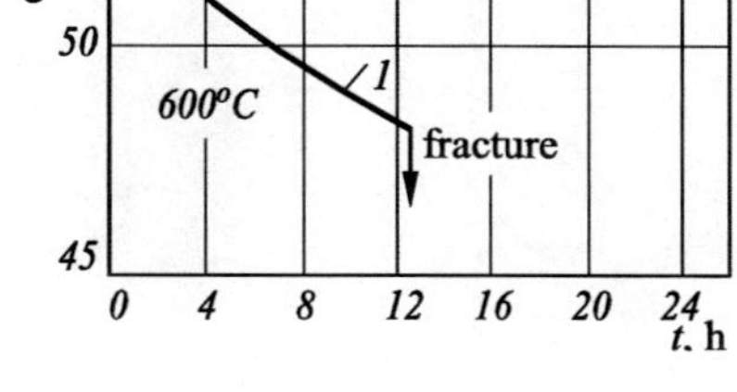

Fig. 3.17 Relaxation curves for 37Ch12N8G8MFB (EI481) steel in a low-plastic state at $\sigma_0 = 600$ MPa: *1* - first loading; *2* - second loading

The first such case is the formation of cracks in the area of the hole for the circular damper connection of the blades of the old design of the low-pressure gas turbine GT-100 (Figs. 1.10 and 3.18), made of the heat-resistant alloy EI893. The blade temperature was 650°C. The calculated coefficient of stress concentration in the blade at the hole in point A was 4.3 ($\alpha = 30°$). It is known that in the course of long holdings of an alloy thermally treated at low temperatures of quenching and aging, at 600-700°C, a finely dispersed γ'-phase precipitates in it, accompanied by an increase in hardness and a decrease in deformation capacity. Tests for long-term strength of this alloy at 650°C revealed a multiple decrease in the time to fracture of samples with an annular notch in comparison with smooth samples. Thus, at a stress of 490 MPa, the time to fracture of smooth specimens was 348 h, and of specimens with an annular notch, 34 h. Brittle fractures of GTU blades made of EI893 alloy were also observed in the zone of stress concentration on the lower flanges under conditions of long-term operation at temperatures of about 650°C.

Thus, it can be stated that during the operation of the blades with a hole for the damper bond, the long-term plasticity of the material under stress relaxation conditions was exhausted, which entailed the formation of cracks and failure of the blades. Failure of turbine blades under conditions of stress relaxation was also observed under conditions when the redistribution of stresses due to the geometric features of the section occurs to a much lesser extent than in tests of specimens under uniaxial stress state. And, finally, the fracture during stress relaxation can conditionally include the cases of formation of thermal fatigue cracks under conditions when thermal stresses do not exceed the elastic limits at the maximum cycle temperature.

Another typical case of failure under stress relaxation conditions is the formation of cracks during tempering of welded impellers of centrifugal compressors made of chromium-nickel-molybdenum steel with $\sigma_{0.2}$ more than 1000 MPa. A typical feature of these designs is the welding of relatively thin wheel blades to more massive outer and inner rims with fillet welds. During the heating process, significant thermal stresses are generated in the wheels. To find the causes of cracking, the characteristics of the short-term creep of the welded joint were determined on a time base of 7 h, corresponding to the tempering time. The tests were carried out at three temperatures

Fig. 3.18 Crack formation in gas turbine blades under stress relaxation conditions

(400, 590, and 620°C) under conditions of constant tensile load (Fig. 3.19 and Table 3.9). Deformations were measured during the first hour every 5 min, then every hour.

For all specimens, the fracture was localized in the material of the welded joint - weld metal or metal at the fusion boundary (exact localization is difficult due to the multiplicity of fracture centers and an uneven fracture surface). The test results showed that on the accepted test base at 400°C, no creep phenomena at stresses below 600 MPa were revealed, all the samples did not fail under test conditions. In tests at 590°C, it was found that the long-term strength on the basis of 7 h (σ_7) is 200 MPa, and in tests at 620°C σ_7 = 140 MPa. In this case, it should be noted the extremely low deformation capacity of the material during fracture, reaching for a number of samples a value of the order of 1% even without taking into account the localization of deformations. To determine the mechanism of this failure, metallographic studies were carried out. It was found that at elevated temperatures, in contrast to tests at temperatures close to room temperature, an intergranular fracture is observed regardless of the location of fracture. Failures of fasteners under relaxation conditions can also be observed with a decrease in the long-term plasticity of the material under conditions of long-term operation. Thus, the considered examples of the failure of power plant components (fasteners, blades, impellers of a centrifugal compressor) allow us to state that fractures under stress relaxation conditions are a fairly common case of static failure of such structures, which manifests itself mainly at a low deformation capacity of the material.

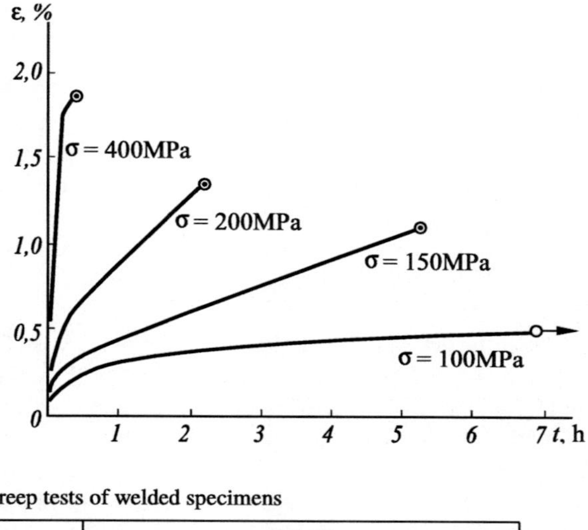

Fig. 3.19 Creep curves of material for welded joints made of chromium-nickel-molybdenum steel at T = 620°C. Points ⊙ correspond to fracture

Table 3.9 Results of short-term creep tests of welded specimens

T, °C	590 °C				620 °C				
σ, MPa	300	250	220	200	400	300	200	150	100
time to rupture	3.5 h	1.5 h	7h, no fracture		12 min	5 min	2 h 50 min	5.5 h	7h, no fracture
δ_f, %	3.0	1.4	0.33	0.26	6.0	2.0	2.7	1.3	0.16

3.2.4.1 Fracture Models Under Stress Relaxation Conditions

The features of the damage development in conditions of stress relaxation are

- low deformation capacity of the material
- localization of the area of occurrence of creep processes
- the development in time of damage processes in the material, which is manifested in the time dependence of strength
- the presence of a stored elastic energy in the component (in the loading system)

The theoretical possibility of crack formation during stress relaxation for materials with low deformation capacity was considered in Birger (1977).

In Lanin (1983); Zemzin et al. (1984); Chizhik and Lanin (1986), the laws of crack propagation under conditions of stress relaxation were established. The greatest number of fractures under stress relaxation conditions with the formation of technological cracks occurs in welded joints of power equipment at the manufacturing stages. Technological cracks in welded structures refer to brittle fractures without traces of visible plastic deformation and have an intergranular character. Hot cracks and cracks during heat treatment of welded structures along the entire length of the grain boundaries. With the development of cold cracks in the focus, the fracture proceeds along the grain boundaries, then it either persists or turns into mixed and intragranular failure (Zemzin et al., 1984). Cracks occur without the application of external loads due to changes in the internal energy of the welded structure. Hot cracks (HC) develop under the action of increasing stresses during welding, while cold cracks (CC) and cracks during heat treatment (CHT) are formed in the process of relaxation (reduction) of residual stresses after welding and during heat treatment (tempering), see Zemzin et al. (1984).

For all types of cracks under consideration, a time dependence of strength is observed. Similar to fractures during creep, there is an explicit dependence of plasticity on the rate of deformation under conditions of loading at increasing stresses, at a constant force, or at decreasing stresses (stress relaxation). In this case, a decrease in the rate of deformation and an increase in the duration of loading leads to a decrease in the fracture stress by 30-50% and the strain of fracture decreases by a factor of 5-10 (Lanin, 1983; Chizhik and Lanin, 1986; Altenbach et al., 2008; Lanin and Getsov, 2009).

The time dependence of strength and plasticity for the conditions of formation of HT is confirmed by many researchers. Similar dependencies are established for CC and CHT. The change in the strength and ductility of the joint, which is prone to the formation of CC and CHT, takes place in time only immediately after welding. The processes (unloading, recovery, etc.), causing the ordering of the microstructure, lead to the stabilization of the properties of the product material and to the emergence of a local stable equilibrium with a minimum of free energy. The release of excess free energy can be associated with the ordering of the microstructural state of the material due to diffusion processes, with the development of plastic deformation during creep and relaxation, as well as with the formation of technological cracks. Relaxation of system stresses under the conditions of one way or another of the release of free energy is due to specific physical processes of self-organization of the microstructural state. The higher the value of excess free energy, the more opportunities the system has for its release and the greater the probability of formation of local minima of free energy in the system. This regularity manifests itself in the occurrence of minima of strength, plasticity and crack resistance in different temperature ranges. Local minima are characterized by threshold values of metal properties.

Taking into account the above, we can point out the main phenomenological regularities of the formation of technological cracks

- the presence of structurally unstable zones with an excess level of free energy
- development of fracture in time under conditions of creep and stress relaxation with a decrease in plasticity and strength
- intergranular fracture with low deformation capacity of nonequilibrium zones of welded joints
- formation of CC and CHT cracks under the influence of relaxation of residual welding stresses

Physical micromechanisms of failure for different technological cracks are different. However, the kinetic features of these processes are described using a system of simple kinetic dependencies within the framework of the general phenomenological theory of creep and fracture. The use of the known kinetic relations of creep (stress relaxation) makes it possible to determine the time before the formation of cracks, their location in the welded joint, as well as to take into account the influence of the dimensions of the structure and the stored of elastic energy in it and the features of the stress state of the welded joint.

3.2.4.2 Methods for Predicting the Conditions of Fracture

Considering the issues of crack propagation under stress relaxation conditions, we introduce definitions to describe relaxation in bodies with cracks (Lanin and Getsov, 2009). The definition of stress relaxation as a process of unsteady creep, accepted in continuum mechanics, under the condition of constant displacements on a part of the body surface as applied to practical problems of fracture mechanics, is insufficient. Indeed, in the general case, one can consider three types of loading of bodies with cracks under conditions of stress relaxation (Fig. 3.20). The first type of loading corresponds to the model case of pure relaxation. In this case, by pure relaxation we

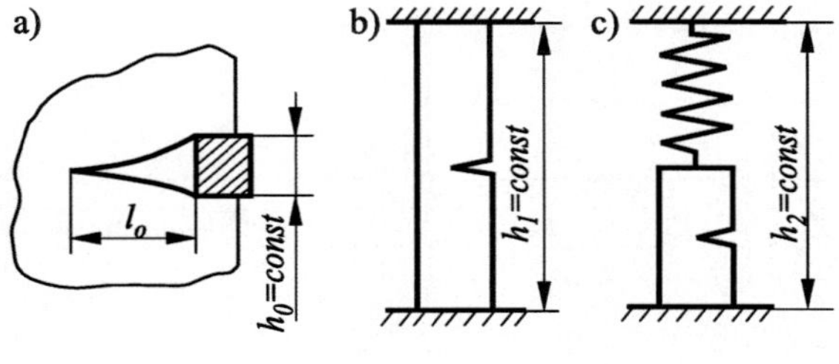

Fig. 3.20 Model representations of relaxation in bodies with cracks: a - "pure" relaxation; b - "pinched specimen" system; c - "specimen-machine" system

mean the relaxation of an infinite body with a crack when its edges are opened by an absolutely rigid thin wedge (Fig. 3.20a). In practice, this case occurs when cracks are formed near rigid inclusions. At the interfaces of the precipitated phases, during flock formation, etc., stress relaxation in real bodies with cracks should be attributed to the case of loading with a stored of elastic energy in the body, when loading is carried out by fixed displacements on a part of the body surface (or a specimen with a crack, Fig. 3.20b). We conventionally classify such loading as a clamped specimen system. Loading with constant specified displacements at the boundaries of the system (microstructure) and localization of fractures in separate zones will be considered as a system with an elastic branch and classified as a sample-machine (Fig. 3.20c).

Some cases of crack propagation in parts and structures in accordance with the classification of relaxation in bodies with cracks are shown in Fig. 3.21.

To describe the relaxation of the stress intensity factor K_{I}, one can apply the theory of heredity in the form proposed by Rabotnov (1969). In the case of pure relaxation, the value of K_{I} for a linearly elastic body is determined by the relation

$$K_{\mathrm{I}0} = \sqrt{\frac{2}{\pi}} \frac{E h_0}{\sqrt{l_0}}, \tag{3.45}$$

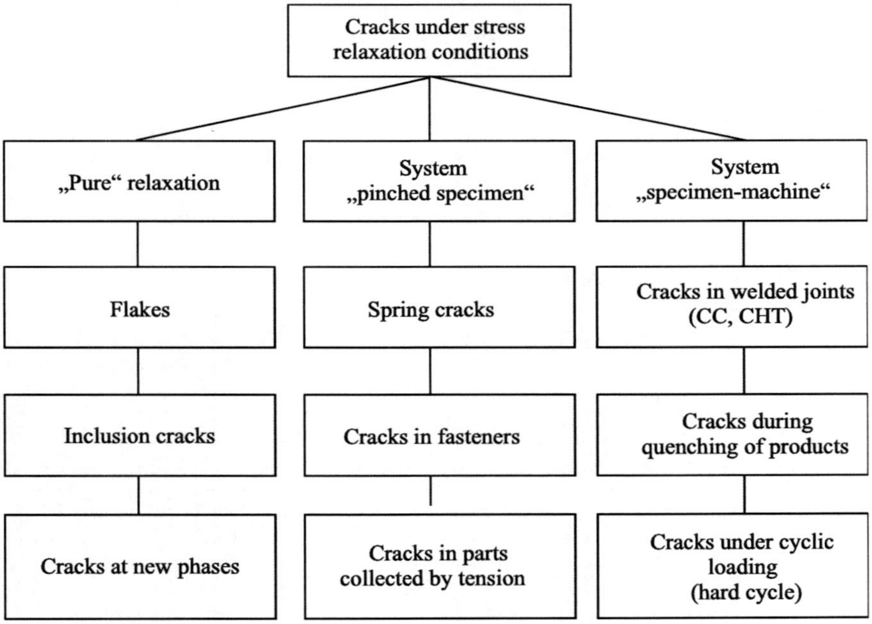

Fig. 3.21 Cracks that develop under stress relaxation conditions

where E is the Young's modulus, h_0 is the thickness of the wedge, and l_0 is the initial crack length. In accordance with the principle of Rabotnov (Getsov et al., 2004; Getsov and Turgenev, 2006), when describing relaxation processes in bodies with cracks, according to the theory of heredity, we replace the value of the Young's modulus with a hereditary Volterra-type operator

$$\tilde{E} = \frac{E}{1 + q\tilde{\Gamma}(t)}, \tag{3.46}$$

where

$$\tilde{\Gamma}(t) = \int_0^{\tau} \Gamma(t - s)$$

$\Gamma(t - s)$ is heredity and q is material constant.

From (3.45) and (3.46) we obtain

$$K_I(t) = \frac{K_{I0}}{1 + q\tilde{\Gamma}(t)}, \tag{3.47}$$

where $K_I(t)$ is the current value of the stress intensity factor, K_{I0} is the initial value of K_I (at $t = 0$).

According to Lanin and Getsov (2009), for relaxation with an elastic branch (in the "specimen-machine" system) we have

$$K_I(t) = \frac{K_{I0}}{1 + qm\tilde{\Gamma}(t)}, \tag{3.48}$$

where $m = (l + p)^{-l}$ and

$$p = \frac{U_1 - U_0}{U_0}. \tag{3.49}$$

Here U_1 is the elastic energy stored in a body (structure) with a crack, U_0 is the elastic energy stored in the body during loading under conditions of pure relaxation K_I.

Equation (3.48) is general. Thus, for $U_1 = U_0$, we have $m = 1$ and we obtain relation (3.48), which characterizes the case of pure relaxation in a body with a crack. As $U_0 \to \infty$, $m \to 0$ follows, and the case of an open (loadable) system is realized with $K_I(t) = K_{I0} = \text{const}$, i.e. creep loading conditions are realized. To describe the relaxation K_I, various types of kernels of the integral operator can be used (Table 3.10). The considered aftereffect kernels do not require complex mathematical calculations for the estimation of the relaxation curves of the stress intensity factor using Eqs. (3.47) and (3.48).

Comparison of the calculated values of the K_I relaxation curves for a number of different aftereffect kernels with the experimental values for hardened steel 38ChN3MFA ($V_{cool} = 20°C/s$, without subsequent recovery) showed their fairly good agreement (Fig. 3.22). For short test times, the best result is obtained by using kernels of power-law and fractional-exponential type. At long times, the choice of

Table 3.10 Variants of the description of the kernels of the integral operator

Type of kernel for $\tilde{E}$	Kernel equation
Maxwell kernel	$\Gamma(t-s)=\dfrac{G}{q\mu}=\lambda$ (3.50)
Exponential (standard Kelvin) kernel	$\Gamma(t-s)=\lambda\exp[-\beta(t-s)]$ (3.51)
Power law kernel	$\Gamma(t-s)=\dfrac{\lambda}{(t-s)^{1-\beta}}$ (3.52)
Fractional exponential kernel	$\Gamma(t-s)=\dfrac{\exp[-\beta(t-s)]}{t-s}\sum_{n=1}^{\infty}\dfrac{A\Gamma^n(\alpha)(t-s)^{\alpha n}}{\Gamma(\alpha n)}$ (3.53)

$G, q, \mu, \lambda, \beta, \alpha$ are material parameters.

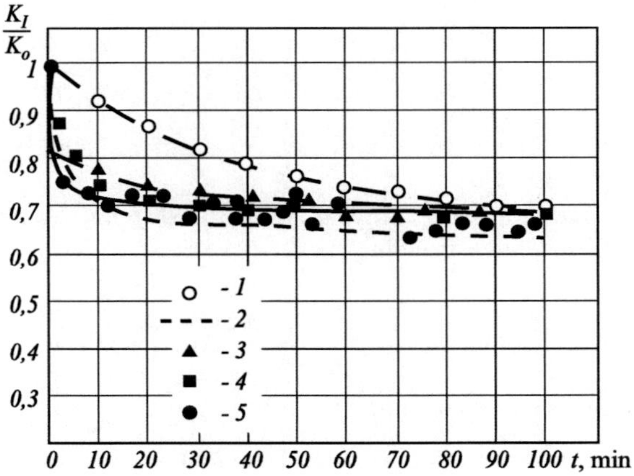

Fig. 3.22 Comparison of calculated and experimental relaxation curves of the stress intensity coefficient K_I: *1* - Maxwell kernel; *2* - power law kernel; *3* - exponential kernel; *4* - fractional-exponential; *5* - experimental results

the type of aftereffect kernel has practically no significant effect on the accuracy of determining the relaxation curves. To analyze the applied issues of brittle fracture when estimating the propagation conditions of creep cracks of relatively small sizes, the simplest operators can be used, and in some cases, under conditions of brittle intergranular fracture, the effect of a change in the crack length on stress relaxation can be neglected. In this case, the relaxation curves of the stress intensity factor K_I are determined based on the results of stress relaxation tests in bodies without cracks.

In Lanin (1983); Zemzin et al. (1984); Chizhik and Lanin (1986), a model of fracture of bodies with cracks during relaxation was proposed. According to this, materials that are not crack-resistant under stress relaxation conditions are characterized by time dependences of the threshold $K_{If\,th}$ and critical values of fracture toughness K_{Ic} (Fig. 3.23).

The fracture process can be divided into two stages. The first stage is the incubation period, which corresponds to the time of accumulation of damage in the material

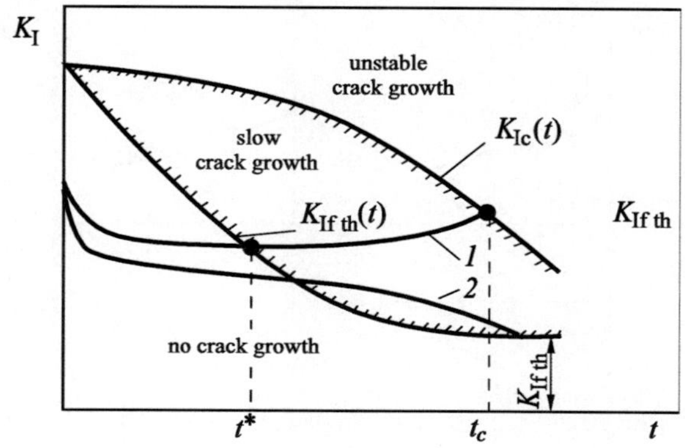

Fig. 3.23 Scheme of bodies fracture with cracks at relaxation K_I

under the action of nominal stresses. The second stage of crack propagation is slow growth under the action of local damage accumulation processes at the crack tip and rapid spontaneous fracture when global damage is reached at a distance from the crack tip. Local and global critical conditions for crack propagation are characterized by threshold $K_{If\,th}$ and critical values of crack resistance K_{Ic}. The kinetic analysis of local damage (Lanin, 2007) made it possible to establish the dependence of the incubation period duration on the rheological characteristics of the material and the values of $K_{Ir\,th}$ and K_{Ic}

$$t_* = \frac{1}{C_0}\left[\frac{1-\bar{K}_I(t)}{\bar{K}_I(t)-\beta(t)}\right]^n, \tag{3.54}$$

where $C_0 = A(n+1)$, A and n are constants. In addition, $\beta(t) = K_{Irth}(t)/K_{Ic}(t)$, $\bar{K}_I(t) = K_I(t)/K_{Ic}(t)$ and

$$K_I(t) = \frac{K_{I0}}{1 + gm\tilde{\Gamma}(t)}$$

From (3.54) it follows that with a decrease in the $K_{If\,th}(t)/K_{Ic}$ ratio, the time before the start of crack growth decreases. An increase in the reserve of elastic energy, characterized by a decrease in the parameter m, leads to an increase in $K_I(t)$ and a decrease in the time to local failure. For $m \to 0$ and $K_I(t) \to K_{I0}$, we have the case of creep loading with a minimum time t^*. Note that an increase in creep resistance causes a decrease in the time before the crack starts growing.

Let us analyze the features of the crack growth stage under stress relaxation conditions. We will use the simplest Maxwell kernel for the model of a linear-viscoelastic medium, since in this case there is a reduced stress relaxation rate and, in accordance with the fracture scheme, there is a shorter time until the crack starts to grow (Fig. 3.24). We write expression (3.45) in the following form

$$\sqrt{l_0} = n\frac{Eh_0}{K_{I0}}, \quad n = \sqrt{\frac{2}{\pi}} \tag{3.55}$$

By replacing E to $\tilde{E}$ we obtain

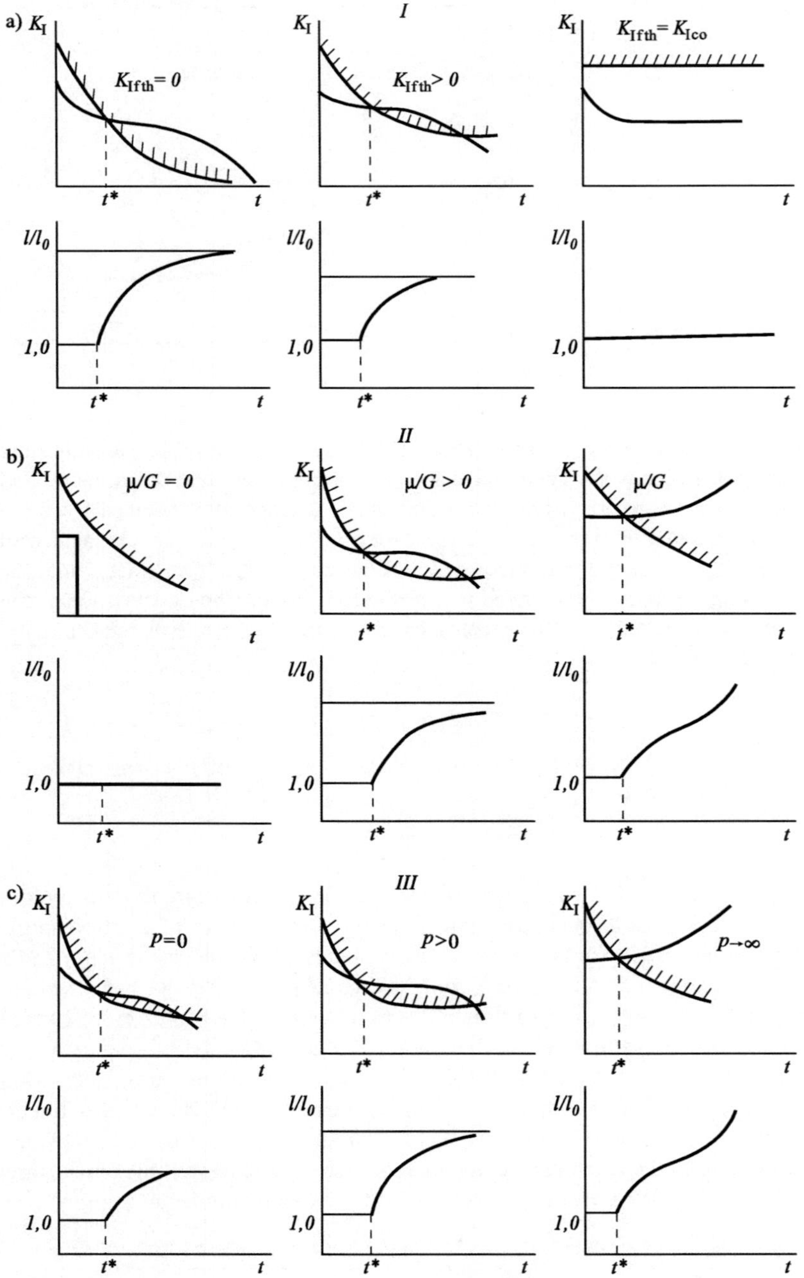

Fig. 3.24 Analysis of influence of the main factors on crack growth kinetics under stress relaxation conditions: a) influence of material crack resistance; b) influence of rheological properties; c) influence of elastic energy reserve

$$\sqrt{l} + q\sqrt{l_0}\tilde{\Gamma}(t_*) + q \int_{t_*}^{t} (t-s)\sqrt{l}\mathrm{d}s = \frac{nE}{K_{\mathrm{Ic0}}} h_{\mathrm{f}}, \tag{3.56}$$

where $h_{\mathrm{f}} = h_0 K_{\mathrm{Ic0}}/K_{\mathrm{Ic}}$. The time dependence of the threshold values of fracture toughness is approximated as follows

$$K_{\mathrm{Ic}}(t) = K_{\mathrm{Ic\,th}} + (K_{\mathrm{Ic0}} - K_{\mathrm{Ic\,th}}) \exp(-ct), \tag{3.57}$$

where c ia a material property. Expanding (3.57) in a power series t and restricting to the first term of the expansion, we have

$$h_{\mathrm{f}} = h_0 \left[1 + c\left(1 - \frac{K_{\mathrm{Ic\,th}}}{K_{\mathrm{Ic0}}}\right) t + \ldots \right] \tag{3.58}$$

Differentiating expression (3.56) and solving the differential equation taking into account (3.58), we obtain the kinetic equation for crack growth in Maxwell's viscoelastic medium under conditions of pure relaxation

$$\sqrt{\frac{a(t)}{l_0}} = \mu^* + (1 - \mu^*) \exp\left[-\frac{G}{\mu}(t - t_*)\right], \tag{3.59}$$

where

$$\mu^* = \frac{c\mu}{G}\left(1 - \frac{K_{\mathrm{Ic\,th}}}{K_{\mathrm{Ic0}}}\right)$$

For relaxation conditions with an elastic branch (sample - testing machine system), we obtain

$$\sqrt{\frac{a(t)}{l_0}} = \mu^*(1 + p) + [1 - \mu(1 + p)] \exp\left[-\frac{G}{\mu}(t - t_*)\right] \tag{3.60}$$

Let us analyze the obtained relations (3.59) and (3.60), considering the influence of various factors on the development of cracks during stress relaxation.

3.2.4.3 Development of Cracks During Stress Relaxation. Influence of Crack Resistance

In accordance with relations (3.59) and (3.60), slow growth of cracks during stress relaxation is possible only for materials prone to damage accumulation over time. In this case, according to (3.59), the process of slow crack growth under conditions of pure relaxation is self-decelerating, which essentially distinguishes the development of cracks during creep and during stress relaxation. As can be seen from relations (3.59) and (3.60), the crack size depends on the ratio of $K_{\mathrm{Ic\,th}}$ and K_{Ic0}. For materials in which a decrease in fracture toughness $K_{\mathrm{Ic\,th}}(t)$ is observed over time, the absolute crack growth before its deceleration is the greater than lower the ratio $K_{\mathrm{Ic\,th}}$ and

K_{Ic0} is (Fig. 3.24). At $K_{Ic\,th}(t) = K_{Ic0}$ $(\dot{\omega}_k = 0)$, slow crack growth is not observed (Lanin, 2007; Lanin and Getsov, 2009). $\dot{\omega}_k$ is a function of K_I and time. This is the kinetic equation of the the rate of damage accumulation at the crack tip under the action of the singular part of the stress state.

3.2.4.4 Influence of the Rheological Properties of the Material

The influence of the rheological properties of the material can be established by analyzing the dependence of the length of a self-inhibiting crack a on the ratio μ/G. Let us consider three particular cases of rheological media (Lanin and Getsov, 2009).

1. Medium of the ideal fluid type ($\mu/G = 0$).
   In this case, the material has a very low creep resistance ($\varepsilon \to \infty$). From (3.59) and (3.60) it follows that for the medium under consideration, any initial loading leads to rapid relaxation of K_I to zero (Fig. 3.22). In this case, slow crack growth does not occur.
2. Medium with moderate creep resistance ($\mu/G > 0$).
   For such a material, a slow crack growth is observed. The length of a self-inhibiting crack is the greater, the higher the creep resistance (the higher μ/G) is (Fig. 3.24).
3. Medium with high creep resistance ($\mu/G \to \infty$). In materials of this type, according to (3.59) and (3.60), unlimited crack growth is observed under conditions of stress relaxation (Fig. 3.24).

Analysis shows that an increase in creep resistance without a simultaneous increase in crack resistance can lead to an intensification of the kinetics of crack growth during stress relaxation, increasing the length of a self-inhibiting crack.

3.2.4.5 Influence of the Supply of Elastic Energy

According to relation (3.60), an increase in the reserve of elastic energy in the loading system is equivalent to an increase in the material's creep resistance (i.e., an increase in μ/G). In the absence of a reserve of elastic energy ($p = 0$), the crack propagation is rapidly inhibited. An increase in the supply of elastic energy leads to a more intense crack growth (Fig. 3.24). At $p \to \infty$, the K_I relaxation is practically absent. Loading corresponds to creep. At the same time, the crack develops until the body is completely failed. The developed theoretical concepts of fracture under stress relaxation conditions have been used to solve numerous practical problems of preventing the formation and development of technological and operational cracks in the elements of power equipment (Lanin, 1983; Zemzin et al., 1984; Chizhik and Lanin, 1986; Lanin, 2007; Lanin and Getsov, 2009).

3.2.4.6 Fracture Under Conditions of Relaxation and Repeated Loading

Under relaxation conditions under single loading, fracture can occur when the creep deformation $(\sigma_0 - \sigma_{res})/E$ reaches the limiting deformation corresponding to that shown in Fig. 3.25. In this case the material is brittle. With a large number of repeated loadings, fracture under conditions of relaxation (hard loading) can also occur in plastic materials, regardless of their level of ductility.

Cases of failure of fasteners in gas turbines are often associated with the exhaustion of the deformation capacity of the material during repeated pull-ups. In Borzdyka and Getsov (1978), experimental data on the behavior of various materials under repeated loading are presented. Figure 3.17 shows the relaxation curves of a low-ductility material, which ended in fracture under the conditions of one loading (at 600°C) and two loads (at 550°C). Examples of curves of cyclic relaxation and fracture of specimens made of ductile alloys ChN35WMT (EI692) at 750°C and ChN70WMYuT (EI765) at 700°C are shown in Figs. 3.26 and 3.27. It can be seen that the values of deformation before fracture of the EI765 alloy (Fig. 3.26) at constant stress and during relaxation under conditions of repeated pulls practically coincide.

For a given material, depending on the initial stress, temperature, number of loadings and time intervals between them, the nature of the cyclic relaxation curves, and, consequently, the fracture conditions can change (see Fig. 3.28). Tables 3.11 and 3.12 show some results of specimen fracture under conditions of repeated loading. To predict the conditions of fracture, the hypothesis of linear summation of damage

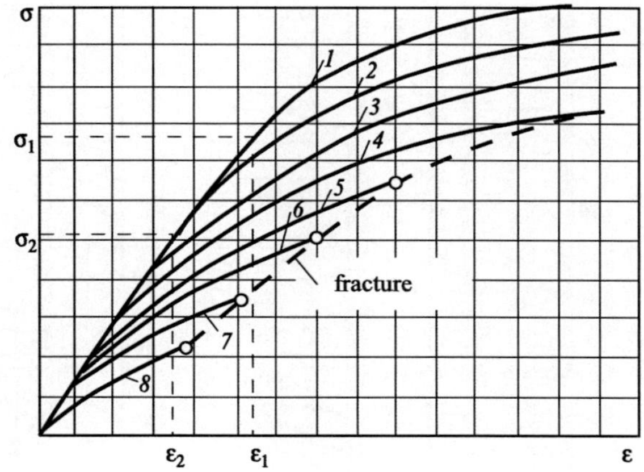

Fig. 3.25 Deformation curves of low plastic material over time, h: *1* – 0, *2* – 10, *3* – 100, *4* – 1000, *5* – 5000, *6* – 10000, *7* – 50000, *8* – 100000

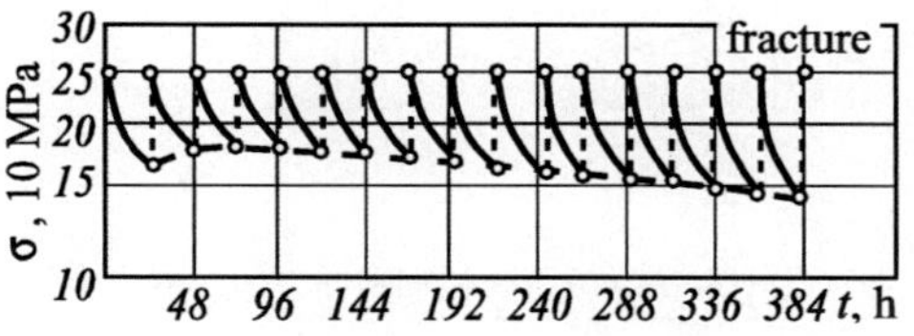

Fig. 3.26 Cyclic relaxation curves of ChN35WMT alloy (EI692) at 750°C

Fig. 3.27 Creep curves of ChN70WMYuT (EI765) alloy at 700°C; *1* - repeated relaxation test at n = 700 MPa, t_{cycl} = 30 min and 20 h (t_f = 97 h); 2 - creep test at σ = 500 MPa (t_f = 100 h)

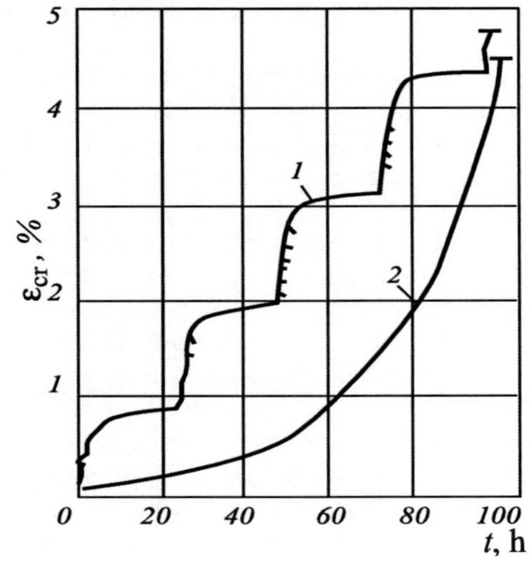

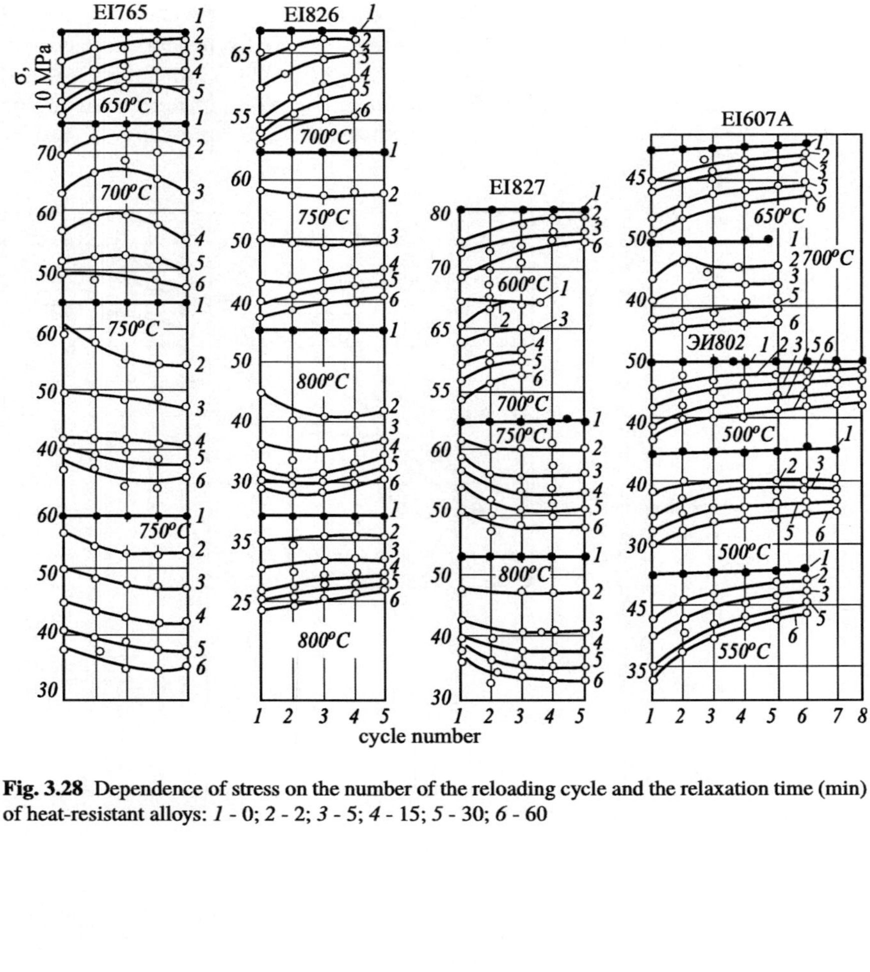

Fig. 3.28 Dependence of stress on the number of the reloading cycle and the relaxation time (min) of heat-resistant alloys: *1* - 0; *2* - 2; *3* - 5; *4* - 15; *5* - 30; *6* - 60

Table 3.11 Results of long-term strength tests under stress relaxation conditions (Borzdyka and Getsov, 1978)

Material	T, °C	σ_0, MPa	t_{aver}, h between loadings	t_f, h	σ_{eq}, MPa, computed by sum of relative durability	σ_{eq}, MPa, computed by ultimate strain
25Ch2M1F (EI723)	580	500	24	371	375	385
			24	408	355	362
			24	416	338	347
			216	1080	278	292
			216	1305	273	288
		400	96	862	267	275
			96	969	288	293
			96	1054	271	278
			96	1092	289	294
			96	1156	260	268
			96	1255	272	280
			96	1482	272	279
			1000	2368	227	239
			1000	2528	220	232
20Ch3MWF (EI415)	550	350	1(173)*	496	342	-
			2(13)*	-	-	-
			4(74)*	-	-	-
			1	707	337	-
		400	1	96	376	-
		450	0.1	12.5	417	-
			1	41	398	-
			10	120	352	-
			100	1066	336	-
			200	1145	336	-
	650	200	11.5 (56)*	168	171	-
			10(7)*	-	-	-
			24	1157	126	-
ChH80TBYuA (EI607A)	650	420	100	3149	346	-
		450	1	337	425	-
			10	750	370	-
			100	2093	349	-
		480	1	181	455	-
			10	640	389	-
			100	896	438	-
			500	7103	278	-

*The number of cycles is indicated in brackets.

Table 3.12 Long-term strength of heat-resistant alloys under cyclic relaxation conditions

Material	T, °C	σ_0, MPa	$\sum \Delta p_i$, %	N	t_f^{exp}, h	DC	FC
ChN70WMYuT	700	700	4.83	34	98	1.07	0.864
(EI765)		720	3.33	25	96.5	0.70	-
EI698	750	550	10.85	79	85	0.99	-
ChN35WMT	750	200	7.30	26	600	0.45	-
(EI692)		250	8.70	16	382	0.55	-
20Ch3MWF	550	350	2.00	649	707	1.50	1.00
(EI415)		350	2.10	260	496	1.20	0.72
		400	1.90	96	96	1.10	0.77
		400	3.70	124	12.5	2.20	0.12
		450	2.20	41	41	1.30	0.27
		450	1.20	12	120	0.70	0.80
		450	1.40	12	1066	2.00	-
		450	0.70	6	1145	1.00	0.46
37Ch12N8G8MFB	550	600	0.07	2	43	1.00	-
(EI481)	600	600	0.08	1	13	1.00	-
ChN80TBXuA	650	420	2.60	34	3149	0.75	-
(EI607A)		450	7.10	339	337	1.10	1.00
		450	5.00	76	750	1.10	0.83
		450	1.50	21	2093	0.43	1.22
		480	3.20	182	181	0.50	1.00
		480	4.30	66	640	0.95	1.00
		480	0.80	11	896	0.18	0.69
		480	2.90	15	7103	0.83	-

DC deformation criterion, FC force failure criterion

in a temporary or, more preferable, deformation interpretation can be used

$$\sum_{i=1}^{n} \frac{t_i}{t_{fi}} = 1, \quad \sum \frac{\Delta p_i}{\varepsilon_{dci}} = 1 \tag{3.61}$$

Here t_i is the operating time of the material at stress σ_i, t_{fi} is the time to fracture at stress σ_1, Δp_i is the creep strain accumulated during the ith loading and ε_{dci} is the deformation capacity of the material under these conditions.

The condition of brittle fracture under relaxation conditions can be described on the basis of strength criteria and the principle of additivity (linear summation) of damage. If we assume the following relation

$$\sigma_{\text{eq}} = \left(\frac{1}{t} \int_{0}^{t} \sigma^{n} \text{d}\tilde{t} \right)^{1/n} = \sigma_{\text{lt s}}, \tag{3.62}$$

where σ changes according to the relaxation curve, then fracture is possible when the equivalent stress σ_{eq} becomes equal to the ultimate long-term strength $\sigma_{\text{lt s}}$ during the same time. Here n is the slope of the long-term strength curve.

Using the second expression in (3.61), the number of loadings N leading to fracture under relaxation conditions and repeated loading can be calculated from the following relation

$$\varepsilon_{\text{dc}} = \sum_{i=1}^{N} \frac{\sigma_0 - \sigma_{t\,i}}{E} \tag{3.63}$$

Here σ_0 is the initial stress and $\sigma_{t\text{I}}$ are the final (residual) stresses before the next loadings.

3.2.5 Long-term Strength Under Complex Stress State

3.2.5.1 Fracture Criteria

To assess the resistance to long-term fracture under conditions of a complex stress state, a number of criteria have been proposed that establish a relationship between the long-term strength limit under uniaxial tension $\sigma_{\text{lt s}}$ with stresses under the action of which a component (sample) fails under a complex stress state. It is advisable to consider the conditions for this in Table 3.1, replacing σ_{B} in the table by $\sigma_{\text{lt s}}$.

The most widespread is the Rankine-Johnson criterion $\sigma_{\text{lt s}} = \sigma_I$ (maximum tensile stress), confirmed in experiments (Pisarenko and Lebedev, 1976; Lebedev et al., 2003). For ductile materials, good agreement between the experimental and calculated values is obtained using the Katz criterion, which is similar to the von Mises criterion ($\sigma_{\text{lt s}} = \sigma_{\text{vM}}$), and for materials with limited ductility, calculations using the Sdobyrev, Pisarenko-Lebedev (see Table 3.1) and Trunin (3.36) criteria show good agreement with experimental data.

Work on the establishment of the fracture criterion most adequate to the experimental data under a complex stress state has continued in recent years. So in Trunin (1994), the time to failure is determined by the formula (3.36), where the equivalent stress is assumed as follows

$$\sigma_{\text{eq}} = \sigma_{\text{k}} \exp[\lambda(\eta - 1)], \tag{3.64}$$

where

$$\sigma_{\text{k}} = \frac{1}{2}[(\sigma_1 + |\sigma_1|)] + \sigma_{\text{vM}}, \tag{3.65}$$

and

$$\eta = \frac{6\sigma_{\rm m}}{\frac{1}{2}[(\sigma_1 + |\sigma_1|)] + \sigma_{\rm vM}}. \tag{3.66}$$

For the EI698 alloy at a temperature of 700°C, a good agreement was obtained between the experimental data and the calculation results (Table 3.13) (experiments at temperatures of 650 and 750°C were used to determine the constants).

The following expressions for the equivalent stress are proposed in Pisarenko and Lebedev (1976); Lebedev et al. (2003)

$$\sigma_{\rm eq} = (\tau_{\rm vM}^m + \tau_1^m)^{1/m}, \qquad \sigma_{\rm eq} = (\sigma_1^n + \chi\sigma_{\rm vM}^n)^{1/n}, \qquad \sigma_{\rm eq} = \frac{1}{2}(\sigma_1 + |\sigma_1|)\omega_1 + \sigma_{\rm vM}\omega_2$$

In Pavlov and Kurilovich (1978); Pavlov (1980), to determine the conditions of failure under complex stress states, the following relation is proposed

$$\frac{{\rm d}\Pi}{{\rm d}t} = f(\sigma_{\rm vM})\,\psi\left(\frac{\sigma_0}{\sigma_{\rm vM}}\right), \tag{3.67}$$

where $\sigma_{\rm vM}$ is the von Mises equivalent stresses, $\psi(\sigma_0/\sigma_{\rm vM}$ is the stress state function, which, according to the test data for the plane stress state of the EI765 alloy at 800°C, has the form $\psi(\sigma_0/\sigma_{\rm vM}) = 7.8(\sigma_0/\sigma_{\rm vM})^2 + 0.145$. Fracture takes place at $t_{\rm f}$, for which the condition

$$\int_0^1 {\rm d}\Pi = \int_0^{t_{\rm f}} f(\sigma_{\rm vM})\psi\left(\frac{\sigma_0}{\sigma_{\rm vM}}\right){\rm d}t \tag{3.68}$$

is valid. The fields of application of different fracture criteria for a large number of different materials are discussed in Lebedev et al. (2003); Naumenko and Altenbach (2007, 2016). Examples for long-term strength analysis of components applying different creep damage equivalent stress formulations are presented in Altenbach and Naumenko (1997); Altenbach et al. (1997, 2000, 2001, 2002), among others.

Table 3.13 Test results of alloy EI698 under complex stress state

σ_1, MPa	σ_3, MPa	$t_{\rm exp}$, h	$t_{\rm comp}$, h
520	-18	13; 22; 28	26
430	-155	135; 273; 329	217
380	-14	919; 1364	746
450	-360	14,5; 18; 22,5	15
300	-240	637,5; 740; 752	1267
350	-280	122; 173; 247	270

3.2.5.2 Effect of Stress Concentration

The characteristics of the long-term strength of notched specimens are of interest as a criterion for the strength of the metal of parts with stress concentrators (holes, notches, fillets, etc.). In the general case, they cannot be analytically characterized by dependencies of the type (3.18), (3.19), and (3.21), which provide the relationship between stress and time to failure only for smooth samples. This is due, firstly, to the fact that the stress state in specimens with stress concentrators is inhomogeneous over the cross section and changes during the test with an intensity that depends on the creep characteristics, which, in turn, depend on the stress, test temperature and degree of work hardening.

The second reason for the nonlinearity of the long-term strength curves of notched specimens in logarithmic and semi-logarithmic coordinates is that the deformation capacity of the material under creep conditions, determining the degree of possible stress redistribution, on the one hand, and the tendency to develop the formed microcracks, on the other, also depends on the stress, temperature and degree of work hardening.

Speaking about the deformation capacity of the material in relation to long-term strength tests of specimens with stress concentrators, it should be borne in mind that the ductility under conditions of a complex stress state p^* can differ significantly (downward) from the ductility of the material under uniaxial tension. For the dependence of p^* on the stiffness of the stress state, relations are usually used that correspond to the concepts of mixed deformation-force criteria proposed by Rabinovich (1970) and Novozhilov (1975). According to Makhutov (1981), we have (3.43), where K_e is the characteristic of the state of the material (in the brittle state of the material, $K_e = 1$, in the viscous state, $K_e = 1.2$).

Analysis of the experimental data available in the literature does not allow establishing a strict relationship between the sensitivity to notch (the ratio of the long-term strength of notched specimens and smooth specimens), the type of alloy, and its plastic characteristics. A number of researchers believe that there is an unambiguous relationship between the ductility of smooth specimens with prolonged rupture and the sensitivity to notch. The author came to the same conclusion when testing samples made of steels EI481 and EI612, made from blanks of various sections, heat-treated for different hardness. At the same time, the results of a number of other studies refute this conclusion. All that has been said applies only to deformable materials. Casting alloys containing many discontinuities of various types are insensitive to notch in creep rupture tests.

Whether a material is sensitive to a notch at a given test temperature under conditions of prolonged static loading is often judged by comparing the time to fracture of smooth specimens tested at an arbitrarily chosen initial stress $\sigma_{0\text{sm}}$ and specimens with an annular notch at normal stress $\sigma_{0\text{n}} = \sigma_{0\text{sm}}$. However, the initial stress of smooth specimens and the nominal stress for specimens with a notch do not reflect the actual picture of their stress state, but only allow comparing the results by a qualitative criterion - whether the material is sensitive to a notch at a given stress concentration and temperature or not. Therefore, the results of such tests cannot be generalized to other temperatures, stresses and stress concentrators. Thus, the conventionality of the notch sensitivity criterion based on the results of testing standard samples is obvious. Since the rate of stress redistribution is determined not only by the properties of the material, but also by the geometry of the component, for real parts made of the same material, a different degree of influence of the notches can take place. To explain the results of long-term strength tests of specimens with circular notches obtained by various researchers and to establish the dependencies

$$\lg t_f = F(\lg \sigma) \quad \text{or} \quad \lg t_f = F(\sigma) \tag{3.69}$$

for specimens with stress concentrators in Getsov (1962), the stress state in the notch and the stress redistribution occurring in time were considered. Specification of the assumptions was discussed in Golubovskii and Trunin (1978); Grubin (1970); Malinin and Nigin (1974); Gariboldi et al. (2016), who proposed various methods for calculating stresses in the minimum section of an axisymmetric notched specimen, both taking into account the initial elasto-plastic deformations and creep deformations. Thus, Ohnami et al. (1976), using the finite element method (FEM) and flow theory, calculated the changes in stresses and strains with time in specimens with annular notches made of Cr-Mo-V-steel, tested at 600°C under conditions of nominal stresses that caused hardening ($\sigma_0 > \sigma_{pr}$) and softening ($\sigma_0 < \sigma_{pr}$) of steel under the action of a notch. In Grubin (1970), the theory of aging was used for calculations, and in Malinin and Nigin (1974) the FEM and the theory of anisotropic hardening. The calculations showed that in a cylindrical specimen with a notch, the maximum stresses are the axial stresses σ_z, the minimum are the radial σ_r (Fig. 3.29a). Maximum axial stresses, however, in some cases can be at some distance from the base of the notch or correspond to the radius $R = R_0$ (R_0 - radius of the sample in the base of the notch). Besides, the position of the maximum changes slightly with time.

To determine the notch effect, either the equivalent stress σ_{eq} or the equivalent strain ε_{eq} are calculated, which are then compared with the corresponding values

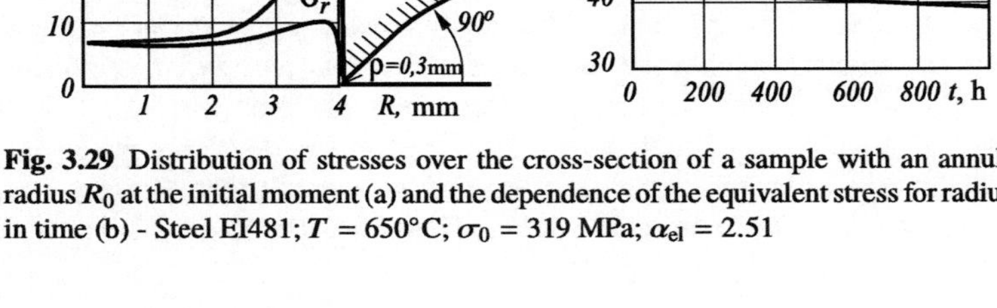

Fig. 3.29 Distribution of stresses over the cross-section of a sample with an annular incision of radius R_0 at the initial moment (a) and the dependence of the equivalent stress for radius $R = 0.95R_0$ in time (b) - Steel EI481; $T = 650°C$; $\sigma_0 = 319$ MPa; $\alpha_{el} = 2.51$

obtained when testing smooth samples. As σ_{eq} in Ohnami et al. (1976) $\sigma_1 = \sigma_z$ was assumed, in Malinin and Nigin (1974), σ_{vM}, and in Golubovskii and Trunin (1978), the value of σ_{eq} according to Eq. (3.3). The values of the maximum stresses, related to the notch surface in Cr-Mo-V-steel, and to a certain distance from the surface for EI481 steel, decrease with time (Fig. 3.29b).

To assess the conditions for the initiation and propagation of cracks, the stress on the surface, averaged over the entire test time, or at some distance from it, is estimated as follows

$$\sigma_{eq}^{(1)} = \left(\frac{1}{t_f} \int_0^{t_f} \sigma^{m_{sm}}(t, R = \text{const}) dt \right)^{1/m_{sm}}, \quad (3.70)$$

where m_{sm} is the slope of the long-term strength curve for smooth specimens.

Expression (3.70) is based on the hypothesis of linear summation of damage and the power-law dependence of the time to fracture on stress. For low-ductility materials, in which the development of cracks occurs predominantly relatively quickly (as is the case, for example, in the EI481 and EI612 austenitic steels), the equivalent stress should take into account only changes in the maximum stress. If the initiation of a crack is not followed by its rapid development (as, for example, in nickel-based alloys EI607A, EI827), then averaging not only in time, but also over the radius is apparently valid. In this case, the equivalent stress is of course less important. If the value of the elastic concentration coefficient is known, the stress redistribution can be calculated by a method based on the use of Eqs. (2.37) by Neuber or (2.38) by Makhutov by extending them to case of creep.

To answer the question about the sensitivity of the material to a notch under these conditions, the value of the equivalent stress for specimens with a notch $\sigma_{eq\,n}$ should be compared with the value of the equivalent stress of a smooth specimen $\sigma_{eq\,sm}$, which takes into account the change in the initial stress during deformation, especially at the tertiary creep stage. For materials that fracture under conditions of a given temperature with low ductility ψ_{pl}, $\sigma_{eq\,sm}$ almost coincides with $\sigma_{0\,sm}$. To calculate $\sigma_{eq\,sm}$, it is necessary to know the change in the relative contraction during the long-term strength test. However, the contraction ψ is not measured over time, but only the creep elongation p is available. It is possible to directly calculate $\psi(p)$ only for samples that have not formed a neck. To recalculate $\psi = \psi[p(t)]$ in the presence of a neck, one can use the dependence obtained by approximating the experimental data.

Let us consider three possible cases of relations between σ_{eq} and $\sigma_{eq\,sm}$ (Figs. 3.30 and 3.31): 1) $\sigma_{eq} > \sigma_{eq\,sm}$, 2) $\sigma_{eq} = \sigma_{eq\,sm}$ and 3) $\sigma_{eq} < \sigma_{eq\,sm}$. In the first case, the material is considered notch sensitive, in the second and third, insensitive. The condition $\sigma_{eq} > \sigma_{eq\,sm}$ usually takes place at low temperatures and normal stresses, at which redistribution also occurs slowly (curves for T_1, see Fig. 3.30a, b), and plastic properties with long-term rupture are small (curve for T_1, see Fig. 3.31). The equality $\sigma_{eq} = \sigma_{eq\,sm}$ is observed at medium temperatures, when an increase in stresses due to a decrease in the diameter of smooth samples during averaging compensates for

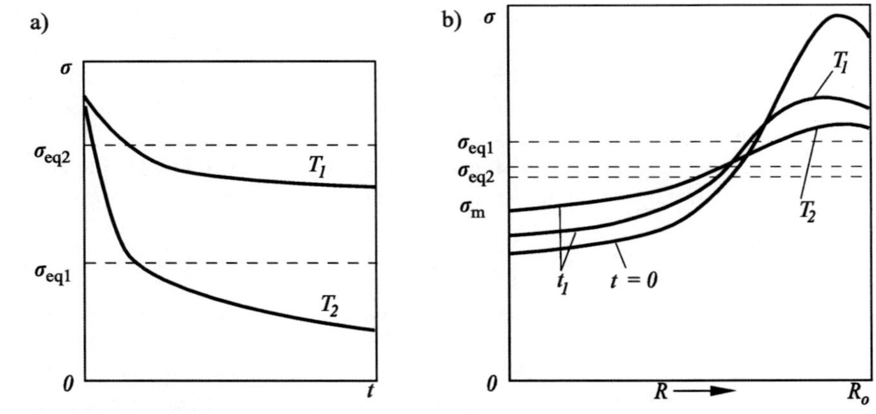

Fig. 3.30 Dependence of stresses σ in a sample with an annular notch on time t (a) and radius R (b) at different temperatures ($T_1 < T_2$), σ_m - average stress

Fig. 3.31 Dependence of relative contraction ψ (a), tensile stress σ (b) on time t (σ_0 =const; $T_1 < T_2$): solid lines - ε_{pl}; dashed lines - p_2 at T_2

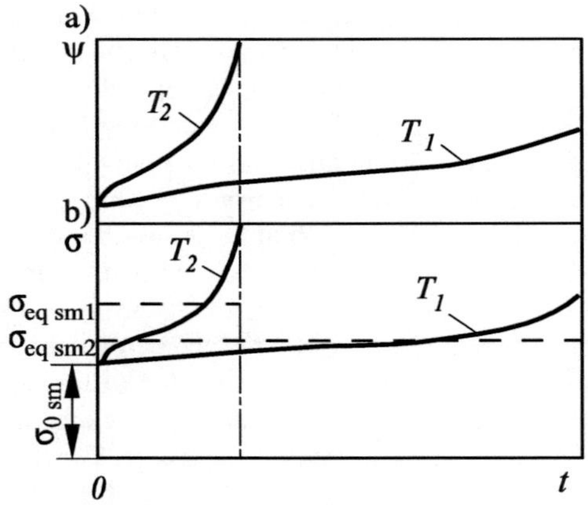

the difference between σ_{eq} and nominal stresses. The ratio $\sigma_{eq} < \sigma_{eq\,sm}$ is usually observed at high temperatures, when the redistribution of stresses occurs extremely rapidly (curves for T_2, Fig. 3.30a, b), and the plastic properties during long-term rupture are high (curves for T_2, Fig. 3.31a).

Certain adjustments to the scheme can be made by the work hardening of the material, which is formed both in the process of machining during the manufacture of the notch, and in connection with plastic deformation during the redistribution of stresses in the concentration zone. In the case of a decrease in the long-term strength of the work-hardened material in comparison with the virgin one at $\sigma_{eq} \le \sigma_{eq\,sm}$, there may be a sensitivity to the notch, and vice versa, with an increase in the long-term strength of the work-hardened material in comparison with the non-riveted one at $\sigma_{eq} > \sigma_{eq\,sm}$, the material can be insensitive to the notch. Using the method described above for assessing the long-term strength of notched specimens, let us analyze the effect of the type of material, test temperature, and notch manufacturing technology on the notch sensitivity for $\sigma_0 < \sigma_{0.2}$ ($\sigma_{0.2}$ is the yield stress at the corresponding temperature).

For heat-resistant deformable alloys, four cases of the location of the long-term strength curves of notched specimens relative to the curves for smooth specimens of the same material are typical (Fig. 3.32). In the case of a ductile material at the considered temperature, having a low creep (relaxation) resistance and not embrittling with an increase in the test time, the stress redistribution occurs extremely rapidly, and, therefore, σ_{eq} for notched specimens at all stresses practically coincides with σ_{0n}. Due to the high ductility at long-term rupture, $\sigma_{eq\,sm}$ is significantly less than σ_{eq}. Therefore, the long-term strength curve of smooth specimens is located below the long-term strength curve of notched specimens (Fig. 3.32a), and at this temperature the material is not sensitive to the notch.

In a ductile material that is embrittled over time and has low creep resistance, ductility with prolonged rupture decreases with an increase in the test base; therefore, at high stresses, $\sigma_{eq\,sm}$ is significantly greater than σ_{eq}, at low stresses, σ_{esmth} practically coincides with σ_{eq} (Fig. 3.32b) or less σ_{eq} (Fig. 3.32d). A certain role in this case can be played by work hardening, which occurs at high stresses and is almost completely removed at a long test duration. If the dependence of ductility at long-term fracture on the test base has a minimum, the location of curves *1* and *2* will also change.

For specimens made of notch-sensitive low-ductility materials, σ_{eq} almost coincides with σ_{0sm} and σ_{eq} is calculated by Eq. (3.70). The long-term strength curve of notched specimens in the general case has three sections with different slopes kl, lm, and mn (Fig. 3.32c). The kl section has a very large slope. This is due to the fact that the maximum conditional elastic stresses ($\sigma_{el} = \sigma_{0n}\alpha_n$) at the initial moment exceed the yield strength, while real elasto-plastic stresses change little with an increase in σ_{0n}. Thus, the effective stress and, consequently, the time for fracture are weakly dependent on σ_{0n}. The lm section is rather flat. This is determined by the fact that with increasing test duration, the proportion of time required for stress redistribution decreases. The mn section has a slope close to the slope of the long-term strength

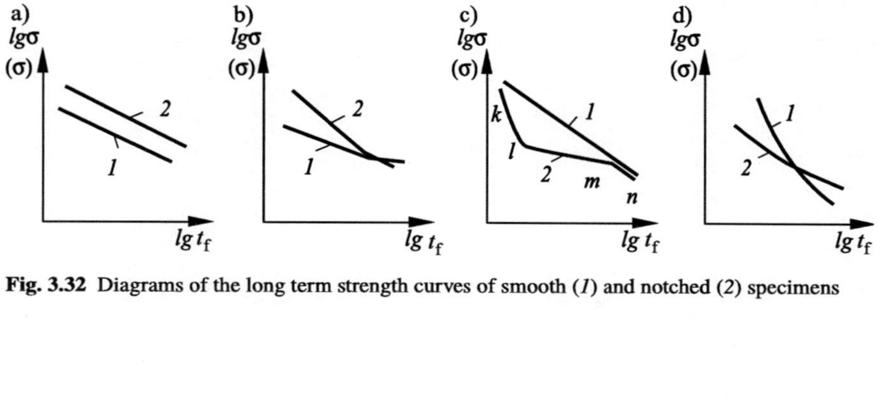

Fig. 3.32 Diagrams of the long term strength curves of smooth (*1*) and notched (*2*) specimens

curve of smooth samples. For this section, the role of exhaustion of durability at the initial moment of testing is extremely small, and σ_{eqn} almost coincides with σ_{0n}.

The considered cases of the location of the long-term strength curves, of course, do not exhaust all the possibilities encountered in the practice of testing different materials. Analysis of the literature data and the results of the tests conducted by the author confirms the existence of all the cases considered. Curves for notched specimens of different heats of the same material may differ (for example, curves A and B in Fig. 3.33).

As the test temperature decreases, the notch sensitivity increases (Table 3.14). However, in a number of cases there are violations of this regularity, which can be caused by: the presence of a minimum deformation capacity in a certain temperature range (Stanyukovich, 1967) the absence of a significant difference between σ_{eq} and σ_{0n} in tests with $\sigma_{0n} \geq \sigma_y$. The effect of work hardening, the removal of which at low temperatures occurs very slowly.

The characteristics of the long-term strength of notched specimens, especially in tests of short duration, depend on the technology of manufacturing blanks to a much greater extent than is observed for smooth specimens. The reason for this is

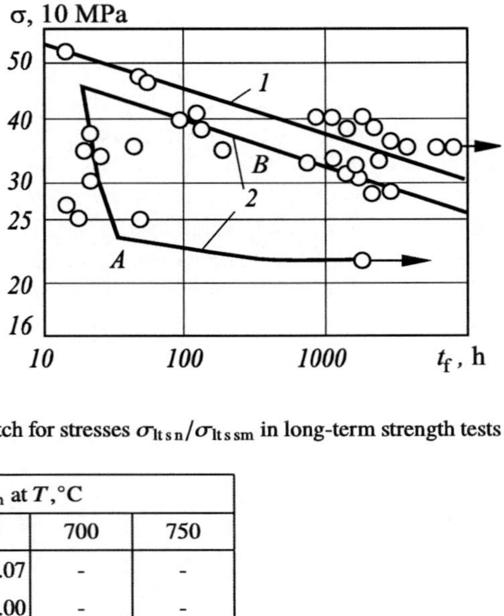

Fig. 3.33 Long term strength curves at 600°C of smooth (*1*) and notched (2) specimens made of EI612 steel

Table 3.14 Coefficients of sensitivity to notch for stresses $\sigma_{lt\,s\,n}/\sigma_{lt\,s\,sm}$ in long-term strength tests of some materials for disks

Forging brand	$\sigma_{lt\,s\,n}/\sigma_{lt\,s\,sm}$ at T,°C				
	500	600	650	700	750
EI481	0.79	0.70-1.00	0.69-1.07	-	-
EI612	0.84	0.54-1.00	0.74-1.00	-	-
EI787	-	-	0.80	0.76	-
EI607A	-	-	1.05	0.98	-
EI437B	-	-	0.77-0.98	1.00	-
EI765	-	0.72	0.86	0.86	1.00
EI698	-	-	1.10	1.10	-

that ductility with prolonged rupture and relaxation resistance in the first period, which determine the fracture of notched specimens, depend very significantly on the modes of heat treatment, forging, rolling, and melting conditions. The deformation of specimens with an annular notch in comparison with the deformation of smooth specimens accumulates in a relatively small volume. Therefore, the statistical probability of cracking in a notched specimen, other things being equal, is less than in smooth specimens. In some cases, this can make some adjustments to the scheme discussed above.

A certain influence on the long-term strength of notched specimens is exerted by the method of making the notch, which is associated with the depth of the work-hardened layer. If this depth exceeds the distance from the root of the notch to the radius corresponding to the maximum stresses, or is equal to it, then work hardening, reducing the plasticity and reducing the ability to redistribute stresses, increases $\sigma_{\text{eq n}}$ and thereby increases the sensitivity to the notch. This is illustrated by the data of S.E. Belyaev and N.G. Plekhanov on increasing the sensitivity to the notch in the case of making an incision with a cutter (instead of grinding). If work hardening increases the creep rate, the effect may be the opposite.

When testing for long-term strength of specimens with different stress concentrators made of notch-sensitive materials, in some cases, an increase in the elastic concentration coefficient α_{N} is not accompanied by a decrease in the time to fracture. An explanation of this behavior can be found by comparing the values of the strain concentration coefficient α_{ε} corresponding to different values α_{N}. An increase in α_{N} leads to an increase in α_{ε}, and, consequently, to the initial plastic deformation, which in turn causes (see Subsect. 2.2.4) an increase in the relaxation rate. Thus, an increase in α_{N} can cause a more intense decrease in stresses in the notch, in connection with which the value of the equivalent stress may be even lower for specimens with a large α_{N} than for specimens with a lower α_{N}.

To determine the coefficient of sensitivity to notch by deformation

$$K_{\delta} = (\delta_{\text{f}} - \delta_0)_{\text{n}}/(\delta_{\text{f}} - \delta_0)_{\text{sm}}$$

Stanyukovich and Chizhik suggested testing specimens with a spiral double-start notch (Stanyukovich, 1967). This sample has two symmetrically located small grooves in each of its sections, the position of which is shifted from section to section. The presence of two undercuts prevents the specimen from bending. The most widespread is the sample with $\alpha_{\text{N}} = 3.7$. When testing such specimens, it was found that all materials are sensitive to notch deformation ($K_{\delta} < 1$). The plasticity of metal, determined on specimens with a spiral cut, obeys the same laws as the plasticity of smooth specimens. The K_{δ} value for different materials can vary significantly.

Ugorskii and Khein (1986) criticize the concept of sensitivity of stress concentration, which relates the behavior of the materials of components with the results of testing samples with an annular notch. One or another sensitivity to stress concentration for a sample with an annular notch does not determine the nature of the notch sensitivity for a sample (part) at a different stress state due to differences in the stiffness of the stress state (displacement of the stress peak). In this work, the guaranteed values of the effective concentration coefficient under static loading under conditions of a plane stressed state k_e are related to the characteristics of long-term plasticity.

It was found that for ductile materials (with $\varepsilon_{pl} > 5\%$) $k_{eq} = \sigma_B(t)/\sigma_n(t)$ depends on α_N according to the formula

$$k_{eq} = (1 \div 1.3)\alpha_N^{2/(m+1)}, \tag{3.71}$$

while for materials with low ductility we obtain

$$k_{eq} = \frac{(1 \div 1.3)\alpha_N}{\sqrt{(\varepsilon_{pl} + \varepsilon_{el})/\varepsilon_{el}}}, \tag{3.72}$$

where $\varepsilon_{el} = \sigma_B(t)/E$, m is the exponent in the power law of the creep rate.

A convincing experimental confirmation of the validity of such estimates was obtained using k_{eq} for plates with a central beveled hole ($\alpha_N = 4$) made of alloy EI893 in an embrittled state at $T < 600\,°C$, for flat models of disc rims made of brittle 25Ch1M1F steel with forked attachment of the blade tails ($\alpha_N = 2.06$), for T-shaped tails of blades ($\alpha_N = 2.9$) made of embrittled steel 25Ch2M1F and for flat models of disc rims made of the same steel with T-shaped attachment of tails ($\alpha_N = 2.55$).

3.2.6 Long-term Strength at Variable Temperatures and Stresses

Evaluation of the strength of materials at variable temperatures is of great importance for components of various heat-power plants (Getsov, 1973; Naumenko et al., 2011b; Kostenko et al., 2013), and especially gas turbines. For example, the temperature of the rotor blades of a 1.5 MW gas turbine operating at a power plant at a compressor station of a main gas pipeline in the presence of external consumers varies during the day from 600 to 760°C (Fig. 3.34).

Considerable interest in the issue of heat resistance of materials at variable, cyclically changing temperatures is caused by the desire of designers to maximize the maximum working temperature of parts and thereby increase the efficiency of machines. At the same time, if the choice of material is made in relation to the maximum temperature and full motor life, then this will cause significant difficulties.

Among the calculation methods for determining the cyclic heat resistance, the most widespread methods are based on the use of the hypothesis of damage additivity in the temporary and deformation interpretation, according to which failure occurs when the sum of the relative durability or relative deformations is equal to one

$$\sum_{j=1}^{n} \frac{t_j}{t_{fj}} = 1, \qquad \int_0^{t_f} \frac{dt}{t_f} = 1, \tag{3.73}$$

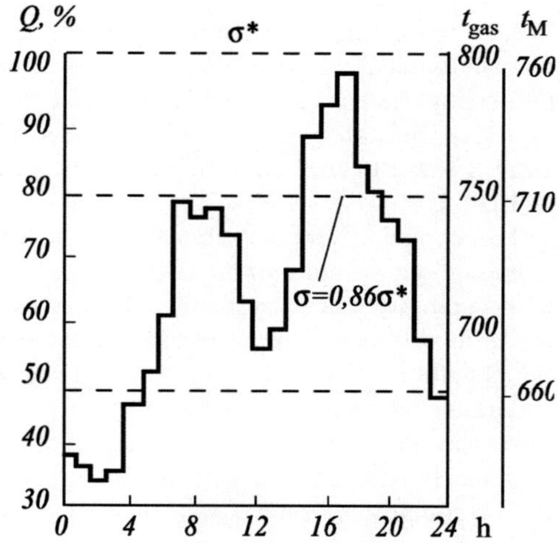

Fig. 3.34 Change during the day in the winter months of load Q, gas temperature t_{gas}, and temperature of rotor blades t_M of a gas turbine with a capacity of 1.5 MW

$$\sum_{j=1}^{n} \frac{p_j}{\varepsilon_{\mathrm{cr\,f}j}} = 1, \qquad \int_0^{\varepsilon_{\mathrm{cr\,f}}} \frac{\mathrm{d}p}{\varepsilon_{\mathrm{cr\,f}}} = 1. \tag{3.74}$$

Here t_j, $t_{\mathrm{f}j}$ are total holding time and time to failure of the sample or component at a given jth temperature level T_j and stress σ_j. $p_j, \varepsilon_{\mathrm{cr\,f}j}$ - creep strain accumulated during t_j and deformation at fracture under conditions t_j and σ_j.

It can be seen from Eqs. (3.73) that the assessment of the durability of the material as the sum of the relative durability assumes, firstly, the absence of irreversible structural transformations occurring at different temperature levels, and, secondly, the absence of the dependence of the time to fracture on the number of cycles. The second assumption makes it possible to assess the durability of materials at cyclically changing temperatures or stresses based on the results of temperature-step tests or tests under a step-changing load. The effect of a change in the test time t and stress σ is usually expressed using conditional characteristics: the equivalent stress σ_{eq}, the equivalent time to fracture $t_{\mathrm{f\,eq}}$ and the equivalent temperature T_{eq}, the formulas for which can be derived from Eq. (3.73).

Using the hypothesis of linear damage summation (3.73), Kinasoshvili (1959) proposed a graphical method for assessing the safety margins for stress k for multi-stage changes in temperature and stress in a component, based on the assumption that when calculating t_f, the stress is equal to $\sigma_{\mathrm{fict}} = k\sigma_{\mathrm{real}}$, ($\sigma_{\mathrm{fict}}$ and σ_{real} are fictitious and real stresses). The time to fracture at a cyclically changing temperature can be determined from the formula

$$t_{\mathrm{f\,eq}} = \frac{\sum t_j}{\sum \frac{t_j}{t_{\mathrm{f}j}}} \quad \text{or} \quad t_{\mathrm{f\,eq}} = \frac{t}{\int_0^{t_{\mathrm{f}}} \frac{\mathrm{d}t}{t_{\mathrm{f}}[\sigma(t)]}}. \tag{3.75}$$

In order to experimentally verify the validity of the model at variable temperatures, a number of industrial alloys on an iron and nickel base, as well as aluminum and titanium alloys were tested in rectangular, triangular, trapezoidal and multistage heating cycles, as well as temperature-step tests (Getsov and Taubina, 1960; Pisarenko, 1965, 1967, 1969; Birger et al., 1975) the cycle varied within wide limits (from 20 s to 24 h).

Predominantly smooth samples were tested. The exhaustion of durability at the maximum temperature of the two- and multi-stage cyclic tests was 30-90% of the total durability. For each mode, 5-8 samples were tested at different stresses and the results of these tests were used to construct long-term strength curves. Under the conditions of the accepted test regimes with a duration of 2000-4000 h, the materials under study can be practically considered structurally stable. To compare the experimental and calculated values of the long-term strength during cyclic tests, as a rule, the characteristics of the long-term strength of the metal of the same heats as at constant temperatures were studied (Figs. 3.35 and 3.36). It should be noted that all failures occurred at temperatures corresponding to the maximum exhaustion of durability.

Analysis of the test results shows the following. The long-term strength of refractory materials at cyclically changing temperatures in logarithmic coordinates is described by straight-line dependencies in a time range of at least two orders of magnitude. Satisfactory agreement between the calculated and experimental values

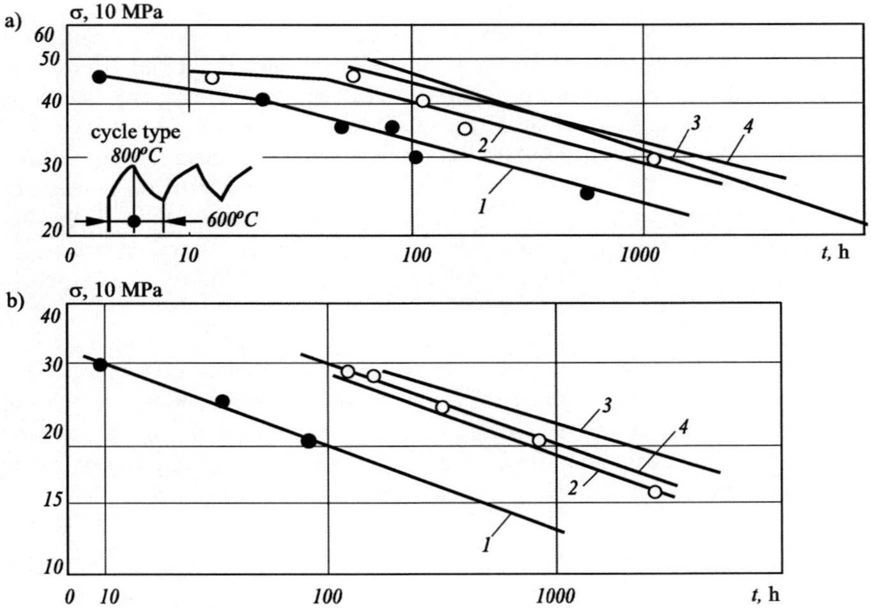

Fig. 3.35 Long-term strength of alloys EI827 (a) and EI765 (b): *1* - at a temperature of 800°C; *2, 3, 4* - cyclically changing temperature; *2* - experimental data; *3, 4* - calculated data

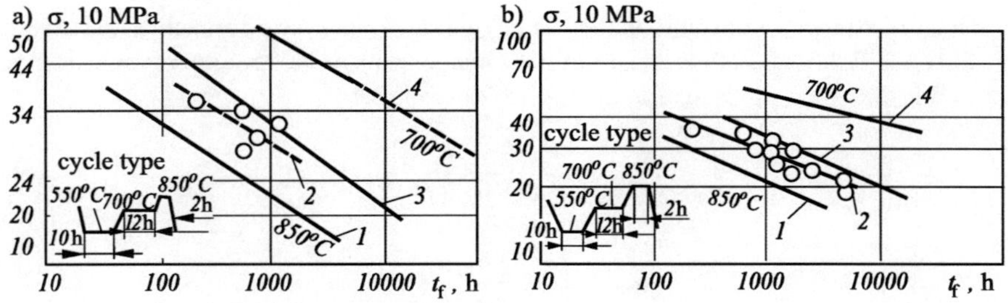

Fig. 3.36 Long-term strength of alloys EI661 (a) and EI827 (b): *1, 4* - constant temperature; *2, 3* - cyclically changing temperature in three-stage modes; *2* - experimental data; *3* - calculated data

of durability is observed. The root-mean-square deviation in time to failure is 10%, in stress - 2% (Fig. 3.37). The absence of influence on the durability of the frequency of cyclic temperature changes (in the range from 3 to 1/1440 cycles/min) was established. No specific effect of the notch was found in testing specimens at cyclically varying temperatures (compared to test results at constant temperatures).

The long-term strength of heat-resistant alloys at cyclically varying stresses was studied in Guarnieri and Yerkovich (1952); Zhurkov and Tomashevskii (1959); Getsov and Taubina (1960); Pisarenko (1965) among others. The processing of the results of these tests showed that in the event of failure, the sum of the relative fatigue life $(t_j/t_{fj}) = 0.2 \ldots 1.5$, and the value of (t_j/t_{fj}) depends on the duration time at the maximum stress and on the number of overloads. One of the reasons for the discrepancy between the calculated and experimental values of t_f is the structural instability of materials in the studied temperature ranges. The variety of structural changes that can take place in a material at elevated temperatures, and the possibility of their different effect on the behavior of the material at temperatures cyclically changing in various modes, require special approaches to solving the problem of

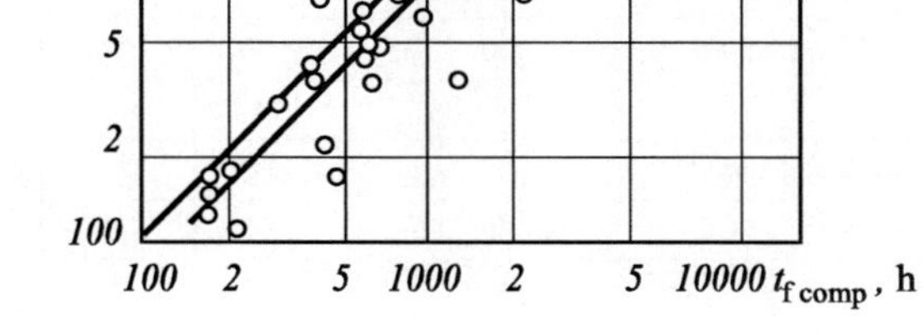

Fig. 3.37 Comparison of calculated ($a = 1$) and experimental ($a_{avg} = 0.8$) fracture time values at cyclically changing temperature: $a \sum(t_i/t_{fi})$

the heat resistance of unstable materials in general. Using the approaches discussed in Subsect. 2.4.5, the effect of structural changes can be taken into account using formulas that are a modification of Eqs. (3.73) and (3.74)

$$t_{\mathrm{f\,eq}} = \sum \frac{t_j}{\sum(t_j k_j / t_{\mathrm{f}j})}, \tag{3.76}$$

$$t_{\mathrm{f\,eq}} = \left(\sum \frac{\dot{p}_{\mathrm{avg}j} \alpha_j}{\varepsilon_{\mathrm{pl\,f}j} \beta_j} \right)^{-1}, \tag{3.77}$$

where k_j, β_j and α_j are correction factors that take into account the history of temperature effects on the properties of the alloy (Tables 2.7 and 3.15). The values of k_j and β_j are determined from experimental data at σ = const as follows

$$k_j = t_{\mathrm{f\,init}} / t_{\mathrm{f\,age}}, \quad \beta_j = \varepsilon_{\mathrm{pl\,f\,init}} / \varepsilon_{\mathrm{pl\,f\,age}}, \tag{3.78}$$

where, the indices "init" and "age" refer to the values of t_{f} and $\dot{p}_{\min}$ of the original and aged metal.

As an example of a calculation using the proposed method, let us assess the performance of a gas turbine disk made of EI787 steel under the following operating conditions:

Turbine operating mode for load Q,%	80	100	125
Motor resource, h	10000	500	20
Maximum temperature, °C	650	700	750

If we do not take into account the influence of microstructural changes in the metal during operation, then the estimated equivalent time to fracture for σ = 350 MPa is $t_{\mathrm{f\,eq}}$ = 3840 h. Taking into account the effect on the properties of only prolonged aging

Table 3.15 Coefficients of the microstructure influence on durability and creep rate

Material	Low temperature aging mode		k_j	β_j
	T,°C	T,°C		
EI481	600	10000	1	-
		20000	3	
	650	10000	2	0.50
EI703	750	2000	1.25-4.00	0.35-0.42
		5000	1.50	0.59-1.08
EP126	750	5000	5	-
EI787	650	10000	9-40	-
		30000-50000	23	0.10
EI827	700	10000	1	-
	800	10000	4	-

at a temperature of 650°C ($k_j = 20$), the equivalent time to failure is $t_{f\,eq} = 292$ h, which gives a difference in the equivalent stress of about 35%.

3.3 Resistance to Fracture for Cyclically Varying Stresses

Cyclic loading of a material depending on temperature, frequency, and cycle asymmetry can cause both quasi-static failure[10] and fatigue failure[11]. With constant sign cyclic stresses, fatigue fracture processes are also observed, caused by the fact that during cyclic deformation processes of reverse elastic aftereffect occur - the width of the hysteresis loop is not zero.

As already noted, creep processes develop in structural materials at temperatures close to room temperature. Hence, the fracture stresses under cyclic constant-sign loading can be significantly less than the ultimate strength. In the presence of stress concentrators, unloading causes the formation of residual stresses, which can be so significant that irreversible deformations of the opposite sign are formed. Under conditions of alternating loading, fatigue fracture is, as a rule, deformation-free from a macroscopic point of view: cyclic deformations during a symmetric cycle are localized only within microvolumes. Similarly to low-temperature creep, the determining mechanism of fatigue at low temperatures is the mechanism of intersection of dislocations: fatigue cracks form and grow in those places where the concentration of defects reaches a critical value.

The fatigue resistance of materials is characterized by the fatigue limit related to the given asymmetry of the cycle $r = \sigma_{\min}/\sigma_{\max}$, temperature, number of cycles, type of stress state and frequency. In this section, we will consider the features of the behavior of materials under high-frequency and low-frequency loading under conditions of constant and varying temperatures. The basic concepts and designations used when considering fatigue processes are defined in GOST 23207-78.

3.3.1 High Frequency Fatigue

Fatigue damage is usually assessed based on the assumption of a linear law of damage accumulation $\Pi = n/N_0$, where n is the number of cycles and N_0 is the number of cycles to failure. For nonstationary conditions $\Pi = \sum n_i/N_{0i}$. At the moment of failure, $\Pi = 1$. In real situations, $\Pi \neq 1$. Thus, in two-stage tests, when the stress amplitude at the first stage is higher than at the second one, accelerated damage is usually observed, which reduces the overall life ($\Pi < 1$). It has been established in NN (1968, 1970) that when passing from a high stress level $\sigma_{a,1}$ through n_1 cycles to

[10] One of the types of quasi-static fracture is fracture from cyclic creep.

[11] Usually we distinguish between high-frequency fatigue ($10^7 - 10^8$ cycles), low-cycle (low-frequency) fatigue ($10^3 - 10^4$ cycles), fatigue in the intermediate region ($10^5 - 10^6$ cycles).

a lower stress level $\sigma_{a,2}$, cyclic microplastic deformations develop more intensively than in tests at a constant stress level $\sigma_{a,2}$ (Fig. 3.38).

When moving from low to high levels, a training effect can be observed that increases overall durability. To assess fatigue damage under a complex loading program, a number of dependencies based on nonlinear summation of damage were proposed: the formulas of Gerber (1874); Dolan and Brown (1952); Henry (1955); Corten and Dolan (1956); Levy (1957); Freudenthal and Heller (1959); Serensen (1975a); Wilkins (1999) and Kogaev (1993). In Troshchenko (1979), a relationship was established between the energy dissipated per cycle (upon reaching a stabilized state) $D = k_{\phi}\delta\sigma_a$ and the number of cycles before failure. Here k_{ϕ} is the form factor.

3.3.1.1 Influence of Cycle Asymmetry

Various authors have proposed formulas for estimates of the influence of cycle asymmetry on the fatigue resistance of materials (Table 3.16). Comparison of the results of calculations using different formulas showed that the most conservative for steels and nickel-based alloys are calculations by the formulas of Soderberg (1930) and Serensen-Kinasoshvili. The most universal for high-frequency fatigue at

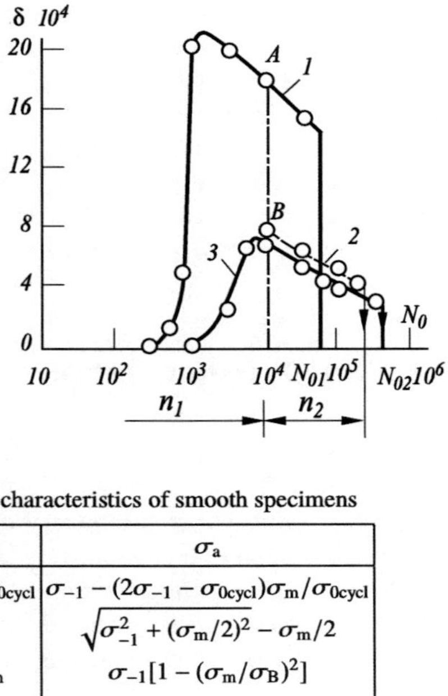

Fig. 3.38 Change in the width of the hysteresis loop δ during the test with $\sigma_a = \sigma_{a,1}$, (*1*), $\sigma_a = \sigma_{a,2}$, (*3*) and with $\sigma_a = \sigma_{a,2}$, after preliminary test with $\sigma_a = \sigma_{a,1}$, (*2*)

Table 3.16 Analytical relationships between fatigue characteristics of smooth specimens

Author(s)	σ_{max}	σ_a
Serensen-Kinasoshvili	$\sigma_{-1} + 2\sigma_m(\sigma_{0cycl} + \sigma_{-1})/\sigma_{0cycl}$	$\sigma_{-1} - (2\sigma_{-1} - \sigma_{0cycl})\sigma_m/\sigma_{0cycl}$
Oding	-	$\sqrt{\sigma_{-1}^2 + (\sigma_m/2)^2} - \sigma_m/2$
Gerber (1874)	$\sigma_{-1}[1 - (\sigma_m/\sigma_B)^2] + \sigma_m$	$\sigma_{-1}[1 - (\sigma_m/\sigma_B)^2]$
Goodman (1899)	$\sigma_m(1 - \sigma_{-1}/\sigma_B) + \sigma_{-1}$	$\sigma_{-1}(1 - \sigma_m/\sigma_B)$
Soderberg	$\sigma_m(1 - \sigma_{-1}/\sigma_y) + \sigma_{-1}$	$\sigma_{-1}(1 - \sigma_m/\sigma_y)$

σ_{0cycl} - fatigue strength for pulsating cycle ($\sigma_m = \sigma_a$)

elevated temperatures is the Alishoev (1968) dependence $\sigma_a = \sigma_{-1} - \psi_0\sigma_m$. The coefficient ψ_0 depends to a very large extent on the creep rate that the material has at a given static stress and temperature. It was found in Alishoev (1968) that for low creep rates ($\dot{p}_{min} < 10^{-7}$%/h), as a rule, $\psi_0 = 0.1 \ldots 0.15$, and at creep rates of $10^{-5} < \dot{p}_{min} < 10^{-4}$%/h for most austenitic steels and nickel alloys the value of ψ_0 is in the range of $0.6 \leq \psi_0 \leq 1.0$. At a high level of static stresses (when $\dot{p}_{min} > 10^{-4}$% / h), they can, as shown in a number of studies, cause an increase in the material's endurance limit. In this case, the coefficient $\psi_0 < 0$.

To determine the fatigue resistance at room temperature, Smith (1910) and Haigh (1917) diagrams are also used (Fig. 3.39). With a constant number of cycles, with an increase in the average stress, the maximum stress increases, and the limiting amplitude decreases (Fig. 3.39 and Table 3.16). Since the dependence of the fatigue limits on the average cycle stresses in the compression region has been studied very little, the possibility of constructing Smith and Haigh diagrams in the compression region from the data of tensile experiments is debatable. Substantial development was obtained by Medekshas, who studied the problems of crack closure in a compression half-cycle and published the results in Balina and Myadekshas (1994).

3.3.1.2 Statistical Processing of Test Results

When assessing fatigue strength, statistical methods are used that are similar to those applied for assessing long-term strength (see Sect. 3.2). The results of statistical processing of experimental data of fatigue tests have shown that the scatter of fatigue resistance characteristics is well described by the law of normal distribution (Serensen and Kogaev, 1969). Thus, in Akimov (1977), the results of processing the experimental data of heat-resistant alloys of various grades are presented. A linear dependence of the form $\lg N = a - b \lg \sigma$ is accepted. The values of the coefficients a and b, as well as the standard deviation S and the correlation coefficient r were determined depending on temperature, environment, cycle asymmetry and sample shape (Table 3.17).

It has recently become common to represent fatigue curves in terms of the probability of failure. These curves can have a constant slope, kinks in different directions, including those with an exit to a horizontal section (fatigue limit at $N > N_0$, independent of the number of cycles). The scatter of the durability values depends on

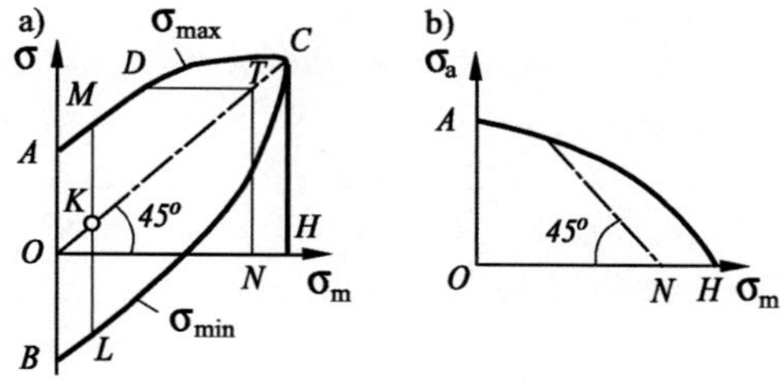

Fig. 3.39 Smith (a) and Haigh (b) charts: *HC* - tensile strength: *HC* = *OH*; *ON* - yield strength, *ON* = *NT*; *OA* - symmetrical cycle fatigue limit; *OA* = *OB*; *KL* = *KM*; *DT* - line limiting the plastic deformation region

Table 3.17 Fatigue resistance parameters of round specimens in air at symmetric cycle. The values of the coefficients are given for stresses σ, 10 MPa

Material	T,°C	a	b	S	r
EI437B	700	31.2	16.4	0.28	0. 94
	600	25.2	11.7	0.36	0.80
EI893	800	26.1	12.6	0.43	0.89
	700	30.8	16.3	0.31	0.92
	600	20.9	9.2	0.29	0.92
EI867	900	22.4	11.2	0.32	0.96
ZhS6K	900	23.2	11.6	0.48	0.95

the acting shear mechanisms of the vacancy-diffusion type. For the statistical description of fatigue fracture processes under the action of both mechanisms, instead of a normal distribution, the concept of bimodal distributions is used, according to which the distribution function of the mixture is determined through the distribution functions of the components using weight coefficients (Zakharova, 1974).

3.3.1.3 Effect of Stress Concentration

The extent to which a particular metal is sensitive to stress concentration under cyclic loading is usually judged by the values of the effective stress concentration coefficient $k_{\text{eff}} = \sigma_{\text{a sm}}/\sigma_{\text{a}N}$, where $\sigma_{\text{a sm}}$ is the limiting stress amplitude for a smooth sample and $\sigma_{\text{a}N}$ is nominal limiting amplitude for a sample with a stress concentrator. The coefficient k_{eff} can be classified for amplitude, average stress, asymmetric cycle with a given degree of asymmetry, different numbers of cycles until failure N_0 of stress states (tension-compression and bending).

Under cyclic loading, the effective concentration coefficient is almost always less than the theoretical one ($k_{\text{eff}} < \alpha_N$). This is due to the redistribution of stresses caused by material flow in the plastic zone and creep, the influence of the stress gradient and the scale factor. In the general case, the fracture conditions depend on the combination of principal stresses at the surface of the concentrator and over the section (see Subsect. 3.1.1). Which of the following factors is decisive depends on the material, stress level and temperature. With an increase in σ_{b}, the value of k_{eff} increases. So, for example, for titanium alloy WT22M E.A. Borisova, A.F. Matveenko et al. obtained the following data:

σ_{B}, MPa	1010	1030	1120	1150	1310	1430	1610
ψ, %	22.5	22.5	14.5	7	21.5	18.5	19
k_{eff}	1.84	1.8	2.46	3.36	2.7	2.68	2.98

A decrease in ductility leads to an increase in the value of k_{eff}. To take into account the mutual influence of the stress concentration caused by the inhomogeneity of the metal structure and the geometrical dimensions of the concentrator, Heywood (1962) proposed the following formula

$$k_{\text{eff}} = \alpha_N/[1 + 2\sqrt{a/\rho}(\alpha_N - 1)/\alpha_N], \tag{3.79}$$

where a is the coefficient of attenuation of the stress concentration in the material in the presence of its inherent equivalent defects.

A decrease in the size of the defect, the radius of the hole or the radius at the base of a stress concentrator of a different shape (ρ) leads to a decrease in the value of the effective stress concentrator k_{eff} during fatigue. This is due to the fact that k_{eff} is influenced not only by the value α_N, but also by the size of the plastic region near the concentrator, which decreases with decreasing ρ (Pickett, 1968).

With increasing temperature and decreasing α_N, the value of k_{eff} decreases. For cast alloys, k_{eff} is less than for forged ones. As a characteristic of the sensitivity of a material to a notch, the following sensitivity coefficient ranging from 0 to 1 is proposed

$$q = (k_{\text{eff}} - 1)/(\alpha_N - 1), \tag{3.80}$$

3.3.1.4 Effect of Temperature

The temperature has different effects on the fatigue characteristics of different metals (Table 3.18). This is due to differences in the temperature dependencies of such material characteristics that affect fatigue resistance, such as yield strength, creep resistance, deformation capacity, the nature of cyclic instability, etc. A decrease in the yield stress of a material with an increase in temperature causes an increase in the width of the hysteresis loop of a cyclically stable material tested at a constant stress amplitude, which, with a constant deformation capacity in the considered temperature range, leads to a decrease in its durability. The nature of the dependence of the creep resistance on temperature is decisive when comparing the durability, especially under conditions of low-frequency tests with stress amplitudes that are less than the cyclic proportionality limit. The foregoing is well illustrated by experiments carried out under conditions excluding the occurrence of creep processes. Analysis of the results of studies of the temperature dependence of the fatigue strength of titanium and iron shows that in the temperature range causing strain aging, the fatigue limit increases to values equal to the cyclic proportionality (elasticity) limit. The intensification of the processes of cyclic softening of the metal with an increase in temperature leads to an accelerated depletion of the plasticity resource and premature failure. The decrease in the fatigue resistance of high-temperature alloys with an increase in temperature and a decrease in the high-temperature strength of the material is illustrated by the data given in Table 3.17, and the curve in Fig. 3.40.

It should also be noted that as the temperature rises, the concept of the fatigue limit without specifying the test base loses its meaning, since the fatigue resistance under these conditions depends on the number of cycles, no matter how large it may be. At very high temperatures, the slope of the fatigue curves for heat-resistant alloys can sharply increase upon reaching a large number of cycles, which is associated with

Table 3.18 Fatigue resistance of heat-resistant nickel alloys at high temperatures (according to Pogrebnyak's data). I, II - corset and cylindrical shapes of the working part of the sample, cb - cantilever bending with rotation, pb - pure bending with rotation

Alloy	Endurance limits, MPa (2 10^7 cycles)			*f*, Hz	Type of loading	Type of specimen
	600°C	700°C	800°C			
EI437BVA	517.5	345.8	286.8	50	pb	I
EI437B (melt 1)	-	335.7	-	50	cb	I
EI437B (melt 2)	-	321.8	-	50	cb	I
EI437B (melt 2)	-	400.5	-	100	pb	II
EI607A (melt 1)	360.0	371.5	-	50	pb	I
EI607A (melt 2)	-	350.8	-	100	pb	II
EI929VA	348.0	374.6	377.7	50	pb	I
EI617	320.8	346.6	313.0	50	pb	I
EI783	350.0	371.6	356.0	50	pb	I
EI698VA	340.0	380.0	324.4	50	pb	I
EI867	320.0	330.0	389.1	50	pb	I
ZhS3DK	-	295.1	-	50	pb	I
VZh105 (melt 1)	559.7	-	-	50	pb	I
VZh105 (melt 2)	460.0	-	-	50	pb	I
VZhL12U	200.0	-	-	50	pb	I
ZhS6K	254.7	-	313.6	50	pb	I
EI661 (melt 1)	371.8	382.4	386.3	50	cb	I
EI661 (melt 2)	307.3	299.0	311.2	50	cb	I
EI893	327.0	-	-	100	pb	II
EI826	-	371.0	337.9	100	pb	II
EI827	-	281.0	-	100	pb	II
EI539	-	365.0	383.0	100	pb	II
EI617	-	360.0	-	100	pb	II

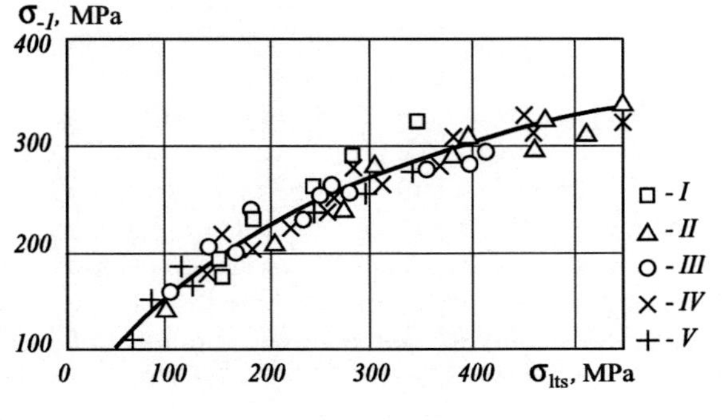

Fig. 3.40 Dependence of the fatigue limit σ_{-1} under tension-compression on the long-term strength σ_{lts}, corresponding to the same time to failure (100-1000 h) for alloys of the Nimonic type: *I, II, III* - Nimonic 80; *IV, V* - Nimonic 90 (Betteridge and Heslop, 1974)

softening at these temperatures. In high-heat-resistant casting alloys of the ZhS6K, WZhL12 type, during solidification and cooling from the quenching temperature, significant stresses are formed at the interphase boundaries (up to 400 MPa). Under loading conditions, the acting stresses are added to the stresses created by internal defects (large carbides, micropores, etc.), which turns out to be equivalent to the action of external concentrators. As a result, the fatigue resistance of such alloys is reduced, especially at low temperatures. With an increase in temperature, σ_{-1} increases due to the processes of stress relaxation and an increase in plasticity (Fig. 3.41). The improvement in the morphology of the carbide phases, achieved by microalloying with hafnium, and the partial relaxation of interfacial stresses during annealing at 950°C lead to an increase in the fatigue strength (Fig. 3.42).

The effect of the loading frequency on the fatigue strength under high-frequency loading is much weaker than under low-frequency loading. Nevertheless, the direct use of the data obtained at frequencies of 3000-5000 Hz in relation to the conditions of cyclic loading with frequencies of 50-20 Hz is considered unreasonable. The nature of the frequency dependence of σ_{-1} is usually determined under conditions where heating of the samples during the test is excluded. According to N.D. Zhukov, with an increase in the frequency from 500 to 2000 Hz, the fatigue resistance of the EI617 alloy at 850°C increases from 300 to 400 MPa for 107 cycles and from 260 to 320 MPa for 108 cycles. Upon reaching a certain critical value (1000-2000 Hz), the dependence of σ_{-1} on frequency has a maximum.

3.3.1.5 Influence of the Type of Loading

The type of loading influences the fatigue strength. A large number of studies are known that indicate that under conditions of an uneven stress state during bending, the fatigue limit σ^{b}_{-1} has greater values than under tension-compression σ^{t-c}_{-1}. So, for the Nimonic 90 alloy at 750°C (see Table 3.19) $\sigma^{b}_{-1} = 359$ MPa and $\sigma^{t-c}_{-1} = 241$ MPa.

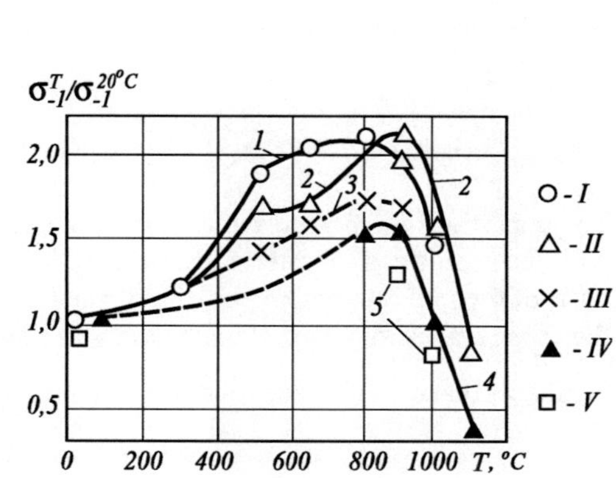

Fig. 3.41 Dependence of the endurance limit of cast heat-resistant alloys $\sigma^{T}_{-1}/\sigma^{20°C}_{-1}$ on temperature *T*:
1, *I* - ZhS6K;
2, *II* - ZhS6U;
3, *III* - WZhL12U;
4, *IV* - ZhS6UDC;
5, *V* - WZhL12U

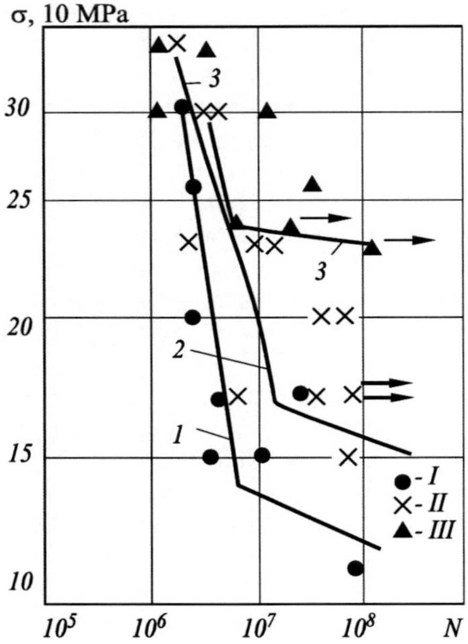

Fig. 3.42 Fatigue curves of the ZhS6U alloy at a temperature of 20 C after various heat treatment: *I, I* - 1210°C, 4 h, air; *2, II* - 1210°C, 4 hours, air; 950°C, 4 h, argon; *3, III* - microalloying with Hafnium, 1210°C, 4 h, air

Table 3.19 Fatigue resistance of Nimonic alloys for $(0.45\ldots1.2)10^8$ cycles at different temperatures in a symmetric cycle (Betteridge and Heslop, 1974)

Material	Type of test	σ_{-1}, 10 MPa, at T, °C								
		600	650	700	750	800	815	870	940	980
Nimonic 80	Tension-compression	29.9	28.8	26.3	19.5	14.2	-	-	-	-
Nimonic 90	Bending with rotation	-	-	-	35.9	23.6	16.1	-	-	-
Nimonic 90	Tension-compression	-	46.9	-	29.3	24.1	-	17.2	10.4	-
Nimonic 95	Bending with rotation	-	-	-	-	41.3	-	32.1	26.5	-
Nimonic 100	Bending with rotation	-	-	-	44.1	-	37.8	30.7	18.1	15.5

3.3.1.6 Fatigue Under Complex Stress State

The conditions for fatigue failure for a complex stress state are usually associated with the conditions for failure in a uniaxial stress state by relations obtained on the basis of the same strength theories as for static fracture (Serensen, 1975b). Consider these conditions for proportional loading in a symmetric cycle

- in accordance with the hypothesis of the maximum shear stress

$$\sigma_{eq} = \sigma_{a1} - \sigma_{a3} = \sigma_{-1}$$

- in accordance with the hypothesis of the octahedral stress

$$\sigma_{eq} = \sigma_a = \sigma_{-1},$$

where $\sigma_a = (\sigma_{a1} + \sigma_{a2} + \sigma_{a3})/3$

A dependence was also proposed with one material parameter equal to the ratio of the endurance limits under variable tension-compression and variable shear σ_{-1}/τ_{-1} (Troshchenko, 1977)

$$\sigma_{eq} = \frac{2}{\sqrt{3}-1}\left(\frac{\sigma_{-1}}{\tau_{-1}} - 1\right)\sigma_a + \frac{1}{\sqrt{3}-1}\left(\sqrt{3} - \frac{\sigma_{-1}}{\tau_{-1}}\right)\sigma_{1a} = \sigma_{-1} \tag{3.81}$$

For conditions of combined action of biaxial static tension and alternating torsional and bending stresses Shkanov et al. (1979) proposed the following criterion

$$\frac{\chi_a\sigma_a + (1-\chi_a)\sigma_{1a}}{\sigma_{-1}} \pm \left[\frac{\chi_m\sigma_m + (1-\chi_m)\sigma_{1m}}{\sigma_B}\right]^n = 1, \tag{3.82}$$

where $\chi_a = (\sigma_{-1}/\tau_{-1})/(\sqrt{3}-1)$, $\chi_m = (\sigma_B/\tau_B - 1)/(\sqrt{3}-1)$, σ_B and τ_B are ultimate strength (long-term strength) in tension and shear, respectively, n is a coefficient. For high-strength steels and alloys, the value of n is found from the ratio

$$n = \left(\frac{2\sigma_{-1}}{\sigma_B}\right)^2 + \frac{(\sigma_B/\sigma_{-1} + 1)}{(\sigma_B/\sigma_{-1})}\left[\frac{\chi_m\sigma_m + (1-\chi_m)\sigma_{1m}}{\sigma_B}\right]^2. \tag{3.83}$$

Satisfactory agreement was obtained between the calculated and experimental values of the fatigue life for a number of structural and heat-resistant alloys with variable bending and biaxial static tension, with alternating torsion and biaxial static tension, with joint alternating bending and torsion in combination with biaxial static tension (Shkanov et al., 1979).

3.3.1.7 Complex Influence of Various Factors

How strong the decrease in the fatigue limit can be under the action of various factors can be judged from the data in Table 3.20. Its consideration shows that in the case when it is not possible to eliminate the corrosive environment and stress concentration, with a small asymmetry of the cycle ($\sigma_m < 150$ MPa), the permissible stress amplitude should not exceed $[\sigma_a] = 50$ MPa, and with large σ_m - $[\sigma_a] = 10$ MPa.

Methods for calculating the characteristics of fatigue resistance of machine components and structural elements made of steel in the high- and low-cycle elastic and elasto-plastic regions are regulated by GOST 25.504-82. The determined characteristics include: median values of endurance limits, endurance limits for given fracture probabilities and component sizes, variation coefficients, coefficients of sensitivity to cycle asymmetry, stress concentration coefficients, etc.

To assess the complex effect of various factors on the endurance of alloys, Kononchuk (1966) proposed an empirical coefficient

Table 3.20 Comparison of calculated and experimental values of fatigue limits under conditions of action of various factors

Material	Specimen	Test conditions					σ_a,	σ_a, MPa
		T, °C	σ_m, MPa	Corrosive environment	Notched	Specimen diameter, mm	MPa	(air, smooth specimen)
20Ch13	Round	80	245	20% NaCl	-	-	50	450
08Ch21NM3D	Round	80	245	20% NaCl	-	-	70	-
EI961	Lock connection	20	340	Air	Special	-	150	550
14Ch17N2	Round	20	0	Sea water	$r = 0.25$ mm	-	70	450-575
EI481	Round	650	320	Air	$r = 0.5$ mm	-	65	320
EI481	Round	650	320	Air	Special	-	20	320
15Ch5MF	Round	20	600	Sea water	-	-	15	-
EI893	Shank	550	100	Air	Special	-	105	-
35	With steel sleeve	20	-	Sea water	-	5	30	270
						50	60	205
35	with brass sleeve	20	-	Sea water	-	5	15	270
						50	80	205

$$k_{sum} = (\sigma_{-1oper}/\sigma_{-1sm} = k_1, k_2, k_3, \ldots, k_{n0}(\sqrt{q_k})^{n_0-1}, \qquad (3.84)$$

where σ_{-1oper} is the blade endurance limit under conditions close to operation, σ_{-1sm} is the endurance limit of smooth samples in air, $k_1, k_2, k_3, \ldots, k_{n0}$ are influence coefficients and $q_k = n_0/(k_1 + k_2 + \ldots + k_{n0})$.

For experimental verification of Eq. (3.84), tests were carried out on the alloys EI437A, EI617, EI867 (Akimov, 1977). In separate experiments for the probability of non-failure of 0.5, the coefficients k_j were determined, characterizing the following effects

- dispersion of properties (when testing profile samples) $k_1 = \sigma_{-1prof}/\sigma_{-1\,S=0.5}$;
- sample shapes $k_2 = \sigma_{-1prof}/\sigma_{-1cycl}$;
- environments (for profile samples) $k_3 = \sigma_{-1prof\,warm}/\sigma_{-1prof\,air}$;
- heat change (for profile samples) $k_4 = \sigma_{-1prof\,warm}/\sigma_{-1prof\,air}$

Comparison of the coefficients k_{sum}, obtained experimentally $k_{sum}^{exp} = \sigma_{-1exploit}/\sigma_{-1sm}$ and calculated by the formula $k_{sum}^{comp} = k_1k_2k_3k_4(\sqrt{q_k})^3$, showed (Table 3.21) the satisfactory accuracy of the estimation method using relation (3.84).

Table 3.21 Comparison of calculated and experimental values of fatigue limits under conditions of action of various factors

Material	T, °C	t, h	k_{sum}^{exp}	k_{sum}^{comp}
EI437	700	0.5	0.86	0.59
	700	5	0.56	0.47
	700	50	0.38	0.38
EI617	800	0.5	0.85	0.70
	800	5	0.70	0.57
	800	50	0.52	0.47
EI867	900	0.5	0.66	0.53
	900	5	0.57	0.55
	900	50	0.47	0.48

3.3.2 Low Cycle Fatigue

Low-cycle fatigue is the main factor that determines the service life of the main parts of aviation gas turbine engines[12]. Therefore, it is customary to use the concept of a flight cycle as a unit of resource in aircraft engine manufacturing. In contrast to high-frequency fatigue processes, in which fractures are observed at a large number of cycles, low-frequency loading leads to the formation of cracks in a number of cases after several cycles. Determination of the fracture characteristics in relation to the number of cycles characteristic of tests under high-frequency loading ($10^7 - 10^8$) is experimentally impracticable under these conditions. That is why fracture processes under conditions of low-frequency loading are usually called low-cycle fatigue. The conditions of low-frequency fracture largely depend on whether the loading cycle is sign-constant or sign-alternating and whether the processes of cyclic creep are realized or not.

3.3.2.1 Basic Features

The main condition that indicates the accumulation of damage under low-frequency loading and the ability of the material to fail is the presence of cyclic irreversible deformations (the range of inelastic deformations, hysteresis loops). The possibility of the formation of such deformations exists not only under alternating sign, but also under constant sign loading. Under these conditions, they are formed either in the presence of concentrators that contribute to the formation of significant residual stresses (higher than σ_y), or due to the reverse aftereffect under creep conditions,

[12] In the former Soviet Union 5 times were organized symposia on low cycle fatigue problems: first in 1971 in Kaunas (Kagan, 1971), second in 1974 in Chelyabinsk (Gokhfeld and Chernyavsky, 1974), third in 1979 in Palanga, fourth in 1983 in Krasnodar, fifth in 1987 in Volgograd (Gadenin et al., 1979; Makhutov et al., 1983, 1987), and sixth in Kishinev (Makhutov and Gadenin, 1991).

or in connection with microplastic deformations. For example, in tests of titanium alloy VT8 at cyclic constant stresses at 20 and 450°C (temperatures at which creep processes develop intensively), fatigue fractures were observed at stresses lower than the elastic limit, while at 350°C, which causes strain aging alloy, sharply inhibiting creep, failure at stresses lower than σ_{pl} was not observed.

3.3.2.2 Influence of Frequency

At elevated temperatures, an increase in frequency in most cases leads to an increase in durability (Fig. 3.43). The degree of influence of the loading frequency on the endurance characteristics depends on the material, the nature of the loading, the stress level (the ratio of the maximum cycle stress to the elastic limit) and temperature.

A number of authors use Eckel's formula (Eckel, 1951) to describe the effect of frequency ω on the durability of materials t_f

$$\lg t_f = \lg b_e - m_e \lg \omega \tag{3.85}$$

where b_e and m_e are material constants. In those cases when the time to fracture does not depend on the frequency, $m_e = 0$.

The influence of frequency on fatigue strength is due to its effect on the elastic hysteresis (unloading modulus) of a given metal, the elastic limit and plastic deformation per cycle, and the accumulated creep deformation per cycle, which in turn depends on the degree of hardening during short-term creep. The latter is significant for stresses below the elastic limit and temperatures at which the short-term creep curves have a hardening stage.

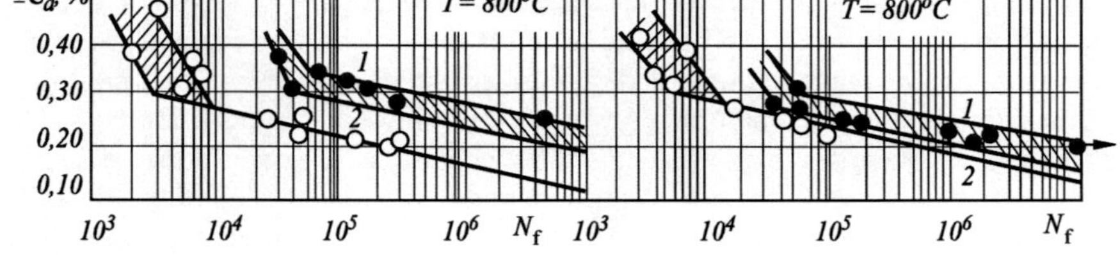

Fig. 3.43 Dependence of the number of cycles before fracture on the amplitude of deformation ε_a and the frequency of fatigue tests ω at an asymmetric cycle (σ_m = 120 MPa) for nickel-based alloys of the EI765, EI826, EI827: *1* - ω = 1200 cycle/min; 2 - ω = 10 cycle/min

The significantly different effect of frequency under cyclic loading under conditions of stresses σ_a, greater and lower than the elastic limit σ_y, is due to the fact that at $\sigma_a > \sigma_y$, the durability is determined mainly by the range of short-term plastic deformation $\Delta\varepsilon_{pl}$, which is affected by the loading time much weaker than the range of creep deformation determining the width of the hysteresis loop for $\sigma_a < \sigma_y$. In this regard, for the same values of $\Delta\varepsilon$, a change in the cycle period leads to significantly different results for materials with high and low values of yield strength (for example, high-temperature nickel-based alloys with $\sigma_{0.2} > 750$ MPa and austenitic steels with $\sigma_{0.2} \approx 200$ MPa).

Based on the deformation criterion of fatigue failure for the elastic region in the form of the following dependence (Getsov, 1971; Manson, 1973)

$$(\Delta p^{(1)})^{n_0} N_0 = C_2, \quad (3.86)$$

where C_2 and n_0 are material constants, Δp is the range of creep deformation per cycle, N_0 is the number of cycles to failure. Taking the hardening theory in the formulation (3.23), we obtain for ε_{max} = const and σ_a = const an expression that allows describing the effect of frequency ω at $\sigma_{max} < \sigma_y$ for durability in conditions of rectangular and with some approximation of sinusoidal cycle forms

$$\frac{N_{f1}}{N_{f2}} = \left(\frac{\omega_1}{\omega_2}\right)^{n_0/(\lambda+1)}, \quad (3.87)$$

where $\omega_1 > \omega_2$. When deriving formula (3.87), it was assumed that the degree of cyclic softening (or hardening) does not depend on frequency. Comparing (3.85) and (3.87), we see that

$$m_e = n_0/(\lambda + 1) - 1 \quad (3.88)$$

Since λ and n_0 depend on the stress level and temperature, expression (3.87) is more general than (3.85). Comparison of the experimental values of the EP220 alloy life, tested at 800 and 900°C and frequencies of 0.5 and 4 cycles/min, with the calculated values according to (3.86) (n_0 and A were determined from the data of other experiments) indicates their satisfactory agreement with respect to the assumptions that were made when deriving (3.87). Indeed we have

$(N_{\omega_1}/N_{\omega_2})_{exp}$	2.5	4.8	12.3
$(N_{\omega_1}/N_{\omega_2})_{calc}$	2.2	5.4	5.4

To describe the dependence of the durability of high-temperature alloys on the stress amplitude and hold time in the half-cycle of tension, Orlov and Balashov (1979) proposed the relation $\lg N_f = C - A \lg t_h - B\sigma_a$, the coefficients of which for the EP220VA and ZhS6KP alloys have the values given in Table 3.22 (t_h is the hold time in the half-cycle of tension).

Table 3.22 Values of coefficients *A*, *B* and *C*

Coefficients	EP220VA for		ZhS6KP for	
	900°C	950°C	900°C	950°C
A	0.5530	0.5924	0.6840	0.719
B	0.0622	0.075	0.0587	0.0401
C	5.6691	5.5561	6.2271	4.891

3.3.2.3 Influence of Temperature, Cycle Asymmetry, Stress Concentration and Type of Loading

The nature of the effect of temperature on low-frequency fatigue resistance depends on the extent to which creep processes affect fracture. If there is no creep, then the temperature-dependent characteristics of the material affect the durability for a given range of deformations: modulus of elasticity, ultimate strength, deformation capacity, proportionality (yield) limit.

The influence of the asymmetry of the cycle under low-frequency loading on the durability is significant, as in the case of high-frequency loading. As an example of the influence of cycle asymmetry, Fig. 3.44 shows the corresponding dependencies for EI826 and EI437B alloys, tested under conditions of alternating bending and tension as well as tension-compression. The effect of stress concentration on durability under low-frequency loading is identical to the effect under high-frequency loading. The effective stress concentration factor depends on the size of the plastic zone at the concentrator. Thus, in experiments with the EP220 alloy and a temperature of 900°C, carried out by A.G. Kostyuk, A.D. Trukhny and Y.D. Martynov (Moscow Energetic Institute), the number of cycles before cracking at $\Delta\varepsilon$ = const decreased with an increase in the diameter of a round hole in the samples.

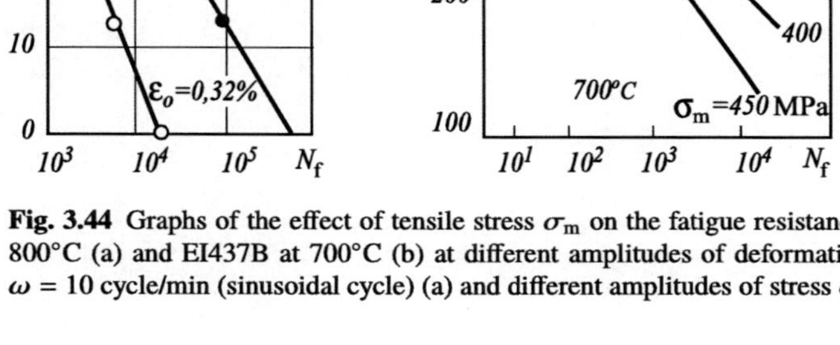

Fig. 3.44 Graphs of the effect of tensile stress σ_m on the fatigue resistance of the EI826 alloy at 800°C (a) and EI437B at 700°C (b) at different amplitudes of deformation ε_a under conditions ω = 10 cycle/min (sinusoidal cycle) (a) and different amplitudes of stress σ_a (sawtooth cycle) (b)

To study the effect of the type of loading on the resistance to low-frequency fatigue at elevated temperatures, tests of the EP220 alloy were carried out at cyclic tension-compression, bending, torsion, as well as at constant variable temperatures. The shape of the cycle under different types of loading was different. A fairly good agreement was found between the values of the durability obtained in tension-compression and torsion (Fig. 3.45), while the flexural fatigue resistance is characterized by higher values of durability and a smaller slope of the fatigue curve.

In Coutsouradis et al. (1994), the fatigue life of the SRR99 alloy was compared for various cycle forms without holding ($N_{0/0}$), with holding at σ_{max} and σ_{min} ($N_{t/t}$), with holding at σ_{max} ($N_{t/t}$), with holding at σ_{min} ($N_{0/t}$) at temperatures of 750 and 1050°C. It was found that due to changes in the values of σ_{max} and σ_{min} during cyclic loading at a temperature of 750 °C, the effect of holding at $\Delta\varepsilon$ = const is characterized by an unconventional relationship

$$N_{t/0} > N_{0/0} \approx N_{t/t} > N_{0/t}$$

At the same time, at a temperature of 1050°C, the influence of the hold time on damage is prevalent and the traditional relationship

$$N_{0/0} > N_{0/t} > N_{t/0} > N_{t/t}$$

can be applied. The standard for the test method for low-cycle fatigue under thermomechanical loading GOST 25.505-85 provides for the determination of the statistical characteristics of scatter of fatigue values at each loading level. Such a determination can be made by the methods described in GOST 25.502-79. In a number of cases, it is of interest to reveal how statistically the characteristics determined from different places of large forging, as well as those obtained on samples of different shapes, differ from each other. In this case, it is advisable to use Fisher's criterion (Fisher, 1934).

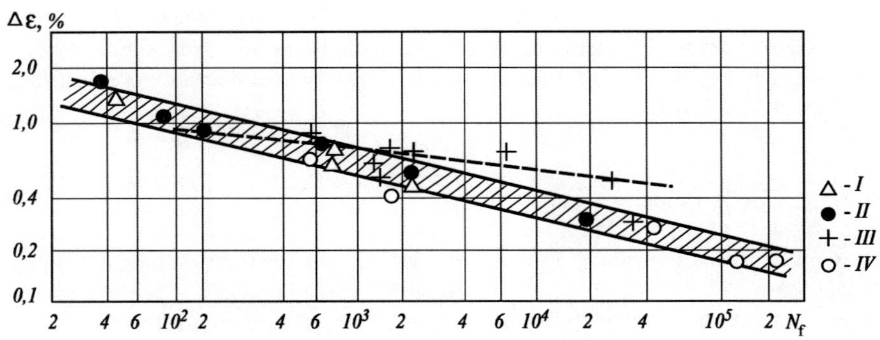

Fig. 3.45 Fatigue curves for EP220VA alloy samples: *I, II, III* - tension-compression, torsion and bending, respectively (T = 900°C = const); *IV* - tension-compression, thermal fatigue (T_{max} = 900°C)

3.3.3 Thermal Fatigue

The phenomenon of thermal fatigue of metals attracted attention back in 1912, when D.K. Chernov, investigating the causes of the cracking of gun barrels, showed the role of cyclically changing temperatures in the formation of cracks. Interest in the issues of resistance of materials to the action of thermal cycles has especially increased in recent years. This is due to the increased maneuverability and operating temperatures of modern power plants and the use of materials with low thermal conductivity. The occurrence of thermal stresses in parts is usually associated with conditions under which the geometric dimensions of the part and (or) the conditions of its operation prevent the free deformation of the metal. Thermal stresses can also arise due to the anisotropy of the linear expansion coefficient a of the metal, the difference in a for different phases, and, finally, due to phase transformations accompanied by changes in volume.

Under cyclic thermal action, metals can undergo irreversible changes in shape caused both by internal structural stresses (Davidenkov and Likhachev, 1962) and by stresses resulting from a temperature gradient over the section of the part (Balandin and Zolotukhina, 1961; Getsov, 1973; Dul'nev and Kotov, 1980; Naumenko et al., 2011b; Kostenko et al., 2013; Längler et al., 2014; Kauss et al., 2019). The processes of irreversible deformation of the second type prevail in the parts of the gas turbine unit made of heat-resistant alloys. Under the influence of cyclic thermal effects in parts, the following four cases of material behavior can occur

1. Thermal stresses are elastic, do not lead to the formation of irreversible deformations and do not cause failure
2. Irreversible plastic deformations are formed in the component during the first few cycles. However, the resulting elastic residual stresses prevent the appearance of plastic deformations in the process of further cyclic deformation - adaptability occurs
3. Alternating plastic deformations are observed during the entire process of cyclic temperature changes (thermal fatigue fractures may occur). The process of accumulation of repeated plastic one-sided deformations develops from cycle to cycle, leading to irreversible deformation. When the resource of plasticity is exhausted, the part may collapse.

3.3.3.1 Test Methods

Table 3.23 shows the typical design forms of samples for thermal fatigue testing, which are most widely used. The test methods for these samples can be divided into two groups. In the methods of the first group, thermal stresses are formed during heating and cooling of rigidly clamped specimens (see Table 3.23, specimens of types 1, 2), at all points of the working part of which the stresses are constant at a given time. The nominal stresses in this case do not depend on whether cracks have formed in the sample or not. These stresses σ in specimens of type 1 (see Table 3.23)

Table 3.23 Shape of specimens for thermal fatigue tests

Sample section	Sample type	Author(s)
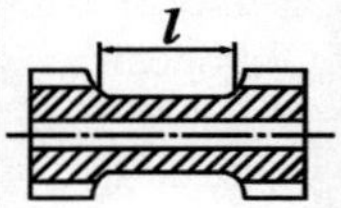	Tubular Coffin	Frey (1954)
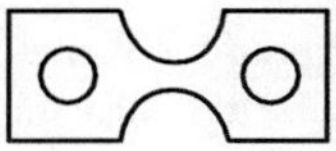	Plane with corset fillets	Gugelev and Getsov (1976)
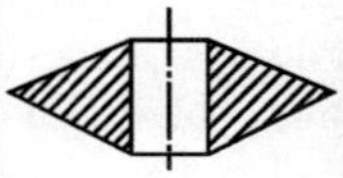	Disk	Littler (1971)
	Disk model with concentrators	Kabelevskii and Stolyarova (1976)
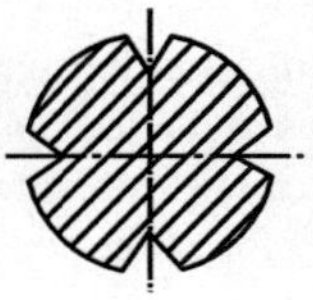	Cylindrical with grooves	Balandin and Zolotukhina (1961)
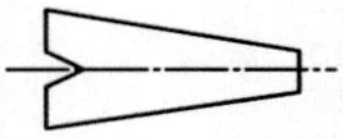	Plane	Lvovsky and Smiyan (1958)
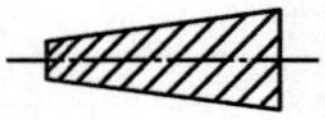	Wedge-shape	Tretyachenko et al. (1975)
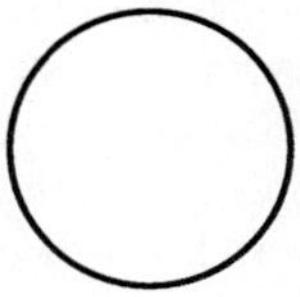	Cylindrical	Ermakov et al. (1972)

are measured during testing on machines of the Coffin type and can vary from cycle to cycle. Thus, for the condition $\Delta\sigma < \sigma_{y1} + \sigma_{y2}$ (where $\Delta\sigma$ is the stress range, σ_{y1} and σ_{y2} are the elastic limits at the maximum and minimum cycle temperatures), the corresponding deformations $\varepsilon = \sigma/E$ are known. At $\Delta\sigma > \sigma_{y1} + \sigma_{y2}$, the deformations can be determined by relatively simple calculation methods. In the test method for specimens of type 2 (Gugelev and Getsov, 1976), the total deformations ε are measured directly during the test. Knowledge of the characteristics of the thermal fatigue strength of the material, determined by the methods of this group of samples (1 and 2), allows for a calculated assessment of the thermal fatigue strength of parts. Both of these methods can be used to assess the thermal fatigue strength of macro-inhomogeneous metals. However, the test method for specimens of type 2 has certain advantages due to the small dimensions of the working part of the specimens and the ability to observe the kinetics of crack propagation.

The methods of the second group (Table 3.23, specimens of types 3, 4, 5, 6, 7 and 8) make it possible to simulate the stress state and conditions of fracture of parts close in shape and configuration, to study the processes of propagation of the formed surface cracks under conditions that decrease with distance from surface of stresses, as well as to study the effect on the number of cycles before cracking of stress concentrators of various shapes, manufactured using different technologies. In these methods, thermal stresses over time not only change during heating, cooling, and holding, but are also different over the cross section of the samples, and in the course of thermal cycling these stresses are redistributed at different points of the sample. All this makes the problem of calculating the values of σ and ε quite difficult. The values of σ and ε turn out to be largely related to the accuracy of the determination or calculation of temperature fields and the accepted hypotheses of plasticity and creep[13]. Therefore, such methods cannot be used as the simplest ones - the basic ones for determining the characteristics of materials required for calculating the strength of components. They can be used to solve the problem of determining the thermal stability of samples with a surface layer having mechanical properties and chemical composition that differ from the core, as well as with various kinds of non-metallic inclusions. Let us consider in more detail the features of test methods for specimens of types 1, 2 and 7.

3.3.3.2 Test Procedure for Installations of the Coffin Type

The samples fixed in the installation (Fig. 3.46) are periodically heated by passing current and cooled. During the tests, strains are recorded using dial indicators or wire strain gauges, the deformations of the support columns. The maximum and minimum cycle temperatures measured by a thermocouple welded in the middle of the sample are maintained in the process tests are constant. The number of cycles to failure is considered to be the number before the formation of a through crack in a thin-walled specimen. Using the characteristics of thermal fatigue resistance obtained during tests on a device with heating by passing a current to assess the strength of parts operated under conditions of another heating method, in the case of intergranular formation cracks in thin-walled specimens can give underestimated results. This is due to the possibility of local heating of the damaged metal. A certain role can be played by surface hardening formed during the manufacture of tubular specimens. Material has a tendency to decrease long-term strength with decreasing dimensions (section), it should be expected that as a result of using thermal fatigue test data for tubular specimens for parts with a larger section, underestimated durability

[13] This problem is most easily solved for cylindrical specimens (Ermakov et al., 1972).

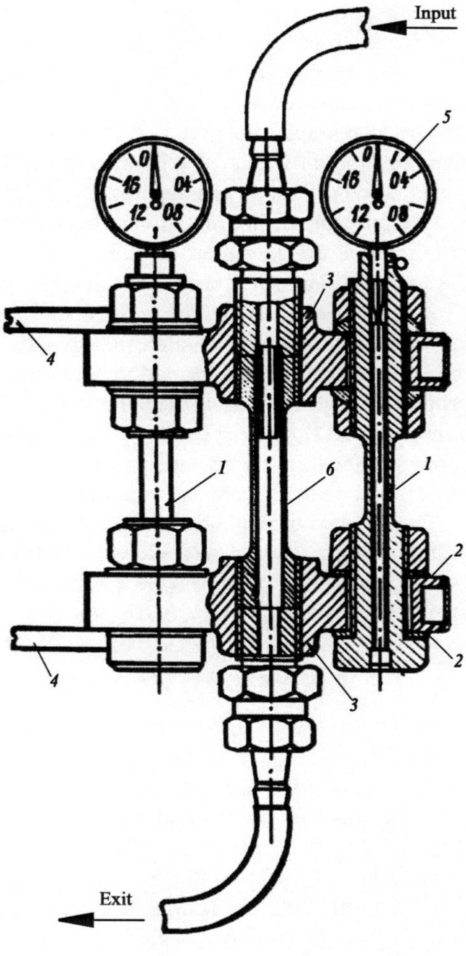

Fig. 3.46 Installation diagram of Coffin type: *1* - supporting columns; *2* - insulators; *3* - flanges; *4* - current supply buses; *5* - indicators; *6* - sample

characteristics will be obtained. Thus, when tested on a Coffin-type installation, we obtain conservative estimates of durability. Tests according to this technique can be carried out with the supply of an aggressive medium both inside the sample and outside. Thermal fatigue curves are constructed on the basis of a series of tests, in which they vary: 1) T_{max} and T_{min} values (at T_{mean} = const); 2) T_{min} value (at T_{max} = const); 3) the value of stiffness K. The first method has a number of disadvantages associated with differences in the nature of the cyclic instability of materials at different T_{max} and different creep contribution made to $\Delta\varepsilon$ with increasing T_{max}. When using the third method, T_{min} should be much less than T_{max} to obtain the number of experimental points required for plotting the curve. Since at low values of T_{max}, a number of materials exhibit cyclic hardening, the degree of which depends on temperature, the extrapolated values of the numbers of cycles to failure for the second and third methods may differ.

Due to the inhomogeneity of the temperature field along the length of the samples, the localization of cyclic plastic deformation and creep deformation takes place in the zone of maximum temperatures. The distribution of deformations along the length changes both during one cycle and with a change in the cycle number. One-sided deformations of a sign can accumulate in different areas.

Methods for the computational determination of test parameters on Coffin-type installations are described in (Getsov and Malygin, 1969). A method based on the use of a microstructural model of the environment has also been proposed in (Gokhfeld et al., 1975). These methods allow one to take into account the non-uniformity of temperature along the length and can take into account the change in resistance to cyclic deformation under conditions of cyclic instability of material properties. Calculations of the deformation diagrams of specimens fixed at different temperatures (T_{min}, T_{max}, $0.5T_{max}$), carried out by the method (Gokhfeld et al., 1975), show that the loop width and, consequently, the number of cycles to failure should not depend on the specimen clamping temperature.

The absence of the influence of the specimen fixing temperature in tests without holding at $\Delta\varepsilon > \varepsilon_{pr1} + \varepsilon_{pr2}$ is evidenced by the data given in Balandin and Zolotukhina (1961), and at $\Delta\varepsilon < \varepsilon_{pr1} + \varepsilon_{pr2}$ - for example, the results of tests of the EI765 alloy at 800°C ↔ 400°C at $\Delta\varepsilon = 0.7(\varepsilon_{pr1} + \varepsilon_{pr2})$

$T_{overcrit}$	20	800
N_0	2233, 2797, 4532	2684, 3331

The effect of the temperature of the samples $T_{overcrit}$ is manifested only in tests with holding at T_{max} and $\Delta\varepsilon < \varepsilon_{pr1} + \varepsilon_{pr2}$. As calculations show (Gokhfeld et al., 1975), in tests without holding time in the center of the sample, plastic compression deformations can accumulate, in tests with holding times - tensile deformations (Fig. 3.47).

Relaxation of stresses during holding at T_{max} not only causes the resulting cyclic creep deformations, but is also accompanied by an increase in the range of inelastic deformations (the absence of changes in $\Delta\varepsilon$ under these conditions is realized only in a uniformly heated sample with absolutely rigid fixation). These features of specimen deformation in tests with holdings at T_{max} (expansion of the loop and the formation of one-sided tensile deformations in the central part) should cause a decrease in the

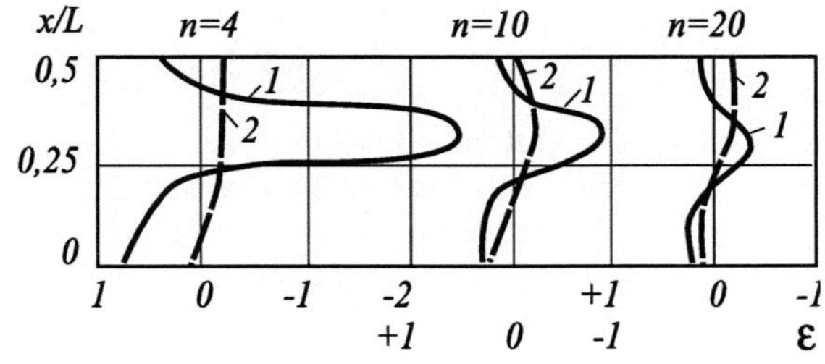

Fig. 3.47 Accumulation of irreversible deformations ε along the length x of a specimen made of steel Ch16N11M3, depending on the number of cycles n: *1* - 800⇄50°C, t_{exp} = 1 min; 2 - 800⇄400°C, t_{exp} = 0 min

durability of specimens tested at $\Delta\varepsilon > \varepsilon_{pr1} + \varepsilon_{pr2}$, in comparison with the durability in tests without holding, as is observed in experiments. The requirements for this test method are set out in GOST 25505-85, which also allows testing solid section samples, including those with a corset working part.

3.3.3.3 Test Procedure for Installations of the ALATOO Type (IMASH-5S-65)

To test specimens of type 2 (see Table 3.24) in the vacuum chambers of IMASH installations, a device is used (Gugelev and Getsov, 1976; Getsov and Rybnikov, 1994), which, at maximum internal forces (up to 10^4 N) arising during thermal cycling, leads to displacement of the sample heads no more than 2 μm. Two dies of the device 1 (Fig. 3.48) are preliminarily compressed on a press with a load of 10^5 N and tightened in this state with two bolts 5. The dies are isolated from each other and from the bolts by mica gaskets 3, washers 6 and fluoroplastic bushings 4. In each of the strips a bushing 8 made of steel ChWG with a hardness of RC 54-56 is pressed, at the upper end of which, perpendicular to the axis of the specimen, there are three teeth with the height of 0.3 mm and an apex angle of 90°. A thread is cut inside the bushings. After installing the specimen on the device, the bolts 7 are passed through the holes in the heads 2 of the sample and are screwed in with sufficient force until the bushing stops. In this case, the teeth of the bushings are introduced into the head of the sample, providing high rigidity fixing the sample in the device. The dimensions of the device allow it to be installed in the working chamber of the ALATOO.

Deformation per cycle is assessed by measuring the displacement of the reference points on the sample surface (marking in the working area is applied before testing on the PMT-3 device) using a micron indicator. Figure 3.49 shows a sketch of a sample in the version with a coating and with a stress concentrator in the form of a hole with a diameter of 0.5 mm, the results of temperature measurements along the length of the sample made of EI607A alloy at different points in the cycle time, as well as two variants of temperature changes in the working part of the sample during cycle time. It should be noted that in the case of testing samples of high-

Table 3.24 Results of tests of various heat-resistant alloys for thermal stability with T_{max} = 900°C

T_{min}, °C	Number of cycles to failure					
	EI868	EI607A	EI765	EI827	EP220	ZhS6UDC
300	-	-	-	67, 324	53, 157	267
500	334, 463	-	692	1778, 558, 1333	106, 117, 991	5295, 7775, 9989
600	-	-	-	-	-	15409
700	2687, 3232	1253	40449, 40899	7100, 6478	12664, 20000	134253, 136000
	4720	1300	10314	10287	37859	
800	41770, 48672	10922	67045	-	114451	
	69990	15251	45503			

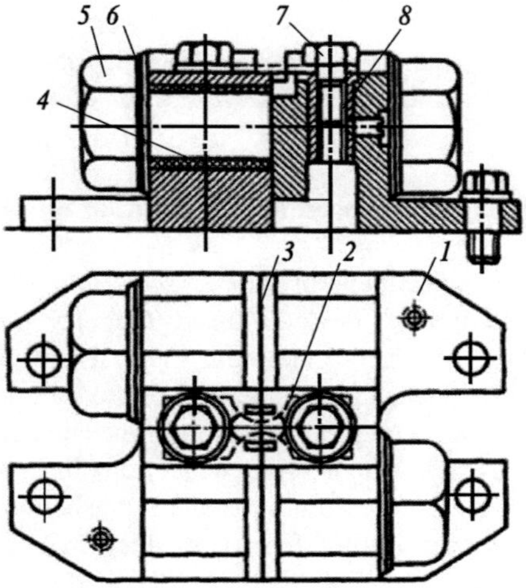

Fig. 3.48 Sample holder type 2 (option I), see Table 3.23

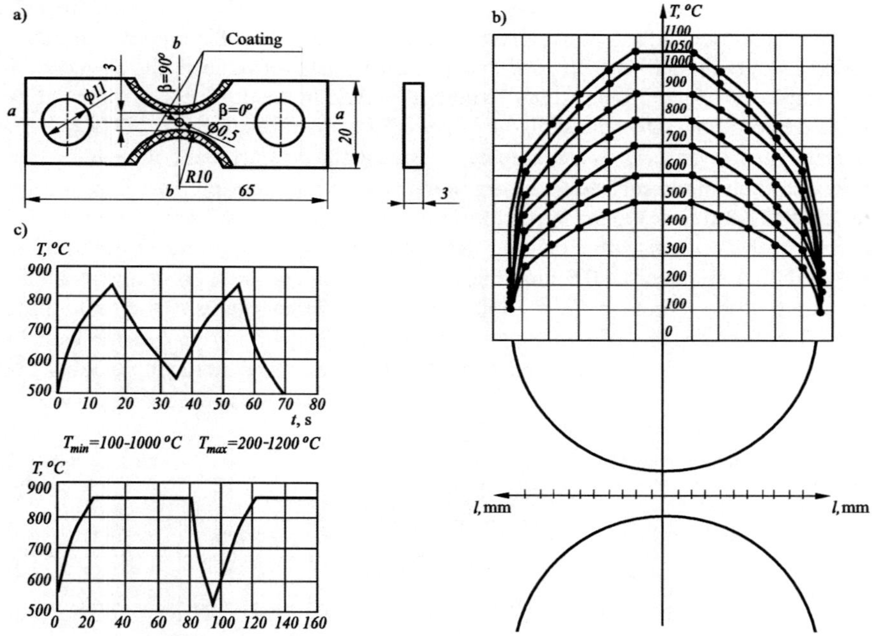

Fig. 3.49 Thermal fatigue test specimen (a), temperature distribution along length in the specimen made of the alloy EI607A (b) and cycle shape (c)

heat-resistant alloys with lower values of the thermal conductivity coefficients, the temperature distribution will differ not more then 10° from that shown in Fig. 3.49b. It should be noted that the curves of temperature distribution along the specimen, due to differences in the coefficient of thermal conductivity, depend on the alloying of steel and the alloy from which the samples are made.

The magnitude of the ranges of conditionally elastic stresses $\Delta\sigma$ and total strains $\Delta\varepsilon$ in the working part of the sample was determined by the following formulas

$$\Delta\sigma = (E_{st1}\alpha_1 T_{max} - E_{st2}\alpha_2 T_{min})\varphi, \quad \varphi = 1 - \Delta k/\Delta l, \quad \Delta\varepsilon = (\alpha_1 T_{max} - \alpha_2 T_{min})\varphi$$

Here E_{st} is the static modulus of elasticity, Δl is the free displacement of the control points during heating from T_{min} to T_{max} and Δk is the measured value of the displacement per cycle of the control microhardness imprints applied to the sample surface along the edges of its working part.

A feature of this technique, like the Coffin technique, is the constancy of conditionally elastic stresses over the cross section of the sample. Both methods use heating the sample by passing a current. At the same time, in recent years, the test procedure with heating the working part of the samples by other methods has been widely used (Bychkov et al., 1991). In these tests, no differences in fracture conditions were found for different heating methods.

The values of the durability to fracture under thermocyclic loading of flat specimens in vacuum and tubular specimens in air, previously tested on the Coffin installations, for a number of investigated high-temperature alloys on a nickel basis turned out to be identical. Comparison of the test results of the samples (the number of cycles to failure) on this installation with the test data on the Coffin installation indicates a good agreement of the results obtained (see below graphs and experimental points in Figs. 3.50-3.52 and 3.53.

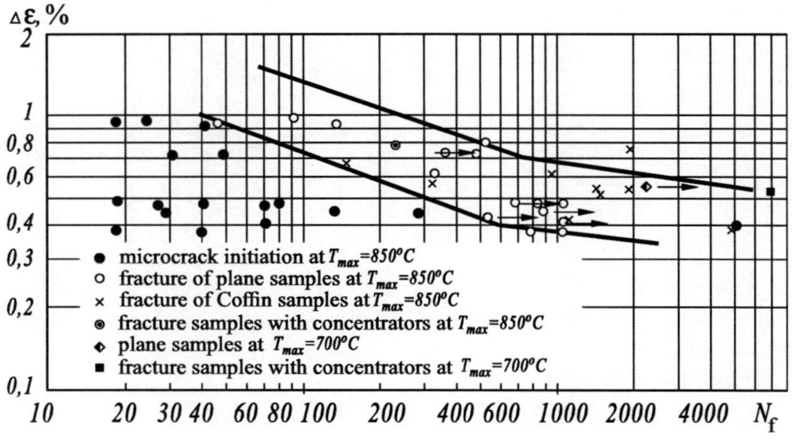

Fig. 3.50 Thermal fatigue resistance of ZhS6K alloy

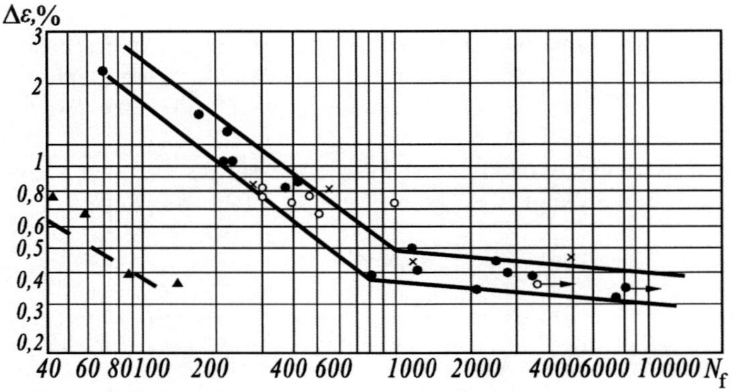

Fig. 3.51 Thermal fatigue resistance of EI893VA and EI893(L) alloys at T_{max} = 800°C: • - fracture of flat specimens made of EI893 steel, ○ - fracture of flat specimens from EI893 alloy with CoCrAlY coating, x - fracture of Coffin's specimens from EI893(L) alloy, ▲ - number of cycles before the formation of microcracks

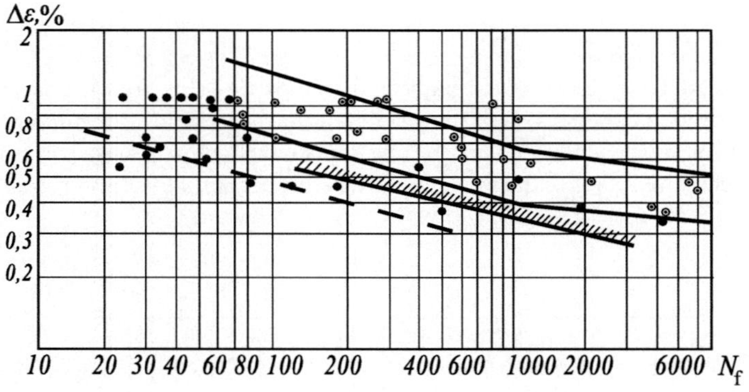

Fig. 3.52 Thermal fatigue resistance of EP220VA alloy at T_{max} = 850°C: ○ -fracture of flat specimens, //// - lower limit of the conditions of destruction of Coffin specimens, • -number of cycles before the formation of microcracks

Methods have been developed for the calculation of the characteristics of the stress-strain state of the working part of a corset specimen (Fig. 3.54), the parameter of which is the loading rigidity of the specimen-fixture system λ (Madudin, 1979; Getsov et al., 2009c).

The described technique allows us

- to study the features of the mechanism of thermal fatigue of the material, depending on the parameters of the cycle
- to determine the weakest structural elements, including those in composite materials, single-layer and multi-layer coatings and welded joints

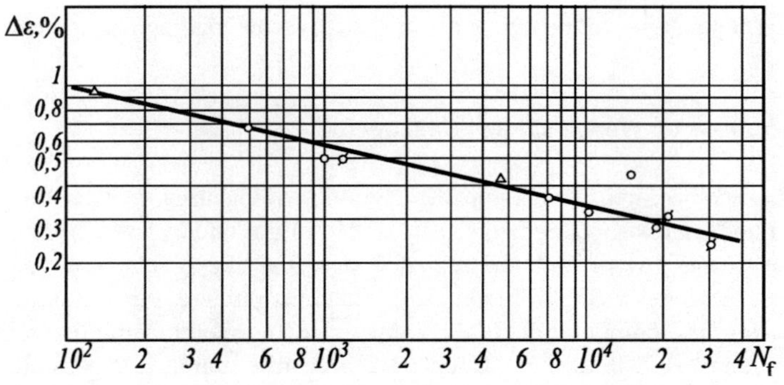

Fig. 3.53 Thermal fatigue resistance of EP539LMU alloys at T_{max} = 850°C: Δ - destruction of flat samples, o, ∅ - destruction of Coffin's samples from blades and billets, respectively

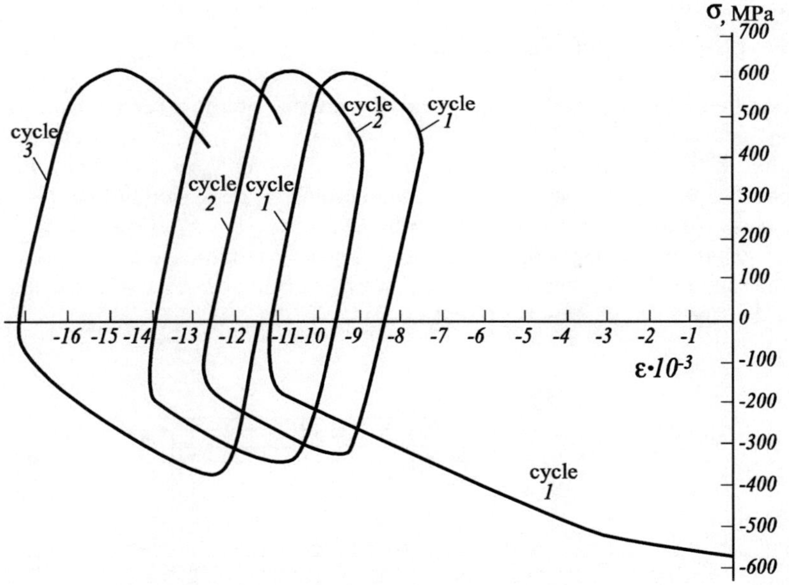

Fig. 3.54 Kinetics of cyclic deformation of a sample made of alloy EI868 when tested under the mode of 150 – 850°C ($\lambda = 5 \cdot 10^{-5}$)

- to carry out thermal fatigue tests of small-sized samples cut from turbine blades after operation
- determine the orientation of the propagation of microcracks
- to study the effect of stress concentrators, for which a hole with a diameter of 0.5 mm is made in the center of the sample
- to compare the thermal fatigue resistance of various technological processes for the manufacture of parts, their repair and coating

- to determine the dependence of the rate of crack growth on the loading parameters

3.3.3.4 Test Procedure for Wedge-shaped Specimens

Tests of wedge-shaped specimens (see Table 3.23, type 7) with heating in a gas flow (GOST 9.910-88) and pseudo-boiling baths (Littler, 1971) are widely used among the methods of the second group. With the help of wedge-shaped specimens of type 7, the conditions of thermal loading of the blade edges are simulated. Furthermore, methods of modeling and computational determination of temperature fields and elastic stresses are developed. In the conditions of gas-dynamic stands, it is possible to simulate the effect on the thermal strength of corrosive media, typical for the operating conditions of the blades.

3.3.3.5 Features of Thermal Fatigue Failure

Analysis of the results of thermal fatigue tests of a number of alloys carried out in a wide range of cycle numbers ($20 \ldots 2 \cdot 10^5$ cycles) showed that the most acceptable presentation of the test results in the coordinates $\lg \Delta\varepsilon - \lg N_p$ or $\Delta\varepsilon - \lg N_p$ for T_{max} = const. Thermal fatigue curves in these coordinates are generally lines with fractures observed for cyclically stable materials at $\Delta\varepsilon \gg \varepsilon_{pr1} + \varepsilon_{pr2}$ (Fig. 3.55).

Typically, gas turbine parts operate in non-stationary modes and experience a range of thermal loads. For these conditions, cyclic durability can be assessed using the linear damage accumulation hypothesis

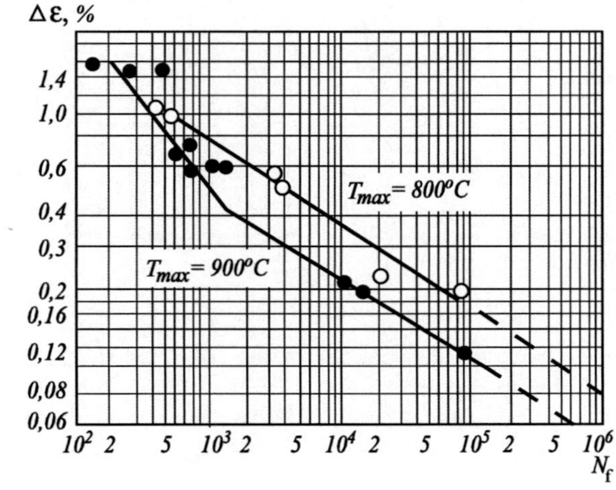

Fig. 3.55 Graphs of changes in the number of cycles to failure N_f depending on the range of elastoplastic deformation $\Delta\varepsilon$ of EI607A alloy in thermal fatigue tests

$$\sum_{i=1}^{m} \frac{n_i}{N_{fi}} = 1, \tag{3.89}$$

where n_i and N_{fi} are the effective and destructive number of loading cycles at each of the existing loads and i is the number of the load. Depending on the sequence of the action of the loads, the value of the total accumulated damage under small-block loading may differ from unity (Dul'nev and Kotov, 1980), while under multi-block loading, it can approach unity (Getsov et al., 1974). In Volkov and Korotkikh (2014), a method of nonlinear summation of fatigue damages of the Π type was proposed

$$\Pi = 1 - \left[1 - \left(\frac{N}{N_f}\right)^{\alpha+1}\right]^{1/(r+1)} \tag{3.90}$$

Thermal fatigue tests conducted by Sobolev and Egorov with steel EI847, made it possible to examine the validity of the hypothesis of linear summation of damage in the case of thermal fatigue. The tests were carried out at $T_{min} = 200°C$, $T_{max} = 750°C$ with variable loading rigidity. Combinations of modes were made up of tests with $\Delta\varepsilon_{pl}$ equal to 0.715, 0.48 and 0.245% under conditions of n_i/N_{fi} ratios equal to 0.2, 0.3, 0.5 and 0.7. Analysis of the experimental data showed that during the transition from a more severe regime to a less severe one we obtain $\sum n_i/N_{fi} < 1$, while when changing from less rigid to more rigid it follows $\sum n_i/N_{fi} > 1$.

In experiments (Serensen and Shneiderovich, 1967) on the accumulation of damage during thermal fatigue $\sum n_i/N_{pi} = 0.8 \ldots 1.5$. Separate experiments carried out with the EP220 alloy under two-stage loading ($T_{max} = 900°C$, $T_{min} = 700$ and 500° C, $t_c = 15$s and 15 min, $n_1 = 2000$, $n_2 = 20$) showed a significant deviation from the linear summation hypothesis damage (the ratio $\sum n_i/N_{pi}$ was equal to 0.37, 0.39, 0.54; Getsov et al., 1974).

3.3.3.6 Influence of Maximum Temperature and Cycle Period

An increase in the maximum cycle temperature, as a rule, leads to a decrease in durability (Fig. 3.56). A slight increase in the heat resistance of the EI765 alloy with an increase in temperature at a small number of cycles to failure, corresponding to tests at $\Delta\varepsilon > \varepsilon_{pr1} + \varepsilon_{pr2}$, is associated with increased short-term plasticity at 900°C, which determines fracture with a relatively small contribution of creep. In tests with holdings at T_{max}, an increase in T_{max} has a significantly greater effect on N_f than in tests without holdings.

Numerous studies of domestic and foreign authors have established that the holding time at T_{max} has a great influence on the durability. However, this effect manifests itself most sharply at $\Delta\varepsilon < \varepsilon_{pr1} + \varepsilon_{pr2}$, i.e. when the thermal fatigue strength is completely determined by the range of creep deformation per cycle Δp.

To describe the dependence of the thermal fatigue resistance on the cycle period t_c on the basis of studies carried out with steels Ch18N9 and 12Ch18N10T, the following formula similar to Eckel's formula (3.85) was proposed in Balandin and

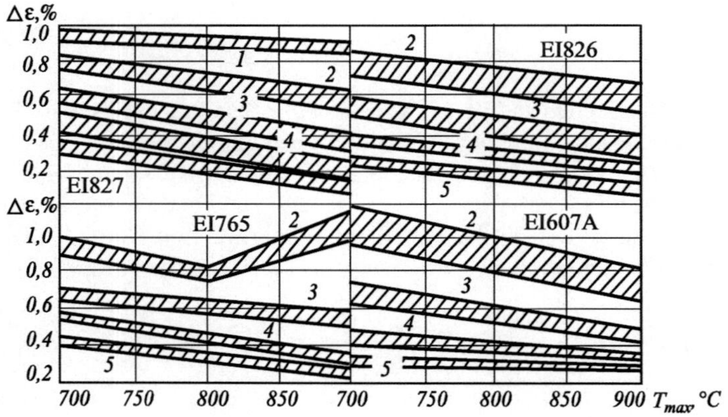

Fig. 3.56 Dependence of the thermal fatigue resistance $\Delta\varepsilon$ of alloys (tests on a Coffin-type installation) EI607A, EI765, EI826 and EI827 on the maximum cycle temperature $T_{\max}$ in tests without holding: *1* - $N_f = 10^2$ cycles; *2* - $N_f = 10^3$ cycles; *3* - $N_f = 10^4$ cycles; *4* - $N_f = 10^5$ cycles; *5* - $N_f = 10^6$ cycles

Zolotukhina (1961),

$$\lg N_f = B_0 - b_c \lg t_{cycl}, \tag{3.91}$$

where B_0 is the parameter depending on the material and deformation range per cycle and b_c is a material parameter.

As Serensen et al. (1969) have shown, expression (3.91) can be used only for very rough estimates (Fig. 3.57). Analysis of the experimental results for different $\Delta\varepsilon$ and t_{cycl}, shown in Fig. 3.58, also confirms the very low accuracy of calculations

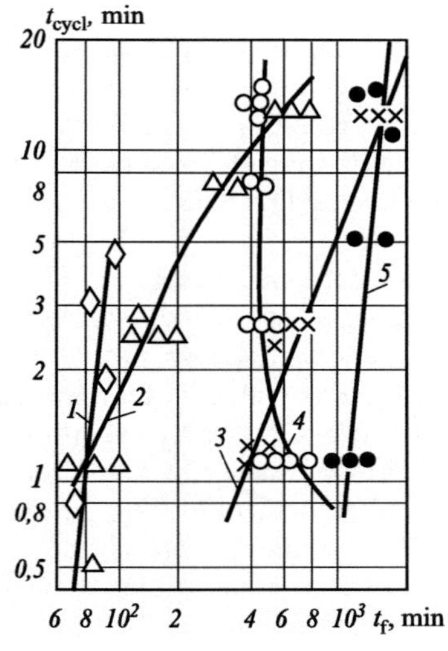

Fig. 3.57 Graphs of changes in thermal fatigue resistance t_f depending on the holding time t_{cycl} for various materials: *1* - Nimonic 90, $T_{\max} = 920°C$; *2* - ChN60WT (EI868), $T_{\max} = 950°C$; *3* - EI481, $T_{\max} = 700°C$; *4* - EI437B, $T_{\max} = 800°C$; *5* - 12Ch18N10T, $T_{\max} = 700°C$

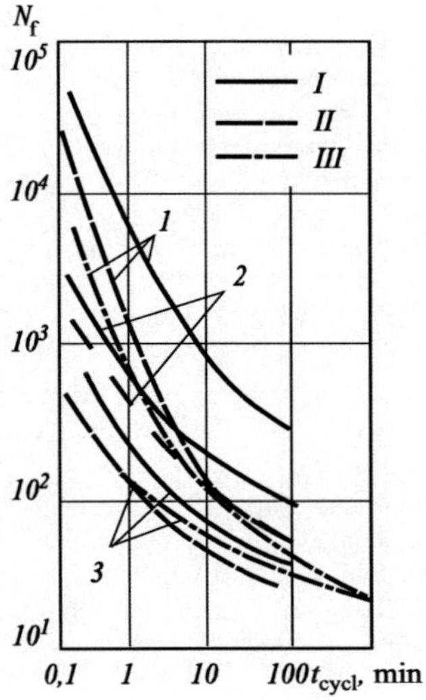

Fig. 3.58 Dependence of the thermal fatigue strength N_f of the EP220 (*I*), EI826 (*II*), EI765 (*III*) alloys on the cycle period t_{cycl} at $T_{max} = 900°C$:
1 - $\Delta T = 200°C$ ($\Delta\varepsilon = 0.3 - 0.35\%$);
2 - $\Delta T = 400°C$ ($\Delta\varepsilon = 0.5 - 0.55\%$);
3 - $\Delta T = 600°C$ ($\Delta\varepsilon = 0.6 - 0.65\%$)

according to Eq. (3.91). The value of the coefficient b_c for one series of tests may differ by 3 times.

Certain advantages in the accuracy of determining the effect on the durability of the cycle period for the conditions $\Delta\varepsilon < \varepsilon_{pr1} + \varepsilon_{pr2}$ are possessed by the dependences obtained from the criterion (3.86). Thus, the results of calculations of the creep deformation per cycle, accumulated in the process of stress relaxation over time at various holding times at T_{max}, indicates a linear relationship between N_f and Δp in logarithmic coordinates. For conditions when, at T_{max}, creep occurs at a constant rate. Using the solution of the relaxation equation (2.26), see Borzdyka and Getsov (1978), and dependence (3.36), we obtain the following expression that establishes the relationship between N and t_{cycl} at $T_{max} = \text{const}$

$$\frac{t_{cycl1}}{t_{cycl2}} = \frac{[1 - f(\Delta t)N_1^{-1/k_0}]^{1-l} - 1}{[1 - f(\Delta t)N_2^{-1/k_0}]^{1-l} - 1}, \tag{3.92}$$

where $f(\Delta t) = C_2^{1/k_0} E/\sigma_0(\Delta t)$ and l is the exponent for creep rate $\dot{p}_{min} = A\sigma^l$. N_1 and N_2 are the numbers of cycles to failure at $t_{cycl} = t_{cycl1}$ and $t_{cycl} = t_{cycl2}$, respectively (Fig. 3.59). The possibility of using relation (3.92) for calculating the thermal fatigue life is limited, on the one hand, by the values of t_{cycl} that exceed the heating and cooling time, and on the other hand, by the values corresponding to static damage comparable to thermal fatigue. As an example illustrating the accuracy of using expression (3.92), Fig. 3.60 shows the calculated curve is compared with the experimental points.

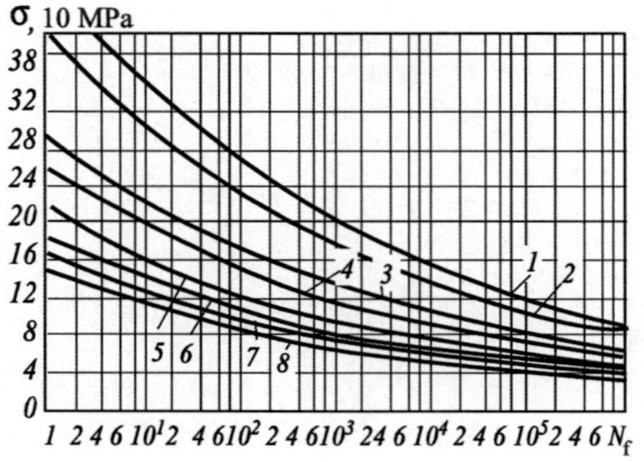

Fig. 3.59 Graphs of changes in the number of cycles before fracture N_f of the EI826 alloy depending on the load σ acting at the maximum cycle temperature T_{max} = 800°C and the cycle period t_{cycl}: *1* - 15 s, *2* - 1 min, *3* - 5 min, *4* - 15 min, *5* - 1 h, *6* - 4 h, *7* - 12 h and *8* - 24 h, respectively

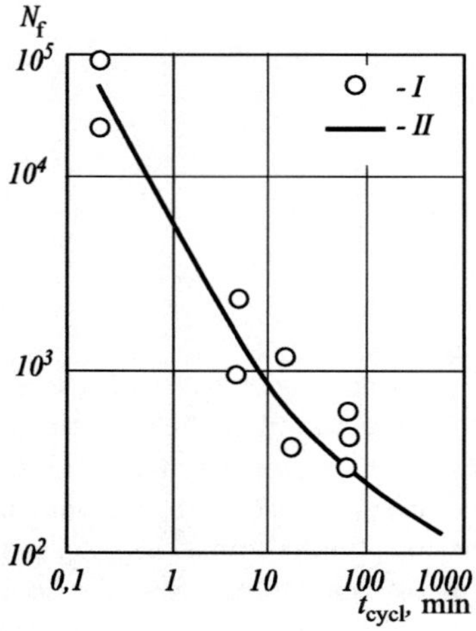

Fig. 3.60 Effect of holding time at T_{max} = 800°C on thermal fatigue strength of EP220 alloy: *I* - experimental points; *II* - computed curve

3.3.3.7 Influence of Non-metallic Inclusions

Under thermal cyclic loading, in contrast to isothermal cyclic loading, high-strength non-metallic inclusions (oxides, alumina, nitrides, silicates, sulfides) in steels and alloys cause, even in those cases when they practically do not affect the mechanical properties under tension, a very significant change in durability (Fig. 3.61). Such a change in the thermal fatigue resistance is associated with the formation of microstructural cyclic stresses (Berezhko et al., 1980)[14]. The value of these stresses is determined both by the difference in the coefficients of linear expansion of the base and inclusions, and by the temperature variation interval.

The holding time and the minimum cycle temperature practically do not affect the degree of reduction in durability, since the temperature dependence of the strength of inclusions at temperatures up to 800°C is very weak. It was also found that the

[14] The tests were carried out on cylindrical specimens heated at different rates and cooled to the minimum cycle temperature in water or sodium.

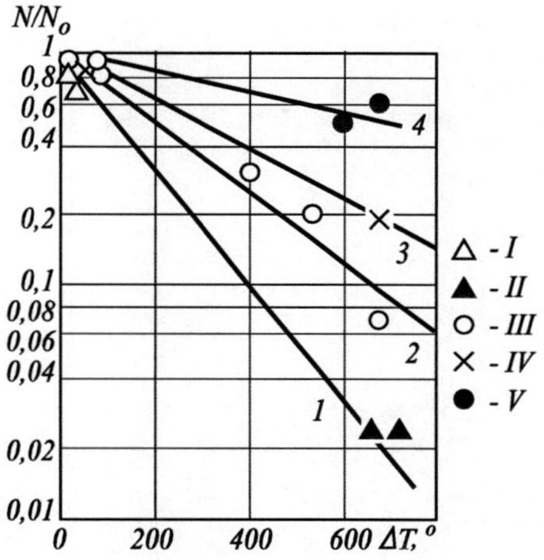

Fig. 3.61 Influence of non-metallic inclusions on the thermal fatigue strength of steels 12Ch18N9 (*1*, *4*), 12Ch2M (*2*), Ch15N3 (*3*), according to Berezhko et al. (1980): *I* - nitrides; *II* - silicates; *III* - alumina; *IV* - oxides; *V* - sulfides (N_0 - number of cycles in the absence of non-metallic inclusions in steel)

presence of pores in connection with the local stress concentration due to differences in the thermal conductivity coefficients negatively affects the thermal fatigue strength. In Berezhko et al. (1980), a method is proposed for calculating the effect of various inclusions and defects on the thermal fatigue resistance. To high-alloyed heat-resistant alloys with a high degree of heterogeneity, the above considerations may apparently not apply. As for homogeneous heat-resistant alloys, the effect of microstructural stresses caused by non-metallic inclusions can significantly affect the thermal shock resistance, similar to phase precipitates during long-term aging (μ- and σ-phases).

All of the above indicates that the characteristics of heat resistance of heat-resistant materials, apparently, have a dispersion that is noticeably higher than that observed in static tests. With a change in the manufacturing technology of workpieces, causing a change in the content of non-metallic inclusions, a noticeable change in the thermal fatigue resistance should be expected.

3.3.3.8 Thermal Fatigue of High Temperature Alloys

Thermal fatigue tests make it possible to obtain comparative characteristics of the thermal cyclic strength of various alloys, which in some cases can be used in calculating the strength of parts. Table 3.24 shows the data of tests of turbofan alloys on a nickel basis, carried out on the Coffin apparatus at T_{max} = const under conditions of $t_c = 15$ s. It is seen that with an increase in the heat resistance (Ti + Al content), the thermocyclic life increases. The exception is the highly plastic alloy EI868. Figures 3.50-3.65 show the results of tests carried out on flat corset and coffin samples (Getsov et al., 2009a).

The relationship between heat resistance and thermal stability is especially strong in tests with exposure at T_{max}. Thermal resistance is greatly influenced by plasticity

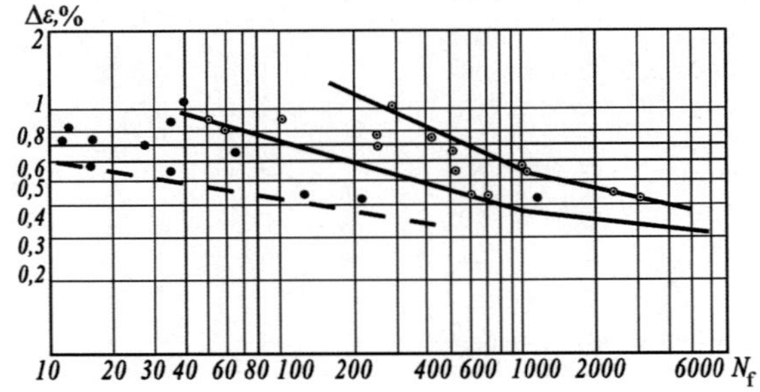

Fig. 3.62 Thermal fatigue resistance of the EI929VA alloy at T_{max} = 850°C:
o - fracture of flat samples,
• - number of cycles before microcracking

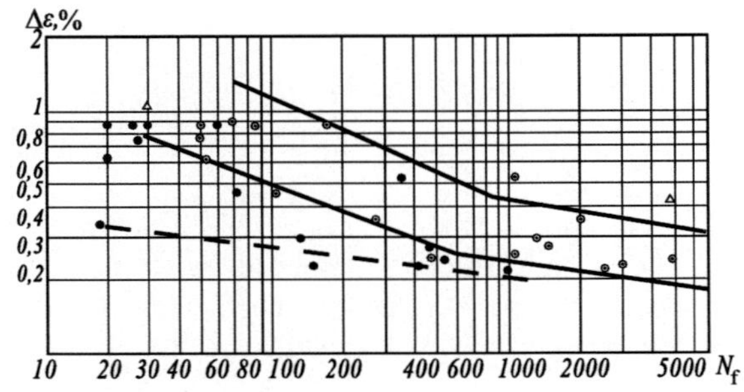

Fig. 3.63 Thermal fatigue resistance of EP539VA alloys at T_{max} = 850°C

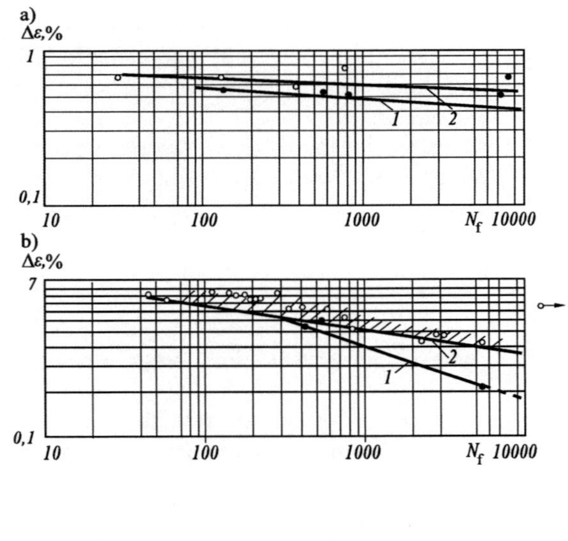

Fig. 3.64 Thermal fatigue resistance of alloys CNK7 (a) and ZMI3U (b) in tests at T_{max} = 750°C (1) and T_{max} = 850°C (2)

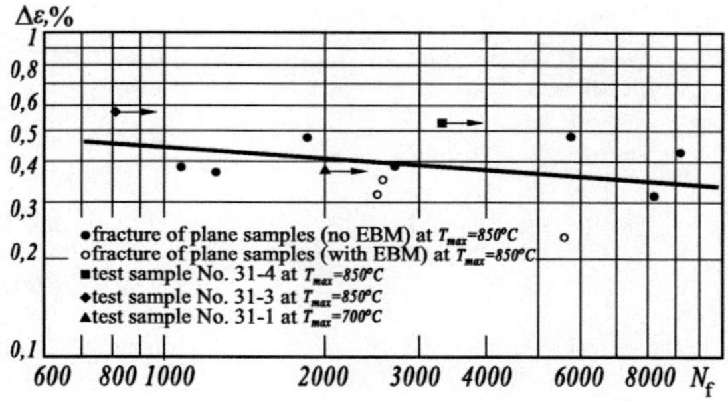

Fig. 3.65 Thermal fatigue resistance of EP957 alloy specimens (EBM - electron-beam melting)

(Fig. 3.66). The influence of holding at T_{max} and $\Delta\varepsilon < \varepsilon_{pr1} + \varepsilon_{pr2}$, in contrast to tests at $\Delta\varepsilon > \varepsilon_{pr1} + \varepsilon_{pr2}$, for heat-resistant alloys of different grades is very significant (Fig. 3.67). Tests of flat corset samples according to the method (Getsov and Rybnikov, 1994) made it possible to determine the durability before the initiation of thermal fatigue cracks in a number of alloys (Table 3.25).

Comparative data on heat resistance of different alloys have been obtained abroad as well. The described tests were carried out on prismatic samples with heating and cooling in pseudo-boiling baths (Al_2O_3, blown by argon) at temperatures of 1088-315°C under conditions t_c = 6min. Maximum durability was obtained for alloys TAZ-8A, MAR-M200, NX188 with directional crystallization, which have the highest heat resistance. Cobalt-based alloys are not preferred over medium-alloyed nickel-based alloys.

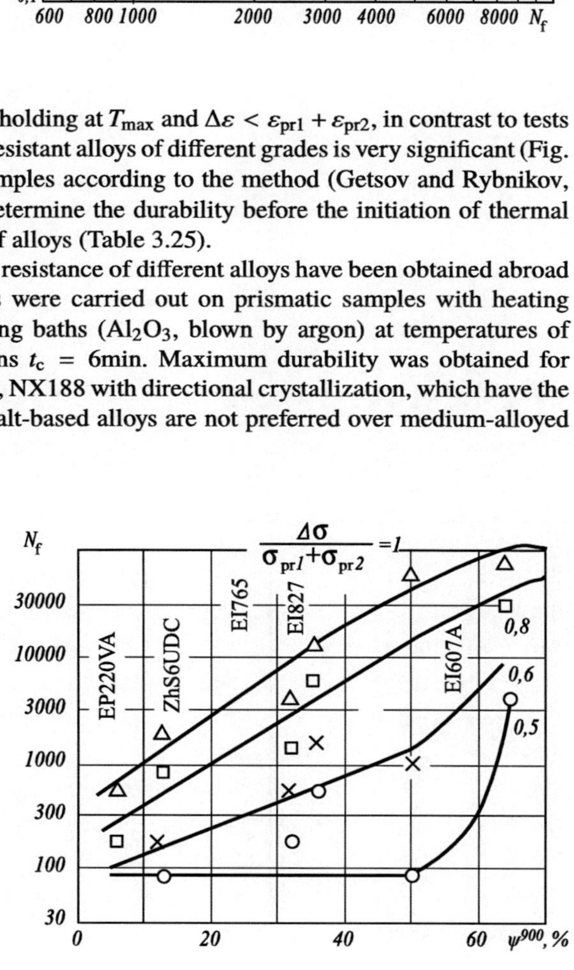

Fig. 3.66 Graphs of the dependence of the thermal fatigue life of heat-resistant alloys on plasticity

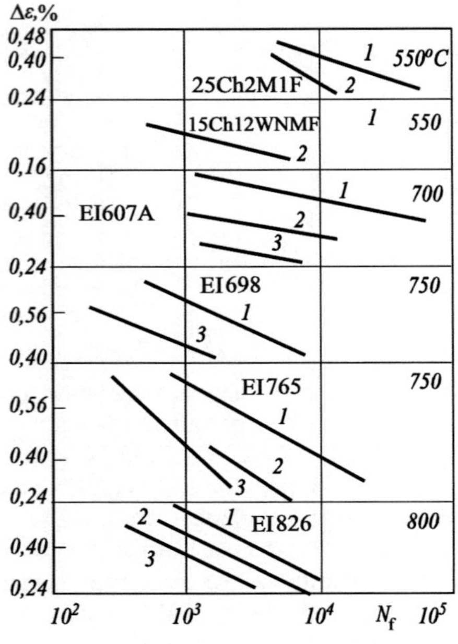

Fig. 3.67 Effect of holding time at T_{max} on the thermal fatigue life of heat-resistant alloys: *1* - t = 0 min, *2* - t = 5 min, *3* - t = 15 min

Table 3.25 Durability to crack initiation N_c and to fracture N_f in high temperature alloys

Alloy	Testing regime			
	150-850°C		500-850°C	
	N_c	N_f	N_c	N_f
EI607A	10	31	-	-
EI826	11	71	10	3540
EI868	17-40	62-93	66-82	2678-2930
ZMI-2: cast +1050°C	1	2-9	64-87	441-643
cast +1180°C	1-2	17-18	2-13	122-1722
cast +1180°+850°	1	6-20	1-15	7-851
EI893	1-30*	296-998*	-	-
EI929	6-40	60-430	10-100	539-1058
EP220	30-68	72-1072	82-265	702-5752
ZhS6K	3-48	15-495	25-303	518-2263
EP539	8-72	53-176	138-864	176-4538

*cycle 100° ⇔800°C

3.3.3.9 Progressive Deformation of Materials Under Thermal Cyclic Loading

One-sided accumulation of irreversible deformations leads to metal damage under thermal cyclic loading, in some cases comparable to damage from cyclic irreversible deformations determined by the width of the hysteresis loop. The safety margins associated with these damages are usually estimated by calculations using the theory of adaptability (Gokhfeld and Cherniavsky, 1979; Chernyavsky and Getsov, 2009). A large number of works in recent years have been devoted to the development of plasticity models that allow describing progressive deformation (in the English-language literature, denoted by the term ratchetting) under cyclic loading (Chaboche, 1991, 1994; Lin et al., 1999; Abdel-Karim and Ohno, 2000; Bocher et al., 2001; Bari and Hassan, 2002; Yoshida and Uemori, 2002; Feaugas and Gaudin, 2004; Vincent et al., 2004; Xu and Jiang, 2004; Besser et al., 2005; Yaguchi and Takahashi, 2005; Kang et al., 2006; Marchal et al., 2006; Naumenko and Altenbach, 2016). Particularly noteworthy is Kang et al. (2006), which provides a detailed analysis of publications by different authors. Unfortunately, with the exception of Shenoy et al. (2005), the test results of mainly austenitic steels are considered but the issues of damage accumulation and fracture under conditions of progressive deformation are not analysed. In addition, these works do not give an answer to the question of how to determine the safety margins under conditions when there are no data on the properties of the material necessary for calculating both cyclic deformation and ratchetting with their arbitrary combination.

An experimental study of the processes of one-sided accumulation of irreversible deformations under thermal cyclic loading of cast heat-resistant alloys was carried out in Getsov et al. (2009b). The experiments were carried out using the technique proposed in Getsov and Rybnikov (1994) under the conditions of cycles with different maximum temperatures on flat corset specimens without concentrators and with a stress concentrator in the form of a hole with a diameter of 0.5mm in the center. The experimental results are shown in Tables 3.26-3.28. Here ε_1 and ε_2 are irreversible compressive strains in the fracture zone of thermo-fatigue specimens, measured in the central zone after their fracture in directions perpendicular and parallel to the plane of the thin section.

The maximum plastic deformation in compression under conditions of thermal cyclic loading is comparable to the deformation capacity of the tested materials in tension. Thus, it can be argued that the contribution of this deformation to damage to the material under thermal cyclic loading can be significant. It should be noted that, even for alloys with low ductility under tension, the value of the deformation under compression turns out to be much higher than the values under tension. This is known to be used in hot deformation of hard-to-form alloys. According to the data obtained by Sizova, long-term ductility under compression is on average 2.5 times greater than under tension. Therefore, damage caused by progressive deformation (accumulated deformations ε_{max}) should be evaluated within the framework of the deformation criterion of fracture by the following equations

$$\Pi_3 = 0.4\varepsilon_{max}/\varepsilon_r, \qquad \Pi_4 = 0.4\varepsilon_{max}/\varepsilon_{cr} \tag{3.93}$$

Table 3.26 Thermal fatigue test results for alloy ZMI3U. Here ε_1 and ε_2 are irreversible deformations in the fracture zone of the samples, measured after their fracture in the directions perpendicular and parallel to the plane of the thin section, "ci" means crack initiation.

Minimum cycle temperature, °C	Maximum cycle temperature, °C	N_{ci}	N_f	$\Delta\varepsilon$, %	ε_1, %	ε_2, %
150	850	4	291	0.98	14.3	8.7
		17	141	0.99	8.7	16.3
		16	153	0.92	19.7	4
		16	183	0.91	3.7	22.3
		3	195	0.86	11.3	20.3
		3	405	0.76	6.7	32.7
		2	58	0.84	2.0	8.7
		2	33	–	11	15.7
250	850	20	690	0.71	10.3	15.7
350	850	45	755	0.62	5.3	10.3
500	850	32	5221	0.43	6.3	22
		2	33	–	16.7	33
		25	2262	0.45	13.3	27.3
		8	560	0.63	4.3	8.7
		4	415	0.59	6.7	5.7
400	750	50	5568	–	15.7	9

Table 3.27 Thermal fatigue tests of CNK7 alloy

Minimum cycle temperature, °C	Maximum cycle temperature, °C	N_{ci}	N_f	$\Delta\varepsilon$, %	ε_1, %	ε_2, %
150	850	1	7	–	1.7	23.3
		13	124	0.7	0	14.7
		27	750	0.8	2.7	7.0
		1	10	–	11	13.3
150	750	10	137	0.6	3.3	11.3
		20	808	0.6	2.3	9.7
		150	7960	0.6	1,0	32,7
		2	30	–	3.3	15
		22	370	0.6	0.7	19
		20	585	0.6	0	18

where ε_r and ε_{cr} are deformations during short-term rupture and in creep tests at T_{max} under tension conditions, respectively. ε_{max} is the maximum of the two values ε_1 and ε_2.

Table 3.28 Thermal fatigue test results for ChS70 alloy

Type of specimen	Thermal cycles regime, °C	t_c, s	N_{ci}	N_f	$\Delta\varepsilon$, %	ε_1, %	ε_2, %
no concentrator	100-750	100	130	5870	0.68	3	8.3
	150-850	115	30	612	0.83	1.6	19.3
	100-750	96	260	1160	0.76	9.3	12.3
	200-750	63	280	10276	0.69	5.3	9
	150-850	98	57	458	0.84	6.3	11
	350-850	59	83	2904	0.72	5.6	8.6
	350-850	47	82	3330	0.68	4.7	14.7
concentrator	100-750	65	20	80	0.675	5	10.7
	200-750	60	147	350	0.603	6	17.3
	100-750	66	2	54	≈0.76	3.7	7.7
	200-750	50	5	117	0.63	6.3	15
	200-750	58	14	901	0.66	5.6	16
	500-850	42	2	446	0.4	5.8	15

3.4 Failure Criteria for Complex Loading Programs

3.4.1 Damage Under Static and Low-cycle Loading

Previously, the terms damage and damaging were used repeatedly. For different loading conditions, different parameters are used as a measure of damage. Some of them are purely phenomenological in nature, for example, time, the number of cycles, while others claim to be associated with some physical processes (irreversible deformation, density, electrical resistance). For complex loading programs, the analytical description of the test results is based on the following assumptions.

Let us assume that failure occurs when the accepted damage parameter q is equal to q_*. Then $\Pi = q/q_*$ is a measure of damage. For multimode loading, the differential notation $\mathrm{d}n = (l/q_*)\mathrm{d}q$ should be used, since both q and q_* change. The fracture condition in this case is given as follows

$$\Pi = \int \frac{1}{q^*}\mathrm{d}q = 1. \tag{3.94}$$

We thus have an expression corresponding to the linear summation hypothesis in the case of one parameter. When failure can occur for several reasons, then these reasons can be both independent and influencing each other. For simultaneously acting n factors capable of separately causing failure, the conditions of failure can be determined from the following conditions

$$\max(\Pi_i) = 1, \quad i = 1, 2, \ldots, n, \tag{3.95}$$

$$\sum_{i=1}^{n} \Pi_i = 1 \quad \text{linear summation hypothesis,} \tag{3.96}$$

$$\sum_{i=1}^{n} \frac{q_i}{q^*(q_j)} = 1, \; i \neq j \quad \text{nonlinear summation hypothesis} \tag{3.97}$$

It seems promising to summarize damages expressed in vector form similar to that described in Chizhik and Petrenya (1978); Petrenya (1997) for static damages by different micromechanisms.

The factors that can cause the failure of materials are usually considered as follows: unilaterally accumulated irreversible deformations, cyclic irreversible deformations, cyclic loads of different frequencies, loads reaching the values of brittle strength at a given stress state, etc. Depending on the accepted parameters of damage, we have criteria of force, deformation, energy and time type. Table 2.2 lists various expressions for material damage under creep conditions using different parameters. In this section, we consider equations of fracture under multifactor loading based on the use of criteria of different types.

3.4.2 Fracture Criteria for Elasto-plastic Deformation

A number of authors have proposed empirical relationships between the range of deformations per cycle and the number of cycles before material failure under low-frequency elasto-plastic deformation

- Langer (1962)

$$\Delta\varepsilon = 2\sigma_{-1}/E + 0.5\varepsilon_{\mathrm{B}} N_0^{-0.5}, \tag{3.98}$$

- Manson (1966)

$$\Delta\varepsilon = [3.5(\sigma_{\mathrm{B}} - \sigma_{\mathrm{m}})/E] N_0^{-0.12} + \varepsilon_{\mathrm{B}} N_0^{-0.6}, \tag{3.99}$$

  and its modification proposed in Trukhny and Stepanov (1978)

$$\Delta\varepsilon = 0.345(\sigma_{\mathrm{B}}/E)^{0.548} N_0^{-0.12} + 0.618\varepsilon_{\mathrm{B}}^{0.367} N_0^{-0.6}, \tag{3.100}$$

- Marin (1968)

$$(4N_0 - 1)(\Delta\varepsilon/2\varepsilon_{\mathrm{B}})^a + (0.5\Delta\varepsilon/\varepsilon_{\mathrm{B}} + \varepsilon_{\mathrm{m}}/\varepsilon_{\mathrm{B}})^b = 1 \tag{3.101}$$

Here $\varepsilon_{\mathrm{B}} = \ln(1-\psi_{\mathrm{B}})^{-1}$, ε_{m} and σ_{m} are the mean deformation and stress, respectively. The value of ε_{B} is determined either by static tests or by selecting the values that best fit the experimental curve. Manson's formula was obtained on the basis of processing test results at room temperature for 29 materials, and the formula (3.100) for 41 materials. Unlike dependencies (3.99) and (3.101), relations (3.98) and (3.100) do not allow taking into account the change in the asymmetry of the cycle, which may take place under real conditions.

To determine durability at elevated temperatures, Manson proposed the ten percent rule, according to which $N_f = 0.1N_0$, where N_0 is determined by Eq. (3.99). As the analysis of the results of experiments carried out with various high-temperature alloys has shown, the values of the durability calculated according to Eq. (3.99), in a number of cases, differ significantly from the experimental data (Fig. 3.68). For the EP220 alloy, however, good agreement was obtained between N_f^{exp} and N_f^{calc}.

Low cycle fatigue characteristics as well as long term strength and fatigue resistance are inherently variable. To determine the minimum values of the durability for low-cycle fatigue under the conditions $\Delta\varepsilon = \text{const}$, it is advisable to use the relation[15]

$$\lg N_f^{l,\beta} = \lg N_f - S(\lg N_f)\frac{t_\beta}{\sqrt{n}} - (l - 0.5)\frac{S(\lg N_f)}{0.341}\sqrt{1 + \frac{n-1}{\chi^2}}, \tag{3.102}$$

similar to that used for static failure (Shalin et al., 1981), the parameters of which are the probability of non-failure l, the confidence probability β and the standard deviation $S(N_f)$ (t_β and χ^2 are determined from the tables for the given number of samples in the sample n and confidence probability β).

However, due to the fact that at the disposal of researchers, as a rule, there is no possibility of obtaining the results of tests for low-cycle fatigue of many heats, recently the dependencies, the parameters of which are the standard mechanical characteristics σ_B, ψ etc., are applied.

The variance of the quantity $\Delta\varepsilon$ determined by the Manson formula

$$\Delta\varepsilon = f(\sigma_B, \psi, N_f) = \frac{3.5\sigma_B}{E}N_f^{-0.12} + \left(\ln\frac{1}{1-\psi}\right)^{0.6} N_f^{0.6}, \tag{3.103}$$

can be calculated using the method of linearization of a function of several arguments, in this case D_{σ_B} and D_ψ (D_{σ_B} and D_ψ are the variances of σ_B and ψ, respectively)

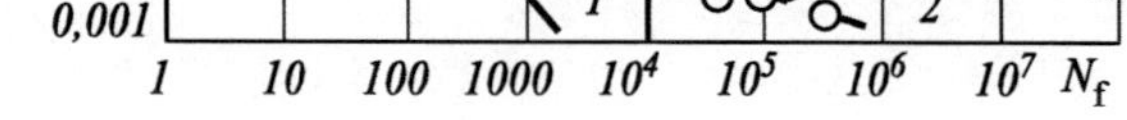

Fig. 3.68 Comparison of experimental values (*I*, *II*) of the fatigue strength of the EI765 alloy at $T = 800°C$ (*1*, 2) with curve (*III*) calculated by the universal slope method: *1* - ω = 10 cycle/min; 2 - ω = 12000 cycle/min

15 The theoretical background is the Pearson's chi-squared test (χ^2) presented, for example, in Wentzel and Ovcharov (1987).

Balepina (1988). It is assumed that there is no correlation between the arguments σ_B and ψ of expression (3.103).

For a fixed N_f, the variance $D_{\Delta\varepsilon}$ is determined from the relation

$$D_{\Delta\varepsilon} = \left(\frac{\partial f}{\partial \sigma_B}\right) D_{\sigma_B} + \left(\frac{\partial f}{\partial \psi}\right) D_{\psi}, \tag{3.104}$$

As a result, we have the relation

$$D_{\Delta\varepsilon} = \left(\frac{3.5 N_f^{-0.12}}{E}\right)^2 D_{\sigma_B} + \left[0.6(1-\psi_m)\left(\ln\frac{1}{1-\psi_m}\right)^{-0.4} N^{-0.6}\right]^2 D_{\psi}, \tag{3.105}$$

As mentioned above, the dispersions D_{σ_B} and D_{ψ} have values depending on the type of alloy, which can be expressed by the dependencies

$$\sqrt{D_{\sigma_B}} = (0.003 \div 0.007)\sigma_{B\,m}; \quad \sqrt{D_{\delta}} = (0.1 \div 0.2)\delta_m. \tag{3.106}$$

The following assumption can be made $\sqrt{D_\psi}/\psi_m = \sqrt{D_\delta}/\delta_m$.

The minimum values of the deformation range corresponding to a given number of cycles can be determined as follows

$$\Delta\varepsilon_{min} = \Delta\varepsilon_m - 3\sqrt{D_{\Delta\varepsilon}}. \tag{3.107}$$

Thus, taking Manson's formula or Langer's formula as a basis, as was done in Makhutov et al. (1989), with known dispersions of σ_{-1} and ψ as well as the low-cycle fatigue curve of an individual melt, one can obtain the calculated probabilistic low-cycle fatigue curves. A further step towards the development of fracture assessment under low-frequency loading is the deformation criteria, the parameters of which are irreversible deformations.

The conditions of material failure at constant and variable temperatures under rigid cyclic loading (tests with constant amplitude deformation) are fairly well described by the Coffin-Manson formula

$$\Delta\varepsilon_{pl}^{k_0} N_0 = C_1, \tag{3.108}$$

where k_0 and C_1 are constants, $\Delta\varepsilon_{pl}$ is the range of plastic deformation (double loop width). In those cases when $\Delta\varepsilon_{pl} \gg \Delta\varepsilon_{pr}$, the cyclic instability of the material during deformation under rigid loading plays a relatively small role, since $\Delta\varepsilon_{pl}^1 \approx \Delta\varepsilon_{pl}^i \approx \Delta\varepsilon$. Under conditions of soft loading, the fracture criterion can be written as

$$\sum_{i=1}^{N_0} (\Delta\varepsilon_{pl}^i)^{k_0} = C_1, \tag{3.109}$$

The validity of Eq. (3.108) has been experimentally confirmed by many researchers. However, if used incorrectly, there may be some contradictions. Indeed, for $\Delta\varepsilon_{pl} = 0$, according to (3.108), $N_0 \to \infty$, i.e. the fatigue limit is equal to the elastic limit. N_0

does not depend on the deformation frequency, which is in agreement with the experimental data only at low temperatures, at which the metal does not creep. In order to check the validity of formula (3.109) with a symmetric cycle ($k_0 = 2$), the values

$$\bar{D} = \sqrt{C_1} = \sqrt{\left[\sum_{i=1}^{N_0} (\Delta\varepsilon^i_{\text{pl}})^2\right]} = \sqrt{\left[\sum_{i=1}^{N_0} (2\delta^i)^2\right]}$$

were computed.

It was found that when using the Coffin-Manson relation (3.108) for testing the EI765 alloy under soft loading, the value of C_1 changes within fairly significant limits (up to 25 times), and the value of $\sqrt{C_1}/\varepsilon_{\text{f}}$ from 0.5 to 6. The value $\bar{D}$ changes relatively little at a constant test temperature. Similar results were obtained when processing data for mild steel at 350°C, pearlitic steel tested at 20°C, and for cast iron tested by A.P. Gusenkov at 550°C. Thus, from constancy of $\bar{D}$ it follows that in the formula (3.109) C_1 can be considered a constant. Material damage accumulated over N cycles of elasto-plastic deformation can be characterized by the value of Π_1, determined in fractions of C_1, i.e.

$$\Pi_1 = \Delta\varepsilon^{k_0}_{\text{pl}} N/C_1 = \left(\Delta\varepsilon_{\text{pl}}/\Delta\varepsilon_{\text{pl f}}\right)^{k_0} (N/N_0),$$

or

$$\Pi_1 = \sum_{i=1}^{N_0} (\Delta\varepsilon^i_{\text{pl}})^{k_0}/C_1, \tag{3.110}$$

The deformation-type equations of the Coffin relation (3.108) for $\Delta\varepsilon > \varepsilon_{\text{p cycl1}} + \varepsilon_{\text{p cycl2}}$ are usually used as criterion dependencies for describing thermal fatigue fracture. The possibility of using criterion (3.109) to assess the durability under thermal cyclic loading in the case of cyclically unstable materials was checked. The values $\bar{D} = \sqrt{\sum[2\delta^{(i)}]^2}$ were calculated as applied to experiments on thermal fatigue of the EI765 alloy at $T_{\max} = 700$°C and $T_{\min} = 20$°C. The values of $\delta^{(i)}$ corresponding to $T_{\max}$ and $T_{\min}$ were found graphically (Fig. 2.22), for some cycle numbers they have the following values

$i = k/2$	1	2	3	4	80	100	200	500	1000
$\delta^{(i)}_{20} + \delta^{(i)}_{700}$	0.8	0.48	0.38	0.45	0.47	0.46	0.48	0.55	0.58

From the values $\Delta\varepsilon^i_{\text{pl}} = \delta^i_{200} + \delta^i_{700}$, the values of D were calculated for $N_{\text{f}} = 2296$, corresponding to the failure of the specimen during the thermal fatigue test. The value $D = 29.7\%$ calculated for conditions of variable temperatures practically coincided with the value of D at $T = 700$°C. The obtained results support the assumption that the mutual influence of deformation processes with varying temperature can be neglected if in this temperature range there are no changes in the structure of the material and the accumulation of residual strains over time (creep).

3.4.3 Adaptability Theory

Among the methods for determining the thermal fatigue strength of parts, the most preferable is the use of calculations based on the application of the theory of adaptability, with the determination of safety margins: a) for an alternating flow (AF) and b) for progressive deformation (PD), and ac) with the determination of the value of a single safety margin associated with the simultaneous manifestation of damage of both types.

The adaptability theory is based on two theorems - Melan's static theorem (1) (Melan, 1936) and Koiter's kinematic theorem (2) (Koiter, 1960) as formulated in Gokhfeld and Cherniavsky (1979)

1. If in a solid such a statical residual stress field is possible, at which, if it were a real residual stress field, then no plastic flow would arise when the loads actually acting in the cycle were applied, then the body will actually adapt to one of these states
2. Adaptability is impossible if there is a field of kinematically possible total (per cycle) increments of plastic deformations $\Delta\varepsilon_{ij}^{n}$, for which the increment in the work of constant, time-independent mass and surface loads exceeds the increment in the work of stresses σ_{ij} on the fictitious yield surface.

Melan's theorem was extended to the creep case in Shorr (1966).

With an alternating flow, the fictitious yield surface degenerates into a sphere, line, or point. Assessments of the state of a component under conditions of a complex cyclic loading program based on the adaptability theory are based on relatively simple computational algorithms, since only conditionally elastic calculations are performed within the framework of this theory.

Using the adaptability theorems, equations are formulated, the solutions of which for the considered structure, subjected to thermal cyclic loads, make it possible to determine the position of the adaptability curves. In addition to the geometry of the component and the conditions of its cyclic loading, in such calculations it is necessary to know the limiting characteristics - some values of the so-called yield stress σ_s.

By setting the values of the cyclic yield stress ($S_{0.4} = 2\sigma_{s}$), including under creep conditions in one of the half cycles ($S_{0.4}^{\text{creep}}$), and evaluating the effect of stress concentration for a given number of cycles using the well-known Neuber formula

$$K_{\mathrm{N}}^{2} = K_{\sigma} K_{\varepsilon}, \tag{3.111}$$

where K_{N} is the theoretical coefficient of stress concentration, K_{σ} and K_{ε} are the elastic-plastic coefficients of stress and strain concentration, we obtain the possibility of an adequate estimate of the reserves before the onset of the adaptability of the structure during thermal cycling. To determine the values of K_{σ} and K_{ε} with a known (from finite element calculations or graphs given in Peterson's handbook, see for example Pilkey, 1997) value of K_{N}, one can use the stress-strain curve (or cyclic stress-strain curve) and the diagram shown in Fig. 3.69.

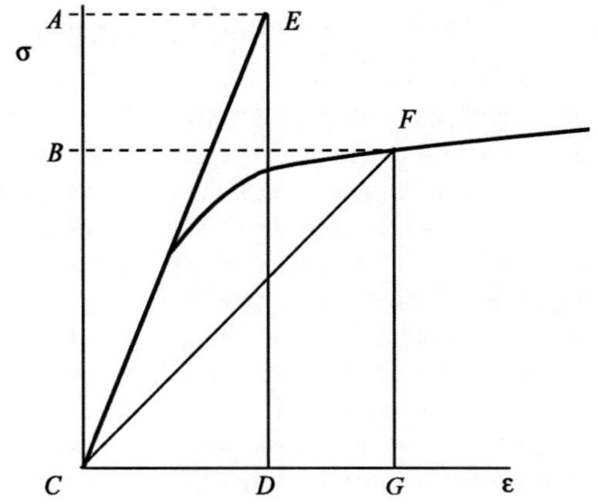

Fig. 3.69 Method for determining the stress and strain concentration factors: areas of the triangles *CDE* and *CFG* are equal. *DE* - stress obtained by elastic calculation ($\sigma = \sigma_m K_N$), *CD* - deformation obtained by elastic calculation ($\varepsilon = \sigma_m/E$). σ_m is the average stress over the cycle ($r = \sigma_{min}/\sigma_{max}$ is the coefficient of asymmetry of the cycle, $r = -1$ for a symmetric cycle)

When estimating the limiting loading cycles corresponding to the alternating plastic flow, the limiting stresses at each temperature are assumed to be $\sigma_s = 0.5S_{0.4}$ during start-up and shut-down and $\sigma_s = S_{0.4}^{cycl} - 0.5S_{0.4}$ during steady operation. For sections of stress concentration, the limiting stresses are taken, in accordance with the Neuber formula (3.111),

$$\sigma_s = \sqrt{E\varepsilon(N)\sigma_{\varepsilon(N)}}$$

where $\varepsilon(N)$ is the amplitude of total deformation corresponding to the appearance of a crack in N cycles, $\sigma_{\varepsilon(N)}$ is the stress corresponding to $\varepsilon(N)$ on the isochronous deformation curve (isochronous cyclic deformation curve). When determining the conditions of adaptability according to the PD, the values of the ultimate strength and long-term strength for the time corresponding to the service life of the component in question are set as the limiting characteristics. The values of the cyclic yield points at different temperatures are determined experimentally (see Getsov et al., 1976). The adequacy of the calculation methods for the adaptability of turbine disks and rims of the guide vanes of gas turbines, as well as elements of metallurgical equipment, is confirmed by the positive experience of operating components for which a safety margin was provided: $K_{AF} = 1.2 - 1.5$ for flow with changing sign and $K_{PD} = 1.9 - 2.2$ for progressive deformation. If the operational model includes several cycles with different parameters, then this method allows us to estimate a single safety margin based on the results of calculations carried out for each type of cycle. At the same time, the method cannot be used for cases when the loading program is not cyclic.

3.4.4 Failure Criterion for Alternating Cyclic Creep

When the creep processes develop quite intensively, they can determine the durability in tests at $\sigma < \sigma_{\text{pl}}$. For the conditions of both hard and soft loading of a cyclically stable or stabilizing material, the fracture criterion can be written in the form (3.86). The validity of dependence (3.86) was confirmed by calculations of the creep strain accumulated during the cycle, carried out in relation to tests of a number of high-temperature alloys. Specimens made of alloys EI765, EI826, EI827 and EP220 were tested under conditions of rigid loading by cyclic bending at 750, 800, and 900°C. To calculate the creep strain for a sinusoidal cycle, the hardening theory was adopted and the characteristics of the short-term creep of materials were used. An analysis of the calculation results showed the existence of linear dependencies in logarithmic coordinates between the range of creep deformation per cycle Δp and the number of cycles to failure N_0 (Fig. 3.70), which confirms the validity of relation (3.86). This also confirms the validity of the relationship between the range of total deformation per cycle $\Delta\varepsilon$ and the range of creep deformation per cycle Δp, i.e.

$$\Delta\varepsilon = A(\Delta p)^{m_*}, \tag{3.112}$$

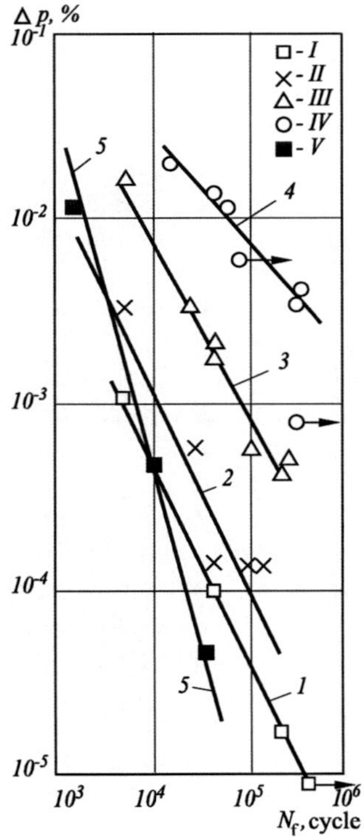

Fig. 3.70 Dependence of the calculated values of creep strains accumulated per cycle on the number of cycles to failure in tests with a frequency $\omega = 10$ cycle/min: *1* - EI765 at temperature 750°C (*I*), *2* - EI765 at temperature 800°C (*II*), *3* - EI826 at temperature 800°C (*III*), *4* - EI827 at temperature 800°C (*IV*), *5* - EP220 at temperature 900°C (*V*)

where A and m_* are constants. Dependence (3.112) can be derived in general form. The validity of (3.86) was also confirmed in experiments with AISI316 steel and AF2-IDA alloy (Fig. 3.71). The possibility of using criterion (3.86) to assess the thermal fatigue life at $\Delta\varepsilon < \varepsilon_{pr1} + \varepsilon_{pr2}$ and conditions when one-sided accumulation of deformations can be neglected is confirmed by calculations carried out in relation to thermal fatigue tests of heat-resistant alloys.

Calculation of Δp for tests without holding at $\Delta\varepsilon < \varepsilon_{pr1} + \varepsilon_{pr2}$ was carried out according to the theory of hardening, extended to the case of variable temperatures and alternating stresses by introducing conditions that allow taking into account the regularities of alternating creep. The obtained dependence between Δp and N_p for alloys EI765 and EI826 confirms the possibility of using criterion (3.86) for conditions of variable temperatures (Fig. 3.72).

In cases where there is a significant decrease in the deformation capacity of the material with an increase in the test base and when the value of Δp changes from cycle to cycle, instead of formula (3.86), we obtain

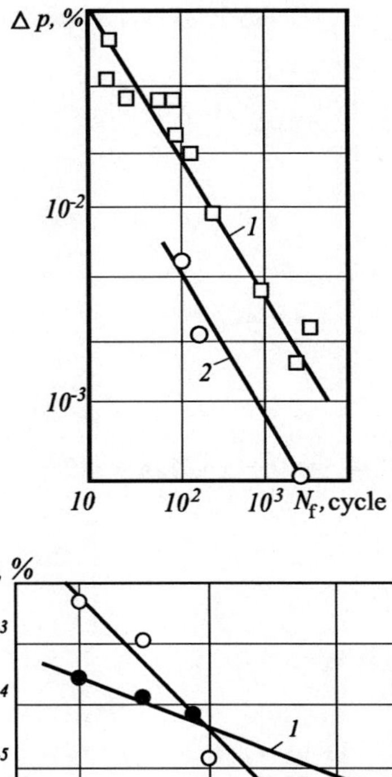

Fig. 3.71 Test results for alternating creep of steel AISI316 at temperature 650°C (*1*) and alloy AF2-1DA at temperature 760°C (2) according to Manson and Hulford

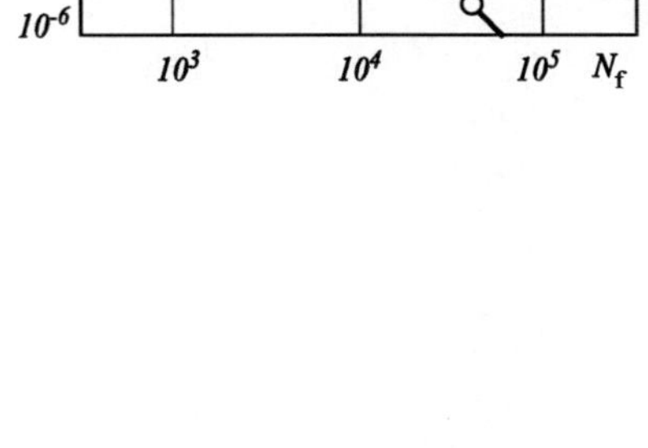

Fig. 3.72 Graphs of changes in the number of cycles to fracture N_f depending on the range of creep deformation Δp per cycle in thermal fatigue tests with T_{max} = 800°C for EI765 (*1*) and EI826 (2) alloys

$$\sum_{i=1}^{N_0} (\Delta p^{(i)})^{n_0} = C_2(N_0), \qquad (3.113)$$

For estimations, the coefficient C_2 can be assumed constant. The degree of damage under conditions of cyclic alternating creep Π_2 can be represented in fractions of C_2, i.e.

$$\left.\begin{aligned} &\Pi_2 = (\Delta p^{(i)})^{n_0} N / C_2 \\ &\text{or} \\ &\Pi_2 = \sum_{i=1}^{N_0} (\Delta p^{(i)})^{n_0} / C_2(N_0) \end{aligned}\right\} \qquad (3.114)$$

3.4.5 Modified Fracture Criteria for Materials Under Cyclic Loading

To determine the durability of components operating at elevated temperatures under low-frequency loading, a number of criteria have been proposed.

- Mixed criteria
  - Kostyuk et al. (1965)

$$\mu N_{\mathrm{f}} \int_0^{t_{\mathrm{cycl}}} (1/t_{\mathrm{f}})\mathrm{d}t + \int_0^{N_{\mathrm{f}}} (1/N_0)\mathrm{d}N = 1, \qquad (3.115)$$

  - Manson (1966)

$$N_{\mathrm{f}} \int_0^{t_{\mathrm{cycl}}} (1/t_{\mathrm{f}})\mathrm{d}t + \int_0^{N_{\mathrm{f}}} (1/N_0)\mathrm{d}N = 1, \qquad (3.116)$$

  - Dul'nev and Kotov (1980)

$$\left[\int_0^{t_{\mathrm{f}}} (1/t_{\mathrm{f}})\mathrm{d}t\right]^a + \left[\int_0^{N_{\mathrm{f}}} (1/N_0)\mathrm{d}N\right]^b = 1, \qquad (3.117)$$

  - Wood (1966)

$$\lg(t/t_{\mathrm{f}}) + a_{\mathrm{f}}[(N_{\mathrm{f}}/N_0)^2 + b_{\mathrm{f}}(N_{\mathrm{f}}/N_0)] = 1, \qquad (3.118)$$

  where μ, a, b, a_{f}, b_{f} are coefficients that take into account the mutual influence of static and cyclic damage and N_0 is the number of cycles to failure in the absence of creep.

- Energetic criteria
  - Kostyuk et al. (1967)

$$MN_\mathrm{f}\left(\oint \sigma \mathrm{d}\varepsilon_\mathrm{pl} + \oint \sigma \mathrm{d}p\right)^{\frac{-\delta_0}{1-\delta_0}}\left(\oint \sigma^{\gamma_k+1}\mathrm{d}\varepsilon_\mathrm{pl} + \oint \sigma^{\gamma_k+1}\mathrm{d}p\right)^{\frac{1}{1-\delta_0}} = 1, \tag{3.119}$$

where M, d_0 and γ_k are constants.

- Deformation criteria
  - Shneiderovich (1968); Gusenkov (1979); Makhutov (1981)

$$\int_1^{N_\mathrm{f}}\left(\frac{\Delta\varepsilon}{\varepsilon_\mathrm{f}(t)}\right)^m \mathrm{d}N + \int_0^{e^*(N_\mathrm{f})}\frac{1}{\varepsilon_\mathrm{f}(t)}\mathrm{d}e = 1, \tag{3.120}$$

  - Getsov (1971, 1978, 2001)

$$\Pi_1 + \Pi_2 + \Pi_3 + \Pi_4 = 1, \tag{3.121}$$

where

$$\Pi_1 = \int_1^{N_\mathrm{f}}\frac{\Delta\varepsilon_\mathrm{pl}^{k_0}\mathrm{d}N}{C_1(\Pi_2,\Pi_3,\Pi_4)}, \qquad \Pi_2 = \int_1^{N_\mathrm{f}}\frac{\Delta p^{n_0}\mathrm{d}N}{C_2(\Pi_1,\Pi_3,\Pi_4)},$$

$$\Pi_3 = \int_1^{\varepsilon_\mathrm{pl}^*(N_\mathrm{f})}\frac{\mathrm{d}\varepsilon_\mathrm{pl}}{\varepsilon_\mathrm{f}}, \qquad \Pi_4 = \int_1^{p^*(N_\mathrm{f})}\frac{\mathrm{d}p}{\varepsilon_\mathrm{cr}(\Pi_1,\Pi_2,\Pi_3)},$$

  - Manson (1973)

$$\Pi_1 + \Pi_2 + \Pi_6 + \Pi_7 = 1, \tag{3.122}$$

where

$$\Pi_6 = \Delta\varepsilon_{\mathrm{pl\,cr}}^{\alpha_1}N/C_6, \qquad \Pi_6 = \Delta\varepsilon_{\mathrm{cr\,pl}}^{\alpha_2}N/C_7, \qquad \alpha_1 = \alpha_2 = 1.25$$

Damage parameters Π_6 and Π_7 and criterion (3.122) itself refer to the conditions of deformation according to the schemes (Figs. 3.73, *V*, a, b, note that δ in the figure means $\varepsilon_{\mathrm{pl\,cr}}$ and $\varepsilon_{\mathrm{cr\,pl}}$, respectively), typical for a rigid loading cycle.

In Gokhfeld et al. (1983); Poroshin (1988), the following criterion was proposed

$$\Pi_1 + \Pi_2 + \Pi_8 = 1, \tag{3.123}$$

where

$$\Pi_8 = \int_1^{\varepsilon}\frac{\mathrm{d}\varepsilon}{\varepsilon_f(T,t,\sigma_0)},$$

and $\sigma_0 = (\sigma_1 + \sigma_2 + \sigma_3)/3$. In criterion (3.123) the total deformation is determined using the microstructural model of the medium, which makes it possible, using its parameters, to better describe the process of damage accumulation within the cycle. Criteria (3.120)-(3.122) do not represent such a possibility. This criterion, like

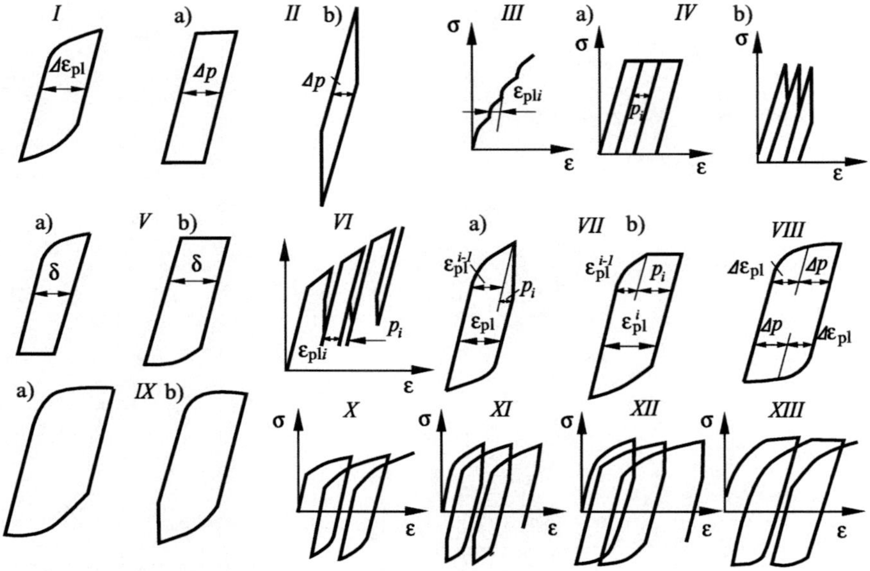

Fig. 3.73 Deformation schemes: *I* - Π_1; *II* - Π_2, *III* - Π_3; *IV* - Π_4; *V, VI* - $\Pi_3+\Pi_4$; *VII* - $\Pi_1+\Pi_3+\Pi_4$; *VIII* - $\Pi_1+\Pi_2$; *IX* - $\Pi_1+\Pi_2+\Pi_3+\Pi_4$; *X* - $\Pi_1+\Pi_3$; *XI, XII* - $\Pi_1+\Pi_3+\Pi_4$; *XIII* - $\Pi_1+\Pi_2+\Pi_3+\Pi_4$

(3.120) and (3.121), allows one to describe damage under cyclic loading accompanied by unilateral accumulation of deformations (soft cycle of cyclic loading).

Let us analyze and compare the criterion equations (3.115)-(3.122). The parameters of the mixed type criteria (3.115)-(3.118) are the time and the number of cycles. Criterion (3.115) is based on the linear summation hypothesis. In criterion (3.115), using the coefficient μ, the mutual influence of damage is taken into account. Criteria (3.117) and (3.118) are based on the assumption of non-linear damage summation. The use of (3.118) is possible only for not too small and not too large values of t/t_f, since according to (3.118) we obtain that $N_f \gg N_0$ at $t \ll t_f$ and $N_f < 0$ at $t \sim t_f$. For certain values of the coefficients, Eq. (3.119) is reduced to (3.116). In contrast to (3.122), the deformation criteria (3.120) and (3.121) relate to the conditions when one-sided accumulation of irreversible deformations can take place. The difference between equations (3.120) and (3.121) is that in (3.121) irreversible deformations are divided into time-dependent and time-independent. For materials in which the values of the deformation capacity at short-term ε_p and long-term ε_l rupture differ significantly, the use of (3.120) to determine the durability under conditions of predominant damage accumulation due to unilaterally accumulated deformations leads to significant differences from experimental data. The possibilities of using criterion (3.121) for various deformation schemes are illustrated by diagrams (see Fig. 3.73). The nature of the $\varepsilon_l(\Pi_i)$ dependences can be established from experiments with different materials, the results of which are summarized in Fig. 3.74. It is seen that

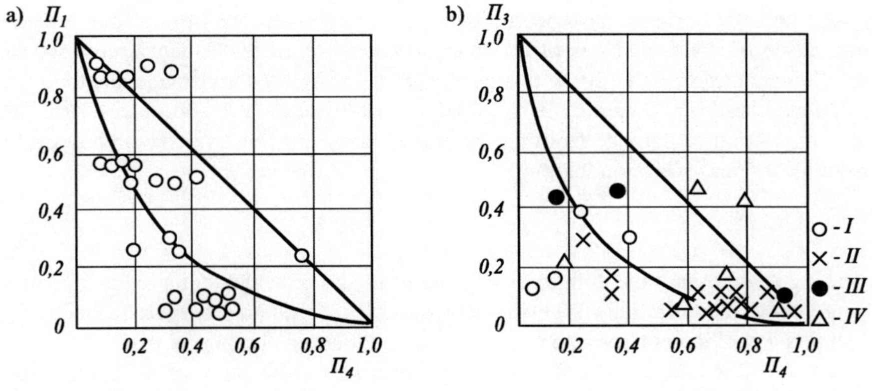

Fig. 3.74 Graphs of the mutual influence of creep and short-term plastic deformation damage for high-temperature alloys: *I* - thermal fatigue under conditions of cyclic plastic deformation and creep; *II* - preliminary plastic deformation; *III* - preliminary creep deformation; IV - cyclic plastic deformation and creep at σ = const

the hypothesis of linear summation, which corresponds to straight lines drawn at an angle of 45°, gives significant differences from the experimental values.

There is a certain connection between criterion (3.117) and a particular case (3.121) of the form $\Pi_2 + \Pi_4 = 1$. If we assume the existence of power-law dependencies between creep strain and time, deformation capacity and time to failure and if damage from cyclic creep is written in the form $(\sum \Delta p_i / C_2)^{1/n_0}$, then (3.122) will be reduced to (3.121).

The generalization of criterion (3.121) to the case of arbitrary cyclic loading programs, proposed in Getsov (1971, 1978, 2001), is a complex of interrelated strain equations and kinetic equations of fracture. To take into account the possibility of brittle fracture under hard stress states, one more term is added to the terms (3.121), associated with brittle strength

$$\Pi_5 = (\sigma_{eq}/S)^m, \tag{3.124}$$

where S is the resistance to all-round stretching, σ_{eq} is the equivalent stress, determined by the formulas for σ_B (see Table 3.1), and m is a constant.

In cases of a complex stress state, it is proposed to correct the values of ductility ε_f^{pl} according to the formulas of Hancock and Mackenzie (1976) and Makhutov (1981). The von Mises strain rate can be used as an equivalent strain measure for polycrystalline materials in Eqs. (3.12) as follows

$$\varepsilon_{eq} = \sqrt{\frac{2}{9}\left[(\varepsilon_{11} - \varepsilon_{22})^2 + (\varepsilon_{22} - \varepsilon_{33})^2 + (\varepsilon_{33} - \varepsilon_{11})^2\right] + \frac{1}{3}\left(\gamma_{12}^2 + \gamma_{23}^2 + \gamma_{31}^2\right)}$$

Similarly, according to Getsov et al. (2004, 2009a), it is necessary to correct and ε_f^{cr}.

A number of criteria have been proposed for the conditions of multimode loading, based on the hypothesis of linear summation. So, M.M. Gadenin proposed to supplement criterion (3.120) with a term $\int(1/N_f)dN$ related to damage due to high-frequency loading (Gadenin, 1976). Tseitlin (1976) proposed a force criterion for the summation of damage from the combined action of low-cycle (with any cycle asymmetry) and vibration loading

$$\sigma_a = \sigma_{-1}[1 + (\sigma_z/B)^{\beta}], \tag{3.125}$$

where σ_a and σ_z are the amplitudes of vibration and repeated-static loads. σ_{-1} and B are endurance limits for high-frequency and low-frequency fatigue, respectively and β is a coefficient depending on material and loading conditions.

Let us illustrate the above criteria by considering the question of summation of long-term static (Π_{st}), low-cycle (Π_{lcf}) and fatigue damages (Π_f). There is no doubt now that with the simultaneous accumulation of Π_{lcf} and Π_{st} the conditions of failure are determined by both components. Only the question of which range of materials and loading conditions can be covered by criteria (3.119)-(3.122) remains controversial. Note, however, that relation (3.121) has certain advantages. Compared to other relations, it has, on the one hand, greater physical validity, versatility with varying materials and loading conditions, on the other hand, it allows one to take into account, if necessary, the mutual effects of damage using relations of the type (3.97).

The number of constants in equations (3.115)-(3.123) can be reduced, because in a number of cases, in a first approximation, they can be considered universal for various materials and loading conditions. However, it should be noted that when determining the constants, it is necessary to use the same methods for determining the stress-strain state, which are supposed to be used in the calculations according to these criteria of safety margins of real parts.

In those cases when, in addition to damage to Π_{lcf} and Π_{st}, Π_f also accumulate, depending on the material and test conditions, failure can take place either in accordance with the criterion max max$(\Pi_i) = 1$, or are described by equations of the type (3.97). It was noted in a number of works that the creep strain accumulated at the first stage does not reduce the fatigue life of the EI698 and EP742 alloys. At the same time, an increase in the level of creep strains leads to a noticeable decrease in fatigue strength, which is satisfactorily described in accordance with the linear summation hypothesis.

For the summation of Π_f and Π_{lcf}, a number of authors also use the hypothesis of nonlinear summation of damage in the form of the relation

$$\Pi_f^{\alpha} + \Pi_{lcf}^{\beta} = 1, \tag{3.126}$$

where Π_{lcf} is damage caused by thermal fatigue.

Let us consider in more detail the general model of fracture of non-brittle heat-resistant materials of deformation type, which allows us to describe the conditions of their fracture under an arbitrarily specified program of unsteady loading and heating and a complex stress state. In this model, we will proceed from the following assumptions (Getsov, 1971, 2001)

1. irreversible deformations can be divided into time-dependent (creep) and time-independent (plastic) deformations (p and ε_{pl}, respectively)
2. the nature of damage caused by cyclic plastic strains Π_1 and creep strains Π_2, unilaterally accumulated plastic strains Π_3 and creep strains Π_4 is different
3. values of damage $\Pi_1, \Pi_2, \Pi_3, \Pi_4$

$$\Pi_i(t) = \int_0^t \frac{\partial \Pi_i}{\partial t} \mathrm{d}t$$

depend respectively on the following quantities

- paths of cyclic plastic deformation λ_1 according to the formula (2.64)
- the path of cyclic creep λ_2 according to the formula (2.65)
- accumulated plastic deformation

$$\begin{aligned} \bar{e}^{pl} &= \int \operatorname{sign}(\boldsymbol{\varepsilon}_{pl} \cdot \cdot \Delta\boldsymbol{\varepsilon}_{pl}) \sqrt{\frac{2}{3} \mathrm{d}\boldsymbol{\varepsilon}_{pl} \cdot \cdot \mathrm{d}\boldsymbol{\varepsilon}_{pl}} \\ &= \int \frac{\boldsymbol{\varepsilon}_{pl} \cdot \cdot \Delta\boldsymbol{\varepsilon}_{pl}}{|\boldsymbol{\varepsilon}_{pl} \cdot \cdot \Delta\boldsymbol{\varepsilon}_{pl}|} \sqrt{\frac{2}{3} \dot{\boldsymbol{\varepsilon}}_{pl} \cdot \cdot \dot{\boldsymbol{\varepsilon}}_{pl}} \mathrm{d}t = \int \mathrm{d}\bar{e}^{pl}, \end{aligned} \tag{3.127}$$

- accumulated creep strain

$$\begin{aligned} \bar{p} &= \int \operatorname{sign}(\boldsymbol{p} \cdot \cdot \Delta\boldsymbol{p}) \sqrt{\frac{2}{3} \mathrm{d}\boldsymbol{p} \cdot \cdot \mathrm{d}\boldsymbol{p}} \\ &= \int \frac{\boldsymbol{p} \cdot \cdot \Delta\boldsymbol{p}}{|\boldsymbol{p} \cdot \cdot \Delta\boldsymbol{p}|} \sqrt{\frac{2}{3} \dot{\boldsymbol{p}} \cdot \cdot \dot{\boldsymbol{p}}} \mathrm{d}t = \int \mathrm{d}\bar{p} \end{aligned} \tag{3.128}$$

4. the rates of accumulation of damage of these four types can be characterized using four mutually dependent functions

$$\frac{\partial \Pi_1}{\partial \lambda_1} = F_1(\Pi_1, \Pi_2, \Pi_3, \Pi_4, t, k_{cor1}) \Delta\lambda_1^{k_0}, \tag{3.129}$$

$$\frac{\partial \Pi_2}{\partial \lambda_2} = F_2(\Pi_1, \Pi_2, \Pi_3, \Pi_4, t, k_{cor}) \Delta\lambda_2^{n_0}, \tag{3.130}$$

$$\frac{\partial \Pi_3}{\partial e} = F_3(\Pi_1, \Pi_2, \Pi_3, \Pi_4, t, k_{cor3}), \tag{3.131}$$

$$\frac{\partial \Pi_3}{\partial p} = F_4(\Pi_1, \Pi_2, \Pi_3, \Pi_4, t, k_{cor4}), \tag{3.132}$$

where τ is some characteristic for a given moment of time, which takes into account the temperature-time prehistory, by the development of aging processes of the material. k_0 and n_0 are constant coefficients, $\Delta\lambda_1$ and $\Delta\lambda_2$ are the increments of the cyclic deformation path for the deformation stages $\Delta\lambda_1 = \lambda_1^{(q)} - \lambda_1^{(q-1)} > 0$,

$\Delta\lambda_2 = \lambda_2^{(r)} - \lambda_2^{(r-1)} > 0$ under the conditions that $\mathrm{d}\lambda_1^{(q)}/\mathrm{d}t > 0$ for $t < t_q$, $\mathrm{d}\lambda_1^{(q)}/\mathrm{d}t = 0$ for $t > t_q$, $\mathrm{d}\lambda_2^{(r)}/\mathrm{d}t > 0$ for $t < t_r$, $\mathrm{d}\lambda_2^{(q)}/\mathrm{d}t = 0$ for $t > t_r$. k_{cori} are parameters that determine the effectiveness of the corrosive effect of the environment

5. failure is determined by the condition

$$\int_0^{\lambda_1^*} \frac{\mathrm{d}\Pi_1}{\mathrm{d}\lambda_1}\mathrm{d}\lambda_1 + \int_0^{\lambda_2^*} \frac{\mathrm{d}\Pi_2}{\mathrm{d}\lambda_2}\mathrm{d}\lambda_2 + \int_0^{e^*} \frac{\mathrm{d}\Pi_3}{\mathrm{d}e}\mathrm{d}e + \int_0^{p^*} \frac{\mathrm{d}\Pi_4}{\mathrm{d}p}\mathrm{d}p = 1, \tag{3.133}$$

where λ_1^*, λ_2^*, e^* and p^* are the limiting parameters of damage corresponding to failure, in the general case, depend on the stress state.

Experimental data obtained under a uniaxial stress state show that not all types of damage, but only some of the parameters of functions F_1, F_2, F_3 and F_4, are decisive (see Fig. 3.74).

Damage increments in one sth cycle can be determined from the relations

$$\Delta\Pi_{1s} = \frac{\left(\oint_s \mathrm{d}\lambda_1\right)^{k_0}}{C_1(\Pi_{3s}, T, t_s)}, \quad \Delta\Pi_{2s} = \frac{\left(\oint_s \mathrm{d}\lambda_2\right)^{n_0}}{C_2(\Pi_{1s}, \Pi_{4s}, T, t_s)},$$
$$\Delta\Pi_{3s} = \frac{\oint_s \mathrm{d}\varepsilon_{\mathrm{pl}}}{\varepsilon_{\mathrm{pl}}(T, t_s)}, \quad \Delta\Pi_{4s} = \frac{\oint_s \mathrm{d}p}{\varepsilon_{\mathrm{cr}}(\Pi_{1s}, \Pi_{3s}, T, t_s)} \tag{3.134}$$

Thus, the number of cycles before failure can be found by solving the kinetic equation

$$\sum_{j=1}^{N_{\mathrm{f}}} \Delta\Pi_{1j} + \sum_{j=1}^{N_{\mathrm{f}}} \Delta\Pi_{2j} + \sum_{j=1}^{N_{\mathrm{f}}} \Delta\Pi_{3j} + \sum_{j=1}^{N_{\mathrm{f}}} \Delta\Pi_{4j} = 1, \tag{3.135}$$

where the terms can be determined by integrating equations (3.134), in which the numerators are calculated based on one or another theory of creep and plasticity, which ensures the adequacy of calculations and experiments for complex cyclic loading programs.

In the particular case, when the mutual effects of different types of damage can be neglected, instead of equation (3.135) we have a fracture condition based on the assumption of linear summation of damage

$$\frac{\sum_j \Delta\varepsilon_{\mathrm{pl}ij}^{k_0}}{C_1} + \frac{\sum_j \Delta p_{ij}^{n_0}}{C_2} + \frac{\sum_j e_{ij}}{C_3} + \frac{\sum_j p_{ij}}{C_4} = 1. \tag{3.136}$$

If in Eq. (3.136) the values of the limiting characteristics are set in the form of dependencies on damages of another type, then it takes on the character of a non-linear summation of damages. Obviously, this assumption reduces the accuracy of assessing the durability, nevertheless, one should not expect this decrease to exceed the accuracy of the determination of deformation parameters in structural elements.

Comparison of models (3.135) and (3.123) shows that each of them has advantages in certain conditions. The first better reflects the cycle-by-cycle kinetics of damage accumulation, which is important for irregular loading. The second has the ability to describe the influence of the shape of the cycle of inelastic deformation on the resistance to low-cycle fatigue.

3.4.6 Experimental Verification of Criteria

The largest number of studies has been devoted to checking the validity of criterion (3.116). Along with a number of data confirming the possibility of obtaining sufficient accuracy when using criteria of the type (3.115) and (3.116), a number of experimental results indicate otherwise. Figure 3.75, for example, it can be seen that based on the hypothesis of linear summation in the temporal interpretation, it is impossible to obtain high accuracy - fractures under low-cycle loading occur earlier than predicted by (3.116). As seen from Fig. 3.76, taking into account the mutual influence of damage using a constant coefficient μ of criterion (3.115) also does not give high accuracy, since the values of μ turn out to be dependent on the test base. Very significant differences from the experiment were obtained in the calculated assessment of the durability of specimens tested for thermal fatigue without holding at stresses lower than the elastic limit (Fig. 3.77). At the same time, in experiments on thermal fatigue with exposure at T_{max}, the difference in durability does not exceed tenfold (Fig. 3.78).

Criterion (3.116) was used to predict fractures under the combined action of static and dynamic loading of the EI617 alloy at 800 °C (Pisarenko, 1965). The tests were carried out in accordance with 10 loading programs. We replace the integrals in

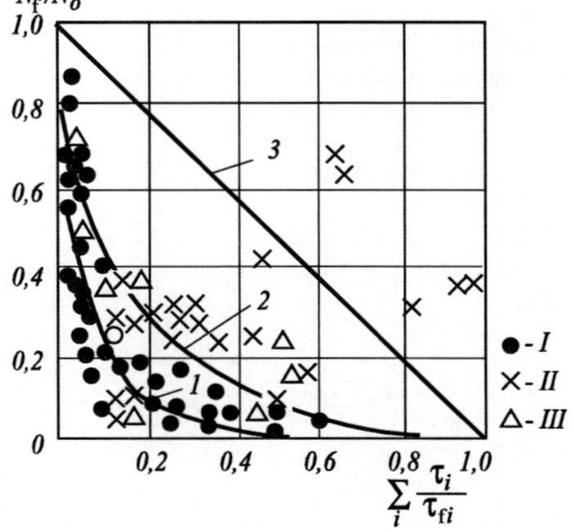

Fig. 3.75 Damage graphs under the combined action of creep and fatigue of alloys: *1* - Ch20N45 and EI437B (*I*); *2* - AISI304 (*II*), steel with 0.14% C (*III*); *3* - curve calculated by the formula (3.116)

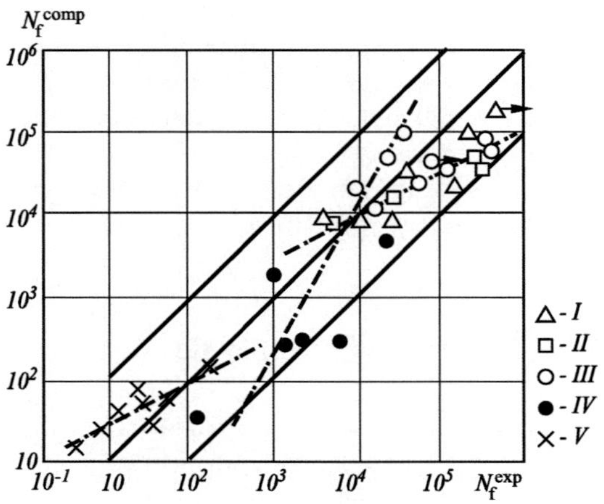

Fig. 3.76 Graphs of the relationship between the calculated by Eq. (3.115) and the experimental numbers of cycles to failure for various alloys: *I*, *II*, *III* - EI765, EI826, EI827 at temperatures of 750, 800 and 850°C; *IV* - EP220 at a temperature of 900°C; *V* - EI612 at a temperature of 650°C

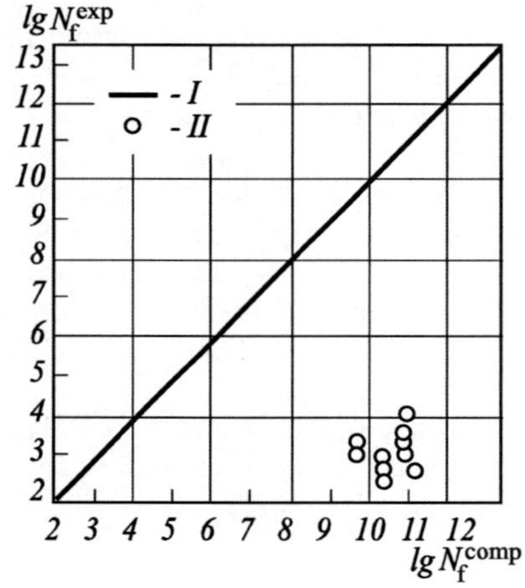

Fig. 3.77 Graph of comparison of the calculated by Eq. (3.116) (*I*) and experimental (*II*) values of the fatigue life of the EI765 and EI826 alloys in thermal fatigue tests without holding at T_{max} = 800°C

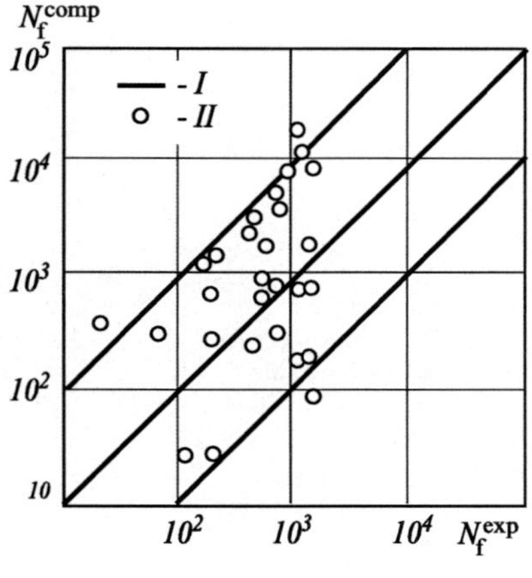

Fig. 3.78 Graphs of the relationship between the calculated by Eq. (3.116) (*I*) and the experimental (*II*) number of cycles to failure in thermal fatigue tests of different alloys in the mode of 900-500°C, t_{cycl} = 2 min (according to L.P. Nikitina's data)

Manson's criterion (3.116) with sums and denote the first term Π_1' and the second Π_2'. Damage ratio at static $\Pi_1' = \sum_{j=1}^{M}(t_j/t_{fj})$ and dynamic $\Pi_2' = \sum_{j=1}^{L}(n_j/N_{0i})$ loading varied in the range $\Pi_1'/\Pi_2' = 0.8\ldots20$. Here M is the number of static loading steps in the block, t_j is the duration of one stage of static loading in the block, L is the number of steps of alternating stresses, n is the number of cycles of changing the variable component of the stress at one stage of the asymmetric cycle within one block and N_0i is the number of cycles to failure in continuous testing. It turned out that an increase in the proportion of fatigue damage to 0.5 does not change the value of accumulated static damage. With an increase in the Π_1'/Π_2' ratio, the value of the total damage at fracture $\Pi_1' + \Pi_2'$ for non-riveted metal decreased from 1.4 to 0.5-0.7, and for cold-worked metal it increased from 0.35 to 0. 8. It was found that for $\Pi_1'/\Pi_2' > 10$ there are fractures of the static type, for $1 \ll \Pi_1'/\Pi_2' < 10$ mixed static-fatigue fractures are observed and, finally, for $\Pi_1'/\Pi_2' < 1$ fatigue-type fractures are detected. Alternating stresses promote crack propagation of permanent static damage.

A good agreement between the calculated and experimental values of the service life according to the criterion $\Pi_2 + \Pi_4 = 1$ (standard deviation 5.8%) was obtained when processing the results of tests for alternating creep of the EI765 alloy reported in Pavlov and Kurilovich (1978). However, it should be noted that under static loading, the fracture conditions $\Pi_3 + \Pi_4 = 1$ are ensured with an accuracy noticeably less than in the case of cyclic modes (Table 3.29). The reasons for the increased scatter of data, apparently, are that under conditions of single deformation, the distribution of deformations along the length of the specimen during creep and instantaneous tension is different, and under conditions of cyclic deformation, the influence of these differences observed in each cycle is leveled.

Comparison of the calculated and experimental values of the fatigue life according to the criteria (3.121) and (3.122) with the experiment carried out in Kononov and Getsov (1984); Pavlov et al. (1989); Getsov (1994) for the cases of uniaxial and complex stress states, showed that in the case of a rigid loading cycle, the root-mean-square deviation N_{calc} from N_{exp} is $S \sim 0.17$ for both criteria. For the case of one-sided accumulation of deformations, criterion (3.121) makes it possible, with

Table 3.29 Influence of the accumulated creep strain on the ductility of heat-resistant alloys

Material	Creep test regime				Tension test temperature, °C	$\frac{p}{\varepsilon_{pl}} + \frac{\delta}{\delta^{(0)}}$
	T, °C	σ, MPa	t, h	p, %		
EI765	700	450	140	0.55	700	0.33
	700	450	106	1.09	700	0.66
	700	300	104	0.21	700	1.0
	700	300	214	1.03	700	0.64
EI723	550	500	62	0.54	550	0.86
EI481	650	370	425	1.93	20	1.1
EI612	650	360	190	2.33	20	1.35
EI607A	700	270	1625	4.91	20	1.77

an error not exceeding two times, to establish what is at the level of the scatter of experimental data for damage of one type (Fig. 3.79). By virtue of its physical validity, criterion (3.121) describes the cycle-by-cycle kinetics of damage better than others.

3.4.7 Fracture Criteria for Complex Stress State

To determine the safety margins in stationary modes of parts in a complex stress state, it is advisable to use the following approaches depending on the plasticity of the material

1. For a plastic material, the margin of safety is determined
   - by viscoelastic-plastic deformations (according to von Mises criterion), determined by the methods described in Chapt. 2, by comparison with the value of the deformation capacity of the material determined by one of the Eqs. (3.12) and (3.43). A comparison of the calculation results for various stress states according to Eqs. (3.12)a and (3.12)b showed which formula gives a more conservative estimate depends on the type of stress state.
   - by the average visco-elastic-plastic stresses in the section of the blade airfoil, determined according to von Mises criterion (see Chapt. 2), by comparing the obtained values with the ultimate strength at the maximum temperature in this section and the duration of operation. This method makes it possible to use the standard safety factors developed in relation to calculations using the bar design scheme.
2. For a brittle material (with ductility less than 3-5%), it is recommended to use either

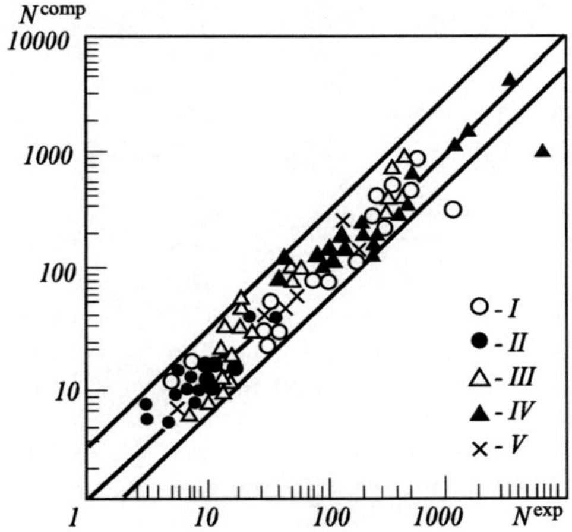

Fig. 3.79 Graphs of comparison of calculated and experimental values of durability (according to test data of pearlite steel AF-2-IDA, austenitic steels Ch16N11M3 and ChN35WT (EI612), nickel-based alloys EI607A, EI765, EP539, EP220): *I* - $\Pi_1 + \Pi_2$; *II* - $\Pi_3 + \Pi_4$; *III* - $\Pi_1 + \Pi_2 + \Pi_4$; *IV* - $\Pi_1 + \Pi_3 + \Pi_4$; *V* - $\Pi_1 + \Pi_2 + \Pi_3 + \Pi_4$

- the criterion of maximum (main) local viscoelastic-plastic deformations by comparing the values obtained in the FEM calculation with the value of the deformation capacity of the material taking into account Eqs. (3.12) and (3.43) or
- the criterion of the maximum (main) local stresses calculated by the FEM in the visco-elastic-plastic formulation, by comparing the obtained values with the ultimate strength at the corresponding operating time and temperature of the element.

To determine the safety margins of the material of the blades in unsteady modes, it is also recommended to use different approaches depending on the plasticity of the material and the loading paths realized in the corresponding finite elements of the part during start-stop.

3.4.7.1 First Approach

Ductile Material

The first approach to determining the material safety margins in unsteady modes assumes that for a ductile (plastic) material, the deformation range $\Delta\varepsilon$ is estimated according to the von Mises criterion for the differences in the ranges of the deformation components (where the ranges are determined as the sum of the maximum tensile deformation and the maximum compression deformation), and the temperature is chosen as the maximum cycle temperature in this point as follows

$$\Delta\epsilon = \Delta\varepsilon^{el} + \Delta\varepsilon^{pl} + \Delta\varepsilon^{cr}, \tag{3.137}$$

where

$$\begin{aligned}
\Delta\varepsilon^{el} &= \frac{\frac{\sqrt{2}}{2(1+\mu)}\sqrt{(\Delta\varepsilon_x^{el} - \Delta\varepsilon_y^{el})^2 + (\Delta\varepsilon_y^{el} - \Delta\varepsilon_z^{el})^2 + (\Delta\varepsilon_z^{el} - \Delta\varepsilon_x^{el})^2}}{+\frac{3}{2}((\Delta\gamma_{xy}^{el})^2 + (\Delta\gamma_{yz}^{el})^2 + (\Delta\gamma_{zx}^{el})^2)}, \\
\Delta\varepsilon^{pl} &= \frac{\frac{\sqrt{2}}{3}\sqrt{(\Delta\varepsilon_x^{pl} - \Delta\varepsilon_y^{pl})^2 + (\Delta\varepsilon_y^{pl} - \Delta\varepsilon_z^{pl})^2 + (\Delta\varepsilon_z^{pl} - \Delta_x^{pl})^2}}{+\frac{3}{2}((\Delta\gamma_{xy}^{pl})^2 + (\Delta\gamma_{yz}^{pl})^2 + (\Delta\gamma_{zx}^{pl})^2)}, \\
\Delta\varepsilon^{cr} &= \frac{\frac{\sqrt{2}}{3}\sqrt{(\Delta\varepsilon_x^{cr} - \Delta\varepsilon_y^{cr})^2 + (\Delta\varepsilon_y^{cr} - \Delta\varepsilon_z^{cr})^2 + (\Delta\varepsilon_z^{cr} - \Delta\varepsilon_x^{cr})^2}}{+\frac{3}{2}((\Delta\gamma_{xy}^{cr})^2 + (\Delta\gamma_{yz}^{cr})^2 + (\Delta\gamma_{zx}^{cr})^2)}
\end{aligned} \tag{3.138}$$

Here, the symbols "el", "pl" and "cr" refer to elastic plastic deformation and creep deformation, respectively. The procedure for separating inelastic strains is described in the literature.

Brittle Material

For a brittle material (with a ductility of less than 3-5%), it is necessary to use the tensile strength criterion under cyclic loading (i.e. crack propagation along a path perpendicular to the direction of the largest principal inelastic strain, and not the method described above. The corresponding method for determining the range of strains is described in NN (2007). In cases where proportional loading takes place, the procedure is simplified and $\Delta\varepsilon = |\sigma_{1\max} - \sigma_{3\min}|/E$. Here $\sigma_{1\max}$ and $\sigma_{3\min}$ are the maximum values of σ_1 and σ_3 in considered point.

In other words, if the loading path during the cycle is close to proportional (see, for example, Fig. 3.80), the moments of time for which the strain components are used in calculating the strain ranges are chosen as giving the maximum magnitudes of tensile or compressive ones (for example, $\varepsilon_1 - \varepsilon_3$). If, at the considered point of the blade, the loading path is not proportional (see, for example, Fig. 3.81), then it is necessary to use the method for determining the range of strains described in NN (2007).

The calculated values of $\Delta\varepsilon$ obtained are compared with the experimental curves of thermal fatigue resistance at the corresponding values of $T_{\max}$ and cycle time. However, due to the fact that at relatively low temperatures (for the materials under consideration, less than 650°C), it is practically impossible to obtain experimental data on thermal fatigue resistance, for such $T_{\max}$ the calculated values of $\Delta\varepsilon$ should be compared with the values estimated using the Manson universal slope formula. At the same time, taking into account the inaccuracy of the durability prediction according to this formula, it is advisable to use sufficiently large values of the permissible safety margins.

Unfortunately, the described calculation method for determining $\Delta\varepsilon$ in components is not free from drawbacks. First, the summation of the ranges $\Delta\varepsilon^{el}$, $\Delta\varepsilon^{pl}$ and

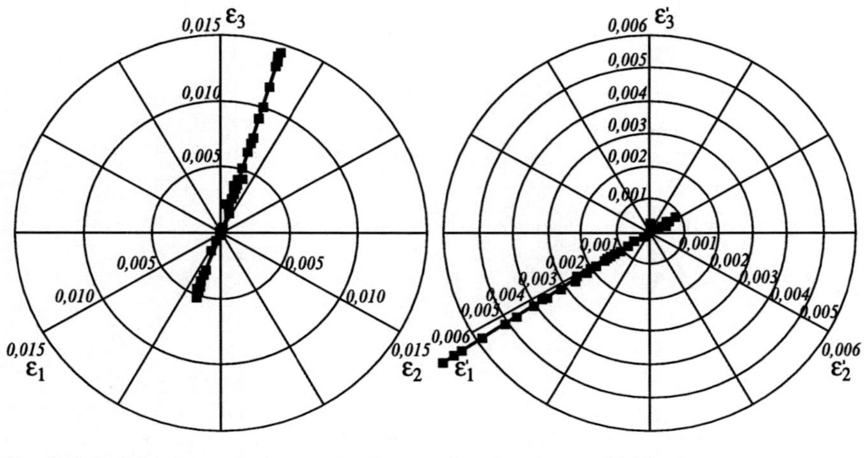

Fig. 3.80 Deformation path trace on the deviator plane (proportional loading)

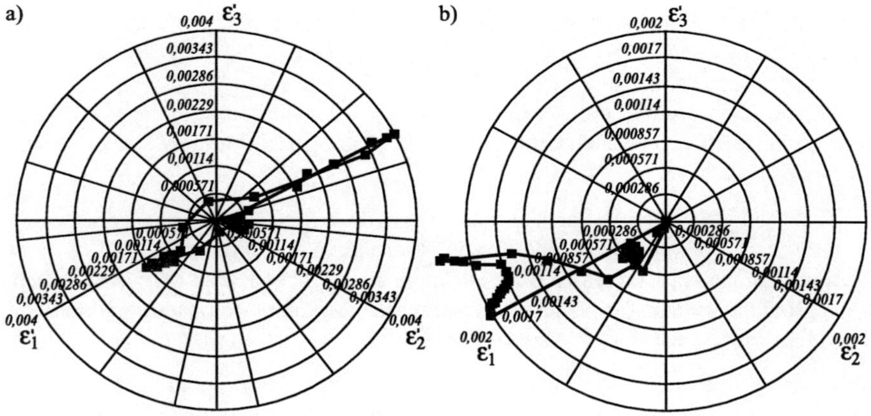

Fig. 3.81 Deformation path trace on the deviator plane: a) proportional loading, b) nonproportional loading

$\Delta\varepsilon^{cr}$ does not take into account the known fact that the contribution of inelastic and elastic deformation to fatigue damage is different. Second, there may be cases when the time moments corresponding to the maxima $\Delta\varepsilon^{el}$, $\Delta\varepsilon^{pl}$ and $\Delta\varepsilon^{cr}$ will be different.

3.4.7.2 Second Approach

The *second approach* is associated with the use of the deformation criterion (3.121), (3.136).

3.5 Features of the Formation and Propagation of Cracks in Heat-resistant Alloys under Static and Dynamic Loading

3.5.1 Conditions and Nature of Cracking

The formation of microcracks along grain boundaries in heat-resistant alloys, causing damage under static loading at elevated temperatures, is associated with the development of intergranular slip processes. Studies presented in Levin and Gugelev (1976) have established that the critical value of intergranular slip, at which microscopically observed cracks appear at grain boundaries, for precipitation hardened austenitic steels and nickel-based alloys in the absence of boundary migration is approximately 0.35μm. The rate of growth of the critical slip value depends on temperature non-monotonically, which is associated with the peculiarities of the changes occurring with the particles of the strengthening phase. For example, for a deformed nickel-based alloy EI607A, the process of intergranular cracking develops at temperatures above 550°C. From 500 to 720°C the slip intensity increases with increasing temperature, in the range 720-850°C it decreases.

The features of intergranular slip were investigated in the course of thermal fatigue tests by the method presented in Gugelev and Getsov (1976) for nickel-based alloys. The analysis showed the following.

1. The value of intergranular slip increases with an increase in the upper cycle temperature and the duration of heating the sample at this temperature.
2. Intergranular slip increases with an increase in the number of cycles and is approximately equal in value to intergranular slip under comparable conditions of tension (temperature and strain rate).

Thus, the process of mutual displacement of grains under conditions of thermal fatigue is much more similar to the same process under conditions of deformation of the same sign than under conditions of alternating isothermal deformation. This conclusion is related to the fact that, in the thermal fatigue test, intergranular slip is not observed at the lower cycle temperature ($T_{\mathrm{min}} < 0.4T_{\mathrm{melt}}$) and develops only in the region of the upper cycle temperature, i.e. under conditions of deformation of one sign. When analyzing the causes of the failure of a part under operating conditions at elevated temperatures, the results of metallographic studies of the nature of the fracture are widely used. Therefore, it is important to know what kind of fracture is characteristic of a particular type of loading.

In what follows, we will consider the intergranular and intragranular nature of fracture, bearing in mind that in the first case, intergranular cracks are found in the fracture zone of the sample. Cases of a mixed nature of fracture are possible, which is associated with an increase in the propagation rate of an already formed crack or, conversely, with a slowdown in the propagation of an intragranular crack, for example, due to the features of the stress state.

Let us consider the specifics of the nature of fracture during high- and low-frequency fatigue, thermal fatigue, creep, instant deformation of specimens, blades and disks made of various heat-resistant alloys at elevated temperatures. In pearlite steels under static loading, the nature of crack propagation depends on the duration of loading. At a temperature of 550°C and a time before fracture of less than 100 h, intragranular cracks are formed, while at $t_f \gg 100$ h - intergranular ones. A change in the nature of crack propagation in these steels leads to a change in the slope of the curves of creep resistance and prolonged fracture (Fig. 3.4).

The change in the slope of these curves is also influenced by a change in the micromechanism of damage - the formation of intergranular micropores or wedge-shaped cracks (Chizhik and Petrenya, 1978). In steels of the ferrite-martensitic class, cracks form and develop transcrystalline at operating temperatures. Cracks in austenitic steels are mainly intergranular.

For deformed nickel-based alloys EI607A, EI765, EI826, EI698, EI893, EI868, during static deformation, an intergranular fracture is observed both during prolonged and short-term loading (deformation). For the complex alloy EP220, intergranular cracks take place even in the case of instant deformation at a strain rate of 2400%/h at 700-900°C. For the low-alloy EI607A alloy, intergranular fracture was observed during short-term creep with $t_f = 2$ min. With creep, microcracks originate on the surface of the samples already at early stages (Fig. 3.82). With a cyclic change in the test temperature, the nature of fracture corresponds to that under static deformation at t_{max}.

The simplest way to determine the features of the propagation of creep cracks is to test specimens with several notches, which, after failure, are analyzed in one of them. So, as a result of tests for long-term strength of specimens with two notches of the same diameter made of steels EI612 and EI481 after long-term fracture along one of the notches in 50% of cases, no visible cracks were found in the second notch, in the zone of which the metal had practically exhausted its durability, while the second notch of the samples from nickel-based alloys EI607A, EI827, EI661 had cracks in all cases. Experimental observation of the growth rate of the formed creep cracks during long-term testing of specimens presents significant difficulties. Figure 3.83 shows the corresponding data for austenitic steel, obtained in Koterazawa and Iwata (1976).

To determine the features of the initiation and development of cracks under conditions of low-frequency alternating loading, metallographic studies of samples and blades that failed at different frequencies, amplitudes, temperatures, types of deformation, and loading conditions were carried out. It was found that at both constant and variable temperatures (up to $T_{max} = 900$°C inclusively), the nature of fracture of the tested samples (Figs. 3.84a, b) and blades (Fig. 3.84c) at loading frequencies up to 4-10 cycles/min is intragranular. However, the tests carried out by Serensen and Meshchaninova (1960) at 900°C showed that when the number of cycles before fracture N_f is more than 109, there is an intergranular fracture under high-frequency loading. With an increase in temperature to 1000°C and more, cracks propagate along grain boundaries.

An increase in the cycle period at $T < 900$°C changes the nature of destruction. An increase in the alloying degree of the alloy somewhat increases the critical value of the loading frequency, which causes a change in the nature of fracture. The ratio

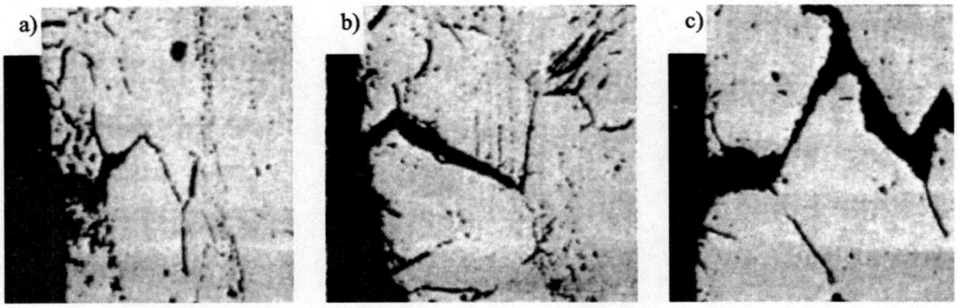

Fig. 3.82 Surface microstructures of samples made of alloy EI607A after long-term strength tests: a) - 600°C, $\sigma = 750$ MPa, $t_f = 6$ min, b) - 650°C; $\sigma = 670$ MPa, $t_f = 7.5$ min; c - 700°C, $\sigma = 250$ MPa, $t_f = 590$ h

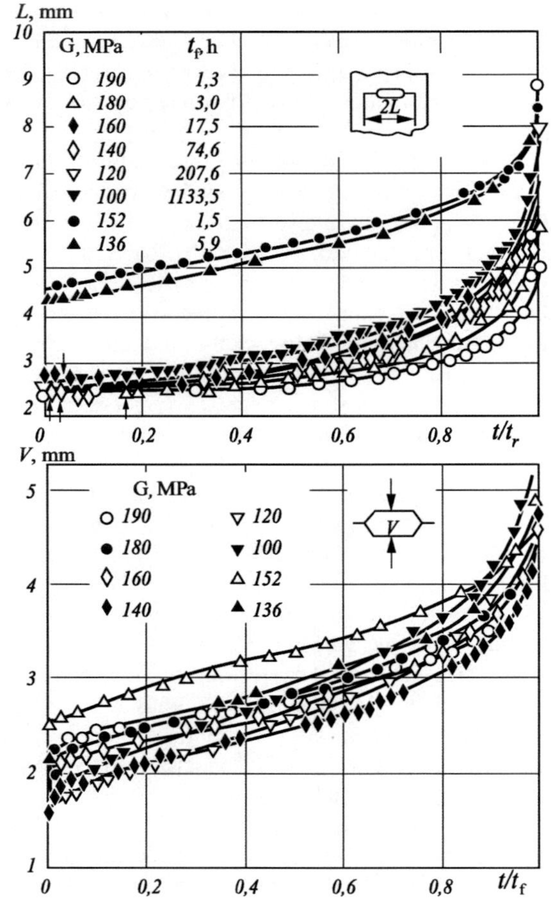

Fig. 3.83 Curves of propagation of creep cracks obtained for steel of the 12Ch18N10T type at 650°C on specimens with a preliminarily applied fatigue crack, *V* - crack center opening

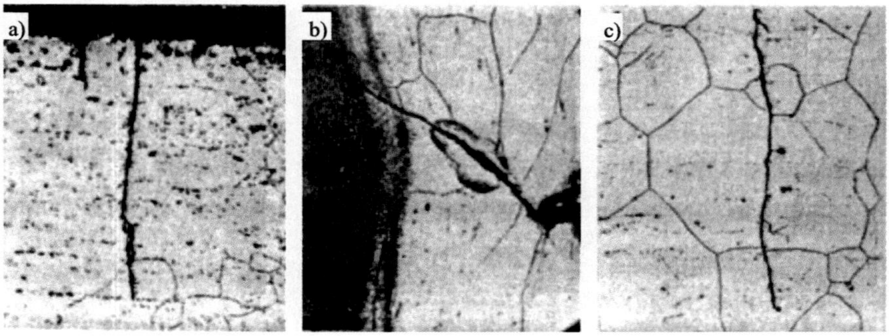

Fig. 3.84 Microstructure of the surface of samples and blades after low-frequency fatigue tests: a) EI765, $\omega = 10$ cycles/min, 800°C, $N_f = 108854$; b) EI827, $\omega = 4$ cycles/min, 800...500°C, $N_f = 10054$; c) EI827, $\omega = 4$ cycles/min, 850...450°C, $N_f = 22000$

between the number of cycles before the initiation of microcracks and before the failure of the sample can be observed from the data in Table 3.25.

An increase in the degree of cycle asymmetry from -1 to -0.3 at constant temperatures and up to approximately 0 at cyclically varying temperatures in thermal fatigue tests does not affect the nature of crack propagation. A decrease in the rigidity of loading promotes the development of processes causing intergranular fracture. The data obtained in the study of the nature of fracture of nickel-based alloys can be summarized using a diagram (Fig. 3.85)[16]. For heat-resistant austenitic steels, as well as for nickel-based alloys, similar dependencies of the nature of cracking during cyclic deformation on the cycle period are observed (holding time at T_{max}). For example, in discs made of steel EI612, tested for thermal fatigue with holding at 650°C, cracks developed along grain boundaries, and in discs tested without holding, inside the grain. Similar results were obtained for steel with 19.4% Cr and 34.4% Ni.

A certain specificity of behavior under cyclic loading was found during thermal fatigue testing of cast specimens of heat-resistant alloys. Tubular specimens made of alloy ZhS6U of directional solidification were tested according to the regime with T_{max} = 900°C without holding at T_{max}. It was found that, in contrast to the data obtained according to Fig. 3.85, the resulting cracks can develop in completely different ways: along the boundaries of crystallites, across crystallites, and also in the form of small tears perpendicular to the crystallites (Fig. 3.86).

The nature of fracture during cyclic alternating loading of heat-resistant cobalt-based alloys also depends on the loading frequency. Thus, for the HS188 alloy (Sahm and Speidel, 1974), a diagram was plotted that separates the regions of intergranular and transgranular fracture depending on the frequency and temperature of tests. As shown by Koterazawa and Iwata (1976), in experiments with austenitic steel 18-8 in the process of cyclic sign-constant loading at high temperatures, the nature of failure

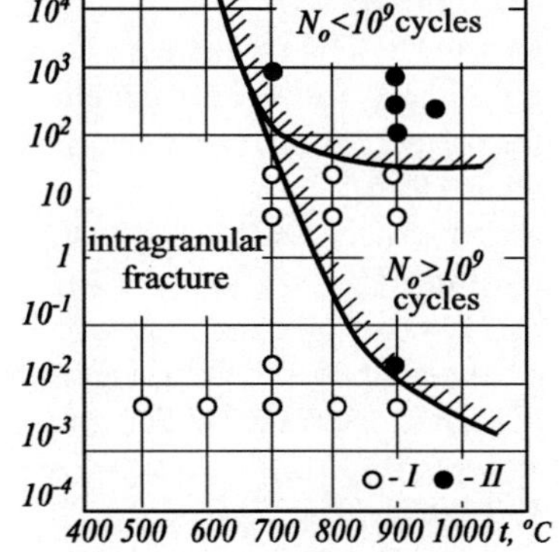

Fig. 3.85 Fracture diagram of nickel-based alloys:
I - transgranular fracture,
II - intragranular fracture

[16] Similar results were obtained in Dul'nev and Kotov (1980).

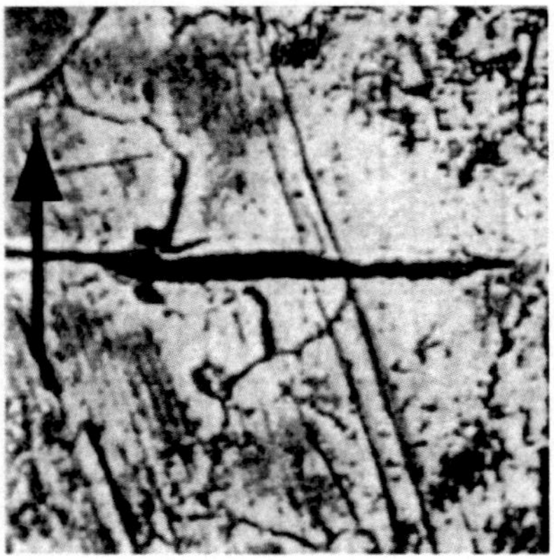

Fig. 3.86 Cracks in a specimen made of ZhS6U alloy tested for thermal fatigue in the 900...700°C mode (the arrow indicates the direction of the load)

depends both on the holding time at σ_{max} and on the value of the stress intensity factor. Cracks originated transgranularly, at a later stage, upon reaching $K = K_{cr}$, propagate along the grain boundaries. With an increase in the exposure time, K_{cr} decreases.

Special consideration should be given to the nature and rate of propagation of stress corrosion cracking cracks. For austenitic steels in water containing NaCl, such cracks propagate transcrystalline at low temperatures, and intergranular at temperatures of ~ 400°C and above.

3.5.2 Crack Propagation Rate

To describe the laws governing the growth of creep cracks, fatigue and low-cycle fatigue, fracture mechanics methods are used Savin (1970); Ivanova and Gurevich (1974); Koterazawa and Iwata (1976); Wasserman et al. (1990); Pokrovskii et al. (1994a,b), based on the fact that the parameter that determines the rate of crack growth (RCG) is the stress intensity factor K_I or the range of stress intensity factors ΔK_I. The K_I values are calculated by the formula $K_I = Y\sqrt{\pi l}$ where Y is the geometry factor, and l is the crack length. The values of K_I are determined for various bodies Savin (1970).

It has been established that the rate of propagation of creep and fatigue cracks has three stages I, II and III (Fig. 3.87). At $K < K_0$ and $K < K_{th}$, an incubation period takes place, during which cracks practically do not grow. So, for values of crack length less than l_{i100} and stresses less than $\sigma_{0.2/100}$, Fig. 3.88, for various disk materials, lateral one-sided cracks on the samples do not propagate and, conversely, at $l_{i100} = (K_{i100}/1.12\sigma_{0.2/100})^2\pi$, fracture occurs after 100 h (Pokrovskii et al., 1994a,b).

The most widespread are two methods for determining the dependence of the crack growth rate (CGR) under creep on the stress intensity factor

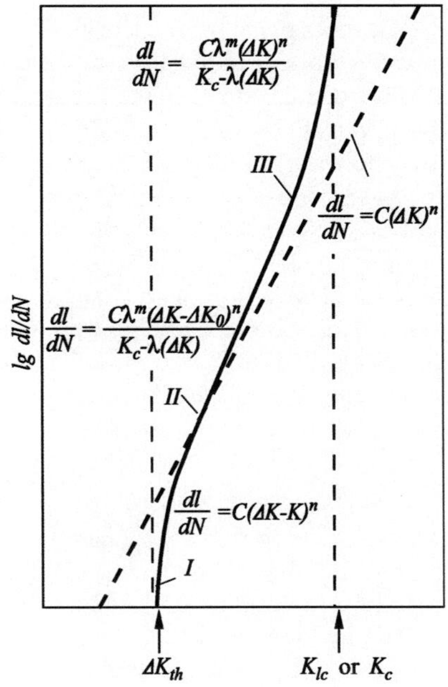

Fig. 3.87 Stages of fatigue crack growth in vacuum (diagrams)

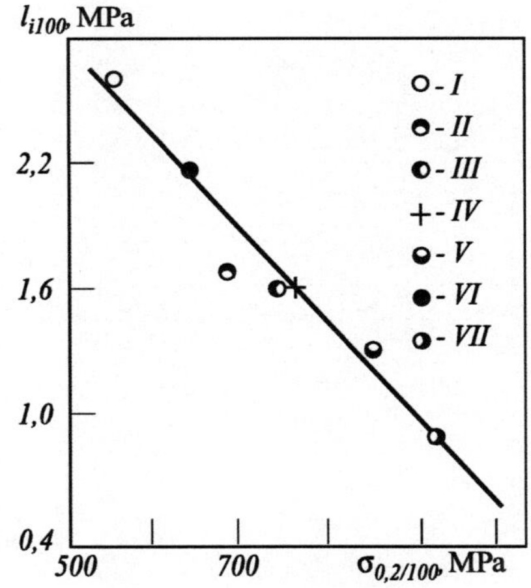

Fig. 3.88 Graph of the dependence of the length of the initial crack, causing the fracture of alloys at a temperature of 650°C for 100 h, on the creep limit $\sigma_{0.2/100}$: *I* - EI698VA; *II* - EP742PVA; *III* –EP741P; *IV* - EP741IA; *V* - EP741NP; *VI* - EP962P; *VII* - EP975P

- test method for compact flat CT specimens for eccentric tension with an induced edge fatigue crack (GOST 25.506-85)
- a method for testing prismatic specimens with an edge crack for bending with the determination of the dependence of the opening at the crack tip on time

To determine the dependence of the CGR under fatigue on K_I, the cracks during the test are observed with an optical microscope using stroboscopic lighting.

The mechanical parameters used to analyze the results of tests for the propagation of creep cracks and to calculate the survivability of parts with formed creep cracks also include the stress in the net section (mean stress in the crack region) σ_{net} and the corrected J-integral (C^*). The C^* -integral is determined by replacing the strain (displacement) in Rice's J-integral with the creep strain rate (or displacement rate). It was found that with a change in the stress level and sample sizes, the parameters K and σ_{net} give an ambiguous determination of the dependence of the rate of growth of a creep crack, while the values of the C^*-integral corresponding to a given value of dl/dt change little with changing test conditions. Thus, in recent years, the use of the C^*-integral is considered preferable (Goldman and Hutchinson, 1975; Landes and Begley, 1976; Margolin et al., 2006). For cylindrical specimens with an annular notch, the value of the C^*-integral is determined by the formula

$$C^* = \frac{2k-1}{2k+1}\sigma_{net}\frac{dV}{dt}, \tag{3.139}$$

where dV/dt is the crack opening rate on the notch surface and k is the creep exponent in the Norton law (2.11). Methods for determining J and C^* integrals are described in detail in Taira and Otani (1986).

Thermal fatigue cracks develop similarly (see Figs. 3.89 and 3.90). The dependencies of the growth rate of fatigue cracks on ΔK for a number of high-temperature nickel-based alloys and stainless steels and alloys for compressor blades are shown in Figs. 3.91-3.93.

The approximation of the experimental data of high-frequency and low-cycle fatigue is carried out using various dependencies (Table 3.30). The most common of these is the following relation

$$i = \frac{dl}{dN} = A(\Delta K)^m. \tag{3.140}$$

For creep and stress corrosion cracking the following Eq. can be applied

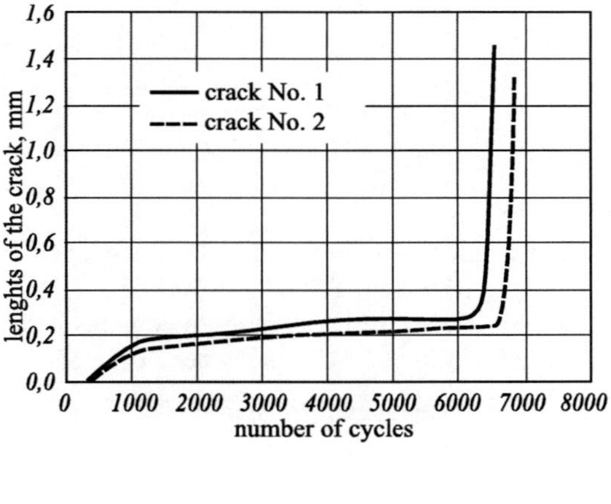

Fig. 3.89 Growth of cracks during thermal fatigue testing of the ZhS6K alloy in the 150-700°C mode of a sample with a concentrator in the form of a hole with a diameter of 0.5 mm

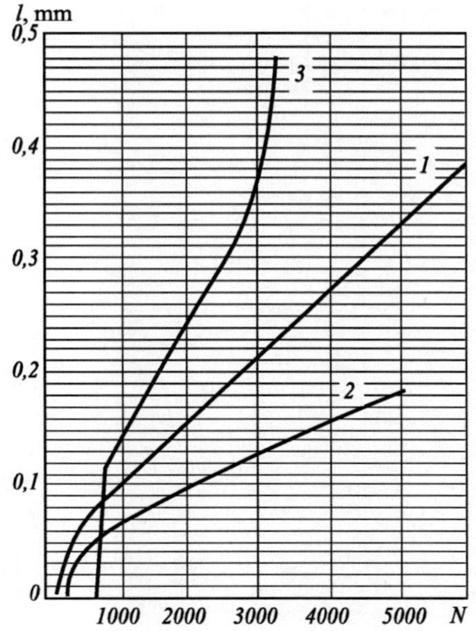

Fig. 3.90 Growth rate of thermal fatigue cracks in alloys EI893 (*1*), EI826 (*2*) under cycle conditions 150-650°C and ZhS6K (*3*) under cycle conditions 150-700°C

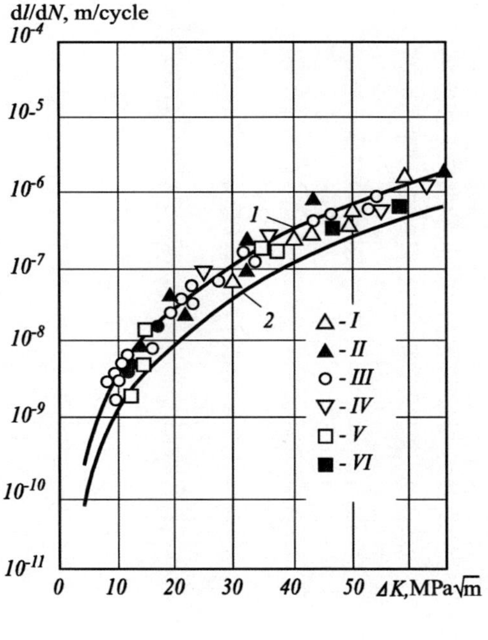

Fig. 3.91 Graphs of the dependence of the growth rate of fatigue cracks dl/dN on the stress intensity factor ΔK of nickel-based alloys in tests in air (*1*) and in vacuum (*2*) at a temperature of 20°C, $\omega = 2.5$ Hz, $R = 0$ (Sahm and Speidel, 1974): *I* - IN718; *II* - IN738; *III* - IN600; *IV* - nimonic 105; *V* - nickel 201; *VI* - nimonic 80A

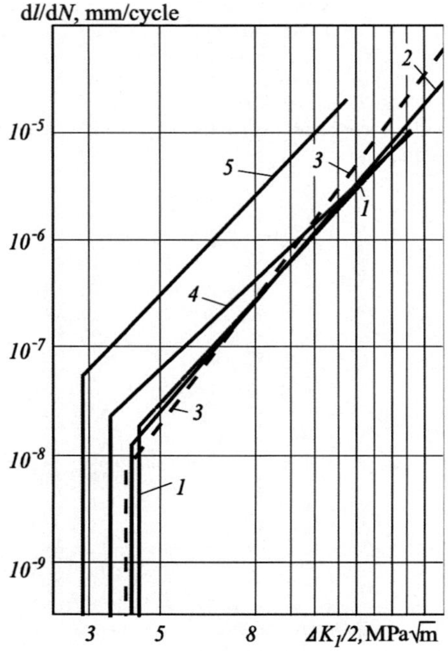

Fig. 3.92 Graphs of the dependence of the growth rate of fatigue cracks dl/dN in seawater of specimens made of steels 08Ch17N6T (*1*), 20Ch13 (2), 14Ch17N2 (*3*), EI961 (*4*), VT-3-1 (*5*) on the range of stress intensity factors $\Delta K_I/2$

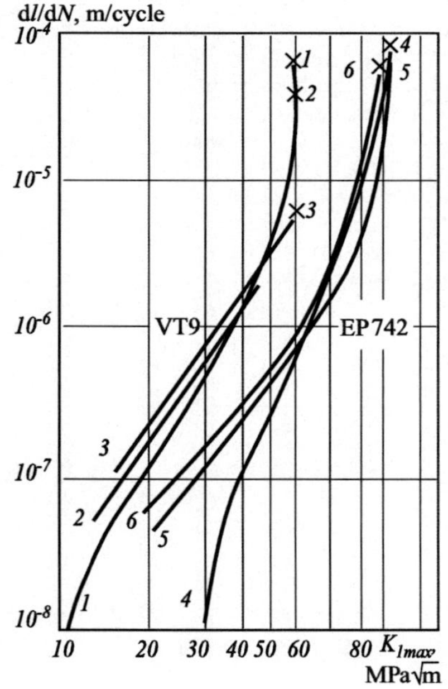

Fig. 3.93 Graphs of the effect of temperature on the GFC (growth of fatigue cracks) speed for VT9 and EP742 alloys: *1*, *4* - T = 293 K; *2* - T = 533 K; *3* - T – 723 K; *5* - T = 773 K; *6* - T = 973 K

Table 3.30 Formulas for fatigue crack propagation rate (Ivanova and Gurevich, 1974)

Equation	Author(s)
$i = \frac{A_1\sigma^2 l}{A_2 - \sigma}$	Head (1953)
$i = Af(\sigma)l$	Frost and Dugdale (1958)
$\lg i = A\sigma_0 - B(\sigma_0 - \sigma_\omega)^{-1} - l; \sigma_0 = K_N\sigma_{nlt}$	McEvely and Illg (1958)
$i = A\sigma^{2.6}l^{1.5}$	Schijve (1962)
$i = A\sigma^2 l$	Upo/Frost
$i = A\sigma^3 l$	Rao Valluri (1963)
$i = A\sigma^4 l^2; i = A\sigma^4 l^3$	McEvily and Boettner (1963)
$i = A\left[\frac{\sigma_\omega}{1 - \frac{l}{W}} - \sigma_\omega\right]^\beta$	Weibull (1963)
$i \cong bl$	Ivanova et al. (1967)
$i = AK^4; i = A(\Delta K)^n$	Paris and Erdogan (1963)
$i = A + \exp(BK)$	Markochev (1966)
$i = A(1 + 2K_m/\Delta K)^{2a}\left(\lambda\frac{\Delta K}{2}\right)^{2(a+b)}$	Roberts and Erdogan (1967)
$i = A\frac{(\Delta K)^n}{(1-R)K_c - \Delta K}$	Forman et al. (1967)
$i = \beta\left(\frac{K_{max}^2 - K_{min}^2}{K_c^2} + \ln\frac{K_c^2 - K_{max}^2}{K_c^2 - K_{min}^2}\right)$	Cherepanov
$i = A(1-R)^2 K^n; n = \frac{2(l+2)}{m+1}$	Pavelko
$i = \varepsilon M_{II}(\Delta K)^n; n = \frac{2\beta}{(1+\beta)mRT}$	Yakobori
$i = \pi^2 K^4/48\sigma_s^2 K_c^2$	Rybakina
$i = A[(0.5 + 0.4R)\Delta K]^n$	Elber (1971)
$i = A\left[\frac{(\Delta K)^2\bar{\gamma}_n}{\pi(\pi_d + 1)t_0^2}\right]^a; \bar{\gamma}_n = \frac{\gamma_n}{(d_0)^{1/(1+n_d)}}$	Gurevich
$i = A(1-R)\frac{K\sqrt{l}}{\sigma_B K_c}$	Morozov (1971)
$i = A[K(1-R)^{0.5}]^n$	Smith (1972)
$i = A[K(1-R)^m]^n$	James (1972)
$i = \frac{A(\Delta K)^n}{(1-R)^m[(1-R)K_c - \Delta K]}$	Nordberg (1972)

In Table 3.30 A, A_1, A_2, B, a, b, λ, n, m, n_d and β are constants. σ_w is the fatigue limit, w is the sample width, γ_n is the octahedral shear strain, $\bar{\gamma}_n/d_0\beta$, d_0 is the size of the plastic zone.

$$\frac{dl}{dt} = BK^n. \tag{3.141}$$

The value of A, according to O.G. Rybakina, is associated with the values of $\sigma_{0.2}$ and K_c by the following dependence[17]

$$A = \frac{\pi^2}{48\sigma_{0.2}^2 K_c^2}$$

However, it should be noted that in a number of cases, the growth rate of stress corrosion cracking cracks turns out to be independent of the values of the stress intensity factor (Fig. 3.94).

Sometimes, to obtain agreement with experiment at small crack sizes, the formulas

$$K = Y\sigma\sqrt{\pi l}; \quad \Delta K = Y\Delta\sigma\sqrt{\pi l} \tag{3.142}$$

are corrected by introducing the initial crack size l_0

$$K = Y\sigma\sqrt{\pi(l + l_0)}; \quad \Delta K = Y\Delta\sigma\sqrt{\pi(l + l_0)} \tag{3.143}$$

This allows one to extend Eqs. (3.140)-(3.141) to the small cracks growth. By integrating the expressions for the rate of crack growth on the values of K_I, it is possible to determine the relationship between the number of cycles from crack initiation until they reach the critical size (determined by the size of the part and the critical value of the intensity factor (for example, K_{Ic} or K_{fc})

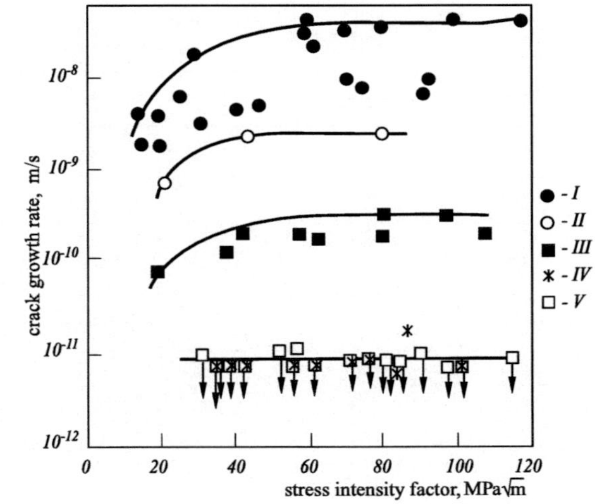

Fig. 3.94 Graphs of the dependence of the growth rate of cracks in stress corrosion cracking of the work-hardened austenitic stainless steel X6MnCrN1818 ($\sigma_{0.2}$ = 1500 MPa) on the stress intensity factor: *I* - 288°C in aerated and deaerated water; *II* - 200°C in deaerated water; *III* - 130°C in deaerated water; *IV* - 90°C in aerated and deaerated water; *V* - 23°C in aerated water

[17] It is shown that A and n are functions of one parameter w_f - specific work of failure under static loading, $A = 1.7191 \cdot 10^{-4}/997^n$, $n = 5.16 - 0.00191w_f$.

$$\int_{N_1}^{N_{cr}} \mathrm{d}N = \int_{l_0}^{l_{cr}} \frac{1}{A(\Delta K_I)^m}\mathrm{d}l; \quad \int_{t_1}^{t_{cr}} \mathrm{d}t = \int_{l_0}^{l_{cr}} \frac{1}{A(\Delta K_I)^m}\mathrm{d}l \tag{3.144}$$

Hence, knowing the characteristics of the crack resistance of the material, it is possible to determine the time or the number of cycles of safe operation of the component, in which cracks may occur during operation, and thus determine the time between the necessary revisions.

The features of the growth of cracks are as follows. At the tip of the crack, plastic deformation processes develop according to the same mechanisms as in the cyclic deformation. As one of the manifestations of these mechanisms, one-sided deformations are formed, leading to the opening of the crack flanges during creep, thermal cycling and quasi-static fatigue loading. With a decrease in stress during the cycle, upon reaching the closing stress σ_{cl}, crack collapse and compression processes occur at loads lower than σ_{cl} (under cycle conditions when $\sigma_{min} < \sigma_{cl}$). These processes are considered in detail in Balina and Myadekshas (1994).

For estimates of the rate of growth of cracks under low-frequency fatigue and cyclic creep, approaches based on deformation criteria developed by Makhutov (1981) can be applied. It is assumed that the propagation of cracks along the length dl in time dt or in one cycle occurs as the deformations in the zone with the size r reach the values of the deformation capacity of the material.

The relations for the radius r in the case of growth of creep cracks can be obtained using the hardening hypothesis by solving (step by step as the crack length l increases) the following system of equations

$$\dot{p}(r, l, \sigma_N) = F_5[p(r, l)K_\varepsilon(r), \sigma_{eq}(r, l)K_\sigma(r), T, \varepsilon_{pl}(r)], \tag{3.145}$$

$$\int_{t}^{t+\Delta t} \dot{p}(r, l, \sigma_N)\mathrm{d}t = \varepsilon_{cr\,f}(t)D_{eq}/J, \tag{3.146}$$

where K_ε and K_σ are strain and stress concentration factors, respectively, σ_{eq} is the equivalent stress, taking into account the effect of stresses along the crack faces (Kabelevsky, 1976) on the level of nominal stresses $\sigma_n = P/F$, $D_{eq}/J = \alpha_0\sigma_{vM}^2/3\sigma_1\sigma_{av}$, $\alpha_0 = 1\ldots1.2$.

Eliminating time from the equations for $r = f(t, l)$ and dr/dt = $f'(t, l)$ the time variable, we obtain the dependence dl/dt = $f(t, \sigma_{eq}, l)$ or dl/dt = $f_7(T, K_{Ieq}, l)$, where K_{Ieq} is the deformation intensity factor, the value of which can be determined if the values of the stress intensity factor and the characteristics of the creep resistance are known. To take into account the influence of damage accumulated by the time Δt at a distance l from the crack tip, it is advisable to introduce into f_7 the dependence on the degree of damage

$$\frac{\mathrm{d}l}{\mathrm{d}t} = \frac{f_7(T, K_{Ieq}, l)}{(1 - \Pi)^{\alpha_0}} = E_7'(T, K_{Ieq}, l, \Pi) \tag{3.147}$$

or

$$\mathrm{d}l/\mathrm{d}t = F_7''(T, \sigma_{\mathrm{eq}}, l, \Pi), \tag{3.148}$$

where α_0 is a coefficient determined from experiments with samples. Similarly, the rate of propagation of cracks caused by cyclic creep can be determined. In this case, instead of equations (3.145) and (3.146), we have

$$\Delta p(r, l, N) = \int_0^{t_{\mathrm{cycl}}} (|\dot{p}|\mathrm{d}t - \dot{p}\mathrm{d}t), \tag{3.149}$$

$$\int_0^{N_{\mathrm{f}}} \Delta p^{n_0} \mathrm{d}N = C_2(t), \tag{3.150}$$

where $\dot{p}$ is determined from Eq. (3.145) such that into the dependence of $\dot{p}$ on N is taken into account. From this we obtain the relations for determining the rate of crack propagation

$$\mathrm{d}l/\mathrm{d}N = F_6'(t, K_{\mathrm{Ieq}}, l, \Pi), \tag{3.151}$$

$$\mathrm{d}l/\mathrm{d}N = F_6''(t, \sigma_{\mathrm{eq}}, l, \Pi), \tag{3.152}$$

The stress and strain concentration coefficients can be determined using the Neuber formula (2.37) or its refinement proposed by Makhutov (2.38).

Taking the approximation of the stress-strain curves of the form (2.4), we obtain the following equation

$$\alpha_N^2 \left[\frac{\sigma_N}{E} + r_0 \left(\frac{\sigma_N}{\sigma_{\mathrm{p}}} \right)^{n_0} + \int p\mathrm{d}t \right] = K_\sigma^2 \left\{ \frac{\sigma_N}{E} + \left[r_0 \left(\frac{\sigma_N}{\sigma_{\mathrm{p}}} \right)^{n_0} + \int \dot{p}\mathrm{d}t \right] K_\sigma^{n_0 - 1} \right\}, \tag{3.153}$$

where α_N is the elastic concentration coefficient, determined on the basis of Irwin's solution by the method described in Makhutov (1981). Having found the value of K_σ from equation (2.34), it is possible to determine K_ε from relations (2.37) or (2.38).

For the case of unilaterally accumulated plastic deformations under the conditions

$$\sum_i \varepsilon_{\mathrm{pl}i} K_\varepsilon = \varepsilon_{\mathrm{p}} D_{\mathrm{eq}} / J, \tag{3.154}$$

the crack growth rate is very high and approaches the speed of sound. Under alternating cyclic loading, the crack growth rate due to cyclic plastic deformations can be determined from the equations:

$$\Delta\varepsilon_{\mathrm{pl}}(r, l, N) = \oint (|\mathrm{d}\varepsilon_{\mathrm{pl}}| - \mathrm{d}\varepsilon_{\mathrm{pl}}), \tag{3.155}$$

$$\int_0^{N_{\mathrm{f}}} \Delta\varepsilon_{\mathrm{pl}}^{k_0} \mathrm{d}N = C_1, \tag{3.156}$$

where $d\varepsilon_{pl} = K_\varepsilon(r)\varepsilon_{pl}(SK_\varepsilon, k, t)$. S is the stress measured from the beginning of unloading. Taking into account the existence of a relationship between ε^t_{cr}, ε^t_{pl}, t, $C_1(t)$, $C_2(t)$ and the characteristics of fracture toughness $\delta^t_c(t)$ and $K^t_{I\delta}$ (δ_c is the critical opening of a crack by the mechanism of breaking its edges, $K^t_{I\delta}$ is the stress intensity factor, calculated from the critical crack opening $K^t_{I\delta} = \sqrt{\sigma_t E \delta_c}$ (Getsov, 1962), the deformation characteristics of the above expressions can be replaced by the parameters $K^t_{I\delta}$ and δ_c that are more usual for fracture mechanics.

For the case of a complex stress state under conditions when it is not obvious whether the crack will develop by separation or shear, and the role of quasi-viscous effects is unknown, the surface points K_{Ic}, K_{IIc}, K_{IIIc} can be used as limiting characteristics (Drozdovskii, 1968; Liebowitz, 1968; Chizhik, 1970).

The rate of crack growth largely depends on the grade of the alloy and the features of its microstructural state (the nature of grain boundaries, precipitates, etc.). Temperature, strain rate, and cycle asymmetry play a significant role. So, as can be seen from Fig. 3.95, an increase in the test temperature of the EP741NP alloy from 650 to 750°C increases the CGR under creep by more than one order of magnitude.

The influence of the asymmetry of the cycle manifests itself in different ways for different materials. Thus, for the Udimet700 high heat-resistant alloy containing Al

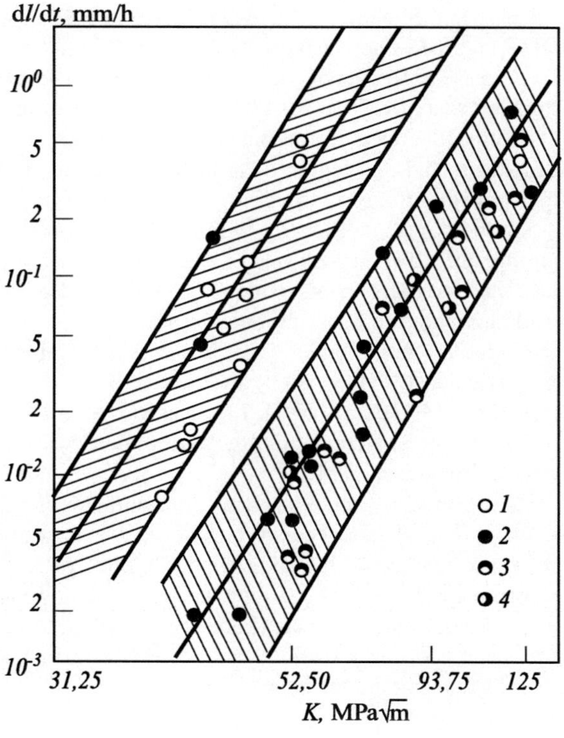

Fig. 3.95 Kinetic diagrams of fracture of EP741NP alloy: *1, 2* - 750°C; *3, 4* - 650°C; *2, 4* - samples with side grooves (according to N.A. Vorobyov)

+ Ti = 8-9%, a change in $R = K_{min}/K_{max}$ in the range from 0 to 0.62 does not affect the dependence of the crack propagation rate on ΔK (Sahm and Speidel, 1974). For pearlitic Cr-Ni-Mo steel, a significant change in the threshold value ΔK_{th} was found with an increase in R from 0 to 0.7. Interruptions in the test usually reduce the rate of crack propagation due to passivation processes at the crack tip. For titanium alloys, the dependence of the position of the curve $dl/dN = f(\Delta K)$ on the stress amplitude was found in Wasserman et al. (1990).

The threshold values of ΔK_{th} depend on the grade of the alloy and the environment. Thus, a comparison of the crack growth curves at room temperature for the group of steels and the VT3-1 titanium alloy showed that high-strength titanium alloys have not only a higher crack growth rate, but also lower threshold values of ΔK_{th} (see Fig. 3.92). The rate of propagation of fatigue cracks in heat-resistant alloys at elevated temperatures is influenced by the frequency of loading and the cycle period (holding time under load and without load) in connection with the processes of oxidation of the crack edges and tip. Thus, for the Inconel718 alloy at 538°C under the conditions of $R = 0.025$, with an increase in the holding time by a factor of 10-20 (up to 2h), an increase in the growth rate was observed at ΔK = const by an order of magnitude (Sahm and Speidel, 1974).

In Pokrovskii et al. (1994a,b), diagrams of the growth of fatigue cracks were plotted in the coordinates dl/dN - K_{Imax} for VT9, EP742, and EP962 alloys based on test data for standard specimens for eccentric tension. It was found that with an increase in the loading frequency from 0.5 to 40Hz, the growth rates of fatigue cracks (FGR) for VT9 alloy increase by 2-4 times, depending on K_{Imax}, and for EP742 alloy, an increase in frequency leads to a decrease in the rate by 2-4 times. Holding under load increases the FGR for VT9 alloy over the entire range of K_{Imax} variation.

With a decrease in the thickness of the sample (from 25 to 12mm), the FGR decreases, and the degree of reduction depends on the type of alloy, K_{Imax} range and temperature. The fraction of the resource of a specimen with a developing crack increases significantly with decreasing stress and decreases with increasing temperature. The growth rate of thermal fatigue cracks depends on the alloy grade and the thermal cycling mode (Fig. 3.90), and the dependence on the maximum cycle temperature is not monotonic. Exposure to an oxidizing environment at high temperatures can blunt the tip of a crack, thereby slowing its growth rate. This, in particular, is evidenced by the results of the experiments of G.A Tulyakov with steel 12Ch18N10T, in which T_{max} was varied at T_{min} = 20°C. It was found that the maximum rate of crack growth in specimens with concentrators occurs at T_{max} = 650°C (Tulyakov, 1974).

In recent years, there has been an accumulation of information on the characteristics of crack resistance and rates of crack growth in domestic materials. However, reference data on such characteristics are not yet available. As an example of such information Table 3.31 presents data on ΔK_{th} for a number of alloys for turbine and compressor blades (Troshchenko et al., 1981, 1987). In Wasserman et al. (1990), the features of the growth of fatigue cracks in gas turbine disks are proposed to be described using the formula

Table 3.31 Threshold ranges of stress intensity factor values and parameters of the Paris equation

Material	T, °C	R	ΔK_{th}, MPa$\sqrt{mm}$	m	C	Application range by ΔK_I, MPa$\sqrt{mm}$, not more than
VT9	20	-1	3.19	3.27	$5.26 \cdot 10^{-9}$	20
		0	2.29	2.47	$8.45 \cdot 10^{-8}$	20
		0.5	1.78	2.17	$3.20 \cdot 10^{-7}$	20
	300	-1	2.81	2.95	$2.09 \cdot 10^{-3}$	20
		0	2.11	2.72	$6.88 \cdot 10^{-8}$	20
		0.5	1.58	2.52	$2.41 \cdot 10^{-7}$	15
	450	-1	2.61	2.38	$7.40 \cdot 10^{-8}$	25
		0	2.00	2.26	$1.54 \cdot 10^{-7}$	20
		0.5	1.39	1.85	$9.51 \cdot 10^{-7}$	15
EI961	20	0	2.24	2.86	$3.82 \cdot 10^{-3}$	-
	20	-1	4.35	4.57	$4.16 \cdot 10^{-10}$	-
20Ch13	20	-1	5.47	4.16	$9.83 \cdot 10^{-11}$	-
14Ch17N2	20	-1	6.10	4.87	$7.59 \cdot 10^{-11}$	-
13Ch11N2W2MF	20	-1	6.16	4.56	$4.58 \cdot 10^{-11}$	-
08Ch17N6T	20	-1	5.54	3.69	$2.05 \cdot 10^{-10}$	-
VT3-1	20	0	3.05	2.57	$6.04 \cdot 10^{-9}$	-
	20	-1	5.69	4.73	$6.43 \cdot 10^{-11}$	-
ZhS6KP	20	-1	5.44	3.61	$2.04 \cdot 10^{-10}$	30
		0	4.20	3.49	$1.04 \cdot 10^{-9}$	30
		0.5	3.14	2.74	$3.00 \cdot 10^{-8}$	25
	800	-1	5.35	4.14	$3.50 \cdot 10^{-10}$	25
		0	4.29	3.32	$5.56 \cdot 10^{-9}$	20
		0.5	2.83	3.00	$8.30 \cdot 10^{-8}$	15
	1000	-1	3.17	4.04	$5.49 \cdot 10^{-9}$	15
		0	2.95	4.22	$2.80 \cdot 10^{-8}$	12
		0.5	2.05	4.20	$1.14 \cdot 10^{-7}$	8

$$\frac{dl}{dN} = \nu_0 \left(\frac{K_{Imax} - K_{t0}}{K_{fc} - K_{Imax}} \right)^m , \tag{3.157}$$

where K_{fc} and K_{to} are critical and minimum threshold values, respectively. K, ν_Q, and m are material constants that depend on the asymmetry of the cycle. The values of the material parameters for VT3-1 and EP742IA alloys are given in Table 3.32.

The growth rate of fatigue cracks depends on the cycle period. Thus, in Pokrovskii et al. (1994a,b), the rate of crack growth at 700°C was investigated for nickel-based alloys EP962 and EP742 under cyclic loading with holding (300 and 3600 s) and

Table 3.32 Parameters of Eq. (3.157)

Parameter	Material	
	VT3-1	EP742IA
K_{fs}, MPa$\sqrt{m}$	1.9	2.13
K_{t0}, MPa$\sqrt{m}$	4.38+10.93R	13.4+18.623R
lg v_0	-5.65-3R^2	-5.195-3.49R
m	1.775+0.474R-1.03R^2	1.941-1.485R

without holding at maximum cycle stress. The growth rates of fatigue cracks in disk materials at loading frequencies according to Pokrovskii et al. (2005) are shown in Fig. 3.96. To describe the observed features, a method was proposed for determining the rate of crack growth, based on the linear summation of the rates of fatigue and creep

$$\left(\frac{dl}{dN}\right)_{CF} = B(K_{max})^m + \dot{A}t_h,$$

where $\left(\frac{dl}{dN}\right)_{CF}$ is the rate of growth of cracks with a trapezoidal shape of the loading cycle, t_h is the holding time, $\dot{A}$ is the average rate of growth of creep cracks in the first section and B is a constant. The second term describes the rate of growth of creep cracks in the first section of creep. Good agreement between the calculated and experimental data has been obtained (see Fig. 3.96).

In a number of publications (for example, Semenov et al., 2015), it was shown that the kinetics of the growth of thermal fatigue cracks is adequately described by the criterion for the growth of thermal fatigue cracks (low-cycle fatigue) of disks and blades with an arbitrary shape of the cycle using the following expression

Fig. 3.96 Experimental lines and calculated points of the $da/dN - K_{max}$ dependence under cyclic loading with holding t_h of EP962 (a) and EP742 (b) alloys: *1* - $f = 0.5$ Hz, $t_h = 0$; 2 - $t_h = 300$ s; $t_h = 3600$ s, ●■ - calculation at exposure $t_h = 300$ and 3600 s, respectively

$$\frac{da}{dN} = B\,(\Delta K_{\text{eff}})^m + \int_0^{t_{\text{cycl}}} A\,[C^*(t)]^q\,dt, \tag{3.158}$$

where C_* integral (Goldman and Hutchinson, 1975; Landes and Begley, 1976; Taira and Otani, 1986; Margolin et al., 2006), the range of the stress intensity factor ΔK_{eff}, the time t, the cycle duration t_{cycl}, the values of the material parameters B, m, A, and q were determined from the fatigue and creep crack resistance tests of materials. The integration was carried out within one cycle (from 0 to t_{cycl}).

3.6 Concluding Remarks

The features of failure considered in this Chapter refer to the materials of components that are in a state after technological heat treatment. At the same time, long-term operation at elevated temperatures changes the microstructure of materials and, consequently, their properties, including strength and plastic characteristics. Thus, the parameters of the material included in the fracture criteria depend on the manufacturing technology of the blanks (casting by various methods, plastic deformation - forging, stamping, rolling, surface treatment, heat treatment) of components and the components themselves and operating conditions. The foregoing, on the one hand, complicates the creation of a database on properties, and on the other hand, it requires the compilation of various technical conditions for a given material in relation to a specific component and a specific manufacturer, regulating a certain number of its mechanical characteristics (usually $\sigma_{0.2}$, σ_{B}, δ, ψ at room and one or two elevated temperatures, as well as long-term strength with a limited test base - 50, 100, 500 or 1000 h). These questions are supposed to be addressed primarily in the next chapter, as well as in other chapters.

Particular attention should be paid to the characteristics of materials with an oriented structure, primarily monocrystalline ones. The features of the properties of such materials will be discussed below. The coverage of a number of issues discussed in this and previous chapters does not claim to be complete. According to the author, they need further research and development of applied programs for implementation. These include the following items

- widespread use of the theory of adaptability in modern calculations of the strength of rotor and stator parts of gas turbines, which would make it possible to adequately assess the safety margins of components without using complex methods, determining the parameters of materials for which requires a large amount of experimental research
- development of a multi-model approach for solving boundary value problems using various theories of thermo-visco-elasto-plasticity in relation to various components operating under non-stationary modes, which would ensure the adequacy of the results of the calculated determination of the stress-strain state of

components to the real picture of the behavior of materials under complex loading programs
- generation of a database on the characteristics of various materials under uniaxial and complex stress states, including r conditions under which various specific effects of inelastic deformation and fracture are manifested
- the development of modern fracture criteria in relation to materials with different levels of ductility, which will make it possible to reasonably choose the criteria in relation to real materials and loading conditions.

At the same time, on the basis of the above consideration, the reader is given the opportunity to critically evaluate the frequently used simplifications in assessing the temperature effects on the properties of materials, the creep laws in terms of excluding the primary stage from consideration, the mutual influence of plastic deformations and creep deformations, and many others.

When considering the selection of fracture criteria that ensure sufficient reliability of operation of GTU parts in non-stationary modes, it is necessary to keep in mind the following. The use of deformation and energy fracture criteria requires a preliminary selection of a method for determining the corresponding parameters, which allows (using the required number of basic experimental data on material characteristics at our disposal) to obtain adequate calculation results with appropriate loading programs. It should be borne in mind that the solution of the problem of determining the basic characteristics of the material in this case seems to be quite complicated and expensive.

On the other hand, the use of force fracture criteria in many cases leads to inadequate results, since the creep processes and especially the cyclic creep processes significantly change not only the parameters of these criteria, but also the degree of cycle asymmetry, making calculations taking into account the asymmetry coefficient meaningless. The foregoing leads to the idea that at the modern level of knowledge, approaches based on the use of fairly simple dependencies, the parameters of which are the values obtained in the calculations of total deformations, are preferable for assessing the strength of parts in unsteady modes.

During the time that has passed since the publication of the book in Russian and the publication of the translation in English by the author in Getsov et al. (2011) new research results are published. Recently published works on the issues discussed in this chapter include Troshchenko (2011); Lokoshchenko (2016).

References

Abdel-Karim M, Ohno N (2000) Kinematic hardening model suitable for ratchetting with steady-state. International Journal of Plasticity 16(3):225–240

Akimov LM (ed) (1977) Vynoslivost' zharoprochnykh splavov (in Russ., Endurance of Heat-resistant Alloys). Metallurgiya, Moscow

Alishoev LR (1968) Issledovaniye nesushchey sposobnosti «yelochnykh» i T-obraznykh khvostovykh soyedineniy lopatok turbin (in Russ., Research of bearing capacity of "herringbone" and T-shaped tail connections of turbine blades). Avtoref. diss. cand. tech. sciences, CNIITMASH, Moscow

Altenbach H, Naumenko K (1997) Creep bending of thin-walled shells and plates by consideration of finite deflections. Computational Mechanics 19(6):490–495

Altenbach H, Zolochevsky A (1996) A generalized failure criterion for three-dimensional behaviour of isotropic materials. Engineering Fracture Mechanics 54(1):75–90

Altenbach H, Lauschke U, Zolochevsky AA (1993) Ein verallgemeinertes Versagenskriterium und seine Gegenüberstellung mit Versuchsergebnissen. ZAMM - Zeitschrift für angewandte Mathematik und Mechanik 73(4):T372–T375

Altenbach H, Altenbach J, Zolochevsky A (1995) Erweiterte Deformationsmodelle und Versagenskriterien der Werkstoffmechanik. Deutscher Verlag für Grundstoffindustrie, Leipzig, Stuttgart

Altenbach H, Morachkovsky O, Naumenko K, Sychov A (1997) Geometrically nonlinear bending of thin-walled shells and plates under creep-damage conditions. Archive of Applied Mechanics 67(5):339–352

Altenbach H, Kolarow G, Morachkovsky OK, Naumenko K (2000) On the accuracy of creep-damage predictions in thinwalled structures using the finite element method. Computational Mechanics 25(1):87–98

Altenbach H, Kushnevsky V, Naumenko K (2001) On the use of solid-and shell-type finite elements in creep-damage predictions of thinwalled structures. Archive of Applied Mechanics 71(2-3):164–181

Altenbach H, Huang C, Naumenko K (2002) Creep-damage predictions in thin-walled structures by use of isotropic and anisotropic damage models. The Journal of Strain Analysis for Engineering Design 37(3):265–275

Altenbach H, Naumenko K, Gorash Y (2008) Creep analysis for a wide stress range based on stress relaxation experiments. International Journal of Modern Physics B 22(31-32):5413–5418

Altenbach H, Bolchoun A, Kolupaev VA (2014) Phenomenological yield and failure criteria. In: Altenbach H, Öchsner A (eds) Plasticity of Pressure-Sensitive Materials, Springer, Berlin, Heidelberg, pp 49–152

Balandin YF, Zolotukhina MA (1961) Novaya metodika ocenki soprotivleniya termicheskoi ustalosti konstrukcionnykh materialov (in. Russ., A new method for evaluating the thermal fatigue resistance of structural materials). Zavodskaya Laboratoriya (1):63–66

Balepina TV (1988) Probability evaluation for exhaustion of the operating capacity of gas-turbine-engine disks with low-cycle fatigue. Strength of Materials 20(7):881–885

Balina VS, Myadekshas GG (1994) Prochnost' i dolgovechnost' i treshchinostoykost' konstruktsiy pri dlitel'nom ciklicheskom nagruzheniy (in Russ., Strength and durability and crack resistance of structures under long-term cyclic loading). Politekhnik, St. Petersburg

Bari S, Hassan T (2002) An advancement in cyclic plasticity modeling for multiaxial ratcheting simulation. International Journal of Plasticity 18(7):873–894

Berezhko BI, Gubin AV, Malygin AF (1980) Vliyaniye skopleniy nemetallicheskikh vklyucheniy na povrezhdayemost' konstrukcionnykh materialov pri teplosmenakh (in Russ., Influence of accumulations of non-metallic inclusions on the damageability of structural materials during thermal cycles). In: Metallovedenie, Rumb, Leningrad, 30, pp 10–21

Besser PR, McKerrow AJ, Iacopi F, Wong CP, Vlassak JJ (eds) (2005) Materials, Technology and Reliability for Advanced Interconnects - 2005, MRS Symposium Proceedings, vol 863. Cambridge University Press, Cambridge

Betteridge W, Heslop J (eds) (1974) The Nimonic Alloys, and Other Nickel-base High-temperature Alloys. Edward Arnold, London

Birger IA (1977) Ob odnom kriterii razrushenija i plastichnosti (in Russ., On a criterion for fracture and plasticity). Izvestija Akademii Nauk SSSR Mekhanika Tverdogo Tela (4):143–150

Birger IA, Shorr BF, Demyanushko IV, Dul'nev RA, Sizova RN (eds) (1975) Termoprochnost' detaley mashin (in Russ., Heat Resistance of Machine Parts). Mashinostroenie, Moscow

Blumenauer H (1996) Werkstoffprüfung. Deutscher Verlag für Grundstoffindustrie, Leipzig

Bocher L, Delobelle P, Robinet P, Feaugas X (2001) Mechanical and microstructural investigations of an austenitic stainless steel under non-proportional loadings in tension–torsion-internal and external pressure. International Journal of Plasticity 17(11):1491–1530

Borodin NA, Stepanov MN (1969) O funkcii raspredeleniya dolgovechnosti pri dlitel'nykh staticheskikh ispytaniyakh (in. Russ., On the distribution function of durability during long-term static tests). Zavodskaya Laboratoriya (1):98–100

Borzdyka AM, Getsov LB (1978) Relaksaciya napryazhenii v metallakh i splavakh (in Russ., Stress Relaxation in Metals and Alloys), 2nd edn. Metallurgiya, Moscow

Bugay NV, Berezina TV, Trunin II (1994) Rabotosposobnost' i dolgovechnost' metalla energeticheskogo oborudovaniya (in Russ., Efficiency and Durability of the Metal of Power Equipment). Energoatomizdat, Moscow

Bulygin IP, Timofeeva LN, Golubovskii ER, Trunin IP, Sizova RN, Boromolova IN (1977) Statistical analysis and prediction of the heat-resistance characteristics of cast alloys for gas-turbine engines (GTD). Strength of Materials 9(11):1337–1343

Bychkov NG, Lavrenteva MA, Feofanova AP (1991) Ustanovka dlya termomekhanicheskogo nagruzheniya obraztsov (in Russ., Installation for thermomechanical loading of samples). Zavodskaya Laboratoriya (12):57–58

Chaboche JL (1991) On some modifications of kinematic hardening to improve the description of ratchetting effects. International Journal of Plasticity 7(7):661–678

Chaboche JL (1994) Modeling of ratcheting evaluation of various approaches. European Journal of Mechanics/Solids 13(7):501–518

Chechulin BB, Khesin YD (1987) Ciklicheskaya i korrozionnaya prochnost' titanovykh splavov (in Russ., Cyclic and Corrosion Resistance of Titanium Alloys). Metallurgiya, Moscow

Cherepanov GP (1979) Brittle Fracture Mechanics. McGraw-Hill International Book Company, New York

Chernyavsky OF, Getsov LB (2009) Raschet na prochnost' detaley GTU s ispol'zovaniyem teorii prisposoblyayemosti (in Russ., Strength analysis of gas turbine parts using the theory of adaptability). In: Petreniya YK, Getsova LB (eds) Prochnost' materialov i resurs elementov energooborudovaniya, St. Petersburg, no. 296 in Trudy CKTI, pp 253–260

Chizhik AA (1970) Soprotivleniye khrupkim i vyazkim razrusheniyam materialov dlya osnovnykh elementov oborudovaniya (in Russ., Resistance to brittle and ductile fracture of materials for basic equipment elements). Avtoref. diss. doct. tech. sciences, CKTI, Leningrad

Chizhik AA, Lanin AA (1986) Inzhenernyy metod ocenki treshchinostoykosti materialov energeticheskogo oborudovaniya v usloviyakh relaksacii napryazheniy (in Russ., Engineering method for assessing the fracture toughness of materials for power equipment under stress relaxation conditions). Trudy CKTI 230:100–106

Chizhik AA, Petrenya YK (1978) O kineticheskikh uravneniyakh povrezhdayemosti pri mezhzerennom razrushenii (in Russ., Kinetic equations of damage at intergranular fracture). Trudy CKTI 169:42–44

Clauss FJ (1960) An examination of high-temperature stress-rupture correlating parameters. In: Proc. AMST, ASTM International, West Conshohocken, PA, vol 60, pp 905–927

Corten HT, Dolan TJ (1956) Cumulative fatigue damage. In: IME-ASME International Conference on Fatigue of Metals, IME, London, pp 235–246

Coutsouradis D, Davidson JH, Ewald J, Greenfield P, Khan T, Malik M, Meadowcroft DB, Regis V, Scarlin RB, Schubert F, Thornton DV (eds) (1994) Materials for Advanced Power Engineering 1994. Springer Netherlands, Proceedings of a Conference held in Liège, Belgium, 3–6 October 1994

Davidenkov NN (1936) Dinamicheskiye ispytaniya metallov (in Russ., Dynamic Testing of Metals). NKTP SSSR, Leningrad

Davidenkov NN, Likhachev VA (1962) Neobratimoye formoizmeneniye metallov pri ciklicheskom teplovom vozdeystvii (in Russ., Irreversible Shape Change of Metals Under Cyclic Thermal Exposure). MashGiz, Leningrad

Demyanushko IV, Birger IA (1978) Deformirovaniye i prochnost' materialov pri slozhnom napryazhennom sostoyanii (in Russ., Calculation of the Strength of Rotating Disks). Mashinostroyeniye, Moscow

Dolan TJ, Brown HF (1952) Effect of prior repeated stressing on the fatigue life of 75S-T aluminum. Proceedings ASTM 52:1–8

Drozdovskii BA (ed) (1968) Prikladnyye voprosy vyazkosti razrusheniya (in Russ., Applied Problems of Fracture Toughness). Mir, Moscow

Dul'nev PA, Kotov PI (1980) Termicheskaya ustalost' metallov (in Russ., Thermal Fatigue of Metals). Mashinostroeni, Moscow

Eckel JF (1951) The influence of frequency on the repeated bending life of acid lead. In: Proceedings - American Society for Testing and Materials, Conshohocken, PA, vol 51, pp 745–760

Elber W (1971) The significance of fatigue crack closure. In: Rosenfeld M (ed) STP486-EB Damage Tolerance in Aircraft Structures, ASTM International, Conshohocken, PA, pp 230–242

Ermakov PI, Malygin AF, Morozov VI (1972) Primeneniye elektronnoy mashiny dlya raschetnoy otsenki neobratimogo izmeneniya razmerov i formy tel pri teplosmenakh (in Russ., The use of an electronic machine for the calculated assessment of irreversible changes in the size and shape of bodies during heat changes). Metallovedenie (16):17–31

Feaugas X, Gaudin C (2004) Ratchetting process in the stainless steel AISI 316L at 300 K: an experimental investigation. International Journal of Plasticity 20(4):643–662

Fisher RA (1934) Statistical Methods for Research Workers, 5th edn. Oliver and Boyd, Edinburgh

Forman RG, Kearney VE, Engle RM (1967) Numerical Analysis of Crack Propagation in Cyclic-Loaded Structures. Journal of Basic Engineering 89(3):459–463

Freudenthal AM, Heller RA (1959) On stress interaction in fatigue and a cumulative damage rule. Journal Aerospace Science 26(7):431–442

Frey D (ed) (1954) Symposium on Effect of Cyclic Heating and Stressing on Metals at Elevated Temperatures, vol 165. ASTM, Philadelphia

Fridman YB (1946) Deformatsiya i razrusheniye metallov pri staticheskikh i udarnykh nagruzkakh (in Russ., Deformation and Destruction of Metals Under Static and Shock Loads). Gos. izdatel'stvo voennoi promyshlennosti, Moscow

Fridman YB (1974) Mekhanicheskie svoistva metallov v dvukh chastyakh. Chast' I: Deformatsiya i razrusheniye (in Russ., Mechanical Properties of Metals in two Parts. Part I: Deformation and Destruction). Mashinostroenie, Moscow

Frost NE, Dugdale DS (1958) The propagation of fatigue cracks in sheet specimens. Journal of the Mechanics and Physics of Solids 6(2):92–110

Gadenin MM (1976) Osobennosti razvitiya deformatsiy i nakopleniya povrezhdeniy pri dvukhchastotnom malociklovom nagruzhenii pri povyshennykh temperaturakh (in Russ., Features of the development of deformations and accumulation of damages during two-frequency low-cycle loading at elevated temperatures). Mashinovedenie (1):69–77

Gadenin MM, Kagan VA, Makhutov NA (eds) (1979) III Vsesoyuznyy rabochiy simpozium po voprosam malociklovoy ustalosti (in Russ., 3rd All-Union Working Symposium on Low Cycle Fatigue). RISO MViSSO Litovskoy SSR - VISI, Palanga

Gariboldi E, Naumenko K, Ozhoga-Maslovskaja O, Zappa E (2016) Analysis of anisotropic damage in forged Al–Cu–Mg–Si alloy based on creep tests, micrographs of fractured specimen and digital image correlations. Materials Science and Engineering: A 652:175–185

Gerber WZ (1874) Bestimmung der zulässigen Spannungen in Eisen-Constructionen (in German, Calculation of the allowable stresses in iron structures). Z Bayer Archit Ing Ver 6(6):101–110

Getsov LB (1962) Ocenka chuvstvitel'nosti k nadrezu pri ispytaniyakh na dlitel'nuyu prochnost' (in Russ., Evaluation of notch sensitivity in long-term strength tests). Zavodskaya Laboratoriya (11):1368–1372

Getsov LB (1971) O kriterii razrusheniya pri slozhnoy programme nagruzheniya (in Russ., Fracture criteria for a complex loading program). In: Vsesoyuznyy rabochiy simpozium po voprosam malotsiklovoy ustalosti, Kaunas, pp 51–55

Getsov LB (1973) Materialy i Prochnost' Detalei Gazovykh Turbin (in Russ., Materials and Strength of Gas Turbine Parts). Mashinostroenie, Leningrad

Getsov LB (1978) Kinetic equations of failure in complex programs of cyclic loading. Strength of Materials 10(7):767–775

Getsov LB (1994) K voprosu o deformacionnykh kriteriyakh razrusheniya materialov pri nestacionarnykh rezhimakh nagruzheniya (in Russ., On the issue of deformation criteria for the destruction of materials under non-stationary loading conditions). Trudy CKTI 260:97–102

Getsov LB (2001) Problemy sozdaniya "universal'noy" teorii razrusheniya materialov (in Russ., Problems of creating a "universal" theory of material destruction). Problemy Mashinostroyeniya i Nadezhnosti Mashin (5):49–55

Getsov LB, Malygin AF (1969) K voprosu o metodike ispytaniy na termicheskuyu ustalost' (in. Russ., Thermal fatigue test procedure). Zavodskaya Laboratoriya (7):836–840

Getsov LB, Rybnikov AI (1994) Metodika i nekotoryye rezul'taty po opredeleniyu usloviy zarozhdeniya i razvitiya termoustalostnykh treshchin v metallakh i zashchitnykh pokrytiyakh (in Russ., Methodology and some results for determining the conditions for the initiation and development of thermal fatigue cracks in metals and protective coatings). Zavodskaya Laboratoriya (9):49–52

Getsov LB, Taubina MG (eds) (1960) Zharoprochnyye splavy pri izmenyayushchikhsya temperaturakh i napryazheniyakh (in Russ., Heat-resistant alloys at varying temperatures and stresses). Gosenergoizdat, Leningrad

Getsov LB, Turgenev KA (2006) Kriterii razrusheniya konstruktsiy pri staticheskom nagruzhenii (in Russ., Criteria for structural failure under static loading). Trudy CKTI 295:204–214

Getsov LB, Malygin AF, Revyakina NN, Balandin IM (1974) Termostoykost' zharoprochnykh materialov dlya lopatok GTU (in Russ., Heat resistance of heat-resistant materials for gas turbine blades). Teploenergetika (3):24–27

Getsov LB, Kononov KM, Rebyakov YN (1976) K opredeleniyu ciklicheskogo predela uprugosti (in Russ., On the definition of the cyclic elastic limit). Zavodskaya Laboratoriya (11):1400–1403

Getsov LB, Margolin BZ, Fedorchenko DG (2004) K voprosu opredeleniya zapasov prochnosti detaley energoustanovok pri ispol'zovanii raschetov MKE (in Russ., On the question of determining the safety margins of parts of power plants when using FEM calculations). In: Proceedings of SPbSPU, St. Petersburg, Mechanics of Materials and Strength of Structures, vol 489, pp 162–177

Getsov LB, Margolin BZ, Fedorchenko DG (2009a) Voprosy opredeleniya zapasov prochnosti elementov mashinostroitel'nykh konstruktsiy pri raschetakh metodom konechnykh elementov (in Russ., Questions of determining the safety margins of elements of machine-building structures in calculations by the finite element method). In: Petreniya YK, Getsova LB (eds) Prochnost'

materialov i resurs elementov energooborudovaniya, St. Petersburg, no. 296 in Trudy CKTI, p 51–67
Getsov LB, Rybnikov AI, Semenov A (2009b) Progressiruyushcheye deformirovaniye materialov pri termociklicheskom nagruzhenii (in Russ., Progressive deformation of materials under thermal cyclic loading). In: Petreniya YK, Getsova LB (eds) Prochnost' materialov i resurs elementov energooborudovaniya, St. Petersburg, no. 296 in Trudy CKTI, p 105–119
Getsov LB, Rybnikov AI, Semenov AS (2009c) Thermal fatigue strength of heat-resistant alloys. Thermal Engineering (5):412–420
Getsov LB, Rybnikov AI, Dobina NI, Dashunin NV (2011) Specific features relating to thermal fatigue rupture of the blade materials of gas turbine units. Thermal Engineering 58(9):736–740
Gokhfeld DA, Cherniavsky OF (1979) Limit Analysis of Structures at Thermal Cycling. Sijthoff & Noordhoff, Alphen aan den Rijn
Gokhfeld DA, Chernyavsky OF (eds) (1974) II Vsesoyuznyy rabochiy simpozium po voprosam malociklovoy ustalosti pri povyshennykh temperaturakh (in Russ., All-Union Working Symposium on Low Cycle Fatigue at Elevated Temperature), vol 1-4. Chelyabinskiy politekhnicheskiy institut imeni Leninskogo komsomola, Chelyabinsk
Gokhfeld DA, Martynenko NE, Sadakov OS (1975) Analiz raspredeleniya deformatsiy v obrazce pri ispytaniyakh na termicheskuyu ustalost' v usloviyakh polzuchesti (in Russ., Analysis of strain distribution in a specimen during thermal fatigue tests under creep conditions). In: Prochnost' Materialov i Konstrukciy, Naukova Dumka, Kiev, pp 93–103
Gokhfeld DA, Kononov KM, Sadakov OS, Poroshin VB (1983) K ocenke dolgovechnosti pri neizotermicheskom malociklovom nagruzhenii s vyderzhkami (in Russ., Assessment of durability under non-isothermal low-cycle loading with holding). Mashinovedenie (4):72–77
Goldman NL, Hutchinson JW (1975) Fully plastic crack problems: The center-cracked strip under plane strain. International Journal of Solids and Structures 11(5):575–591
Golubovskii ER, Trunin II (1978) Temperature-time dependence of long-time structural strength. Strength of Materials (10):1410–1416
Goodman J (1899) Mechanics Applied to Engineering. Longmans, Green & Co., London
Gorokhov MY, Getsov LB, Kutyrev VV, Golubovskii ER (2004) Raschetno-eksperimental'noye issledovaniye nesushchey sposobnosti model'nykh vrashchayushchikhsya diskov (in Russ., Computational and experimental study of the bearing capacity of model rotating disks). In: Proceedings of SPbSPU, St. Petersburg, Mechanics of Materials and Strength of Structures, vol 489, pp 162–177
Grubin AN (1970) Raschet na prochnost' yelochnogo zamka lopatok turbin (in Russ., Calculation of the Strength of the Herringbone Lock of Turbine Blades). Mashinostroenie, Leningrad
Guarnieri GJ, Yerkovich LA (1952) The influence of periodic overstressing on the creep properties of several heart-resistant alloys. In: ASTM Proceeding, vol 52, pp 934–950
Gugelev BM, Getsov LB (1976) Metod mikrostrukturnogo issledovaniya povrezhdeniy v metallakh pri termicheskoy ustalosti (in. Russ., Method of microstructural study of damage in metals during thermal fatigue). Zavodskaya Laboratoriya (1):94–98
Gusenkov AP (1979) Prochnost' pri izotermicheskikh i neizotermicheskikh malociklovom nagruzhenii (in Russ., Strength Under Isothermal and Non-isothermal Low-cycle Loading). Nauka, Moscow
Haigh BP (1917) Experiments on the fatigue of brasses. J Inst Metals 18:55–56
Hancock JW, Mackenzie AC (1976) On the mechanisms of ductile failure in high-strength steels subjected to multi-axial stress-states. Journal of the Mechanics and Physics of Solids 24(2):147–160
Head AK (1953) XCVIII. The growth of fatigue cracks. The London, Edinburgh, and Dublin Philosophical Magazine and Journal of Science 44(356):925–938
Henry DL (1955) A theory of fatigue-damage accumulation in steel. Transactions of the ASME 77:913–918
Heywood RB (1962) Design Against Fatigue. Chapman and Hall, London
Hoff NJ (1953) The necking and the rupture of rods subjected to constant tensile loads. Trans of ASME J Appl Mech 20(1):105 – 108

Irwin G (1957) Analysis of stresses and strains near the end of a crack traversing a plate. Trans ASME Journal Applied Mechanics 24:361–364
Ivanova VS, Gurevich SE (eds) (1974) Ustalost' i vyazkost' razrusheniya metallov (in Russ., Fatigue and Fracture Toughness of Metals). Nauka, Moscow
Ivanova VS, Kudryashov VG, Terent'ev VF (1967) Prochnost' metallov pri tsiklicheskikh nagruzkakh (in Russ., Strength of Metals Under Cyclic Loads). Nauka, Moscow
James LA (1972) The effect of stress ratio on the elevated temperature fatigue-crack propagation of type 304 stainless steel. Nuclear Technology 14(2):163–170
Kabelevskii MG, Stolyarova LI (1976) Estimation of thermal fatigue strength of rotor steel in the presence of stress concentration. Strength of Materials 8(7):832–837
Kabelevsky MG (1976) O koncentracii teplovykh napryazheniy v poluellipticheskoy vytochke (in Russ., On concentration of thermal stresses in a semi-elliptical recess). Mashinovedenie (6):54–59
Kablov EN, Golubovsky ER (1998) Zharoprochnost' nikelevykh splavov (in Russ., Heat Resistance of Nickel Alloys). Mashinostroenie, Moscow
Kachanov LM (1974) Osnovy mekhaniki razrusheniya (in Russ., Foundations of Fracture Mechanics). Nauka, Moscow
Kagan VA (ed) (1971) I Vsesoyuznyy rabochiy simpozium po voprosam malociklovoy ustalosti (in Russ., All-Union Working Symposium on Low Cycle Fatigue). Kaunasskiy politekhnicheskiy institut, Kaunas
Kang G, Kan Q, Zhang J, Sun Y (2006) Time-dependent ratchetting experiments of SS304 stainless steel. International Journal of Plasticity 22(5):858–894
Karzov GP, Margolin BZ, Shvetsova VA (1993) Fiziko-mekhanicheskoye modelirovaniye processov razrusheniya (in Russ., Physical and Mechanical Modeling of Fracture Processes). Polytekhnic, St. Peterburg
Kauss O, Tsybenko H, Naumenko K, Hütter S, Krüger M (2019) Structural analysis of gas turbine blades made of Mo-Si-B under transient thermo-mechanical loads. Computational Materials Science 165:129–136
Khinsky PG (1968) K voprosu otsenki sklonnosti rotorov i diskov k khladnolomkosti (in Russ., To the question of assessing the tendency of rotors and disks to cold brittleness). Trudy CKTI 84:174–185
Kinasoshvili RS (1959) Opredeleniye zapasov prochnosti pri nestatsionarnoy temperature i nestatsionarnoy napryazhennosti (in Russ., Determination of the reserves of strength at non-stationary temperature and of non-stationary tension). Izv AN SSSR Mekhanika i Mashinostroyenie (3):126–128
Kogaev VP (1993) Raschety na prochnost' pri napryazheniyakh, peremennykh vo vremeni (in Russ., Strength Calculations at Stresses, Variable in Time), 2nd edn. Mashinostroenie, Moscow
Koiter WT (1960) General problems for elastic-plastic solids. In: Sneddon IN, Hill R (eds) Progress in Solid Mechanics, North-Holland, Amsterdam, pp 165–221
Kolupaev VA (2018) Equivalent Stress Concept for Limit State Analysis, Advanced Structured Materials, vol 86. Springer, Cham
Kononchuk NI (1966) Metody ocenki vynoslivosti zharoprochnykh splavov (in Russ., Methods for Assessing the Endurance of Heat-resistant Alloys). Metallurgiya, Moscow
Kononov KM, Getsov LB (1984) Failure criteria of materials in cyclic loading. Strength of Materials 16(2):248–253
Kostenko Y, Almstedt H, Naumenko K, Linn S, Scholz A (2013) Robust methods for creep fatigue analysis of power plant components under cyclic transient thermal loading. In: ASME Turbo Expo 2013: Turbine Technical Conference and Exposition, American Society of Mechanical Engineers, pp V05BT25A040 – V05BT25A040
Kostyuk AG, Trukhniy AD, Getsov LB (1965) O prochnosti detaley energeticheskikh ustanovok pri nestacionarnykh rezhimakh (in Russ., On the strength of parts of power plants under non-stationary modes). Teploenergetika (1):48–53

Kostyuk AG, Trukhniy AD, Getsov LB (1967) Deformation and fracture of crystalline materials under complex loading-program conditions. Journal of Applied Mechanics and Technical Physics 8(3):45–49

Koterazawa R, Iwata Y (1976) Fracture Mechanics and Fractography of Creep and Fatigue Crack Propagation at Elevated Temperature. Journal of Engineering Materials and Technology 98(4):296–304

Krisch A (1962) Über die Auswertung von Langzeitstandversuchen. Archiv für das Eisenhüttenwesen 33(2):107–113

Kutyrev VV (2001) Statisticheskiy analiz nesushchey sposobnosti diskov (in Russ., Statistical analysis of the bearing capacity of disks). In: Collection of Reports from a Scientific and Technical Conference Dedicated to the Memory of N.D. Kuznetsova, June 21-22, 2001, Samara, vol I, pp 14–21

Landes JD, Begley JA (1976) A fracture mechanics approach to creep crack growth. In: Rice J, Paris P (eds) Mechanics of Crack Growth, ASTM International, West Conshohocken, PA, STP590, pp 128–148

Langer BF (1962) Design of pressure vessels for low-cycle fatigue. Trans ASME Journal of Basic Engineering 84(3):389–399

Längler F, Naumenko K, Altenbach H, Ievdokymov M (2014) A constitutive model for inelastic behavior of casting materials under thermo-mechanical loading. The Journal of Strain Analysis for Engineering Design 49(6):421–428

Lanin AA (1983) O kineticheskikh uravneniyakh povrezhdayemosti pri mezhzerennom razrushenii (in Russ., Evaluation of the crack resistance of steels during quenching). Trudy CKTI 204:75–81

Lanin AA (2007) Soprotivlyayemost' khrupkim lokal'nym razrusheniyam zharoprochnykh staley i svarnykh soyedineniy elementov energooborudovaniya pri dlitel'nom vysokotemperaturnom nagruzhenii (in Russ., Resistance to brittle local fractures of heat-resistant steels and welded joints of power equipment elements under prolonged high-temperature loading). Avtoref. diss. doct. tech. sciences, NPO CKTI, St. Peterburg

Lanin AA, Getsov LB (2009) Zakonomernosti obrazovaniya i razvitiya treshchin v elementakh energoustanovok v usloviyakh relaksatsii napryazheniy (in Russ., Laws of formation and development of cracks in the elements of power installations under stress relaxation conditions). In: Petreniya YK, Getsova LB (eds) Prochnost' materialov i resurs elementov energooborudovaniya, St. Petersburg, no. 296 in Trudy CKTI, p 162–176

Larson FR, Miller J (1952) Time-temperature relationship for rupture and creep stresses. Trans ASME 74:765–771

Lebedev AA, Kovalchuk BI, Giginyak FF, Lamashevsky VP (2003) Mekhanicheskiye svoystva konstrukcionnykh materialov pri slozhnom napryazhennom sostoyanii (in Russ., Mechanical Properties of Structural Materials Under Complex Stress State). Publishing house "In Yure", Kiev

Levin EE, Gugelev BM (1976) Svyaz' mezhdu obrazovaniyem mikrotreshchin i mezhzerennym proskal'zyvaniyem v zharoprochnykh materialakh (in Russ., Relationship between microcracking and intergranular slippage in refractory materials). Fizika Metallov i Metallovedeniye (3):630–636

Levy JC (1957) Cumulative damage in fatigue - A design method based on the S-N curves. Journal of the Royal Aeronautical Society 61(559):585–591

Liebowitz H (1968) Fracture, An Advanced Treatise, vol II: Mathematical Fundamentals. Academic Press, New York

Lin TH, Wong KKF, Teng NJ (1999) Micromechanics of hysteresis loops of fatigue in a single crystal. Trans ASME Journal of Applied Mechanics 67(2):338–343

Littler DJ (ed) (1971) Thermal stresses and thermal fatigue: Proceedings of the International Conference held at Berkeley Castle, Gloucestershire, England. Butterworths, Waltham, MA

Lokoshchenko AM (ed) (2016) Polzuchest' i dlitel'naya prochnost' metallov (in Russ., Creep and Long-term Strength of Metals). FIZMATLIT, Moscow

Lvovsky MY, Smiyan IA (1958) Novaya metodika ispytaniy listovykh zharostoykikh materialov na termostoykost' (in. Russ., A new method for testing heat-resistant sheet materials for heat resistance). Zavodskaya Laboratoriya (29):202–204

Madudin VN (1979) Analiz kinetiki poley neuprugikh deformatsiy korsetnogo obraztsa pri ispytanii na termicheskuyu ustalost' (in Russ., Analysis of the kinetics of inelastic deformation fields of a corset specimen during thermal fatigue testing). In: Trudy ChPI, NPO CKTI, Chelyabinsk, No. 236, pp 77–83

Makhutov NA (1981) Deformacionnye kriterii razrusheniya i raschet elementov konstrukcii na prochnost' (in Russ., Deformation Criteria of Fracture and Structural Strength Analysis). Mashinostroenie, Moscow

Makhutov NA, Gadenin MM (eds) (1991) VI Vsesoyuznogo simpoziuma "MaloCiklovaya ustalost' – metody uprochneniya i vosstanovleniya materialov i konstrukciy" (in Russ., 6. All-Union Symposium "Low Cycle Fatigue - Methods of Hardening and Restoring Materials and Designs"), vol 1-2. Chisinau Polytechnic Institute, Kishinev

Makhutov NA, Gadenin MM, Blednova ZM (eds) (1983) IV Vsesoyuznyy rabochiy simpozium "Malociklovaya ustalost' – mekhanika razrusheniya, zhivuchest' i materialoyemkost' konstrukciy" (in Russ., 4th All-Union Working Symposium "Low-cycle fatigue - fracture mechanics, survivability and material consumption of structures"), vol 1-2. NTO Mashprom - KPI, Krasnodar

Makhutov NA, Gadenin MM, Gokhberg YA (eds) (1987) V Vsesoyuznyy rabochiy simpozium po voprosam malociklovoy ustalosti (in Russ., 5th All-Union Working Symposium on Low Cycle Fatigue), vol 1-2. NTO Mashprom - VPI, Volgograd

Makhutov NA, Zacarinny VV, Bazaras ZL (1989) Statisticheskiye zakonomernosti malociklovogo razrusheniya (in Russ., Statistical Patterns of Low-cycle Destruction). Nauka, Moscow

Malinin NN, Nigin AA (1974) Dlitel'naya prochnost' obrazcov s koncentratorami napryazheniy pri nestacionarnom rezhime nagruzheniya (in Russ., Long-term strength of specimens with stress concentrators under unsteady loading conditions). Izvestiya Vuzov Mashinostroyeniye (1):28–33

Manson SS (1966) Interfaces between fatigue, creep, and fracture. International Journal of Fracture Mechanics 2(1):327–336

Manson SS (1973) The challenge to unify treatment of high-temperature fatigue - A partisan proposal based on strain range partitioning. In: Carden A, McEvily A, Wells C (eds) STP520-EB Fatigue at Elevated Temperatures, ASTM International, West Conshohocken, PA, pp 744–782

Manson SS, Haferd AM (1953) A linear time-temperature relation for extrapolation of creep and stress-rupture data. TN 2890, NACA, Washington

Manson SS, Succop G (1956) Stress rupture properties of inconel700 and correlation on the basis of several time-temperature parameters. In: Symposium on Metallic Materials for Service at Temperatures Above 1600 F, ASTM International, West Conshohocken, PA, pp 40–46

Marchal N, Forest S, Rémy L, Duvinage S (2006) Simulation of fatigue crack growth in single-crystal superalloys using local approach to fracture. In: Moinereau D, Steglich D, Besson J (eds) Local Approach to Fracture, Presses de l'Ecole des Mines de Paris, no. 9 in EMMC, pp 353–358

Margolin BZ, Gulenko AG, Buchatskii AA, Balakin SM (2006) Modeling for fracture in materials under long-term static creep loading and neutron irradiation. Part 3. Crack growth rate prediction for austenitic materials. Strength of Materials 38(6):565–574

Marin NI (1968) Staticheskaya vynoslivost' elementov aviacionnykh konstrukciy (in Russ., Static Endurance of Elements of Aircraft Structures). Mashinostroenie, Moscow

Markochev VM (1966) Metodika i issledovanie kinetiki makrorazrusheniya listovykh materialov pri odnokratnom i povtornom nagruzhenii (in Russ., Methodology and study of the kinetics of macrofracture of sheet materials under single and repeated loading). Avtoref. diss. cand. tech. sciences, VIAM, Moscow

McEvely AI, Illg W (1958) The rate of crack propagation on two aluminium alloys. Technical Note 4394, NACA, Washington

McEvily AJ, Boettner RG (1963) On fatigue crack propagation in f.c.c. metals. Acta Metallurgica 11(7):725 – 743

Melan E (1936) Theorie statisch unbestimmter Systeme aus ideal-plastischem Baustoff. Sitz-Ber Akad Wiss Wien, Abt IIa 145:195–218

Morozov EM (1971) Calculating the strength of pressure vessels in the presence of cracks. Strength of Materials 3(9):1013–1018

Murry G (1963) Extrapolation of the results of creep tests by means of parametric formulae. In: ASME/ASTM/IMech. Engng. Proceedings Joint Conference on Creep, IMech. Engng., New Yorl-London, pp 6–87–6–100

Naumenko K, Altenbach H (2007) Modeling of Creep for Structural Analysis. Springer Science & Business Media

Naumenko K, Altenbach H (2016) Modeling High Temperature Materials Behavior for Structural Analysis: Part I: Continuum Mechanics Foundations and Constitutive Models, Advanced Structured Materials, vol 28. Springer

Naumenko K, Kostenko Y (2009) Structural analysis of a power plant component using a stress-range-dependent creep-damage constitutive model. Materials Science and Engineering: A 510:169–174

Naumenko K, Altenbach H, Kutschke A (2011a) A combined model for hardening, softening and damage processes in advanced heat resistant steels at elevated temperature. International Journal of Damage Mechanics 20:578–597

Naumenko K, Kutschke A, Kostenko Y, Rudolf T (2011b) Multi-axial thermo-mechanical analysis of power plant components from 9-12%Cr steels at high temperature. Engineering Fracture Mechanics 78:1657 – 1668

Nikitin VI (1959) Metody obrabotki rezultatov ispytanii na dlitel'nuyu prochnost' (in Russ., Methods for processing test results for long-term strength). Zavodskaya Laboratoriya (12):1492–1498

Nikitin VI, Taubina MG (1965) Masshtabnyi effekt pri vysokoy temperature i staticheskoy nagruzke (in Russ., Scaling effect at high temperature and static load). Teploenergetika (4):21–26

NN (ed) (1964) Fracture Toughness Testing and its Applications, ASTM, Philadelphia, PA, symposium presented at the 67th Annual Meeting of the American Society for Testing and Material, Chicago, IL, June 21-26, 1964

NN (1968) Metallovedeniye i prochnost' materialov (in Russ., Metallurgy and Strength of Materials), Trudy Volgogradskogo Politekhnichekogo instituta, vol 1. VPI, Volgograd

NN (1970) Metallovedeniye i prochnost' materialov (in Russ., Metallurgy and Strength of Materials), Trudy Volgogradskogo Politekhnicheskogo instituta, vol 2. VPI, Volgograd

NN (2007) Metodika rascheta prochnosti osnovnykh elementov reaktornykh ustanovok na bystrykh neytronakh s natriyevym teplonositelem (in Russ., Methodology for calculating the strength of the main elements of sodium-cooled fast neutron reactors). Operating organization guidance document RD EO 1.1.2.09.0714-2007, ROSENERGOATOM CONCERN, Moscow

Nordberg H (1972) The effect of microstructure on fatigue-crack growth in a 12% chromium steel. Scandinavian Journal of Metallurgy 1(1):23–25

Novozhilov VV (1975) O perspektivakh fenomenologicheskogo podkhoda k probleme razrusheniya (in Russ., On the prospects of a phenomenological approach to the problem of destruction). In: Mekhanika Deformiruyemykh Tel i Konstruktsiy, Mashinostroyeniye, Moscow, pp 349–359

Ohnami M, Umeda K, Awaya Y, Takada M (1976) Study on creep rupture of notched cylindrical specimens of 1Cr-1Mo-1/4V steel at elevated temperature. Bulletin of JSME 19(136):1100–1106

Orlov AF, Balashov BF (1979) Some features of damage accumulation for GTE under low cycle loading and elevated temperatures. In: Malociklovaya ustalost' elementov konstrukciy - III. Vsesoyuznyi simpozium, Palanga, vol 1, pp 92–94

Orr RL, Sherby OD, Dorn JE (1954) Correlations of rupture data for metals at elevated temperatures. Trans ASM 46:113–128

Paris P, Erdogan F (1963) A Critical Analysis of Crack Propagation Laws. Journal of Basic Engineering 85(4):528–533

Parton VZ, Morozov EM (1974) Mekhanika uprugoplasticheskogo razrusheniya (in Russ., Mechanics of Elastoplastic Fracture). Nauka, Moscow

Paul B (1968) Macroscopic criteria of plastic flow and brittle fracture. In: Liebowitz H (ed) Fracture: An advanced treatise, vol II (Mathematical Fundamentals), Academic Press, New York

Pavlov PA (1980) Mekhanicheskiye sostoyaniya i prochnost' materialov (in Russ., Mechanical States and Strength of Materials). Izdatel'stvo LGU, Leningrad

Pavlov PA, Kurilovich NN (1978) Predel'noye soprotivleniye konstruktsionnykh staley v usloviyakh vysokotemperaturnoy polzuchesti (in Russ., Ultimate resistance of structural steels under conditions of high-temperature creep). In: Prochnost'i ustoychivost' inzhenernykh konstruktsiy, Izdatel'stvo Altayskogo politekhnicheskogo instituta, Barnaul, pp 39–49

Pavlov PA, Getsov LB, Krasnov EG (1989) Use of strain criterion for a computational evaluation of the failure conditions of materials under a cyclic load in a complex stressed state. Strength of Materials 21(3):275–278

Petrenya YK (1997) Fiziko-mekhanicheskie osnovy kontinual'noi mekhaniki povrezhdaemosti (in Russ., Physico-mechanical Foundations of the Continuum Damage Mechanics). NPO CKTI, St. Petersburg

Pickett AG (1968) Prediction of Low-Cycle Fatigue Life of Specimens With Fabrication Flaws. Journal of Engineering for Industry 90(4):620–626

Pilkey WD (1997) Peterson's Stress Concentration Factors, 2nd edn. John Wiley & Sons, New York

Pisarenko GS (ed) (1965) Termoprochnost' materialov i konstruktsionnykh elementov (in Russ., Thermal Strength of Materials and Structural Elements), vol 3. Naukova Dumka, Kiev

Pisarenko GS (ed) (1967) Termoprochnost' materialov i konstruktsionnykh elementov (in Russ., Thermal Strength of Materials and Structural Elements), vol 4. Naukova Dumka, Kiev

Pisarenko GS (ed) (1969) Termoprochnost' materialov i konstruktsionnykh elementov (in Russ., Thermal Strength of Materials and Structural Elements), vol 5. Naukova Dumka, Kiev

Pisarenko GS, Lebedev AA (1976) Raschet na prochnost' vrashchayushchikhsya diskov (in Russ., Deformation and Strength of Materials under Complex Stress Conditions). Metallurgiya, Moscow

Pokrovskii VV, Troshchenko VT, Tseitlin VI, Ezhov VN, Zamotaev VS, Sidyachenko VG (1994a) Estimating working life for aircraft gas turbine engine disks at the stage of fatigue crack development. Part 1. Strength of Materials 26(11):796–800

Pokrovskii VV, Troshchenko VT, Tseitlin VI, Ezhov VN, Zamotaev VS, Sidyachenko VG, Samuleev VV (1994b) Estimating working life for aircraft gas turbine engine disks at the stage of fatigue crack development. Part 2. Strength of Materials 26(12):861–871

Pokrovskii VV, Ezhov VN, Sidyachenko VG (2005) Theoretical-experimental model for predicting crack growth rate in structural alloys of aircraft gas turbine engine disks with regard to fatigue and creep. Strength of Materials 37(6):558–565

Poroshin VB (1988) Effect of the shape of strain cycles on damage cumulation in different types of low-cycle loading with holding. Strength of Materials 1(20):45–52

Rabinovich VP (1966) Prochnost' turbinnykh diskov (in Russ., Turbine Disk Strength). Mashinostroenie, Moscow

Rabinovich VP (1970) Kriteriy sklonnosti k khrupkosti detaley mashin (in Russ., Criterion for the tendency to brittleness of machine parts). Optimizatsiya Metallurgicheskikh Processov (4):426–431

Rabotnov YN (1969) Creep Problems in Structural Members. North-Holland, Amsterdam

Rao Valluri S (1963) Some recent developments at "GALCIT" concerning a theory of metal fatigue. Acta Metallurgica 11(7):759–775

Roberts R, Erdogan F (1967) The Effect of Mean Stress on Fatigue Crack Propagation in Plates Under Extension and Bending. Journal of Basic Engineering 89(4):885–892

Rtishchev VV, Khrabrov PV, Korsov YG (1986) Issledovaniya v oblasti sluzhebnykh kharakteristik materialov - osnova povysheniya nadezhnosti i resursa gazovykh turbin (in Russ., Researches in the field of service characteristics of materials - the basis for increasing the reliability and resource of gas turbines). Energomashinostroenie (6):25–30

Sahm PR, Speidel MO (1974) High Temperature Materials in Gas Turbines. Elsevier Scientific Publishing Company, Amsterdam

Savin GN (1970) Stress Distribution Around Holes. No. F-607 in NASA Technical Translations, NASA, Washington D.C

Schijve J (1962) Fatigue crack propagation in light alloy sheet material and structures. In: Advances in Aeronautical Sciences, Pergamon, pp 387–408, Proceedings of the Second International Congress in the Aeronautical Sciences, Zürich, 12–16 September 1960

Semenov AS, Semenov SG, Getsov LB (2015) Methods of computational determination of growth rates of fatigue, creep, and thermal fatigue cracks in poly- and monocrystalline blades of gas turbine units. Strength of Materials 47(2):268–290

Serensen SV (1975a) Nesushchaya sposobnost' i raschet detaley mashin na prochnost' (in Russ., Bearing Capacity and Strength Calculation of Machine Parts). Mashinostroenie, Moscow

Serensen SV (1975b) Soprotivleniye materialov ustalostnomu i khrupkomu razrusheniyu (in Russ., Resistance of Materials to Fatigue and Brittle Fracture). Atomizdat, Moscow

Serensen SV, Kogaev VP (eds) (1969) Mekhanicheskaya ustalost' v statisticheskom aspekte (in Russ., Mechanical Fatigue in a Statistical Aspect). Nauka, Moscow

Serensen SV, Meshchaninova GP (1960) Nekotoryye osobennosti ustalostnogo razrusheniya splavov pri povyshennykh temperaturakh (in Russ., Some features of fatigue fracture of alloys at elevated temperatures). In: Ustalost' Metallov, AN SSSR, Moscow, pp 19–26

Serensen SV, Shneiderovich RM (eds) (1967) Soprotivleniye deformirovaniyu i razrusheniyu pri malom chisle ciklov nagruzheniya (in Russ., Resistance to Deformation and Fracture at a Small Number of Loading Cycles). Nauka, Moscow

Serensen SV, Dul'nev RA, Bychkov NA (1969) Estimating rupture resistance in thermal fatigue. Strength of Materials 1(1):9–15

Shalin RE, Bulygin IP, Golubovsky ER (1981) Zharoprochnost' splavov dlya gazoturbinnykh dvigateley (in Russ., Heat Resistance of Alloys for Gas Turbine Engines). Metalurgiya, Moscow

Shanley F (1961) Methods for describing the mechanical behavior of materials. In: Dorn JE (ed) Mechanical Behavior of Materials at Elevated Temperatures, University of California Engineering Extension Series, McGraw-Hill, New York

Shenoy MM, Gordon AP, McDowell DL, Neu RW (2005) Thermomechanical Fatigue Behavior of a Directionally Solidified Ni-base Superalloy. Trans ASME Journal of Engineering Materials and Technology 127(3):325–336

Shesterikov SA (ed) (1983) Zakonomernosti polzuchesti i dlitel'noi prochnosti (in Russ., Laws of Creep and Long-term Strength). Mashinostroenie, Moscow

Shkanov IN, Muratov FI, Braude NZ (1979) Ob opredelenii ekvivalentnykh napryazheniy pri ocenke prochnosti konstrukcionnykh materialov v usloviyakh slozhnogo napryazhennogo sostoyaniya (in Russ., On the determination of equivalent stresses when assessing the strength of structural materials under complex stress conditions). In: Tekhnologiya proizvodstva i prochnost' detaley letatel'nykh apparatov i dvigateley, KAI, Kazan', pp 56–68

Shneiderovich RM (1968) Prochnost' pri staticheskom i povtorno-staticheskom nagruzheniyakh (in Russ., Strength Under Static and Repeated-static Loading). Mashinostroenie, Moscow

Shorr BF (1966) Periodicheskiye protsessy i prisposoblyayemost' pri polzuchesti (in Russ., Periodic processes and creep adaptability). Prochnost' i Dinamika Aviacionnykh Dvigateley (4):188–194

Smith JH (1910) Some experiments on fatigue of metals. J Iron Steel Inst 82(2):246–318

Smith KN (1972) A mean stress function for fatigue crack propagation. Canadian Metallurgical Quarterly 11(2):435–438

Soderberg CR (1930) Factor of safety and working stress. Trans ASME 52(Part APM-52-2):13–28

Stanyukovich AV (1967) Khrupkost' i plastichnost' zharoprochnykh materialov (in Russ., Brittleness and plasticity of heat-resistant materials). Metallurgiya, Moscow

Taira S, Otani R (1986) Teoriya vysokotemperaturnoi prochnosti materialov (in Russ., Theory of High Temperature Strength of Materials). Metallurgiya, Moscow

Tretyachenko GN, Kravchuk LV, Kuriat RI, Voloschenko AP (eds) (1975) Nesushchaya sposobnost' lopatok gazovykh turbin pri nestatsionarnom teplovom i silovom vozdeystvii (in Russ., Bearing

Capacity of Gas Turbine Blades Under Non-stationary Thermal and Force Impact). Naukova Dumka, Kiev

Troshchenko VT (ed) (1977) Prognozirovaniye prochnosti materialov i konstruktsionnykh elementov mashin bol'shogo resursa (in Russ., Forecasting the Strength of Materials and Structural Elements of Machines with a Long Service Life). Naukova Dumka, Kiev

Troshchenko VT (1979) Prochnost' metallov pri peremennykh nagruzkakh (in Russ., Strength of Metals Under Variable Loads. Naukova Dumka, Kiev

Troshchenko VT (ed) (2011) Prochnost' materialov i elementov konstrukciy (in Russ., Strength of Materials and Structural Elements). Institute for Problems of Strength "G.S. Pisarenko" NAS of Ukraine, Kiev, Proceedings of the International Scientific and Technical Conference "Strength of Materials and Structural Elements", Kiev, September 28-30, 2010

Troshchenko VT, Prokopenko AV, Torgov VN, Baumshtein MV, Getsov LB (1981) A study of the influence of service operating time and a corrosive medium on the life of gas-turbine engine compressor blades. Strength of Materials 13(4):401–407

Troshchenko VT, Pokrovskiy VV, Prokopenko AV (1987) Treshchinostoykost' metallov pri ciklicheskom nagruzhenii (in Russ., Fracture Toughness of Metals Under Cyclic Loading). Naukova Dumka, Kiev

Trukhny AD, Stepanov YV (1978) Statisticheskiy analiz krivykh malociklovoy ustalosti (in Russ., Statistical analysis of low-cycle fatigue curves). Trudy CKTI 4:82–89

Trunin II (1976) A mechanical equation of the condition of metal materials and prediction of high-temperature strength characteristics. Strength of Materials 8(9):1007–1012

Trunin II (1994) Endurance of high-temperature materials with a complex stressed state. Metal Science and Heat Treatment 36(12):626–631

Trunin II, Fridman YF, Perets IM (1979) Parametric method of evaluating the long-term strength and creep limit of metal lots. Strength of Materials 11(10):1169–1172

Tseitlin VI (1976) Estimating strength under multicomponent loads. Strength of Materials 8(5):515–518

Tulyakov GA (1974) Termicheskaya ustalost' v teploenergetike (in Russ., Thermal Fatigue in Thermal Power Engineering). Mashinostroenie, Moscow

Ugorskii AE, Khein EA (1986) Link between long-term structural strength and long-term ductility in low-strain fracture. Strength of Materials (18):741–747

Vasilchenko GS (1962) Usloviye razrusheniya neravnomerno nagretykh vrashchayushchikhsya diskov (in Russ., Condition for the destruction of non-constant heated rotating disks). Trudy CNIITMASH 30:5–18

Vincent L, Calloch S, Marquis D (2004) A general cyclic plasticity model taking into account yield surface distortion for multiaxial ratchetting. International Journal of Plasticity 20(10):1817–1850

Volkov IA, Korotkikh YG (2014) Modeling of fatigue life of materials and structures under low-cycle loading. Mechanics of Solids 49(3):290–301

Volkov SD (1959) Osnovy statisticheskoj teorii prochnosti (in Russ., Basics of the statistical theory of strength). In: Ioffe AF, Kurdjymov GB, Zhurkov SN (eds) Nekotorye problemy prochnosti tverdogo tela, Izdatel'stvo Akademii Nauk SSSR, Leningrad, Moscow, pp 325–333

Wasserman NN, Sendyurev IV, Pykhtin SA, Smetannilkov YP (1990) Zakonomernosti razvitiya ustalostnykh treshchin materialov diskov GTD (in Russ., Laws of fatigue crack development in GTE disc materials). In: Dinamicheskaya Prochnost' Mekhanicheskikh Sistem, Perm'skii Gosudarstvennyi Tekhnicheskii universitet, Perm', pp 30–35

Weibull W (1963) A theory of fatigue crack propagation in sheet specimens. Acta Metallurgica 11(7):745 – 752

Wentzel E, Ovcharov L (1987) Applied Problems in Probability Theory. Mir Publishers

Wilkins ML (1999) Two-dimensional elastic-plastic flow. In: Computer Simulation of Dynamic Phenomena, Springer, Berlin, Heidelberg, Scientific Computation, pp 83–112

Wood DS (1966) The effect of creep on the high strain fatigue behaviour of pressure vessel steel. Weld J Res Sup 45(2):90–97

Xu B, Jiang Y (2004) A cyclic plasticity model for single crystals. International Journal of Plasticity 20(12):2161–2178

Yaguchi M, Takahashi Y (2005) Ratchetting of viscoplastic material with cyclic softening, Part 1: experiments on modified 9Cr–1Mo steel. International Journal of Plasticity 21(1):43–65

Yoshida F, Uemori T (2002) A model of large-strain cyclic plasticity describing the Bauschinger effect and workhardening stagnation. International Journal of Plasticity 18(5):661–686

Zakharova TP (1974) Ob ocenke minimal'nykh predelov vynoslivosti na osnove predstavleniy o dvukh mekhanizmakh ustalostnogo povrezhdeniya metallov (in Russ., On the assessment of the minimum endurance limits based on the concept of two mechanisms of fatigue damage to metals). Mashinovedenie (6):143–150

Zemzin VN, Chizhik AA, Lanin AA, Shron RZ (1984) Usloviya obrazovaniya treshchin pri svarke i termicheskoy obrabotke. Chast' I. O roli polzuchesti v obrazovanii treshchin (in Russ., C rack formation conditions during welding and heat treatment. Part I. On the role of creep in crack formation). Svarochnoe Proizvostvo (11):1–4

Zhurkov SN, Sanfirova TP (1955) Temperaturno-vremennaya zavisimost' prochnosti chistykh metallov (in Russ., Temperature-time dependence of the strength of pure metals). Doklady Akademii Nauk SSSR 101(2):237–241

Zhurkov SN, Tomashevskii EI (1959) Vremennaya zavisimost' prochnosti pri razlichnykh rezhimakh nagruzheniya (in Russ., Time dependence of strength under different loading conditions). In: Nekotorye Problemy Prochnosti Tverdogo Tela, Izd. AN SSSR, Moscow

Chapter 4
Influence of Technological Factors and Long-term Operation on the Microstructure and Properties of Heat-resistant Materials

4.1 Dependence of Properties on Metallurgical Factors, Size, and Orientation of Grains

4.1.1 Influence of Melting and Casting Methods

In the manufacture of heat-resistant alloys, special melting methods are widely applied: in vacuum, in a protective atmosphere using electroslag and vacuum arc remelting, using various deoxidizers, small additives, including rare earth elements. For parts made by precision casting, the method of filling and crystallization is essential. Melting and casting methods affect the properties of the metal both before and after hot working (forging, rolling, heat treatment). It has been found that the methods of melting and casting affect the content of gases, various oxides in the form of films, non-metallic inclusions, harmful impurities (As, Pb, Bi), usually chemically undetectable in the metal, as well as the size of inclusions, their distribution within the grain and porosity. Vacuum remelting affects the anisotropy of the properties, the amount and nature of the distribution of non-metallic inclusions, hardenability, transition temperature of brittleness, and especially the liquation inhomogeneity of the metal.

The degree of influence on the mechanical properties of the above mentioned methods of melting and filling is different for different alloys: the strength properties of highly plastic unhardened steels (type 12Ch18N10T) are insensitive to the melting method; the ductility of steel after vacuum arc remelting (VAR) increases markedly. For low plastic deforming alloys, the use of melting and casting in vacuum leads to a significant increase of ductility and, as a result, to a decrease in notch sensitivity.

For a number of materials, after remelting, a decrease in long-term strength by 5-10% is observed, due to changes: in the content of elements that form the γ'-phase; the kinetics of precipitation hardening, as well as the content of nitrides, carbides and carbonitrides. It may turn out that after remelting, it is necessary to change the heat treatment regime, which ensures obtaining optimal properties.

L. B. Getsov, *Materials and Strength of Gas Turbine Parts*, Advanced Structured Materials 150, https://doi.org/10.1007/978-981-16-0534-5_4

Studies of the influence of the melting method on the properties of steel EI415 (Table 4.1) were carried out by Yu.A. Starovoitov and the author. The steel was thermally treated in different modes up to hardness HB 269 and HB 363[1]. Vacuum arc remelting of steel has led to a sharp decrease in the number of nonmetallic inclusions compared to the metal of open melting: the score is less than the first for point inclusions (GOST 1778-70) and the absence of line inclusions instead of 3-5 points for both point and line inclusions of oxides, respectively. The oxygen content decreased 50 times (0.00015 instead of 0.008%). Impact toughness after VAR doubled; the ductility of steel slightly increased. After VAR, the metal practically became insensitive to notch in long-term strength tests even at hardness HB 363, while in open melted metal (with hardness HB 269 and HB 363) $\sigma_{\text{lt s}}$ under conditions of stress concentration decreases sharply (Fig. 4.1). The positive effect of melting in vacuum on long-term strength was also established by studies with strongly and weakly aging austenitic steels containing 15% Cr, 32% Ni, 3% Mo and 6% W.

As an example of the beneficial effect of remelting on long-term strength, in Borzdyka (1974) the results of experiments with alloy EI893, in which open-melted ingots of the same rate were subjected to electroslag remelting (ESR) and VAR, after which they were tested at 600°C (Table 4.2) are given.

High-chromium steel 15Ch12WNMF (Generson et al., 1975) is widely used for disks and rotors of stationary GTUs. Forging from this steel is made from deep-vacuum forged ESR ingots. As a result, ingots are obtained in which there is no eccentric segregation, the tendency to hydrogen brittleness is reduced (the hydrogen content is reduced approximately two times), the number of nonmetallic inclusions is reduced (approximately three times) and, as a result, the ductility increases and the transition temperature of brittleness (the tendency to brittle fracture) decreases in comparison with the metal of open melting.

The method of batch evacuation has also become widespread abroad. All General Electric turbine disks are manufactured from VAR materials or from vacuum

Table 4.1 Mechanical properties of steel EI415 at a temperature of 20°C, melted using different technologies

Melting method	Heat treatment mode			Hardness	σ_B	$\sigma_{0.2}$	δ	ψ	KCU
	Oil quenching	Tempering		d_{indent}					
	T,°C	T,°C	t, h	mm	MPa	MPa	%	%	kJ/m^2
Open oven	1000	650	5	3.2	1275	1200	14.8	58.0	230
Vacuum arc remelting	1000	650	5	3.2	1242	1160	17.2	64.2	480
Open oven	1050	700	2	3.7	920	820	16.8	62.5	1010
Vacuum arc remelting	1050	700	2	3.7	900	713	20.4	65.5	2703

d_{indent} - indentation diameter for Brinell hardness test; KCU - impact toughness on samples with rounded notch

[1] The mode with a low tempering temperature (HB 363) is not used in industry, however, the study of steel in this state allows us to reveal the effect of melting in more contrast.

Fig. 4.1 Long-term strength of steel EI415 of open melting (a, b) and after VAR (c) during heat treatment for hardness HB 363 (a, b) and HB 269 (c): *1* - smooth specimens, *2* - notched specimens

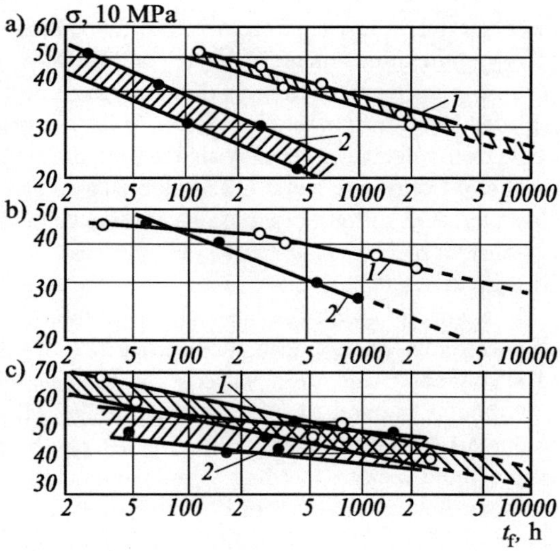

Table 4.2 Influence of the smelting method on the durability of the EI893 alloy at a temperature of 600°C, σ = 550 MPa

Temperature regime of stamping, °C	Melting method	t_f, h	δ_f, %	ψ_f, %
1060 - 960	OA	3237-5212	2,8-4,4	2,8-5,6
	ESR	5785-8483	2,2-4,4	2,8-7,2
	VAR	> 10000	-	-
1130 - 1030	OA	2366-3026	3,6-7,6	3,4-8
	ESR	3182-4672	3,2	7,2
	VAR	> 10000	-	-

OA - open, arc; ESR - electroslag remelting; VAR - vacuum arc remelting

degassing steel with low sulfur content. Vacuum arc remelting of high heat-resistant alloys sharply increases the technological ductility. So, for example, when forging disks from an EI698 alloy of open smelting, surface cracks are often formed on them, which are completely absent on the forged alloy after VAR.

The use of casting in a vacuum makes it possible to obtain a metal free from skin even in cases where the components of its chemical composition have a high affinity for oxygen. It is known, for example, that without melting in a vacuum or protective atmosphere by casting, it is rather difficult to obtain high-quality blade blanks from high-heatresistant nickel-based alloys, reinforced with an intermetallic phase. When receiving large ingots of vacuum-arc melts from heat-resistant alloys, off-axis (spotted) segregation was observed. It was found (Pridantsev, 1968) that the formation of off-axis segregation is associated with increased arc power and remelting rate. When evaluating the properties of cast heat-resistant alloys for blades, one should bear in mind the possible differences in the characteristics of the mechanical properties of the blade metal in its different zones, associated with different feeding conditions and crystallization; metal of blades and trefoil or finger samples, usually used for control tests (from melting or daily casting); metal melted in air, in a vacuum, in a protective atmosphere. The results of testing precision-cast samples and samples made from finger samples may also differ from each other, and even more so from the results of testing the metal of the club-shaped samples and the metal of the blades. However, such differences in properties are sometimes not observed. The properties of polished specimens without and with a casting crust differ significantly (Table 4.3; Kulygin et al., 1971).

In connection with the foregoing, one should be very critical of the data on the characteristics of the mechanical properties of cast alloys obtained during testing of specimens made from various samples, which are given in the literature for a particular method of melting and filling. Reliable data, allowing to judge the characteristics of the metal of the blades, can be established only by the results of tests of blades cast according to the technology adopted in this production. As for the operational characteristics of the blades, their knowledge requires, in addition, studies of the structural strength of these blades, as well as the study of the experience of their operation. Nevertheless, the characteristics of cast alloys obtained as a result of testing the metal of samples can undoubtedly be the basis for choosing a material.

The cooling rate of the castings of the blades after their filling affects both the hot brittleness of the blades (the formation of cracks, especially in the places of transition from the thin-walled feather of the cooled blades to a relatively massive lock), the formation of microporosity, and the structure and properties. At comparatively low solidification rates of castings from ZhS6K-type alloys, the carbide eutectic precipitates in the form of long needles, and sometimes even a continuous network, the primary precipitates of the γ'-phase are large. At high cooling rates, the size of the γ'-phase particles is small, they have a rounded shape; however, under these conditions, cracks and porosity can form. Shpindler and Landa (1976) proposed an optimal technology for producing dense castings with high values of ductility, fatigue resistance and long-term strength, which consists in casting into shell corundum-sillimanite molds heated to 900°C without a support filler, holding the cast block in a furnace for no more than 2-3 minutes and subsequent cooling in air; to combat hot cracks in the pen area, two or three additional thermal insulation layers are applied to the model.

Table 4.3 Ultimate, endurance and long-term strength of ZhS6K alloy at different temperatures, obtained during testing of samples with different surface conditions

Specimen	σ_B, MPa	σ_{-1}, MPa	σ_B, MPa	σ_{100}, MPa	σ_B, MPa	σ_{100}, MPa
	20°C		800°C		975°C	
With a casting crust	970	560-620	880	520	600	190
Polished	810	460-500	760	450	510	150

4.1.2 Influence of Deformation Conditions of Workpiece

It depends on the deformation technology whether the workpiece metal has retained areas with an incompletely broken cast structure, whether zones with a recrystallized uneven-grain structure have formed (Fig. 4.2), whether the fibers are located in the workpiece in a direction favorable for the work of the part under the influence of external forces, or not. In the process of deformation, the metal acquires one or another degree of cold, semi-hot or hot work hardening.

The above factors can have a certain effect on the characteristics of the mechanical properties and the heat resistance of the metal. For example, areas with an unbroken cast structure in a turbine disk, due to the anisotropy of mechanical properties they create, can lead to a redistribution of stresses, a change in safety margins and, in some cases, to an imbalance of the disk under operating conditions. An unfavorable arrangement of fibers (for example, in a preform of a turbine blade) for the same reason causes a decrease in its structural strength.

The mechanical properties and long-term strength of the metal of disk forging depend on the direction of cutting of the samples and the structure. The most sensitive to the conditions of deformation are the characteristics of metal plasticity. Specimens cut in the direction perpendicular to rolling or in the axial direction of the forging, fracture in a shorter time than specimens in which the fiber is located along the axis.

A number of studies are devoted to the analysis of the influence of forging, as well as work hardening, which is not eliminated by subsequent heat treatment, on

Fig. 4.2 Macrostructure of a stamped workpiece of a blade made of alloy EI765

the properties of metals. Interest in the latter issue is associated with the conditions for the manufacture of some workpieces made of heat-resistant materials: work-hardening of the surface of parts (especially thin-walled) during their machining, autofretting of disks, semi-hot hardening of forging of austenitic steel disks in order to increase the yield strength of the hub metal, cold rolling of sheets and profile workpieces of blades, flexible and rolling sheets and pipes; embossing of blades, riveting of blade spikes, mechanical-thermal and thermo-mechanical processing of workpieces, etc.

A number of studies carried out in Russia and abroad have established a positive effect of residual stresses initiated during the scrolling of disks on their performance. One of the reasons for this positive effect is a decrease in the working stresses in the central part of the disk, which is usually contaminated with various kinds of inclusions and sometimes contains various kinds of defects, the development of which into cracks is inhibited with decreasing stress. In this regard, all General Electric turbine disks are subjected to cold and hot scrolling at stresses in the region of the central hole slightly exceeding the yield point, which ensures that there are no brittle fractures during operation.

4.1.3 Effect of Grain Size of Deformed Alloys

The nature and degree of the effect of grain size on the properties of heat-resistant steels and alloys depend on the type of material, the conditions in which the metal with different grain sizes is obtained, the mode of heat treatment after quenching and the test conditions. Changes in grain size can have different effects on creep resistance, creep strength, ductility and fatigue resistance. The deformation ability usually decreases with increasing grain size (Stanyukovich, 1967). As for the creep resistance, along with the large creep limit of metals with a coarse-grained structure as compared to fine-grained metals, in some conditions, grain growth may be accompanied by a decrease in creep resistance.

When studying the effect of grain size on the mechanical properties of alloys, it turns out that the method of obtaining workpieces with different grain sizes is not indifferent. If grain growth (for example, in nickel-based alloys) is achieved by an increase in the hardening temperature, then the obtained data do not characterize the dependence of the creep resistance on the grain size unambiguously, since an increase in the hardening temperature affects not only the intensification of recrystallization processes, but also the processes of dissolution of carbides and γ'-phases before their subsequent isolation and homogenization of the solid solution[2].

In the manufacture of forging from heat-resistant steels and alloys, areas with a recrystallized coarse-crystalline structure are often formed. One of the reasons for the formation of a coarse-grained structure in austenitic heat-resistant steels and nickel-based alloys is the intensive recrystallization process during heat treatment of the

[2] Betteridge believes that the degree of homogenization of the solid solution has a decisive effect on the creep rate (Betteridge, 1959).

deformed metal. The degree of hot deformation causing the intensive development of these processes is called critical ε_{cr} (ε_{cr} - 0.5. . . 5%). Forging cracks often originate at the junction of the boundaries of small and large grains. Failure of specimens during long-term strength tests usually also begins at the border of regions with different structures.

The effect of the recrystallized grain size on the properties of EI481, EI437B, EI765 and EI893 widely used in gas turbine building is considered in Getsov et al. (1960); Belolipetskiy (1970). The effect of grain size on the properties of alloys EI826, EP539, EI929, and EP220 was studied on samples with artificially grown grains by double forging to a critical degree of deformation with intermediate annealing and subsequent heat treatment (Fig. 4.3). It was found that an increase in the grain size leads to a significant increase in the density of precipitation of Me_6C carbides along the grain boundaries in all alloys and, as a consequence, to a decrease in the impact toughness, ultimate strength, elongation and narrowing (Table 4.4). The effect of grain size on yield strength is rather weak. The durability with prolonged rupture of the EP220 alloy does not change noticeably with an increase in the grain size, and for the EI826 alloy it even increases (Table 4.5), while the long-term ductility of both alloys decreases very significantly (by 3-6 times). A sharp decrease in the durability was found in the case of the EP539 alloy with a different grain structure. Thus, in the case of a homogeneous structure under test conditions at 900°C, σ = 200 MPa, the metal of the blades met the technical requirements (t_f > 50 h); in the case of a different-grained structure, the time to fracture varied for different

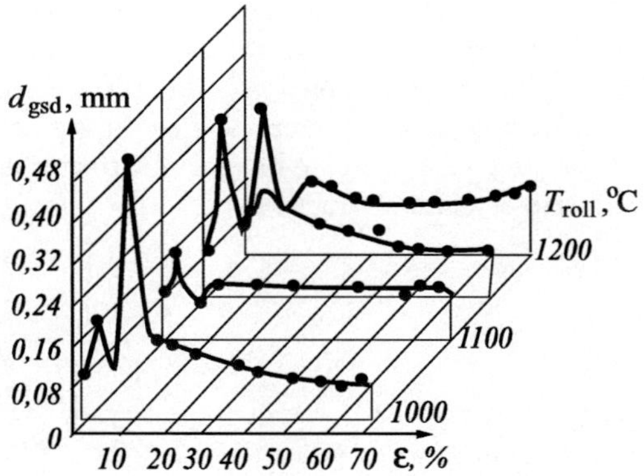

Fig. 4.3 Recrystallization diagram of the EI826 alloy (after complete heat treatment according to the regime: 1180°C (6 h) and 1000°C (4 h) and 900° (8 h), d_{gsd} - nominal grain size diameter, T_{roll} - rolling temperature)

Table 4.4 Effect of grain size on the mechanical properties of alloys

Material	σ_B, MPa	δ, %	ψ, %	KCU, kJ/m^2
EI826	1200/830	13/4	12/3	430/140
EI929	1350/900	30/3	30/5	600/220
EP220	1280/819	30/4.9	24/4.2	500/50

The numerator gives the value for a metal with a homogeneous fine-grained structure and the denominator gives the value for metal with a coarse and uneven grain.

Table 4.5 Results of long-term strength tests of the EI826 alloy of various grain sizes at temperature 800°C and $\sigma = 300$ MPa (Getsov and Rybnikov, 1988)

Grain size, mm	t_f, h	δ_f, %	ψ_f, %
< 0.5	326	13.6	15.8
~ 0.5	370-410	5.2-6.0	6.5-7.5
~ 1.0	187-273	3.9-4.5	4.5-5.2
~ 2.0	498-506	1.6-3.4	2.3-4.5
~ 3.0	845-871	0.8-3.5	1.0-3.6
3 – 4	962-979	2.4-3.8	2.9-4.9

samples in the range from 8 to 24 hours. The ductility was very high ($\delta_f = 5\ldots21\%$; $\psi_f = 8\ldots34\%$).

The value of the limited fatigue limit at 850°C for the EP220 alloy decreases with an increase in the grain size by 15% for a base of 10^5 cycles, by 10% for a base of 10^6 cycles and 5% for a base of 10^7 cycles. The effect of grain size on fatigue resistance is also characterized by the curves in Fig. 4.4. On the occurrence of thermal fatigue macrocracks found on Coffin-type and corset specimens, the grain size has a different effect depending on the alloy composition and loading conditions (Table 4.6). The rate of penetration of macrocracks into the depth of the sample (during thermal cycling of cylindrical samples by the "dipping" method) increases with the coarsening of the grain.

Some alloys show the tendency to the formation of a coarse-crystalline structure in the surface layers of the workpieces, which, in the opinion of a number of researchers, is associated with the depletion of the surface layer with alloying elements under conditions of high-temperature heating during hot deformation and heat treatment in air. Heating in an inert gas or vacuum protects the workpieces from the formation of a coarse-crystalline structure.

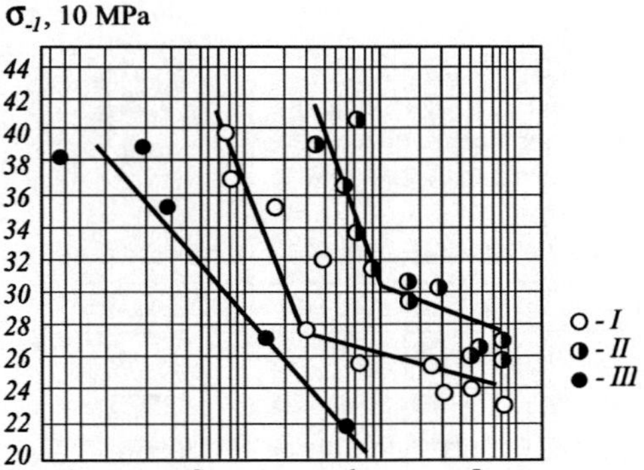

Fig. 4.4 Fatigue curves of EI826 alloy depending on grain size (in mm) at 800°C (according to data of Yu. P. Belolipetskiy): *I* - 0.05-0.1, *II* - 0.25, *III* - 0.1-3)

Table 4.6 Effect of grain size on resistance to thermal fatigue (Getsov and Rybnikov, 1988)

Alloy	Grain size, mm	$\Delta\varepsilon$, %	T_{max}, °C	N_f
EI826	0.05-0.2	0.45	800	1430
	1.0-1.5			17
EP220	0.02-0.05	0.54	850	8700
	0.1-0.3			95, 576, > 615
	1.0-2.5			450
	0.1-0.3	0,64	850	219, 1500, > 2450, 4634
	1-2.5			1230, 1395, 3140
	0.02-0.05	0.83	850	527
	0.1-0.3			346
	1.0-2.5			385

4.2 Influence of Crystal Orientation on the Properties of Cast Alloys

For the manufacture of blades from cast alloys, processes for obtaining an oriented structure (directional crystallization and single-crystal structure) are currently widely used. Usually, the crystallographic and growth anisotropy of the material is distinguished. Growth anisotropy depends on the direction with respect to the crystal growth axis (orientation of the dendritic structure).

The monograph (Shalin et al., 1997) and the proceedings of VIAM (Kablov, 2006; Kishkin, 2006) are devoted to a detailed presentation of issues related to the methods of preparation, structure and properties of single-crystal alloys. In this section, we are forced to concisely dwell on the coverage of these issues with the involvement of data obtained by different researchers (Chan and Leverant, 1987; Dul'nev et al., 1988a; Erickson and Harris, 1994; Li and Smith, 1994; Rtishchev, 1994; Telesman and Ghosn, 1996; Getsov, 1997; Reed et al., 2000; Schubert et al., 2000; Jo and Kim, 2003; Zhou et al., 2003; Golubovskiy et al., 2004, 2005a,b; Marchal et al., 2006; Getsov et al., 2007a,b, 2008a,b,c, 2009; Cormier et al., 2009; Getsov and Semenov, 2009; Nozhnitsky and Golubovsky, 2009).

4.2.1 Microstructure of Monocrystalline Materials

After correctly carried out heat treatment, the microstructure of single crystals of high-temperature alloys consists of a homogeneous solid solution in which cuboid particles of the γ'-phase (~ 60%) form a quasi-periodic three-dimensional macrolattice (Shalin et al., 1997). The dendritic structure with the [001] growth axis, revealed on a longitudinal section, consists of first-order dendritic branches, which correspond to the [001] crystallographic direction and are oriented along the crystallization heat flux. The branches of the second order are located across the heat flow, are weakly developed and are arranged in rows along the (010) and (100) planes. In contrast to single crystals with the [001] orientation with a columnar structure of dendrite branches, single crystals with the [011] orientation have a layered structure.

Single crystals with the [111] orientation, grown under conditions of a macroscopically flat front, do not have dendritic branches, predominantly developed in any one direction [001] - during its growth, the first and second order axes alternate continuously. In the course of long-term high-temperature holdings (in the process of creep), coagulation of the γ'-phase particles occurs, the nature of which depends on the orientation (see Fig. 4.5).

To compare the influence of the orientations <001>, <011>, and <111> on the structure, Golubovskiy et al. (2004) propose to use the parameter $f \sim \delta\Delta\sigma_N$, where δ is the lattice parameter of γ- and γ'-phases, $\Delta\sigma_N$ is the difference between the normal components of the external stress σ_0 on the lattice γ-phase. Then for

- <001> $f \sim \Delta\sigma_0$, because for (100) and (010) $\sigma_N = 0$ and the raft structure is formed along (001),
- <011> $f \sim \Delta\sigma_0$, because $\sigma_N = \sigma_0 \cos^2 \psi$ for (001) and (010), and for (100) $\sigma_N = 0$ and the raft structure is formed at an angle of 45° and
- <111> $f = 0, \sigma_N = \sigma_0/3$ and the raft structure is formed along the direction of the external load.

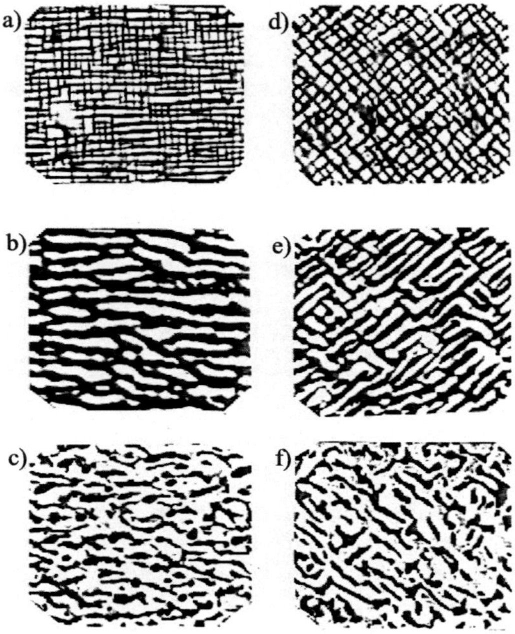

Fig. 4.5 Microstructure of specimens of alloys ZhS32 with the actual orientation <001> (a, b, c) and <111> (d, e, f). Formation of a raft structure at various stages of creep. Test conditions: $T = 1273°\text{K}$, $\sigma = 250$ MPa, a, d - $t = 10$ h, b - $t = 50$ h, e - $t = 120$ h, c - $t = 100$ h, f - $t = 300$ h

4.2.2 Anisotropy of Elastic Moduli

The relationship between deformations and stresses in an anisotropic elastic body is described by the generalized Hooke's law, which in tensor notation has the form:

$$\varepsilon_{ij} = S_{ijkl}\sigma_{kl}, \tag{4.1}$$

where S_{ijkl} are the components of the elastic compliance tensor in the x, y, z coordinate system. In the case of an arbitrary orientation of the single crystal, all tensor components are nonzero. However, if the coordinates x, y, z coincide with the crystallographic axes [100], [010], [001], then the form of the tensor is simplified, since some of its components vanish, and the remaining ones turn out to be equal to one of the three constants $S_{1111}, S_{1122}, S_{2323}$. In the crystallographic coordinate system, the elastic compliance matrix has the form:

$$\begin{bmatrix} S_{11} & S_{12} & S_{12} & 0 & 0 & 0 \\ S_{12} & S_{11} & S_{12} & 0 & 0 & 0 \\ S_{12} & S_{12} & S_{11} & 0 & 0 & 0 \\ 0 & 0 & 0 & S_{44} & 0 & 0 \\ 0 & 0 & 0 & 0 & S_{44} & 0 \\ 0 & 0 & 0 & 0 & 0 & S_{44} \end{bmatrix} \tag{4.2}$$

The following notation is used here: $S_{11} = S_{1111}, S_{12} = S_{1122}, S_{44} = 2S_{2323}$. Thus, to determine the properties of a single crystal in the x, y, z coordinate system, only three constants need to be known.

For the values of elastic compliance in Shalin et al. (1997) for the heat-resistant alloy ZhS6F, the following relations were experimentally obtained, in which T is the temperature in Kelvin:

$$\begin{aligned} S_{11} &= 0.7167 \cdot 10^{-5} - 0.2486 \cdot 10^{-9}T + 2.2848 \cdot 10^{-12}T^2, \\ S_{12} &= -0.2742 \cdot 10^{-5} + 0.1766 \cdot 10^{-9}T + 0.9999 \cdot 10^{-12}T^2 \\ S_{44} &= 0.7709 \cdot 10^{-5} + 0.5500 \cdot 10^{-9}T + 1.5106 \cdot 10^{-12}T^2 \end{aligned} \tag{4.3}$$

Apparently, they can be used for other alloys, because the modulus of elasticity changes little with a change in the chemical composition for alloys of the same class.

The components of the elastic compliance tensor in the x, y, z coordinate system, experimentally obtained in Shalin et al. (1997), were used to determine the values of the elastic moduli and Poisson's ratio of single-crystal alloys with different orientations (see Tables 4.7, 4.8 and Figs. 4.6 and 4.7). For tensile modulus of elasticity $E_{001} = 1/S_{11}$. For the shear modulus of single-crystal alloys, the relation $G = E/2(1 + \nu)$ is not valid. To calculate the temperature-orientation dependencies of G, it is necessary to know the direction cosines of the x, y, z axes relative to the crystallographic coordinate system and the temperature dependence of the components of the elastic compliance matrix S_{11}, S_{12}, S_{44}. In particular, $G_{001} = 1/S_{44}$. The values of the coefficient of linear thermal expansion α are also given according to the data in Shalin et al. (1997).

Table 4.7 Thermoelastic properties of single-crystal alloys at different temperatures

T,°C	E_{001}, 10^5 MPa	ν	G_{001}, 10^5 MPa	α_{001}, 10^{-6} K^{-1}
20	1.372	0.395	1.250	12.4
500	1.199	0.417	1.107	13.8
800	1.049	0.428	0.996	17.1
900	0.998	0.432	0.959	18.1
1000	0.948	0.435	0.921	22.2

Table 4.8 Values of the elastic modulus of single-crystal alloys depending on the orientation

Orientation hkl	E, GPa		
	20°C	750°C	1000°C
001	137.2	107.3	94.8
011	235.0	188.2	169.5
111	308.2	251.2	230.0

For directions [001] and [011], the values of the modulus of elasticity depend on the azimuthal orientation. In Figs. 4.6 and 4.7 are shown the characteristic surfaces E_x and G_{xy}, calculated on the basis of the ratios connecting the elastic moduli with elastic compliance. The calculation (Shalin et al., 1997) was carried out for 40 different orientations, uniformly scattered over the area and sides of the stereographic triangle.

The critical shear stress τ_{cr}, starting from which plastic deformations are formed, does not depend on the orientation of the single crystals and, according to the Schmid law (Schmid and Boas, 1935), is written as:

$$\tau = \sigma \cos \chi \cos \varphi = \sigma F > \tau_{cr}, \tag{4.4}$$

where F is the Schmid factor. The dependence of the elastic modulus for an arbitrary orientation is described by Eqs. (2.5)-(2.8).

4.2.3 Poisson's Ratio Anisotropy

Poisson's ratio is especially sensitive to orientation (Shalin et al., 1997). For orientation 001, Poisson's ratio $\nu = -S_{12}/S_{11}$. When compressed along the 011 direction, the transverse deformation in the 011 direction will always be negative. This explains the region of negative values in Fig. 4.8 for Poisson's ratio ν. For the transverse deformation coefficient - Poisson's ratio at 20°C, dependencies on the azimuthal angle are obtained and shown in Fig. 4.8. The dependencies of Poisson's ratio on the azimuthal angle largely depend on the axial orientations (Fig. 4.9). It is seen that for the 011 orientation, the azimuthal orientation significantly affects the value of ν_{yx}.

Fig. 4.6 Axial-orientational dependence of the modulus of elasticity E_x at temperature 20°C (a), 750°C (b) and 1000°C (c)

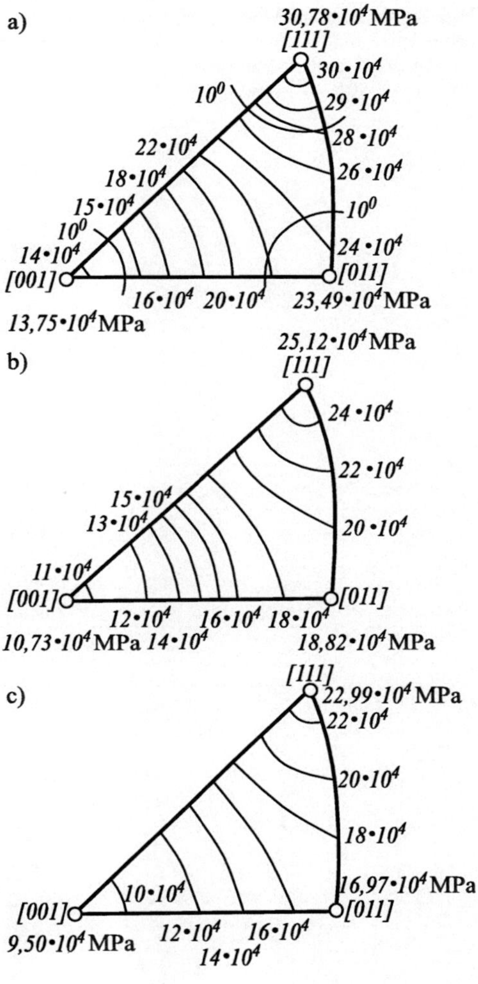

4.2.4 Anisotropy of the Coefficient of Thermal Expansion

As can be seen from Table 4.9, the influence of anisotropy on the coefficient of thermal expansion can be neglected.

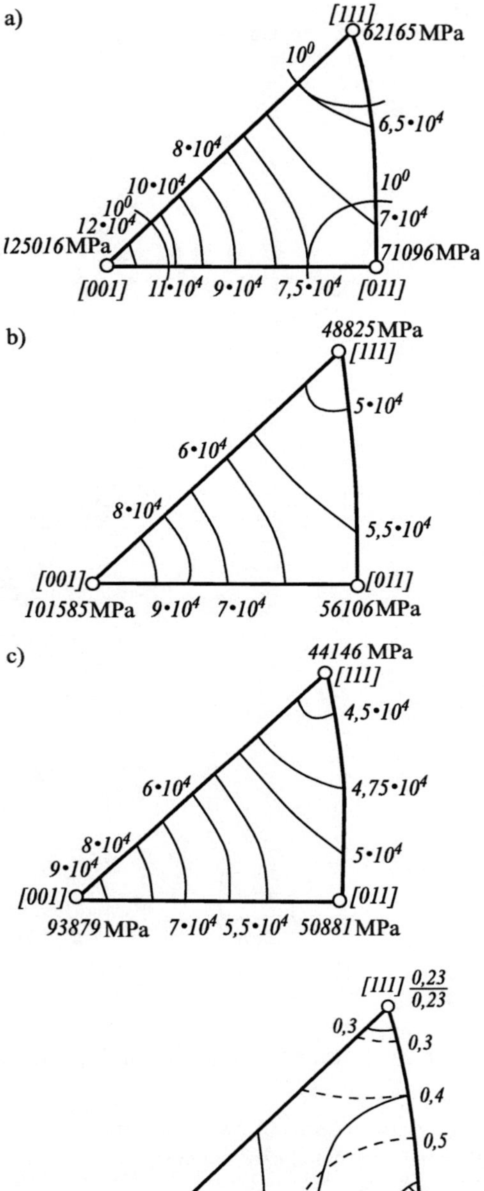

Fig. 4.7 Axial-orientational dependence of the shear modulus G_{xy} at temperature 20°C (a), 750°C (b) and 1000°C (c)

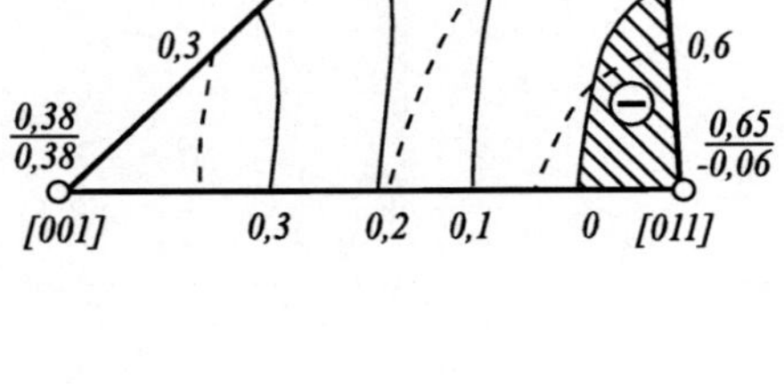

Fig. 4.8 Isolines of the maximum (solid lines) and minimum (dashed lines) values of Poisson's ratio ν_{yx} for directions x belonging to the standard stereographic triangle. The numerator and denominator of fractions in the corners of the triangle correspond to the maximum and minimum values ν_{yx} (shaded - area of negative values).

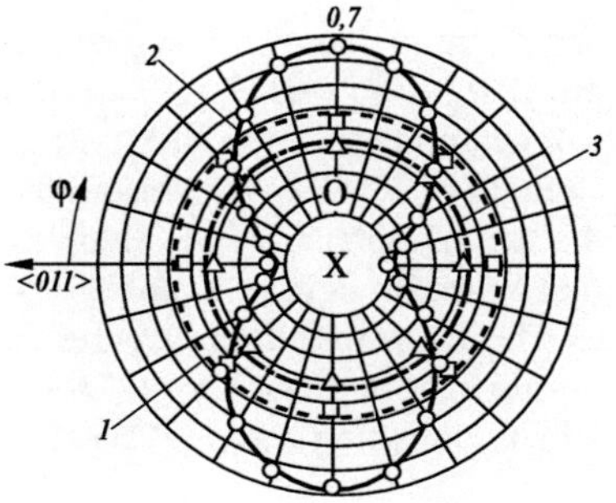

Fig. 4.9 Dependence of the Poisson ratio ν_{yx} on the azimuthal angle for single crystals of different orientations: *1*, *2*, *3* - computed curves, □, ○, △ - experimental points for specimens with orientation [001], [011], [111]

Table 4.9 Temperature dependence of the linear expansion coefficients of single-crystal nickel alloys (Shalin et al., 1981)

Orientation	α, 10^{-6} K^{-1}, in temperature interval °C									
	20... 100	100... 200	200... 300	300... 400	400... 500	500... 600	600... 700	700... 800	800... 900	900... 1000
[001]	12.4	12.8	13.2	13.7	13.8	14.3	16.4	17.1	18.1	22.2
[011]	12.4	12.9	13.4	14.2	14.2	14.4	16.1	17.1	18.4	21.8
[111]	12.3	12.8	13.3	13.8	14.1	14.3	16.1	17.7	18.9	22.6
[112]	12.5	12.9	13.3	13.9	13.9	14.2	16.6	17.2	18.5	21.9

4.2.5 Short-term Mechanical Properties

The effect of crystallite orientation on the mechanical properties of cast directed crystallization alloys is different. So, for example, the data given in Shalin et al. (1997) for the ZhS6U alloy, the tests of which were carried out along and across the crystallites, indicate the absence of differences in σ_B, δ and ψ. At the same time, according to Rtishchev (1994), a significant influence of orientation on $\sigma_{0.2}$ of the CNK7M alloy is observed (Fig. 4.10). The Rene-4 alloy exhibits a dependence on the orientation of the values of the yield point at 760 and 980°C under tension σ_y^t

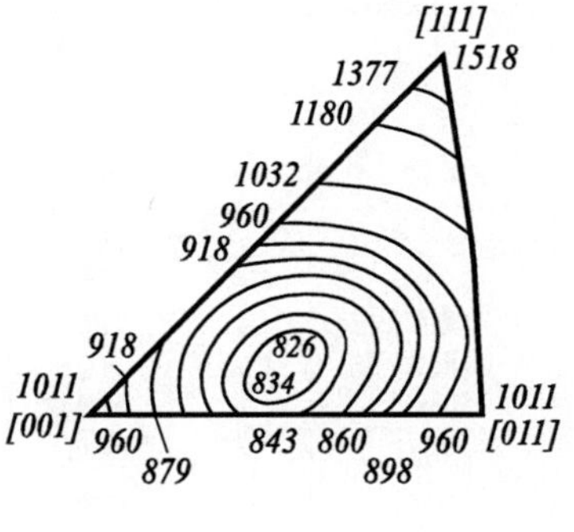

Fig. 4.10 Orientation dependence of the yield stress at temperature 20°C of the alloy CNK7M (Rtishchev, 1994)

and compression σ_y^c: in some cases $\sigma_y^t > \sigma_y^c$ ([001] at 760°C, [111] at 980°C), in others $\sigma_y^t < \sigma_y^c$ ([011], [023], [145] at 760°C) (Gabb et al., 1986).

For an alloy of the ZhS type in Yagodkin et al. (1987) the minimum ultimate strength at 20°C was possessed by specimens whose axis lies in the central part of the stereographic triangle. Samples with the [111] axis and especially [100] have the maximum strength at 20 and 975°C.

The nature of the temperature and orientation dependencies of carbon-free single-crystal heat-resistant alloys (ZhS36, ZhS-40, PWA1480, PWA1484, CMSX-2) and alloys containing carbon (ZhS32, ZhS6F, MAR-M200, MAR-M247) differ significantly. This is evidenced by the data (Shalin et al., 1997; Kablov, 2006; Nozhnitsky and Golubovsky, 2009) given in Tables 4.10 and for non-carbon and carbon alloys, depending on orientation and temperature. From Table 4.10, one can see a nonmonotonic dependence of the characteristics of the ZhS47 alloy on temperature, which is associated with differences in the nature of fracture. The properties shown in Table 4.11 for a number of alloys at two orientations illustrate their significant effect on properties. The nature of plastic deformation of single crystals of various orientations predetermines the type of fractures after short-term tensile tests. If deformation is carried out by multiple sliding along intersecting planes, then the sample is uniformly narrows and remains its original cylindrical shape. Such deformation was observed at all temperatures for the [001] and [111] orientations.

For carbon alloys with orientations [001] and [111], fracture begins in the interdendrite regions. In the range 700-800°C, the fracture of these samples is of a spalling character along the planes of the octahedron. Samples with orientation [011] have the same type of destruction. In the case of deformation along one shear plane, the section acquires the shape of an ellipse (for orientations [112] and [110]). The deformation curves of single crystals with the [111] and [011] orientations have a sharp yield point up to a temperature of 900°C (Fig. 4.11).

The yield point and critical shear stress have a non-monotonic temperature dependence (Fig. 4.12). In single crystals of heat-resistant alloys with a large amount of the γ'-phase, Schmid's law is not fulfilled. In particular, the compressive and tensile yield strengths differ significantly, including with a change in orientation (Fig. 4.13). Figure 4.14 shows similar data for a carbon-free alloy, and Figs. 4.15 and 4.16 show orientation dependencies of the yield stress of single crystals within a standard stereographic triangle.

Table 4.10 Mechanical properties (Kablov, 2006) of ZhS47 alloy with orientation [001]

Temperature, °C	σ_B, MPa	$\sigma_{0,2}$, MPa	δ, %	ψ, %
20	1088	936	29.2	30.7
700	1147.7	908	13.0	23.0
800	1222	922	12.7	18.4
900	1051	1021	26.3	28.1
1000	745	693	36.2	33.3
1100	420	406	13.0	13.0
1200	199	177	24.5	79.0

Table 4.11 Mechanical properties of single-crystal alloys (Shalin et al., 1997)

Tempe-	σ_B, MPa					$\sigma_{0,2}$, MPa					δ,%				
rature,	[001]			[111]		[001]			[111]		[001]			[111]	
°C	ZhS32	ZhS36	ZhS40	ZhS32	ZhS36	ZhS32	ZhS36	ZhS40	ZhS32	ZhS36	ZhS32	ZhS36	ZhS40	ZhS32	ZhS36
20	1310	1080	1170	1650	1600	1000	940	1010	1330	1270	7.5	9.5	9	9.5	9.6
700	1210	1190	1210	-	1230	980	960	940	-	1150	5.5	10	6.7	-	10
800	1310	-	1330	1290	-	1040	-	1260	-	-	6.5	-	13	10	-
900	1030	1000	1030	1020	1020	860	940	970	990	960	13	27	19	17	19.5
1000	750	850	780	730	780	620	680	730	650	730	14.5	19	23	8	21.5
1100	440	600	470	440	520	400	490	410	-	460	15.5	22	28	13	21.5
1150	280	440	340	-	420	250	380	290	-	360	15.5	-	26	-	32

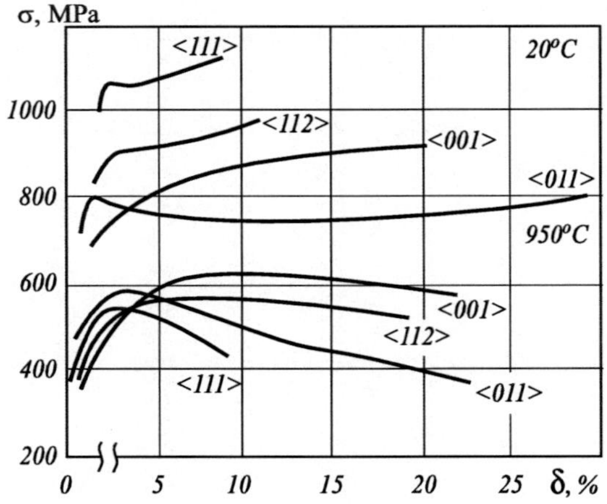

Fig. 4.11 Typical deformation diagram of single crystals at temperatures 20°C and 950°C (Shalin et al., 1997)

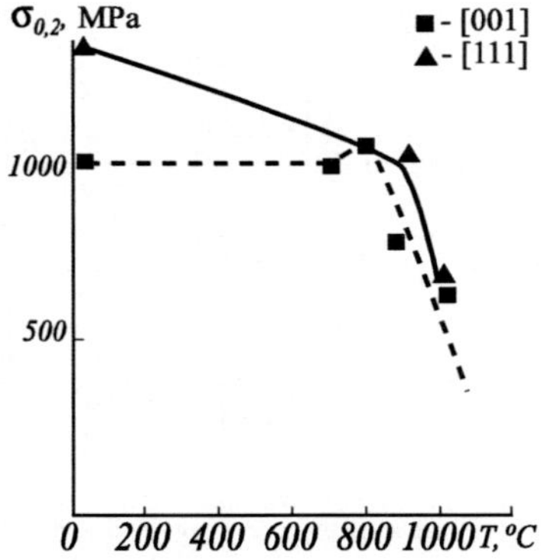

Fig. 4.12 Temperature dependence of the yield point of single crystals from a carbon alloy ZhS32 with orientations [001] and [111]

4.2.6 Resistance to High Frequency Fatigue

The influence of orientation on the resistance to fatigue fracture of the cast ZhS6F alloy with directional crystallization was studied on samples cut in the [001] and [111] directions. The test results (Table 4.12) show that specimens with orientation [111] have increased characteristics of the threshold value ΔK_{th} and lower rates of growth of fatigue cracks at various asymmetries of the cycle.

The influence of orientation on the fatigue resistance of the ZhS32 single crystal alloy was studied at cantilever bending. Table 4.13 shows the values of σ_{-1} corresponding to $2 \cdot 10^7$ cycles, and in Fig. 4.17 values of the anisotropy coefficients of the fatigue limit of a number of alloys.

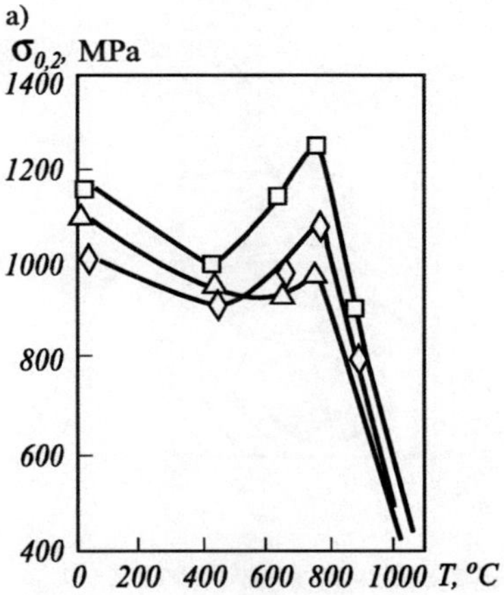

Fig. 4.13 Temperature dependence of the yield point of single crystals from an alloy PWA1480 with orientation [001], [011], [111] under tension (a) and compression (b) (icons - experimental points)

Fig. 4.14 Temperature dependence of the yield point of single crystals from non-carbon heat-resistant alloy AM-1 with orientation [001], [011], [111] (icons - experimental points)

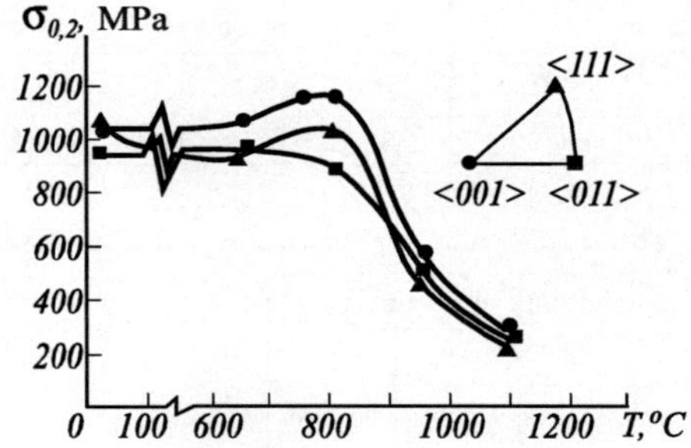

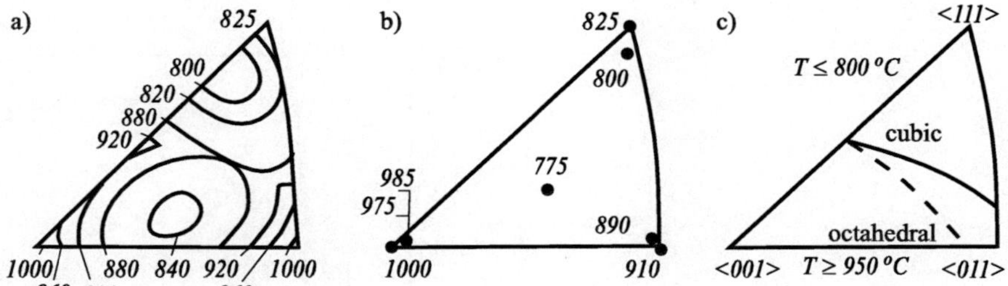

Fig. 4.15 Isoconcentric temperature yield stress contours at temperature 650°C for single crystals from non-carbon alloy AM-1 taking into account the cubic slip system near the angle [111]: a) calculated values of yield points (isocontours), b) experimental data, c) regions of cubic and octahedral slip (calculation)

Fig. 4.16 Orientational dependence of the yield point at 20°C of single crysals from alloy ZhS6K inside the standard stereographic triangle (• - experimental points)

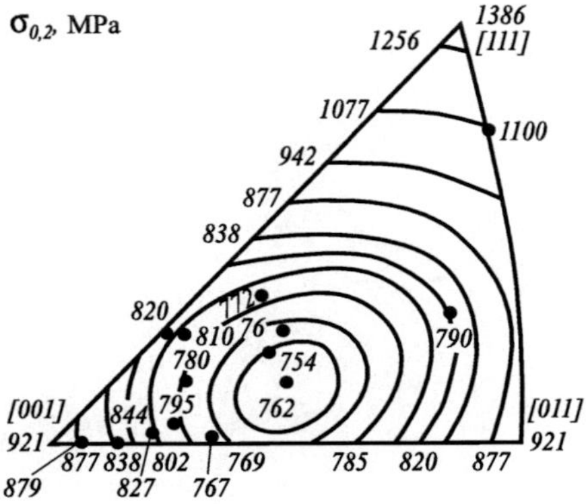

Table 4.12 Crack resistance characteristics of ZhS6F alloy (Troshchenko et al., 1987)

Temperature, °C	Orientation	dl/dN			ΔK_{th}, MPa$\sqrt{m}$			E, GPa
		$R=-1$	$R=0$	$R=0.5$	$R=-1$	$R=0$	$R=0.5$	
20	[001]	$1.48 \cdot 10^{-6}$	$6.49 \cdot 10^{-6}$	$2.53 \cdot 10^{-5}$	5.75	3.99	2.78	115-140
	[111]	$4.26 \cdot 10^{-7}$	$7.21 \cdot 10^{-7}$	$3.65 \cdot 10^{-6}$	8.4	6.96	5.97	190-230
800	[001]	$6.31 \cdot 10^{-6}$	$2.52 \cdot 10^{-5}$	$1.67 \cdot 10^{-4}$	4.73	3.45	2.38	86-105
	[111]	$7.68 \cdot 10^{-7}$	$2.03 \cdot 10^{-6}$	$9.81 \cdot 10^{-6}$	7.65	6.47	5.47	148-180

Table 4.13 Fatigue resistance σ_{-1} (MPa) of ZhS32 alloy

Orientation	Temperature, °C	
	20	900
[001]	340	350
[111]	420	430

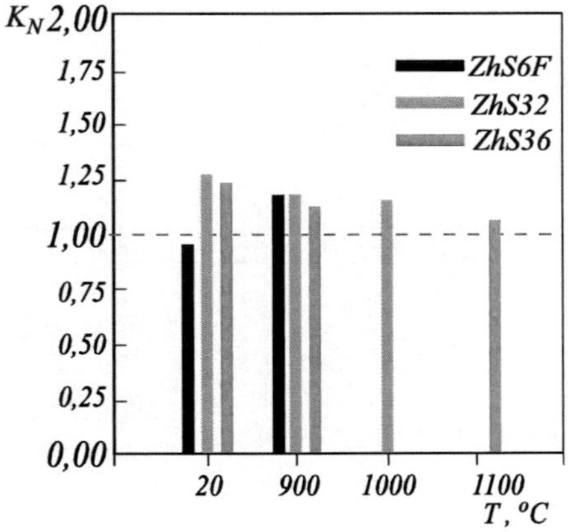

Fig. 4.17 Dependence of the anisotropy coefficient K_N <111> of the fatigue limit ($N = 2 \cdot 10^7$ cycles) on the temperature for the alloys ZhS6F, ZhS32 and ZhS36

The fatigue resistance characteristics depend on the values of the elastic modulus. So at 20°C for the orientation [001] E = 139 GPa, and for [111] E ~ 300 GPa, i.e. as the modulus of elasticity increases, so does the fatigue resistance. However, direct tests of blades made of ZhS6K alloy (Marchal et al., 2006) with edge grain boundaries showed that, even if the stresses are determined using local values of the elastic modulus, the fatigue resistance at crystal orientation at the [011] edge is lower than at [001] (although the value is $E_{011} > E_{001}$).

In Reed et al. (2000); Schubert et al. (2000); Cook and Chatterjee (1999); Zhang and Wang (1999) are devoted to the study of the growth patterns of fatigue cracks in foreign single-crystal alloys. Thus, in Cook and Chatterjee (1999), fatigue tests were carried out at 1038°C for specimens with holes of various diameters from 0.1 to 0.5 mm made of PWA1484 alloy with orientation <001>. A Pt-Al coating was applied to the surface of the samples. The loading frequency was 120 and 300 Hz. In Zhang and Wang (1999), experimental results are presented concerning the influence of the temperature level, loading conditions and the influence of the environment on the growth rate of fatigue cracks.

In Reed et al. (2000), a significant effect of the azimuthal orientation on the direction of crack growth at stage I in the single crystal UDIMET 720 alloy was established. Fatigue tests were carried out at four-point bending of beam specimens at temperatures of 20 (in air and in vacuum) and 600°C (in vacuum) with a frequency of 20 Hz, R = 0.1 and 0.5.

In Schubert et al. (2000), fatigue tests of the CMSX-4 single crystal alloy were carried out in vacuum and in air. Tests of specimens with a cross section of 8 × 8 and 4.5 × 12 with initial cracks of 0.3-0.5 mm were carried out at 750 and 1000°C. Dependencies of the rate of crack growth on ΔK were determined. It was found that when tested in vacuum, the growth rate of fatigue cracks is higher than in air.

4.2.7 Long-term Strength and Creep Resistance

Alloys obtained by directional solidification have significantly higher values of plasticity not only in short-term tests, but also in long-term strength tests; their heat resistance also increases (Table 4.14). As studies by various authors have shown,

Table 4.14 Influence of structure on long-term strength of MAR-M200 alloy

Structure	σ_{100}, MPa	σ = 690 MPa t_f, h	ε_f, %	$\dot{p}$, 10^{-2} %/h	σ_{100}, MPa	σ = 345 MPa t_f, h	ε_f, %	$\dot{p}$, 10^{-2} %/h	σ_{100}, MPa	σ = 208 MPa t_f, h	ε_f, %	$\dot{p}$, 10^{-2} %/h	σ_{100} MPa
	760°C				870°C				927°C	982°C			1037°C
Equiaxial	630	4.9	0.45	7	340	246	2.2	3.4	260	35.6	2.6	2.4	120
Directional	700	366	12.6	1.45	420	280	35.8	7.7	280	67	23.6	2.6	170

samples cut in the [111] and [001] directions have the longest duration, and the shortest — in the [011] direction (Fig. 4.18). The results of testing the ZhS6U alloy with different structure at 975°C (Table 4.15) also indicate a noticeable effect of orientation on durability and long-term ductility.

At the same time, at relatively low temperatures, the effect of orientation on long-term strength, according to Khan (1986), turns out to be more significant (see Table 4.16) and orientation [001] is more preferable. Long-term ductility of single-crystal alloys is usually much higher than that of polycrystalline alloys. It differs little from short-term plasticity at the same temperature. Nevertheless, there is a significant decrease in the long-term plasticity of the samples cut at an angle of 90° to the crystal orientation [001] (Fig. 4.18). Monocrystalline blades and directional solidification blades are usually subjected to 100% X-ray control to determine orientation: the permissible deviation from the [001] axis usually does not exceed 10°.

In eutectic alloys with directional crystallization, the orientation of carbides has a significant effect on the properties. Tests of specimens made of the (Co, Cr) - $(Co, Cr)_7C_3$ alloy (Miles and Mclean, 1977) indicate a decrease in the durability and creep resistance with an increase in the angle between the load application and the carbide orientation:

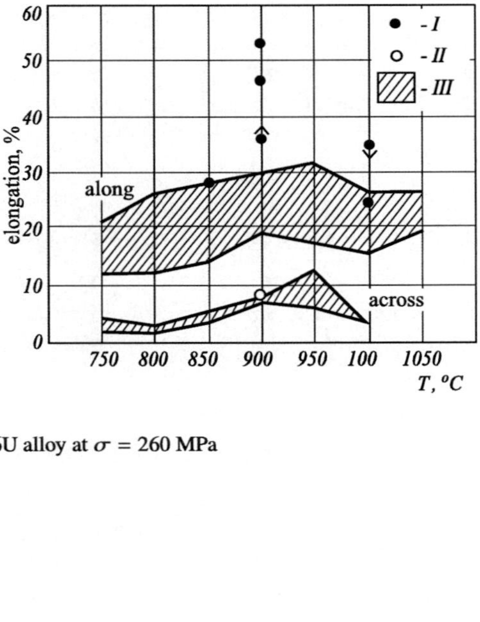

Fig. 4.18 Graphs of the dependence of the long-term plasticity of an DS CM247 LC alloy of different orientations on temperature (Coutsouradis et al., 1994): *I* - samples cut along the crystal axis, *II* - samples cut across the crystal axis, *III* - data from the literature

Table 4.15 Long-term strength of the ZhS6U alloy at σ = 260 MPa

Structure	t_f, h	δ_f, %	ψ_f, %
Large-columnar			
along	64	18	27
across	34	4	6
Thin-columnar			
along	31	10	21
across	12	1.4	2.7

Table 4.16 Influence of orientation on long-term strength of heat-resistant alloys

Alloy	Temperature, °C	Orientation	σ_{10}, MPa	σ_{100}, MPa	σ_{500}, MPa	t_f, h	δ_f, %
						at σ = 750 MPa	
ZhS32	1000	[001]	382	258	190	-	-
	1000	[011]	362	249	180	-	-
	1000	[111]	392	285	222	-	-
CMSX-2	760	6° [001]	-	-	-	759	12.5
	760	13° [001]	-	-	-	64	11.8
	760	22° [001]	-	-	-	484	9.6
	760	8° [111]	-	-	-	32	25.9
	760	19° [111]	-	-	-	21	38.5

Angle, degree	0	10	15	20	40	60	90
σ_{100}, MPa	254	232	202	167	117	107	107
$\sigma_{1\%/100\,h}$, MPa	236	204	140	138	82	61	58
$\sigma_{2,5\,10^{-2},\%/h}$, MPa	313	266	226	178	101	81	78

With an increase in the time to fracture and the angle between the direction of carbides and the load, the ductility of the eutectic alloy increases significantly (from 2 to 40%).

Detailed studies of the long-term strength of single-crystal alloys were carried out in Golubovskiy et al. (2005a); Kablov (2006); Nozhnitsky and Golubovsky (2009). To assess the effect of axial anisotropy on long-term strength, the authors used the following relationship:

$$K_t^{<hkl>} = \frac{\sigma_t^{<hkl>}}{\sigma_t^{<001>}} \tag{4.5}$$

with the long-term strength $\sigma_t^{<hkl>}$ and $\sigma_t^{<001>}$ for the crystallografic orientation <hkl> and <001>. In Fig. 4.19 the curves of the long-term strength of the ZhS36 alloy at 900-1100°C and the dependencies for the coefficient $K_t^{<hkl>}$ are shown. The figure demonstrates the non-monotonic dependence of the value of $K_t^{<hkl>}$ on the time to fracture, which the authors associate with the difference in the nature of fracture.

To determine the time to fracture of single-crystal alloys, as well as polycrystalline ones, the temperature-time dependence (4.6) is used

$$t_f = \xi T^m \sigma^{-n} \exp \frac{U_0 - \chi\sigma}{RT}, \tag{4.6}$$

whose parameters for two single-crystal alloys are given in Table 4.17. During the creep tests, the microstructure of single-crystal alloys changes - the so-called raft structure is formed. Alloys with this microstructure have reduced long-term strength. However, no systematic studies of this effect have been conducted.

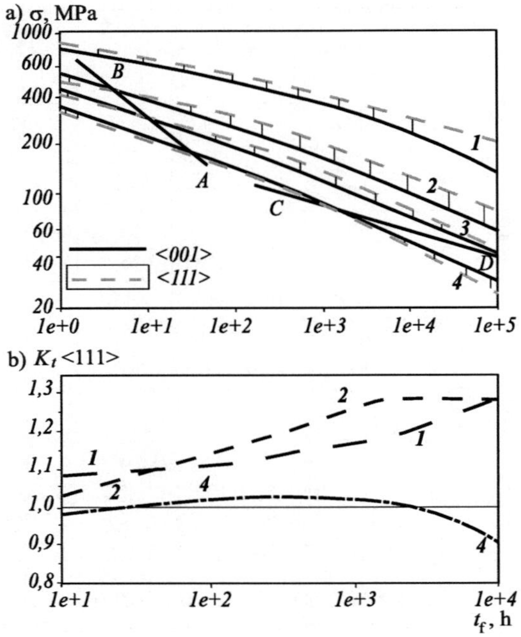

Fig. 4.19 Results of long-term strength tests of ZhS36 single-crystal alloy (*1* - 1173°K, *2* - 1273°K, *3* - 1323°K, *4* - 1373°K); a) time to fracture of the alloy with orientations <001> and <111> (the line AB and CD show the boundaries of dominant either <001> or <111> orientation - an inversion takes place: first <001> to the left of AB, then to the right of AB <111>, and then, to the right of the CD, again the advantage <111>), b) dependence of $K_t^{<hkl>}$ on time to fracture and temperature

Table 4.17 Coefficients of Eq. (4.6)

Alloy	ΔT, °K	<hkl>	m	n	ln ξ	U_0, kJ/mol	χ, J/(mol MPa)
CMSX-4	1023-1073	<001>	0	0	-31.7	440.2	146
	1073-1173	<001>	2	0	-57.6	536.3	130
		<111>	1	4	-39.0	667.9	158
	1173-1273	<001>	2	4	-28.3	441.9	25
		<111>	2	0	-58.9	587.2	236
ZhS36	1173-1237	<001>	1	2	-40.2	546.5	119
		<111>	2	4	-45.4	644.3	91
	1273-1337	<001>	2	4	-29.9	462.9	59
		<111>	2	3	-48.8	653.3	144
ZhS47	1173-1273	<001>	2	4	-29.85	475.7	48.5
	1273-1373	<001>	2	3	-37.36	508.2	93.32

The orientational dependencies of the durability of various single-crystal alloys are shown in Fig. 4.20, 4.21 and 4.22. It can be seen from an examination of the figures that different alloys have different orientational dependencles of long-term strength, which in turn depend on the temperature and duration of tests.

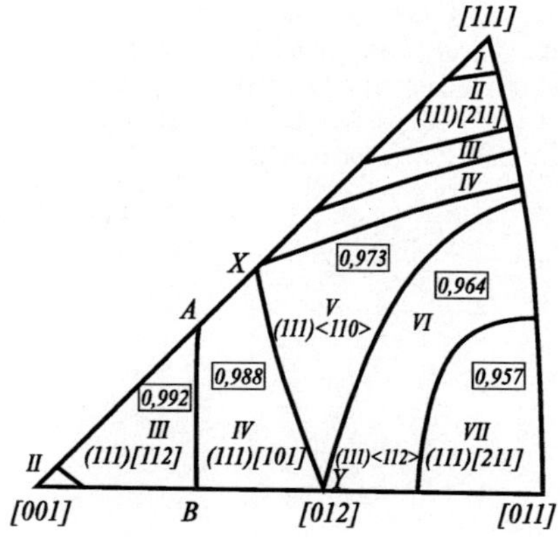

Fig. 4.20 Orientational dependence of the durability of single crystals of the CNK7M alloy at 750-900°C for long-term strength based on 100-25000 h (Rtishchev, 1992). *I* - maximum, *II* - very high, *III* - high, *IV* - medium, *V* - low, *VI* - lower, *VII* - minimum

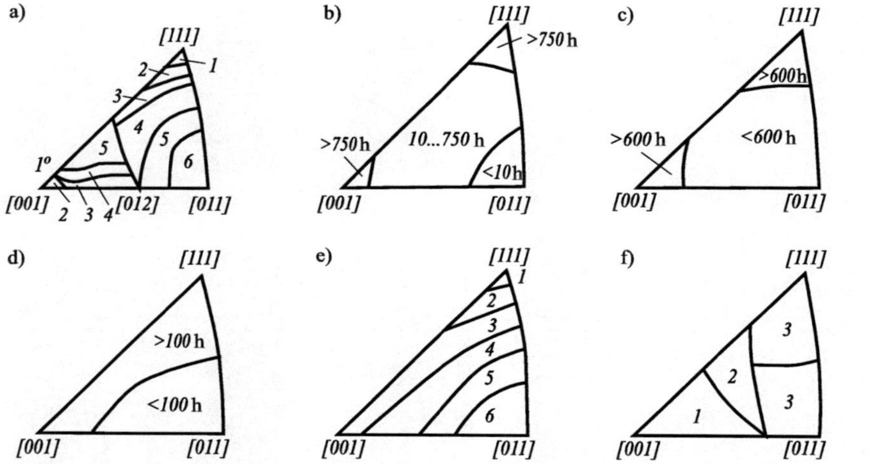

Fig. 4.21 Orientation dependence of the durability of single crystals of high-temperature alloys according to the data of R.A. MacKay (a), B.H. Kear (b-d), G.J.S. Higginbotham (e) and Y.G. Nikigawa (f) (Shalin et al., 1997)

In Golubovskiy et al. (2005b), the authors concluded that the azimuthal orientation does not affect the long-term strength of alloys in the temperature range 1000-1100°C. It has been established that the nature of fracture of single-crystal alloys with different orientations is different (see Figs. 4.23 and 4.24). In all cases, single crystals of carbon and carbon-free alloys with orientations [001] and [111] retain their cylindrical shape, and single crystals with orientation [011], in contrast to fracture upon short-term tension, acquire an elliptical sectional shape. Most single-crystalline superalloys contain relatively little amount of chromium. Therefore, their resistance to oxidation is somewhat reduced. They are usually used with protective coatings applied to them. Creep tests of uncoated single-crystal CMSX-2 alloy at

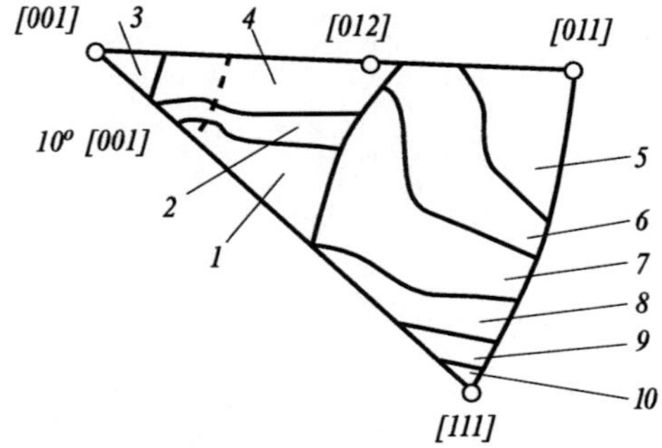

Fig. 4.22 Influence of crystallographic orientation on characteristics of long-term strength of the ZhS26 alloy: *1* - bad, *2, 7* - intermediate, *3, 9* - very good, *4, 8* - good, *5* - especially bad, *6* - bad, *10* - best

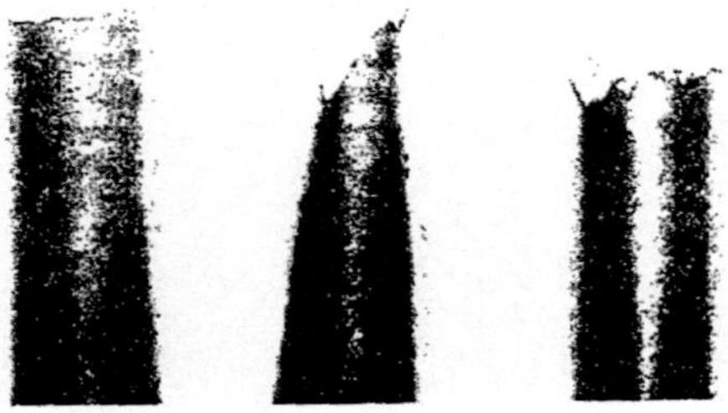

Fig. 4.23 Fracture during creep at 750°C of single crystals with orientations [001], [011] and [111]

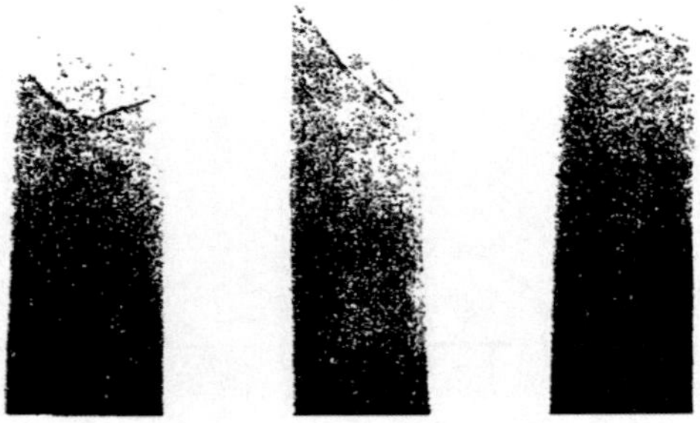

Fig. 4.24 Fracture during creep at 900°C of single crystals with orientations [001], [011] and [111]

982°C (Jo and Kim, 2003) showed that an oxide film formed on the surface of samples (blades) somewhat slows down the propagation of creep cracks.

Data on the creep resistance of ZhS6F and ZhS32 single-crystal alloys with different orientations are given in Shalin et al. (1997, 1981). Separate creep curves for the ZhS36 alloy were obtained for the 001 orientation (Fig. 4.25). The issues of calculating the stress-strain state of blades made of single-crystal alloys and determining the conditions for their fructure are considered below.

Slip systems in single-crystalline alloys are the base of numerical simulations. Table 4.18 shows the possible slip systems and the Schmid factor for single crystals of high-temperature nickel alloys (with an fcc lattice). In order to effectively identify the slip systems, along which the specimen fracture occurred, A.S. Semenov developed a program (programming language C++) that allows evaluating the Schmid factor

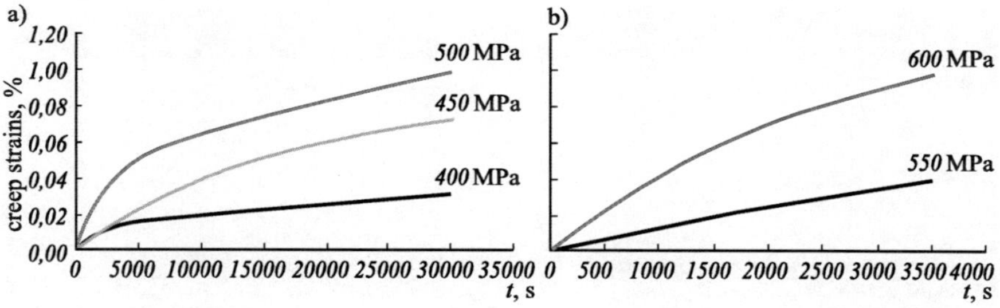

Fig. 4.25 Creep curves of ZhS36 alloy with orientation [001] at 900°C (a) and 850°C (b)

Table 4.18 Possible slip systems and the Schmid factor for single crystals with an fcc lattice

Crystallographic orientation of the loading axis	Schmid factor, M	Slip system
<001>	0.5	{011} <011>
	0.47	{111} <112>; {112} <111>
	0.408	{011} <111>; {111} <011>
<011>	0.5	{001} <001>
	0.47	{111} <211>; {211} <111>
	0.447	{001} <120>; {120} <001>
<111>	0.47	{001} <110>; {011} <100>
	0.447	{100} <012>; {012} <100>

and visualizing the most dangerous slip systems for arbitrary orientations of the specimen and loading axes (see Fig. 4.26).

4.2.8 Thermal and Low-cycle Fatigue

It was shown in Gabb et al. (1986) that the fracture resistance of a single-crystal alloy at a loading frequency of 0.1 Hz, temperatures of 760°C and 980°C for samples cut in the [001] direction are higher than for samples with other orientations of the structure. A similar situation is associated with the fact that the values of the yield stress at the [001] orientation, both static and cyclic, are higher than the values obtained in tests of metal with other orientations. This is evidenced by the combination of low-cycle fatigue test results in the coordinates $\Delta\varepsilon_{pl} - N_f$.

Values of the durability of the ZhS6U alloy of directional solidification during thermal fatigue tests of flat corset specimens tested according to the CKTI method described in Chapt. 3 with $T_{max} = 900$°C, for the orientation parallel and perpendicular to the crystallites differed little from each other (Getsov, 1981). When the boundary was located at an angle of 45° significant decrease in durability was obtained (Fig. 4.27). Similar results were obtained in tests of ZhS6U alloy in the

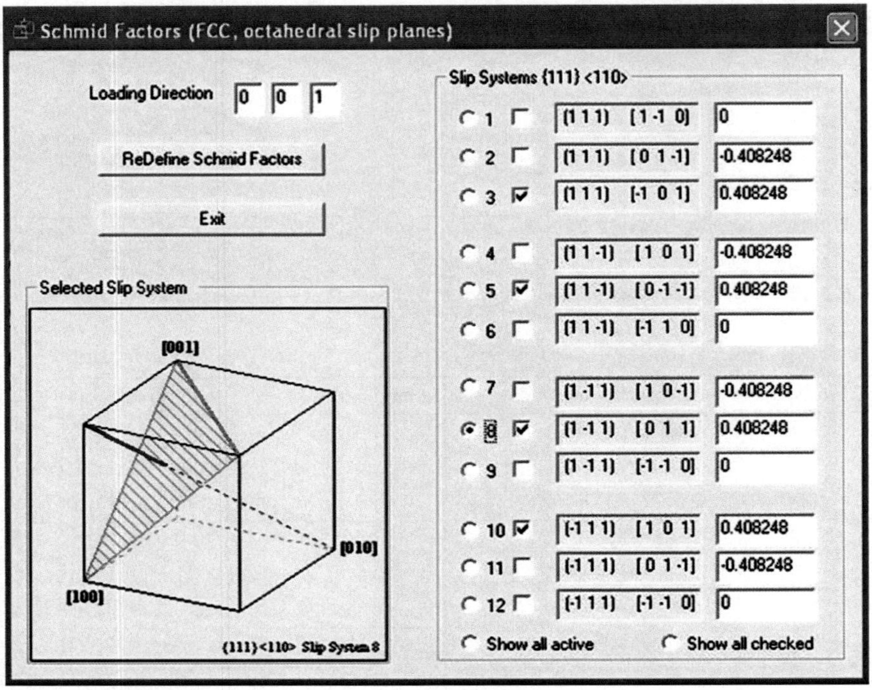

Fig. 4.26 Monitoring of active slip systems, for arbitrary specimen orientations and loading axes

Fig. 4.27 Graph of the influence of the orientation of crystallites in the ZhS6U alloy on the thermal fatigue resistance at T_{max} = 900°:
1 - parallel to loading axis,
2 - transverse, *3* - at angle 45°

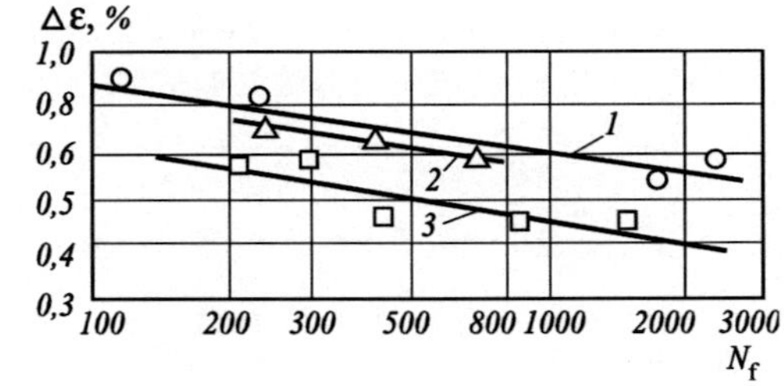

20-900°C mode (Dul'nev et al., 1988b), as well as the DS CM247 alloy at 850°C (see Fig. 4.28; Erickson and Harris, 1994). It has been established that the metal of directional solidification and single-crystal workpieces has significantly higher values of thermal fatigue life compared to metal with an equiaxial structure (Table 4.19; Dul'nev and Kotov, 1980).

In single-crystal alloys under thermal cyclic loading, plastic deformations are concentrated in shear bands, the density of which is higher in the inter-axial spaces (Kulygin et al., 1971) than in the axes of dendrites. Thermal fatigue microcracks originate in the zone of carbides (for carbon alloys) and primary precipitates of the γ'-phase. According to Dul'nev et al. (1988a); Golubovskiy et al. (2004), in single

Fig. 4.28 Dependence of low-cycle fatigue resistance of alloy DS CM247 LC at 850°C on specimen orientation: *I* - 0°, *II* - 45°, *III*, *IV* - 90°, *V* - with equiaxial structure

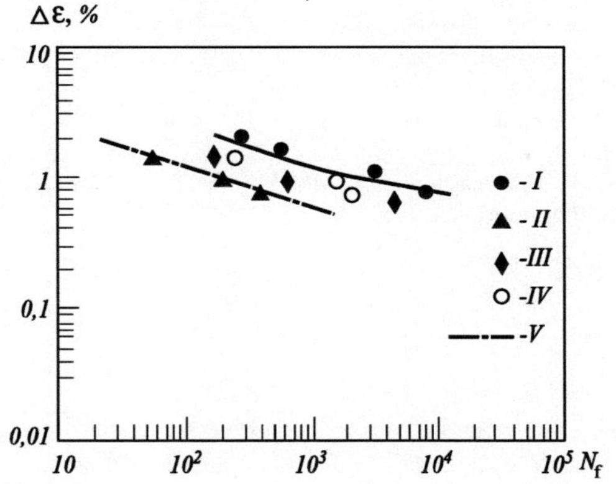

Table 4.19 Resistance to thermal fatigue of materials with different structures

Material	Number of cycles to failure during testing by mode	
	1088 - 315°C	1125 - 375°C
TAZ-8A	$\frac{600-800}{4350-6500}$	$\frac{450}{1200}$
MAR-M200	$\frac{13}{1250-4700}$	$\frac{-}{1200}$
NX-188	$\frac{100-298}{5100}$	$\frac{50}{5125}$
IN-100	$\frac{38}{2400}$	-

The numerator gives the number of cycles before the formation of cracks for alloys with an equiaxial structure, in the denominator - with directional crystallization, obtained by testing wedge-shaped specimens with heating and cooling in pseudo-boiling baths.

crystals with orientations [001] and [111], the formation of thermal fatigue cracks occurs at practically the same elastic-plastic deformations, despite the fact that the modulus ratio E_{111}/E_{011} reaches 2.2-2.4. The thermal stability of single crystals with orientations [011] and [112] differ insignificantly, and their elastic moduli are equal to each other. Figures 4.29-4.31 show the results of the respective tests.

Using SEM methods (Golubovskiy et al., 2004), the fracture structure of samples tested by the Coffin method was investigated. 4 zones (Fig. 4.32) of crack initiation and propagation were identified:

1. a smooth zone adjacent to the source of destruction of a crystallographic nature corresponds to a slow propagation from the source located near precipitates of carbides or eutectics,
2. stage of accelerated crack propagation with fatigue grooves,
3. stage of accelerated development with alternating areas of static destruction, and
4. sample break zone.

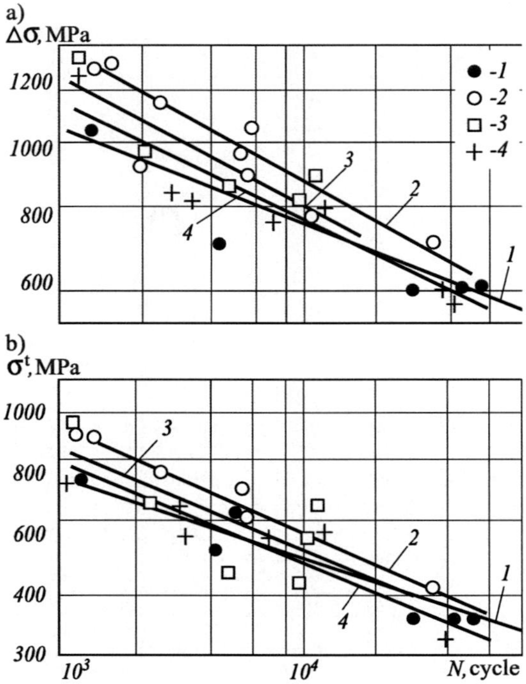

Fig. 4.29 Dependence of the durability (number of cycles N) on the stress $\Delta\sigma$ (a) and tensile stresses σ^t (b) in the cycle for single crystals of the ZhS6F carbon alloy with different orientations: *1* - [001], *2* - [111], *3* - [112] , *4* - [011]

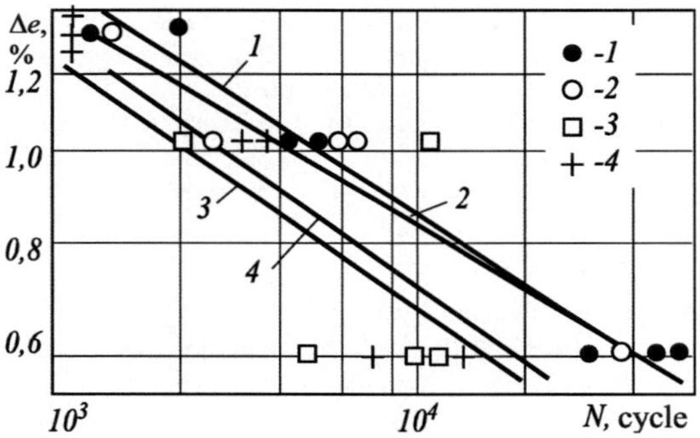

Fig. 4.30 Dependence of durability (number of cycles N) on the range of elastoplastic deformations Δe in a cycle for single crystals with different orientations: *1* - [001], *2* - [111], *3* - [112], *4* - [011]

Tests of the ZhS6U alloy with directional solidification according to the mode 700-900°C (under cycle conditions without holding at T_{max}), carried out on Coffin's samples, revealed the following features of the formation and development of cracks. In the same samples during testing, cracks were found: a) along the boundaries of crystallites (along the axis of the sample), b) across the crystallites (tear-off cracks), and also c) small tears in the direction perpendicular to the crystallites (Shalin et al., 1997). It should be noted that the nature of the destruction depends on the duration of the cycle and T_{max}, as is observed and for alloys with a polycrystalline structure.

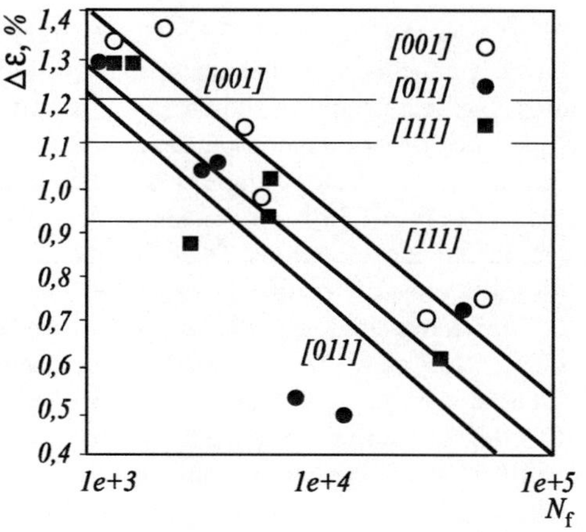

Fig. 4.31 Dependence of the thermal fatigue strength on the axial orientation of the ZhS6F alloy at T_{max} = 1050°C

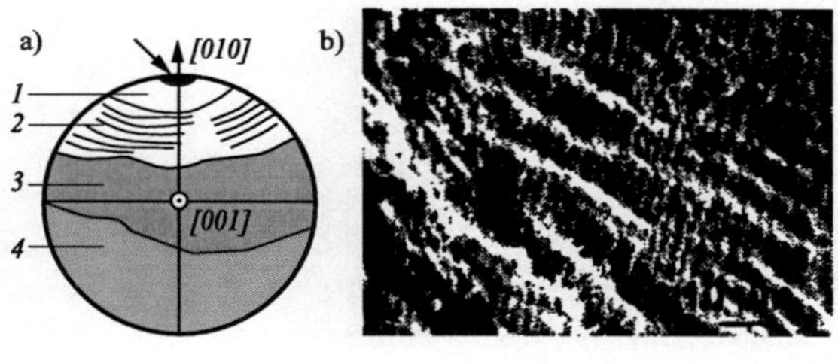

Fig. 4.32 Thermal fatigue fracture pattern. a - Scheme of crack initiation and development, b - fatigue grooves

In Li and Smith (1994), a comparison was made of the results of low-cycle fatigue tests of specimens made of single-crystal SRR99 alloy at different temperatures (750, 950, and 1050°C). Not only the crystallographic orientation was varied, but also the type of cycle (without exposure (0/0), with exposure for 2 min in the half-cycle of tension (t/0), with exposure for 2 min in the half-cycle of compression (0/t), with exposure 2 min in both half-cycles (t/t)). It was found that at 1050°C, the durability to fracture under the conditions of the cycle t/t is minimal. At 750°C, the following relation was observed:

$$(N_f)_{t/0} > (N_f)_{0/0} \approx (N_f)_{t/t} > (N_f)_{0/t}.$$

Under thermal cycling loading, the 0/t cycle is usually realized. In numerous tests for thermal fatigue of polycrystalline materials, the dependence on the holding time at T_{max} (Shalin et al., 1997).

In Zhou et al. (2003), it was established that in thermal fatigue tests without holding and with holding in a half-cycle of compression (at T_{max} = 900°C) single-crystal alloys TMS-75 and TMS-113 have in the second case the durability before cracking is less than in the first. In works carried out by domestic and foreign researchers on polycrystalline materials, the existence of a dependence of their thermal fatigue strength on the amount of the hardening phase in the alloy was established.

In Cook and Chatterjee (1999) the effect of creep under conditions of the presence of stress concentrators under thermal cyclic loading of the CMSX-4 single crystal alloy was analyzed. As shown in Chan and Leverant (1987), the intensity factors K_{I}, K_{II}, K_{III} do not depend on the anisotropy of the elastic constants, but are determined only by the slope angles crack path relative to the loading axis and the sample surface. In Fig. 4.33 shows the dependence of the rate of growth of cracks for metal with different orientations. To determine the effect of crystallographic orientation on the stress intensity factor, the dependence (Fig. 4.34) can be used.

Similar tests for low-cycle fatigue in a zero-cycle were carried out on ST specimens with ZhS6F alloy. The test results are shown in Fig. 4.35. It is seen that the axial orientation has a significant effect on the dependence of the fatigue crack growth rate (FCGR). The influence of the azimuthal orientation was also noted for samples with the [001] axial orientation. The anisotropy of the FCGR is due to the features of the dendritic structure of single crystals of different orientations. Samples with the [001] orientation are destroyed along a plane practically perpendicular to the direction of the first-order axes of the dendrites, with the [011] orientation - at an angle of 45°. and with the [111] orientation - mainly along the interdendritic sections parallel to the dendritic axes of the first order. At high temperatures (~ 980°C), the FCGR does not depend on the orientation of single crystals under conditions of uniaxial and complex stress states (Gabb et al., 1986).

Detailed studies of the FCGR of the PWA1484 single crystal alloy were carried out in Telesman and Ghosn (1996). The tests were carried out on CT samples in vacuum with the cycle asymmetry coefficient $R = 0.1, 0.5$ and constant temperatures

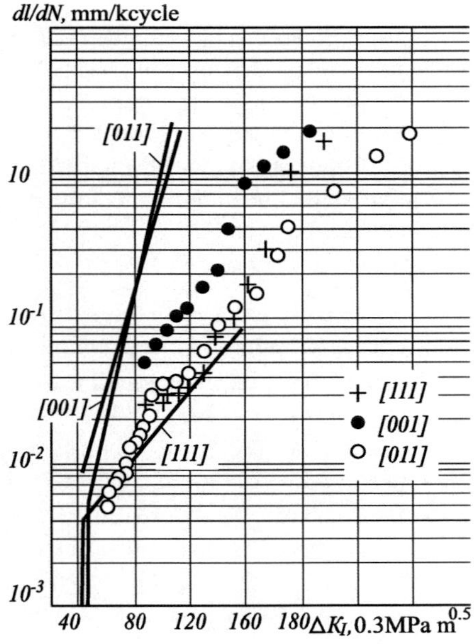

Fig. 4.33 Dependence of the low-cycle fatigue crack propagation rate dl/dN at 20°C on K_I for ZhS6F single crystals with different orientations (points) in comparison with the data for the MAR-M200 alloy (solid lines)

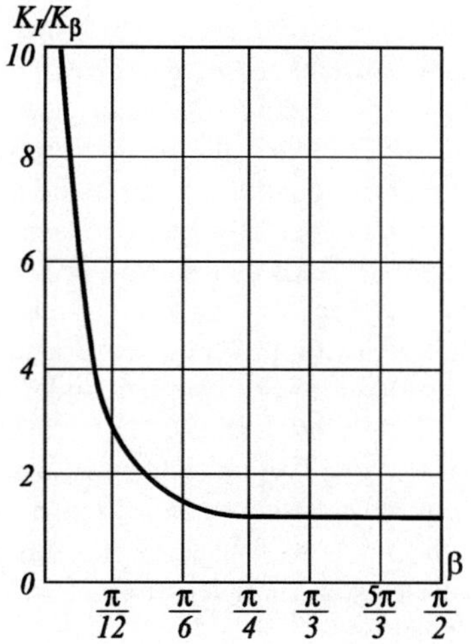

Fig. 4.34 Dependence of the intensity factor on the orientation of the crack (angle β). The depence was constructed by Getsov and Rybnikov (1988) from the data of Savin (1970)

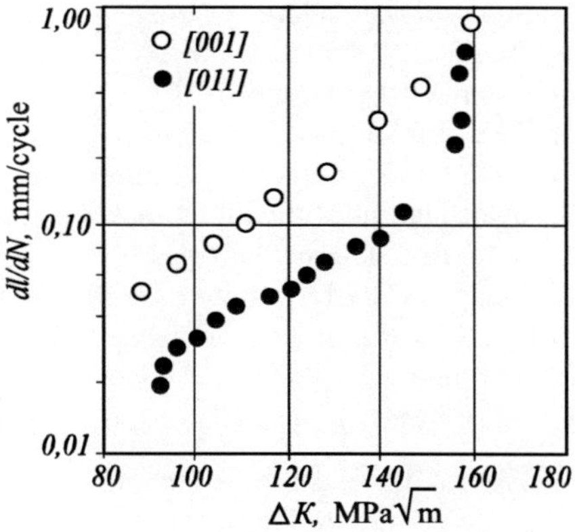

Fig. 4.35 Dependence of the FCGR of the ZhS6F alloy at $T = 293°K$ on the axial orientation (Golubovskiy et al., 2005b)

ranging from 427 to 870°C. The cycle frequency varied from 1 to 15 Hz. In the investigated temperature range, the nature of the destruction was not associated with the crystallographic orientation - it occurred by shear (in accordance with mode 1).

The issues of comparing the results of the resistance of materials to thermal fatigue and low-cycle fatigue at constant temperature deserve special consideration. Unfortunately, at our disposal there are no experimental data characterizing such a comparison for single-crystal materials. Therefore, we will conduct it based on general considerations.

First of all, we note the differences in test conditions and the nature of changes in the stress state as the number of cycles increases. In thermal fatigue (TMF) tests, the cycle gradually tends to become close to symmetric due to cyclic plastic deformation and cyclic creep. The deformation parameters of the cycle change, in connection with the processes of cyclic softening (cyclic hardening) at both the maximum and minimum cycle temperatures. For parts under conditions of both cyclic thermal stresses and constant loads, damage is caused by both cyclic deformations of alternating flow and static damage caused by unilateral accumulation of deformations. For the shoulder blades, the second type of damage usually plays a minor role. Thus, both under test conditions and in the blades, a rigid loading scheme is realized.

Low-cycle fatigue (LCF) tests are typically performed under conditions of a substantially asymmetric cycle under constant-sign stresses, leading to the accumulation of one-sided deformations. Under these conditions, a scheme is realized that is close to soft loading. The stress concentration at LCF plays a noticeably greater role in comparison with its effect under fracture conditions under thermal cyclic loading.

In Getsov et al. (2007a,b, 2008a,b,c, 2009); Getsov and Semenov (2009), a cycle of thermal fatigue tests was carried out at various values of T_{max}, with and without exposure, without concentrators and with stress concentrators. The purpose of these works was to determine the features of thermal fatigue fracture of a single-crystal alloy used for the manufacture of GTU rotor blades. The tasks set in the work included:

1. Study of the features of the deformation relief, the formation of microcracks and the nature of fracture during thermal cycling of an alloy with different crystallographic orientations.
2. Determination of the dependence of the fracture conditions on the maximum cycle temperature, temperature range and stress concentration.
3. Investigation of the effect of pores and recrystallization processes on the degree of thermal fatigue damage.
4. Study of the effect on the durability of holding at the maximum cycle temperature.
5. Study of the processes of one-sided accumulation of deformations during heat pumping.
6. Determination of the modes of destruction of samples.

We applied a complex technique described in Subsect. 3.3.3, based on the use of a special device that provides rigid fixation of flat corset samples (Fig. 3.49a) and their cyclic heating by passing current. The samples were heated according to a preset program, which was automatically maintained during the test. The behavior of the sample material at different points on its surface was observed with a microscope with a magnification of ×250. During the testing, the following were revealed: the features of the deformation relief, which characterize the mechanism of accumulation of thermal fatigue damage; durability until the first microcracks appear in different elements of the metal structure; growth rate of incipient cracks; durability to destruction of the sample. After the test, the directions of slip lines, the direction of propagation of cracks, fractographic studies of fractures were measured, the values of unilaterally accumulated deformations were determined. Calculations of the stress-strain state of the samples were carried out using the finite element package "PANTOCRATOR" (Semenov, 2003) at various modes of thermal cyclic loading in order to differentiate the modes of destruction: crystallographic and non-crystallographic

The concept of a non-crystallographic fracture mode is used in a phenomenological context, as a fracture observed at the macrolevel that does not coincide with any crystallographic slip plane. The magnitude of the range of conditionally elastic stresses $\Delta\sigma$ and total deformations $\Delta\varepsilon$ in the working part of the sample was determined by the formulas of Subsect. 3.3.3:

$$\begin{aligned}\Delta\sigma &= (E_{st1}\alpha_1 T_{max} - E_{st2}\alpha_2 T_{min})\varphi;\\ \Delta\varepsilon &= (\alpha_1 T_{max} - \alpha_2 T_{min})\varphi;\end{aligned} \qquad \varphi = 1 - \frac{\Delta k}{\Delta l}. \tag{4.7}$$

Here E_{st} is the static modulus of elasticity; Δl is the free movement of the control points during heating from T_{min} to T_{max}, Δk is the measured value of the displacement per cycle of the control imprints of microhardness applied to the surface of the sample along the edges of its working part. By calculating with the FEM using the simplest model of viscoelastoplasticity, the kinetics of the stress-strain state of the samples under creep conditions and in the plastic region was determined.

Samples of a single crystal alloy were made from plates of a high temperature single crystal alloy with the orientations [001], [111] and [011], having different azimuthal orientations (Table 4.20). To study the effect of stress concentrators in the working part of the sample, a hole was made perpendicular to the plane of the section with a diameter of 0.5 mm. Before proceeding to the examination of the test results, it should be noted that the used technique and technology for the manufacture of test specimens from a single-crystal alloy made it possible to obtain an extremely small scatter of the experimental results, which made it possible to determine a large number of regularities on a limited number of specimens (28). So, the results of the conducted experiments are as follows (Getsov et al., 2008c).

4.2.8.1 Influence of Maximum Cycle Temperature and Temperature Range

An increase in the maximum cycle temperature from 850 to 1050°C at a constant temperature range leads to a monotonic decrease in the durability of the alloy (Table 4.21). An increase in the temperature range decreases the durability (Table 4.22). Similar results were obtained for samples with stress concentrators.

Table 4.20 Orientation of the tested samples

Series number	Orientation	Deviation from axial orientation, degree	Azimuthal orientation of the sample plane, degree
1	111	5.64	8.26
2	011	4.51	11.27
3	011	8.33	14.43
4	011	9.67	7.86
5	001	5.47	41.97

Table 4.21 Influence of maximum cycle temperature on durability

No. specimen	Orientation	ΔT	T_{min}, °C	T_{max}, °C	N
3-0	[011]	750	150	900	951
3-1			200	950	450
3-2			250	1000	63
3-5		500	450	950	2535
3-3			500	1000	1220
3-4			550	1050	356
5-1	[001]	750	150	900	560
5-2			250	1000	95

Table 4.22 Influence of temperature range on durability

No. specimen	Orientation	T_{min}, °C	T_{max}, °C	N
3-2	011	250	1000	63
3-3		500		1220
5-2	001	250		95
5-3		500		1460

4.2.8.2 Influence of Orientation on Durability Under Thermocyclic Loading

Analysis of the test results shows that, depending on the temperature conditions of the tests, the following relationships can be written: $N_{111} > N_{001} > N_{011}$. At the same time, specimens with a concentrator for orientation <011> have a greater durability than specimens with orientation <111> (Table 4.23). One also studied the effect on the durability of the azimuthal orientation of samples with the axial orientation [011] (series 2, 3, 4). The results obtained (see Table 4.24) indicate the existence of such an effect of azimuthal orientation on durability.

Table 4.23 Order of the durability under thermocyclic loading depending on the orientation and test mode

Sample type	Test mode, °C	Durability ratios
with concentrator	350-850	$N_{011} > N_{111}$
smooth	150-900	$N_{111} > N_{001} > N_{011}$
	250-1000	$N_{001} > N_{011}$
	500-1000	$N_{001} > N_{011}$

Table 4.24 Influence of the azimuthal orientation of the single-crystal alloy [011] on the number of cycles before fracture under thermocyclic loading in the mode T_{min} = 150°C, T_{max} = 900°C

No. specimen	Deviation from exact axial orientation, degree	Azimuth orientation, degree	Cycles to failure
3-0	8.33	14.43	951
2-1	4.51	11.27	100
4-2	9.67	7.86	308

4.2.8.3 Effect of Holding at Maximum Cycle Temperature

Holding at the maximum cycle temperature decreases durability (Table 4.25).

Table 4.25 Influence of holding at T_{max} on durability

Sample series	Test mode, °C	Holding time at T_{max}, min	Cycles to failure
1	150-900	0	823
		2	140
		5	16
4	150-900	0	308
		2	17
		5	26
2 with concentrator	500-1000	0	187
		2	62

4.2.8.4 Effect of Stress Concentration

Under conditions of stress concentration, a significant (10-15 times) decrease in durability is observed (Table 4.26). This decrease in durability is noticeably more significant than in polycrystalline alloys. However, in tests with holdings at the maximum temperature due to stress relaxation processes, the samples with concentrators had a durability only 3 times different from that observed for samples without concentrators.

4.2.8.5 Features of the Deformation Relief

Multiple slip (Fig. 4.36) with TCI was observed for samples of all investigated orientations, in contrast to creep, when there is no multiple slip at orientation [011]. An increase in the maximum cycle temperature and a decrease in the temperature

Table 4.26 Influence of stress concentrator on fracture conditions

No. specimen	orientation	Time at T_{max}, min	T_{min}, °C	T_{max}, °C	N
4-2 without concentrator	011	0	150	900	308
4-1 with concentrator					25
1-2 without concentrator	111	0	150	900	823
1-1 with concentrator					50
2-2 without concentrator	011	0	500	1000	472
2-6 with concentrator		0			187
2-3 with concentrator		5			62

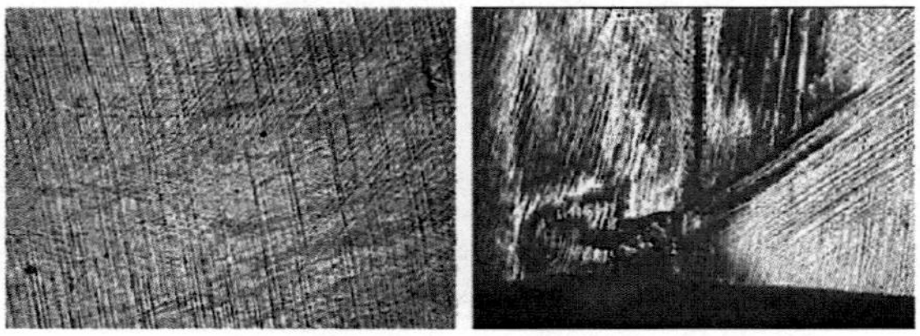

Fig. 4.36 Deformation relief of samples of series 5 after tests under thermocyclic loading

range (increase in durability) lead to an increase in the contribution of creep and intensification of multiple slip processes. the results of measurements and calculations (Table 4.27) of the angles of the slip lines of samples with different orientations, tested under different modes, were compared.

Table 4.27: Comparison of the experimental values of the slopes of the slip lines

No. specimen	T_{max}, °C	Holding time, min	Experimental slope of slip lines, degree	Theoretical slope of slip lines, degree	
				octahedral	cubic
1-1	900	0	15.8	16.4	-
			83.2; 82.5	92.6 (87.4)	-
			128.5; 130.8	130 (50)	130.2
1-2	900	0	22.4; 20.8	16.4	-
			130.5	130 (50)	130.2
1-3	1000	2	15; 16	16.4	-
			129; 130	130 (50)	130.2
1-4	850	5	41; 43	-	38.3
			45; 51	50	-
			81; 85	92.6 (87.4)	-
			160.2; 162	16.4 (163.6)	-
1-5	900	2	14; 18; 16	16.4	-

Continuation on next page

Table 4.27: Continued

No. specimen	T_{max}, °C	Holding time, min	Experimental slope of slip lines, degree	Theoretical slope of slip lines, degree	
				octahedral	cubic
			48; 51 97	16.4 92.6	- -
1-6	850	0	17.8 131.2	16.4 130 (50)	- 130.2
1-7	900	5	61	-	54.9
Theoretical values				50.0; 152.1; 16.4; 92.6	38.3; 130.2; 54.9
2-1	900	0	10 80; 82.5; 76.8; 78 78.2; 82; 83; 84.2 93	6.2 99.1 (80.9) - - 95.7 (84.3) 95.7	- - - - - -
2-2	1000	0	75.8; 80.5; 88.8; 86 99 114; 126	99.1 (80.9) 95.7 (84.3) 99.1	- - - - 47.8(132.2)
2-3	1000	2	5 86 93; 96	6.2 95.7 (84.3) 95.7 (84.3)	- - -
2-4	850	0	0; 1.3 84.5	0.2 95.7 (84.3)	- -
2-5	1000	5	83 94	95.7 (84.3) 95.7 (84.3)	- -
2-6	1000	0	4.2; 3,8 79.8; 81.2; 82; 85.8 93.5; 95.5	6.2 99.1 (80.9) 95.7 (84.3) 95.7 (84.3)	- - - -
Theoretical values				6.2; 99.1; 95.7; 0.2	43.0; 47.8; 161.3
3-0	900	0	15 85 170; 174	- 93.6 (86.4) 4.8 (175.2)	14.8 - -
3-1	950	0	70.5 89	74.8 93.6 (86.4)	
3-2	1000	0	35 71.5	- 74.8	38.6

Continuation on next page

Table 4.27: Continued

No. specimen	T_{max}, °C	Holding time, min	Experimental slope of slip lines, degree	Theoretical slope of slip lines, degree	
				octahedral	cubic
			93.2	93.6 (86.4)	-
3-3	1000	0	91; 95.5	93.6 (86.4)	-
			169.2	167.6	14.8 (165.2)
3-4	1050	0	21; 22.5	167.6 (12.4)	
			65.5	74.8	
			93.2; 92.5; 92	93.6 (86.4)	
3-5	950	0	94	93.6	
			70	74.8	
Theoretical values				74.8; 4.3; 167.6; 93.6	127.3; 38.6; 14.8
4-1	900	0	50	48.8	-
			78.5	-	83.9 (95.2)
			155; 154; 162.4; 158.5	163.2	-
4-2	900	0	48.1; 51.2; 49.2; 48.5	48.8	-
			86.5; 81.5	-	83.9; 84.8 (95.2)
4-4	900	5	96; 99	-	95.2
			45	48.8	-
			66	60.2	-
4-3	950	2	48.5; 47; 50	48.8	-
			121	119.8 (60.2)	-
4-6	950	0			
Theoretical values				5.1; 60.2; 163.2; 48.8	83.9; 5.7; 95.2
5-1	900	0	93.2	90.8	95.5 (84.5)
5-2	1000	0	3.5	-	5.3; 5.6
			92.6	90.8	95.5
			107.1	100.2 (79.8)	-
5-3	1000	0	39.4	40.8 (139.2)	-
			90.6	90.8	-
			149	150.2 (29.8)	-

Continuation on next page

Table 4.27: Continued

No. specimen	T_{max}, °C	Holding time, min	Experimental slope of slip lines, degree	Theoretical slope of slip lines, degree	
				octahedral	cubic
Theoretical values				79.8; 139.2; 90.8; 29.8	5.6; 84.5; 174.7

End of the table

In the experiments, good agreement was obtained between the observed slip lines - deformation relief on the surface of the samples with the calculated values.

4.2.8.6 Features of the Nature of Fracture

The following features of the thermal fatigue fracture of the samples were revealed:

1. The fracture of the samples did not occur through the development of microcracks, which originated after a relatively small number of cycles, but mainly along the newly incipient cracks.
2. Both crystallographic propagation of cracks along the slip lines and predominantly non-crystallographic, oriented at angles to the slip lines were observed.
3. A change in the direction of crack propagation was observed. For orientation [011], the propagation of cracks also occurred along the axis of the specimen (Fig. 4.37).
4. Microcracks at the pores are not, as a rule, a source of destruction (Fig. 4.38).
5. In the process of testing samples with a work-hardened surface during testing at a high temperature, recrystallization processes took place (Fig. 4.39). No slip lines were found in the newly formed grains, which in the case of crystallographic destruction are the source of the destruction processes.

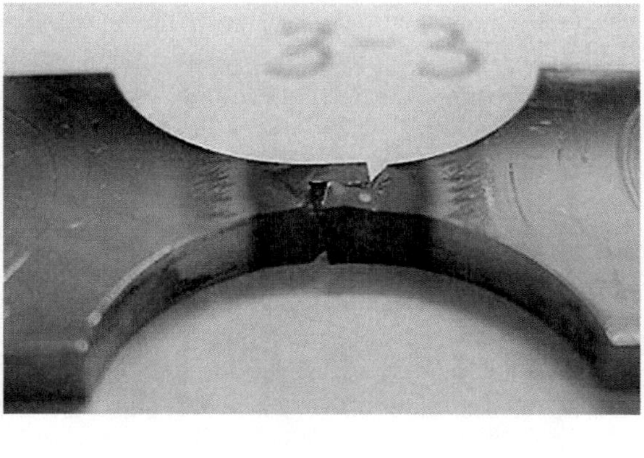

Fig. 4.37 Sample appearance after fracture

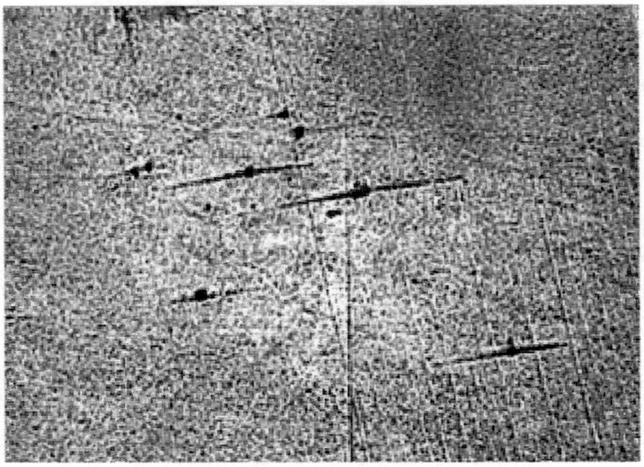

Fig. 4.38 Microcracks near pores after sample fracture

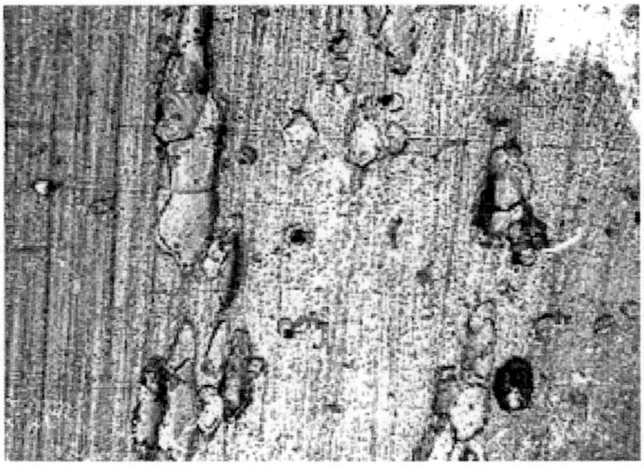

Fig. 4.39 Recrystallization in the process of testing under thermocyclic loading of a sample with a work-hardened surface

4.2.8.7 Processes of Unilateral Progressive Deformation

In the case of thermocyclic loading of single crystals, progressive deformation is observed - the formation of a barrel in the middle part of the samples. The inelastic deformations after fracture are characterized by the values given in Table 4.28. Here $\Delta\varepsilon_1$ and $\Delta\varepsilon_2$ - are irreversible deformations in mutually perpendicular directions in %. A review of the data in Table 4.28 shows the following:

4.2.8.8 Fractographic Features of Thermal Fatigue Cracks

Since during the test, the formation of microcracks and their propagation only in the plane of the thin section were observed, after the fracture of the samples, a fractographic study of the fracture was carried out. It was found that the propagation of cracks from the concentrator across the entire section occurred from the concentrator (see Fig. 4.40). At the same time, the centers of fracture for samples without con-

Table 4.28 Influence of stress concentrator on fracture conditions

No. specimen and orientation	Minimum cycle temperature, °C	Maximum cycle temperature, °C	Holding time, min	Cycles to failure	ε_1, %	ε_2, %
1-1 with concentrator [111]	150	900	0	50	4.0	24.3
1-2 [111]	150	900	0	823	6.7	10.7
1-5 [111]			2	140	16.3	21
1-7 [111]			5	16	18.3	28
1-6 with concentrator [111]	350	850	0	320	2.0	16
1-4 with concentrator [111]			5	118	4.0	11.7
1-3 [111]	500	1000	2	194	17.7	23.7
2-2 [011]	500	1000	0	472	12	-
2-5 [011]			5	317	23	29.7
2-1 [011]	150	900	0	100	15.3	10
2-4 with concentrator [011]	350	850	0	2952	6.0	9.3
2-6 with concentrator [011]	500	1000	0	187	10.6	10.3
2-3 with concentrator [011]			2	62	5.3	13.7
3-0 [011]	150	900	0	951	6.0	10.3
3-1 [011]	200	950	0	450	3.6	10.7
3-2 [011]	250	1000	0	63	1.5	12
3-3 [011]	500	1000	0	1220	6.3	7.3
3-4 [011]	550	1050	0	356	3.0	14.7
3-5 [011]	450	950	0	2535	4.3	-
4-1 with concentrator [011]	150	900	0	25	9.7	6.7
4-2 [011]	150	900	0	308	7.0	10.0
4-6 [011]			2	17	21.3	7.0
4-4 [011]			5	26	18	6.0
4-5 [011]	450	950	0	626	3.7	9.7
4-3 [011]			2	128	6.0	10.3
5-1 [001]	150	900	0	560	9.0	-
5-2 [001]	250	1000	0	95	12	6.3
5-3 [001]	500	1000	0	1460	12	4.7

centrators were formed both in the observation zone (polished surface) and on the opposite side (plane) of the sample (see Fig. 4.41). In some cases, fatigue grooves were observed (see Fig. 4.42).

4.2.8.9 Analysis of the Stress-strain State of the Samples

For the determination of the stress-strain state of the samples, a thermo-viscoelastic micromechanical model was used as a material model, taking into account the presence of slip planes. The considered model is a generalization of the Taylor model (Taylor, 1938). The governing equations in this case are as follows (Semenov and Getsov, 2010, 2014a; Chowdhury et al., 2017, 2018a,b):

Fig. 4.40 Fracture surface of sample No. 1-4 with a concentrator

Fig. 4.41 Fracture surface of sample No. 1-5 (arrows - local direction of crack growth)

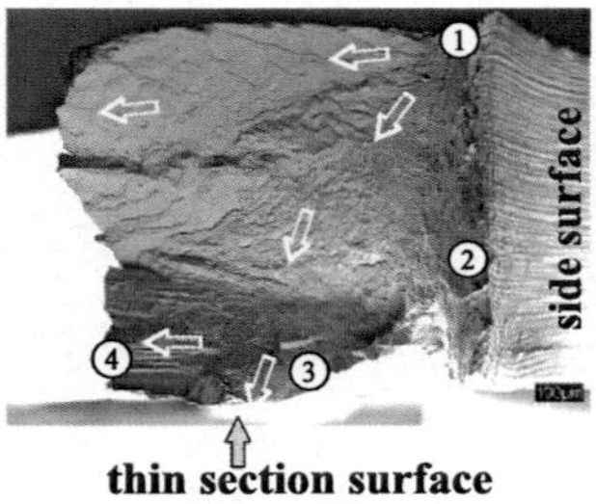

Fig. 4.42 Fracture surface of sample No. 1-6 (fatigue grooves)

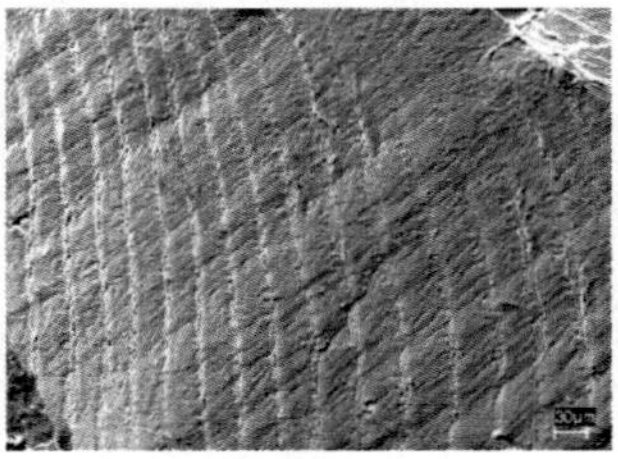

$$
\begin{aligned}
\dot{\boldsymbol{\varepsilon}}^{\mathrm{p}} &= \sum_{\alpha=1}^{N} \dot{\gamma}^{\alpha} \boldsymbol{P}^{\alpha}, \\
\boldsymbol{P}^{\alpha} &= \frac{1}{2} (\boldsymbol{n}^{\alpha} \boldsymbol{l}^{\alpha} + \boldsymbol{l}^{\alpha} \boldsymbol{n}^{\alpha}), \\
\dot{\gamma}^{\alpha} &= A^{\alpha} \left|\tau^{\alpha}\right|^{n} \operatorname{sign}(\tau^{\alpha}), \\
\tau^{\alpha} &= \boldsymbol{\sigma} \cdot\cdot \boldsymbol{P}^{\alpha}, \\
\boldsymbol{\sigma} &= {}^{4}\boldsymbol{C} \cdot\cdot (\boldsymbol{\varepsilon} - \boldsymbol{\varepsilon}^{\mathrm{p}} - \boldsymbol{\varepsilon}^{T}),
\end{aligned} \tag{4.8}
$$

where $\boldsymbol{\varepsilon}^p$ denotes the tensor of the plastic strains, N - number of slip systems, $\boldsymbol{P}^\alpha$ is the Schmid tensor of the slip system, $\boldsymbol{n}^\alpha$ are the normals to the slip plane, $\boldsymbol{l}^\alpha$ are the slip directions, τ^α is the shear stress with respect to the slip system α, ${}^4\boldsymbol{C}$ is the tensor of the elastic moduli.

The experimentally determined material parameters are: the exponent n, the flow law coefficients (similar to Norton's law) A^α. Due to the lack of the necessary detailed experimental data for different slip systems when calculating A^α at this stage of calculations, the same values were used for all slip systems $A^\alpha = A$. A parametric analysis of the effect of A on the results of calculations was realized (see Fig. 4.43). For further analysis, the minimum value A was selected at which a noticeable hysteresis loop is observed. In all the results of finite element calculations considered below, the following values were used: $n = 5.95, A_2 = 3.78 \cdot 10^{-22}\,(\text{MPa})^{-5.95}$. All calculations were performed using the relations (4.2) and (4.3).

At the first stage, in order to conduct a qualitative analysis, preliminary calculations were performed without taking into account the temperature dependence of the mechanical properties of the material. Comparison of the hysteresis curves obtained "with" and "without" taking into account the temperature dependence of the thermoelastic properties of the material is shown in Fig. 4.44. It can be seen that the width of the loop and the magnitude of irreversible deformations increase when the dependence of the elastic modulus on temperature is taken into account.

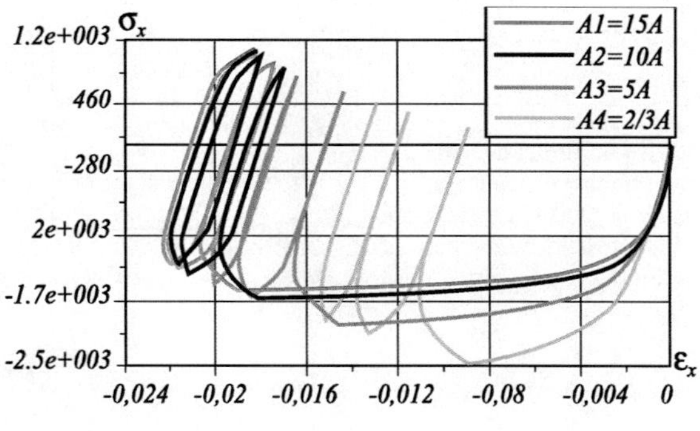

Fig. 4.43 Hysteresis curves for different values of A ($\dot{\gamma} = A\tau^n$) (3rd sample, 3 cycles, center point of center section, node 61)

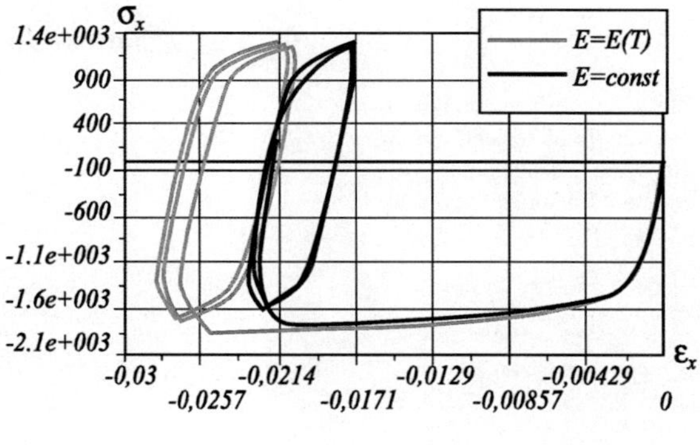

Fig. 4.44 Influence of the temperature dependence of the elastic modulus on the position of the cyclic deformation curves

Comparison of the deformation curves calculated in relation to the test conditions with holding and without holding at T_{max} is illustrated by the curves in Fig. 4.45. It can be seen that in tests with holding, both the width of the hysteresis loop and the value of unilaterally accumulated strain (compression) sharply increase. At the second stage, the deformed state of the samples under thermal cyclic loading was calculated (see Fig. 4.46). As seen in the middle part of the sample, the process of barrel formation is observed. The history of changes in the strain tensor components for the central point of the central section is shown in Fig. 4.47. The performed calculations showed, that ε_z is the dominant strain, but the values ε_x and ε_y are also essential. To compare the deformations of specimens with different orientations, calculations were carried out in relation to a loading cycle of 150 – 900°C (Table 4.29). Analysis of the data obtained indicates that in samples of series 2, 3, and 4 with orientation [011], the accumulation of deformations occurs mainly in the direction ε_z, and in samples of series 5 (with orientation [001] - predominantly in the perpendicular direction, as in the experiment. Some specificity is observed for the samples of series 4. Perhaps it is associated with a change in the results for a larger number of loading cycles.

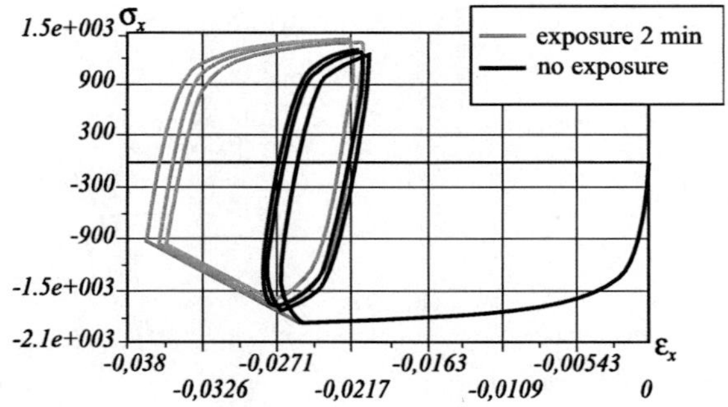

Fig. 4.45 Influence of two-minute exposure at T_{max} = 900°C on the shape of tsteresis loops (sample of series 3, central point)

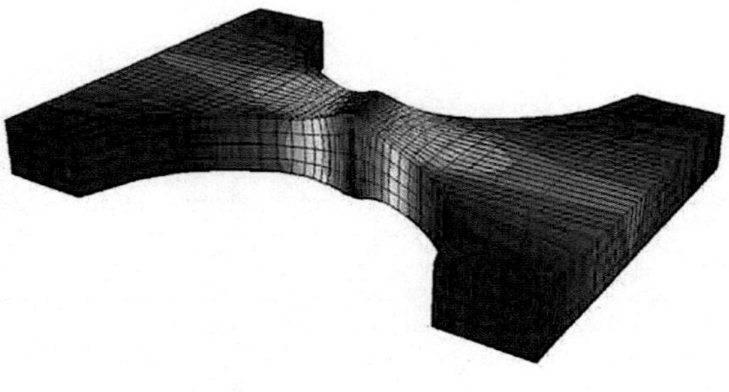

Fig. 4.46 Deformed state of the sample of series 3 on the 3rd cycle. Displacements are enlarged 25 times for clarity of consideration

Table 4.29 Results of calculating the deformed state for various samples (3rd cycle, $T_{min} = 150°C$). The displacements are enlarged 30 times

Sample	3D	2D
1-2	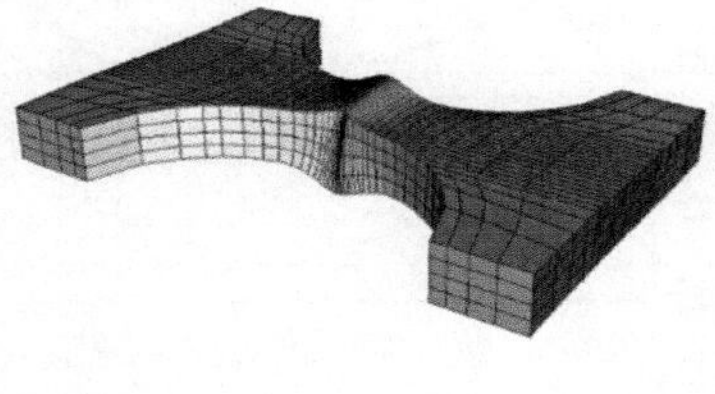	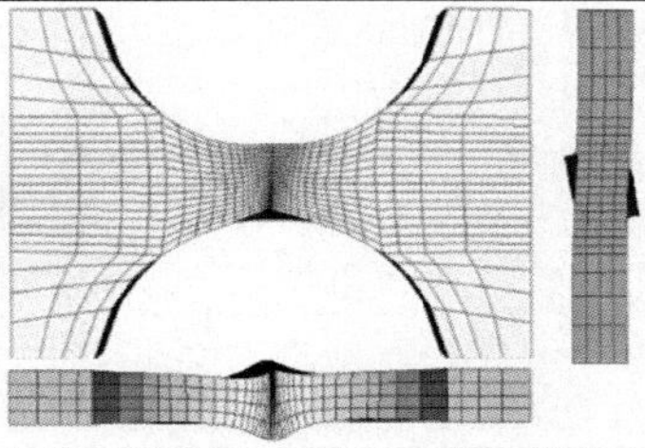
2-1	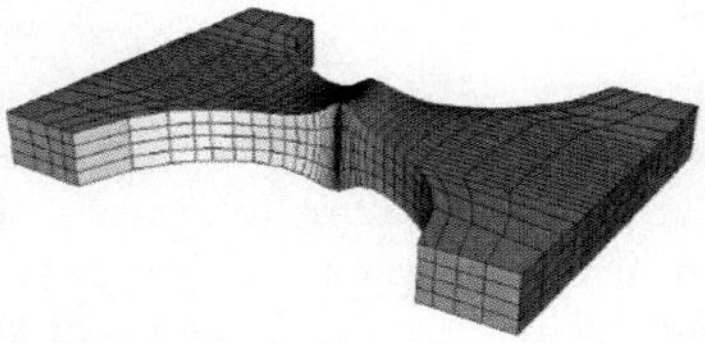	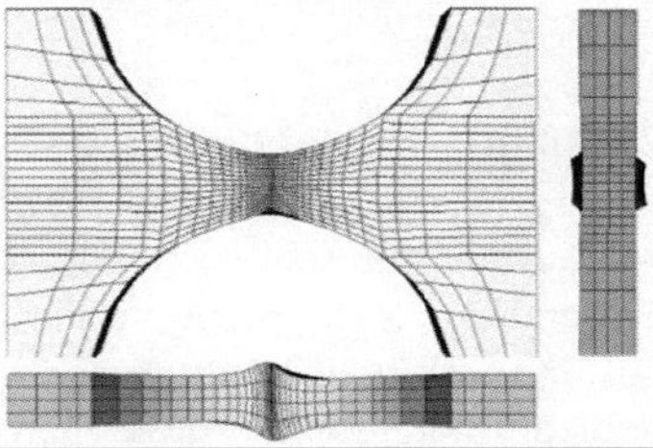
3-0	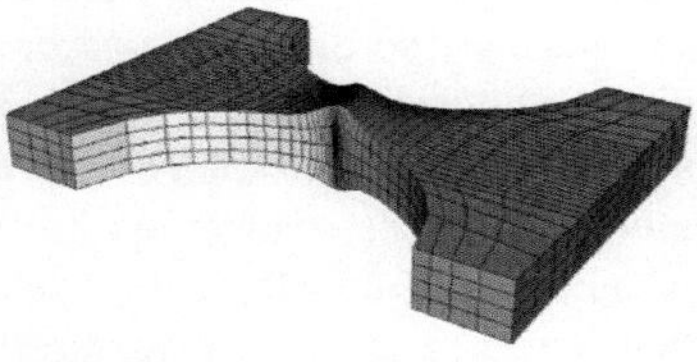	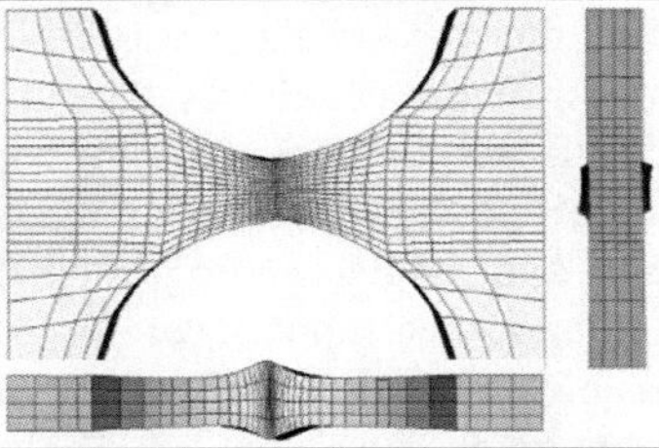
4-2	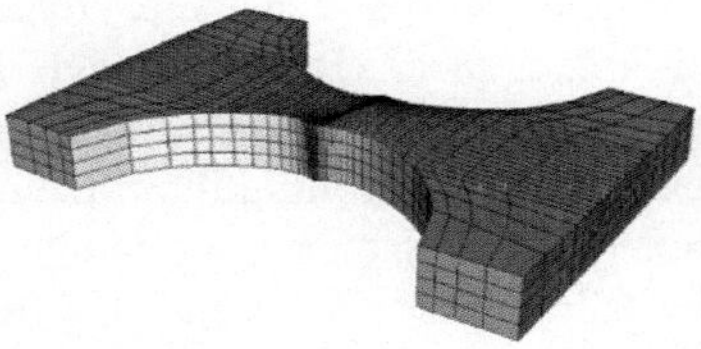	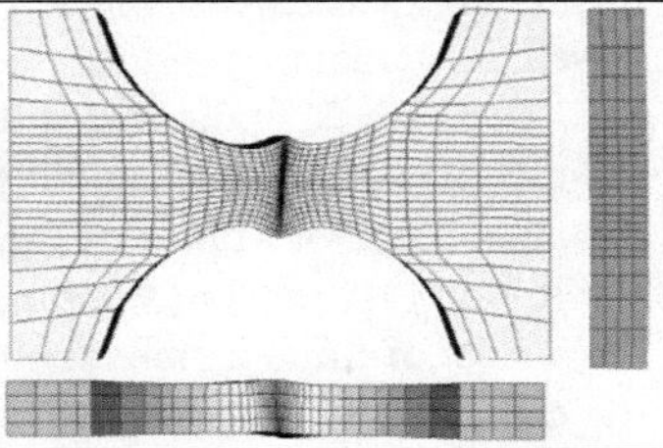
5-1	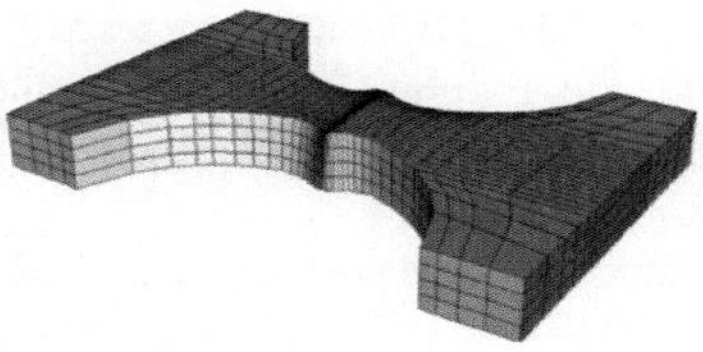	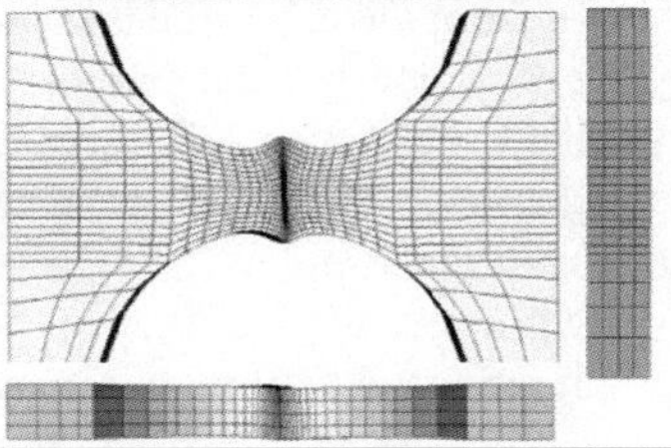

Fig. 4.47 History of changes in the strain tensor components at the central point of the central section of sample 3-0 during thermal cycling

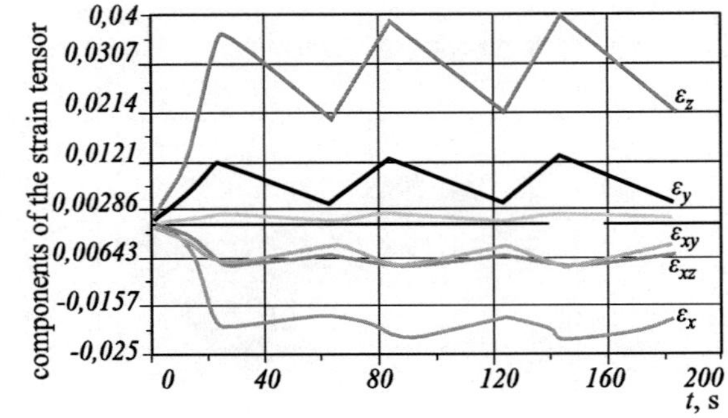

4.2.8.10 Analysis of Fracture Mechanisms

To determine the fracture modes (separation or shear), the analysis of the formed microcracks (in the plane of the thin section - "top") and macrocracks (top and side), the results of measurements and calculations of the angles of slip lines, the results of fractographic studies and calculations of the stress-strain state based on the model thermoviscoelastic material for specimens with different orientations tested under different conditions. The analysis was carried out on the assumption that in the half-cycle of tension (during cooling), the nucleation and development of tear-off cracks is possible, the growth direction of which is determined by the orientation of the area of the maximum principal value of the stress tensor, and in the half-cycle of compression (upon heating), the nucleation and development of shear cracks is possible, the direction of growth which is determined by the orientation of the areas of maximum shear stresses. The concept of a non-crystallographic fracture mode was used as a fracture that does not coincide with any of the crystallographic slip planes. The place of potential initiation of a shear crack was made based on the hypothesis of maximum shear stresses.

A comparison was made of the stress fields controlling the fracture process for different samples (see, for example, Table 4.30), and, most importantly, a comparison of the results of finite element and crystallographic analysis for the fracture angles of the samples. So, for sample 2-5, the crystallographic forecast turned out to be different from the experiment at points A, B, D by a maximum of 6°, and at point C by 13°; of course, the element forecast turned out to be different from the experiment at points A, B, D by a maximum of 4°, and at point C by 7° (see Table 4.31). Here AB is the line in the plane of the thin section, BC and CA are the lines on the lateral surfaces of the sample.

In Dul'nev et al. (1988b), tests were carried out according to the Coffin method for thermal fatigue according to the mode 100 ↔ 950°C at three levels of elastoplastic deformations $\Delta\varepsilon$: 1.3; 1.0 and 0.6% for single crystals of the heat-resistant alloy ZhS6F with the orientations [001], [011] and [111]. The test results are shown in Table 4.31.

Table 4.30 Design stress fields σ_1 and τ_{max} for sample series 1 (3rd cycle, T_{min} = 150°C)

Maximum principal stress σ_1 at T = 150°C	Maximum shear stress τ_{max} at T = 900°C
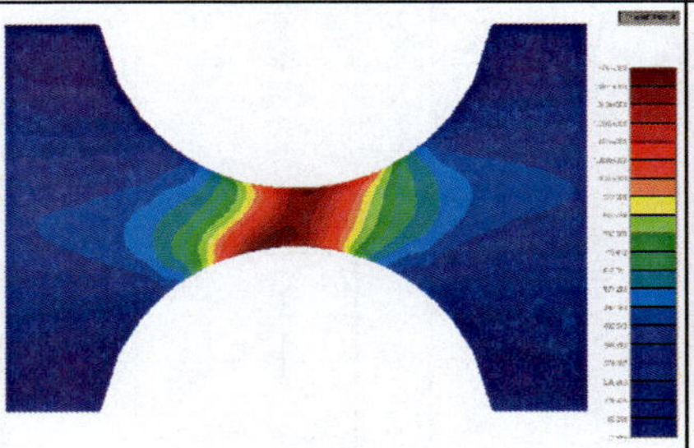	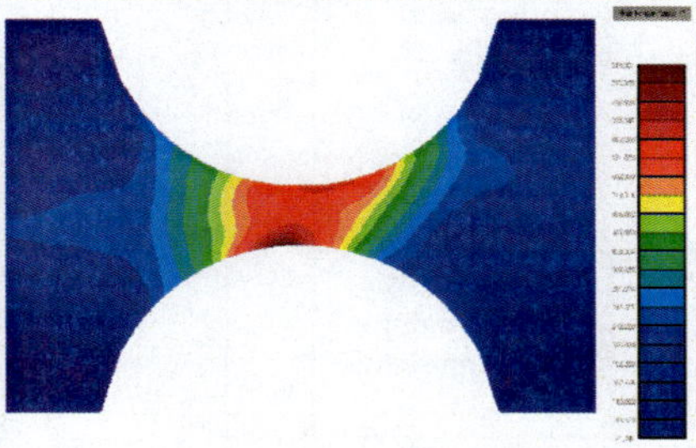
$\sigma_{x\,max}$ = 1559 MPa, τ_{max} = 856 MPa	

Table 4.31 Comparison of experimental data with the results of crystallographic and finite element analysis for sample 2-5

	Test	Crystallography	FEM	
			tear-off mode	shear mode
A	89 / 46	84 / 51 (1-11)	90 / 90	88 / 45
B	89 / 49	84 / 51 (1-11)		
C	90 / 38	84 / 51 (1-11)		
D	90 / 46	84 / 51 (1-11)		

The denominator is the angle in degrees on the top surface, the numerator is the angle on the side surface.

In general, four zones were identified on the fractures. The first zone (I) is directly adjacent to the fracture center, has a **crystallographic** character and corresponds to a slow crack propagation. The second zone (II) corresponds to the stage of **accelerated development** and has a rough structure. In this zone, signs of repeated loading in the form of a folded relief or fatigue grooves are quite clearly detected. The third zone (III) is a zone of sharply accelerated fracture with a structure inherent in repeated loading with alternating single fracture sections. The fourth zone (IV) corresponds to a single dollop. Table 4.32 shows the sizes of these zones (l_I, l_{II}, l_{III}). It can be seen that with an increase in the stress range $\Delta\sigma$ (and, accordingly, the strain range $\Delta\varepsilon$), the fraction of the crystallographic stage l_I decreases. In the case of sample orientation [001] the crystallographic stage is observed at all considered levels $\Delta\varepsilon$. For sample [011] at $\Delta\varepsilon$ = 1.3%, the crystallographic stage is not observed. And for the sample with the [111] axis, the crystallographic stage is not observed at $\Delta\varepsilon$ = 1.0% and $\Delta\varepsilon$ = 1.3%.

As has been established in a number of experimental studies, an increase in temperature (Cunningharn et al., 1994) and a decrease in frequency (Leverant and Gell, 1975) in single crystals lead to a change in the fracture mechanism, at which there is a transition from propagation in **crystallographic** (octahedral) planes to

Table 4.32 Thermal fatigue test results at T_{max} = 950°C without holding single crystals of ZhS6F nickel alloy with different orientations

Crystallographic orientation of the samples				$\Delta\sigma$, MPa	σ^-, MPa	σ^+, MPa	N, cycles	$\Delta\varepsilon$, %	Number and nature of the fracture foci	Length of the zones in the fatigue fracture, mm				K_c, MPa$\sqrt{m}$
Approximate sample axis	Angles (degrees) between sample axis and direction									l_I	l_{II}	l_{III}	l_{max}	
	[001]	[011]	[111]											
[001]	≈0	44.9	54.6	630	280	350	27828	0.6	two primary, multiple secondary	1.5	2.5	0	4.0	223
	2.1	43.5	52.6	970	350	620	5073	1.0	one primary, many secondary	1.05	-	1.7	2.15	285
	3.1	42.8	51.6	940	260	680	1922	1.3	many simultaneously arisen	0.5	1.25	0	1.75	220
[011]	41.0	4.0	33.0	620	290	330	40303	0.6	seven to eight simultaneously arising	1.9	1.0	0	2.9	187
	43.0	8.0	29.0	830	300	530	3162	1.0	-	0.75	1.0	1.25	3	255
	42.0	4.0	32.0	1290	580	710	1033	1.3	-	0	0.55	0.5	1.05	234
[111]	50.0	29.0	7.0	700	300	400	32900	0.6	five consecutive	1.2	0.8	1.2	3.2	222
	49.2	32.3	5.6	1050	370	680	5540	1.0	several simultaneously arising	0	1.4	0.5	1.9	256
	49.3	32.8	5.4	1300	40	900	1323	1.3	same	0	0.6	0.5	1.1	242

growth in directions determined not by the orientation of the crystal, but by the loading conditions (in accordance with the **Mode I**). Non-crystallographic growth is observed mainly along the γ/γ'-phase separation line.

One of the possible explanations (Telesman and Ghosn, 1996) of such a transition is the occurrence of additional damage due to environmental impact. Oxygen embrittlement in the vicinity of the crack tip occurs in accordance with a diffusion mechanism and is sensitive to temperature, time (frequency) and oxygen concentration. Comparison of the results of experiments carried out in air (Cunningharn et al., 1994) and in vacuum (Telesman and Ghosn, 1996) showed that the temperature at which the transition occurs increases significantly when tested in vacuum. As the frequency decreases, the embrittlement time within the cycle increases, which promotes oxygen penetration to a greater distance, which leads to a transition from the crystallographic propagation mode to the non-crystallographic mode I. Also, an increase in temperature accelerates the diffusion process, contributing to the transition from the crystallographic propagation mechanism to the non-crystallographic one.

As shown in Chan et al. (2005), the threshold value for non-crystallographic cracks ΔK_{thI} is lower than the value $\Delta K_{\mathrm{th}(111)}$ for cracks propagating in the crystallographic plane, therefore, when decreasing during testing ΔK_{e} below $\Delta K_{\mathrm{th}(111)}$ the crack cannot continue to grow in accordance with the crystallographic mechanism, and proceeds to propagation in Mode I. Thus, experiments show that the transition from the crystallographic to the non-crystallographic stage is controlled by the following parameters: temperature T, frequency f, and SIF range ΔK.

Thus, maps of fatigue crack growth mechanisms (similar to maps of fracture under static loading; Frost and Ashby, 1982) can be constructed for a set of these parameters, where the boundaries between the regions of crystallographic and non-crystallographic growth are indicated. Depending on the regime of thermal cyclic loading and crystallographic orientation, certain dominant types (modes) of destruction appeared in the samples. The correspondence of fracture modes to the value of the maximum sample temperature in the cycle $T_{\max}$ and the temperature range in the ΔT cycle was presented in the form of maps of fracture mechanisms (see, for example, Fig. 4.48).

Due to the absence of pure modes of crystallographic or non-crystallographic destruction, as well as in a number of cases of ambiguity in the choice between the modes due to the similarity of the predictions of crystallographic and finite-element analysis, when constructing maps of destruction mechanisms, instead of the concepts of modes of "crystallographic" or "non-crystallographic" destruction, we used the classification of modes by the orientation of the mean kink plane. In this case, 4 modes were considered:

- crystallography and stress-strain state (Mode 90 / 45, Mode 90 / 90)
- crystallography (Mode 45 / 45, Mode 45 / 90)

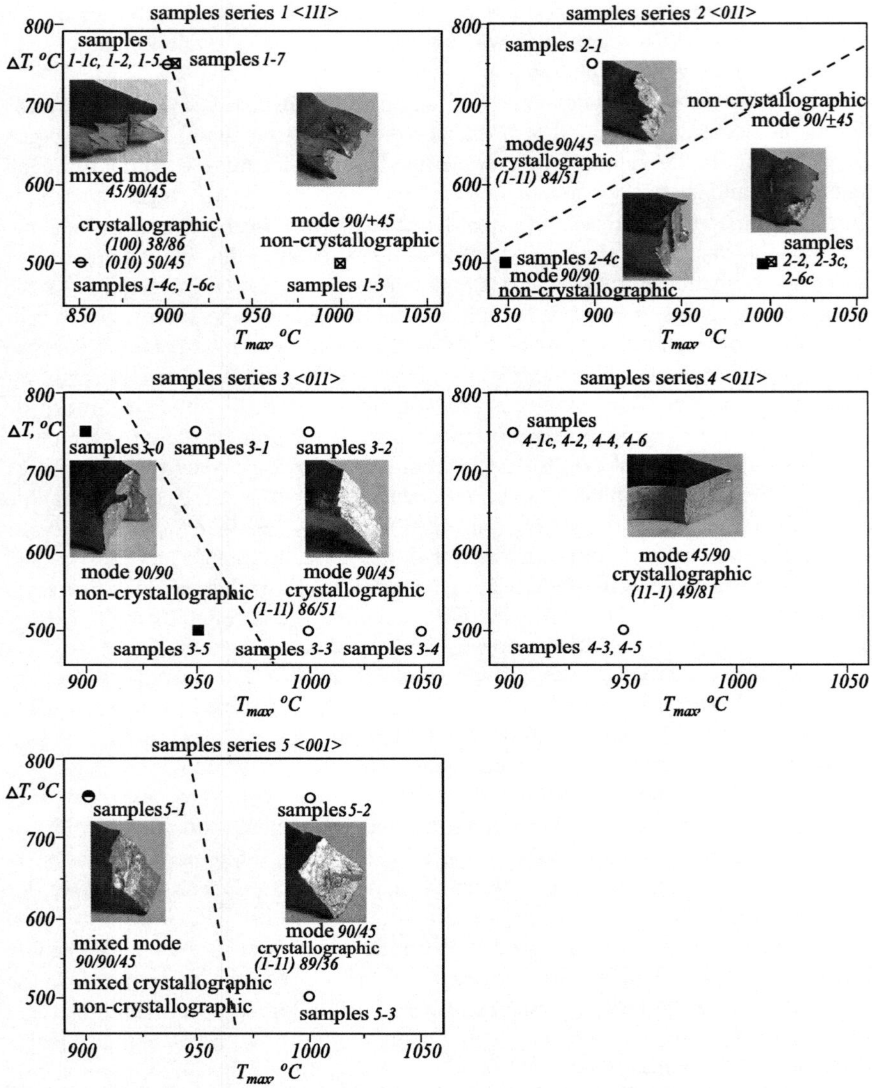

Fig. 4.48 Fracture mechanisms of samples of five series

4.2.9 Fracture Criteria for Monocrystalline Materials under Thermal Cyclic Loading

As shown in Chapt. 3, under low-cycle and thermal-cyclic loading of parts and specimens, two mechanisms of damage accumulation are observed. The first is associated with cyclic inelastic deformations - alternating yielding (AY), the second with one-sided accumulation of inelastic deformations - progressive deformation (PD). The share of these mechanisms in the accumulation of damage to various parts and specimens under various conditions of thermocyclic loading is associated with the rigidity of loading and with the characteristics of plasticity and creep of the material.

The deformation criterion for the fracture of metallic materials under low-cycle and thermal-cyclic loading (see Chapt. 3) has the following form

$$\Pi_1 + \Pi_2 + \Pi_3 + \Pi_4 = 1 \tag{4.9}$$

This criterion is based on the linear summation of damage caused by cyclic plastic deformations

$$\Pi_1 = \sum_1^n (\Delta\varepsilon_{\text{pl}i})^k / C_1,$$

cyclic creep deformations

$$\Pi_2 = \sum_1^n (\Delta\varepsilon_{\text{c}i})^m / C_2,$$

unilaterally accumulated plastic deformations

$$\Pi_3 = \sum_1^n \varepsilon_{\text{pl}i}\varepsilon_{\text{r}},$$

and unilaterally accumulated creep strains

$$\Pi_4 = \sum_1^n \varepsilon_{\text{c}i}\varepsilon_{\text{cr}}.$$

In case of mutual influence of the listed types of damage, the corresponding adjustments are introduced into the limiting characteristics of the material $C_1, C_2, \varepsilon_{\text{r}}, \varepsilon_{\text{cr}}$. The criterion was tested experimentally under conditions of uniaxial and complex stress state as applied to samples and parts of gas turbine plants made of isotropic heat-resistant steels and alloys, the damage of which is associated with low-cycle and thermal fatigue (see Chapt. 3). The quantitative result of determining the durability of parts according to the criterion under consideration depends on the choice and accuracy of determining its deformation parameters. As applied to isotropic materials, is usually taken as such parameters, respectively, the intensities of the deformation ranges (3.137) and (3.138).

The possibility of generalizing the deformation criterion (4.9) with respect to single-crystal alloys was investigated. We used the above-considered test results of five series of samples from a single-crystal alloy of different orientations:

Number of series	Orientation	Eulerian angle α	β	γ
1	111	35°4'	37°10'	94°4'
2	011	108°40'	11°15'	156°4'
3	011	75°14'	14°26'	-113°11'
4	011	-174°50'	37°9'	-179°17'
5	001	-95°19'	41°40'	179°48'

The tests and calculations of the stress-strain state for the investigated single-crystal alloy in comparison with the calculated crystallographic directions of the cracks made it possible to separate the conditions for the propagation of crystallographic cracks (along the slip lines) from the conditions for the propagation of non-crystallographic cracks (in the directions of action of the maximum tensile stresses), see Fig. 4.48.

Thus, in relation to the test results of a single-crystal alloy with different crystallographic orientations, a modification of the criterion (4.9) was proposed, which consists in the fact that instead of the intensity of deformations as the determining parameter of the criterion, the main deformations are used here:

$$\begin{aligned} &\text{Criterion1} - \Pi_1(\Delta\gamma_{pl}) + \Pi_2(\Delta\gamma_c) + \Pi_3(\gamma_{pl}) + \Pi_4(\gamma_c) = 1 - \text{shear},\\ &\text{Criterion2} - \Pi_1(\Delta\varepsilon_{pl}) + \Pi_2(\Delta\varepsilon_c) + \Pi_3(\varepsilon_{pl}) + \Pi_4(\varepsilon_c) = 1 - \text{separation}. \end{aligned} \tag{4.10}$$

The choice of the criterion formulation is carried out on the basis of the experimentally obtained diagrams of the change in the destruction modes when the parameters of the cycle T_{max} and ΔT (Fig. 4.48): crystallographic (in this case, the parameter of the criterion is the maximum shear deformation) and non-crystallographic (in this case, the parameter of the criterion is the maximum deformation upon fracture by separation).

The verification of the criterion for a single-crystal alloy was carried out using the results of tests by R.N. Sizova and E.R. Golubovsky (CIAM) to assess the damage values Π_3 and Π_4. The dependencies (3.92) and (3.93) were used. To determine the parameters of damage Π_1 and Π_2 the following formulas were used:

$$C_1 = (0,5\varepsilon_r)^2, \qquad C_2 = (0,75\varepsilon_{cr})^{1,25}, \tag{4.11}$$

the full deformation range was found from the experiment

$$\Delta e = (\alpha_1 T_{max} - \alpha_2 T_{min})\varphi, \qquad \varphi = 1 - \Delta k/\Delta l.$$

The range of elastic deformation was determined by the formula

$$\Delta\varepsilon_{el} = 2\sigma_{-1}/E \tag{4.12}$$

and the range of plastic deformation $\Delta\varepsilon_{pl} = \Delta e - \Delta\varepsilon_{el}$.

The results of determining the damage of some samples of a single-crystal alloy under thermal cycling loading are shown in Table 4.33. It can be seen from the table that, except for the specimen with orientation 001, tested in the 150-900°C mode, the total damage of all specimens differs from 1.0 by no more than 2 times.

Table 4.33 Results of determination of damage to a single-crystal alloy under conditions of non-crystallographic fracture (separation)

Orientation	T_{min}, °C	T_{max}, °C	t_c, s	Cycles before the formation of a trunk crack, N_m	$\Delta\varepsilon$, %	$2\sigma_{-1}/E$	$\Delta\varepsilon_{el}$, %	$\Delta\varepsilon_{pl,c}$, %	Π_1	Π_2	Π_3	Π_4	$\sum\Pi_i$
[111]	150	900	72	190	0.73	$860/2.4 \cdot 10^5$	0.36	0.37	0.27	0	0.22	0	0.59
			247	50	0.69	$860/2.4 \cdot 10^5$	0.36	0.33	0	0.59	0	0.35	0.94
			378	12	0.73	$860/2.4 \cdot 10^5$	0.36	0.37	0	0.18	0	0.52	0.7
	500	1000	149	80	0.64	$648/2.3 \cdot 10^5$	0.28	0.36	0	0.6	0	0.4	1.0
[001]	150	900	72	500	0.86	$700/9.98 \cdot 10^4$	0.7	0.16	0.01	0	0.1	0	0.11
	250	1000	48	40	0.855	$500/0.95 \cdot 10^5$	0.52	0.335	0.18	0	0.25	0	0.43
	350	1000	28	1400	0.655	$500/0.95 \cdot 10^5$	0.52	0.135	1.13	0	0.25	0	1.38

Recently, calculations of the thermal cyclic durability before the initiation of macrocracks in corset samples from single-crystal alloys have been carried out using the deformation criterion (4.9), see Fig. 4.49. It is obvious that the calculated values of durability differ from the experimental ones by no more than 5 times.

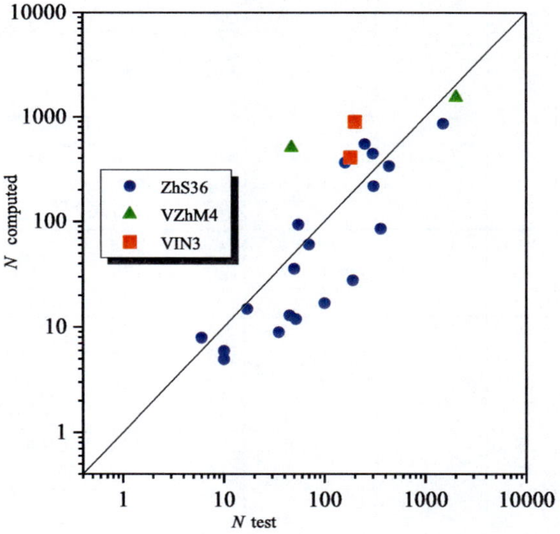

Fig. 4.49 Comparison of the calculated number of cycles before the formation of a main crack using the author's four-component criterion (see Sect. 3.4 and Semenov and Getsov, 2014b) and experimental results

4.2.10 Crystallographic Features of High-cycle Fatigue Fracture of Single Crystals

Table 4.34 shows the average values of the fatigue limits of single-crystal alloys ZhS6F, ZhS32, ZhS36 and ZhS40 with orientations 001 and 111 (Shalin et al., 1997) under pure bending conditions at various temperatures. It is obvious that all alloys exhibit an anomaly in the temperature dependence of the fatigue limit and its sensitivity to stress concentration. This concentration sensitivity is higher for single-crystal alloys than for polycrystalline alloys. In carbon-free alloys, the fatigue limits are higher because they do not contain carbides in which microcracks arise. In Pridorozhny and Sheremetyev (2005), the influence of orientation on the fatigue strength of cooled and uncooled single-crystal blades was investigated (see Figs. 4.50-4.52 showing the conservation factor[3] $K = A[001]/A[hkl]$).

Table 4.34 Fatigue limits of single crystals from various alloys with [001] and [111] orientations at different temperatures

Alloy	Orientation	Sample shape	Fatigue limits σ_{-1} (MPa) at temperature			
			20°C	900°C	1000°C	1100°C
ZhS6F	[001]	smooth	200	340	230	-
		notched K_σ	80	260	180	-
		K_σ	2.5	1.31	1.23	-
	[011]	smooth	240	350	-	-
		notched K_σ	140	220	-	-
		K_σ	1.7	1.6	-	-
	[012]	smooth	220	380	-	-
		notched K_σ	-	-	-	-
		K_σ	-	-	-	-
	[111]	smooth	240	410	320	-
		notched K_σ	120	220	180	-
		K_σ	2.0	1.86	1.78	-
ZhS32	[001]	smooth	340	350	270	-
		notched K_σ	120	220	180	-
		K_σ	3.1	1.35	-	-
ZhS36	[001]	smooth	300	350	270	17
		notched K_σ	190	280	-	16
		K_σ	1.57	1.25	-	-
	[111]	smooth	360	430	-	17
ZhS40	[001]	smooth	320	380	160	-
		notched K_σ	210	280	-	-
		K_σ	1.52	1.35	-	-

[3] The blade is designed according to the passport characteristics for orientation 001. Since they always get with deviations up to ±10 degrees from the axial orientation and it is not known how much from the azimuthal orientation, then it is necessary to know how to take this into account. In Pridorozhny and Sheremetyev (2005), such an attempt was made in relation to fatigue strength, and in Golubovskiy et al. (2005a), in relation to long-term strength.

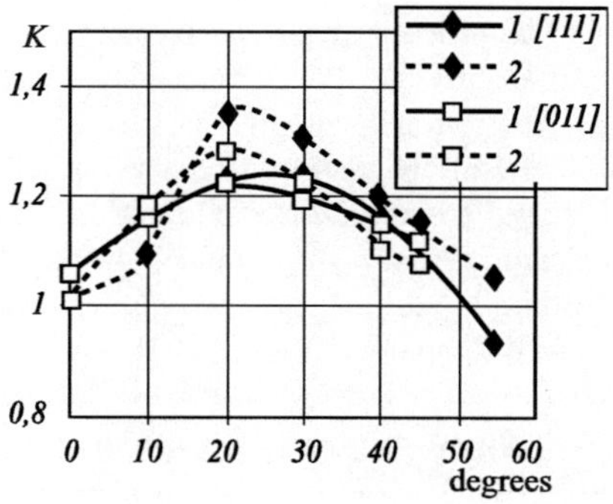

Fig. 4.50 Change in the value of the conversion factor depending on the axial orientation of the uncooled blade

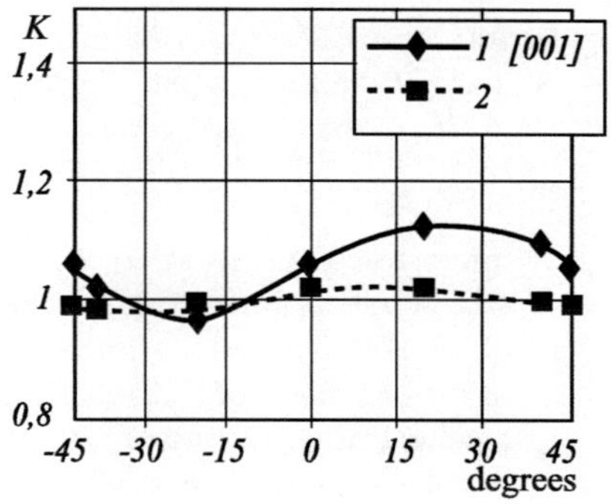

Fig. 4.51 Change in the value of the conversion factor depending on the azimuthal orientation of the uncooled blade. *1* - conversion factor for maximum reduced shear stresses, *2* - conversion factor for maximum shear stresses

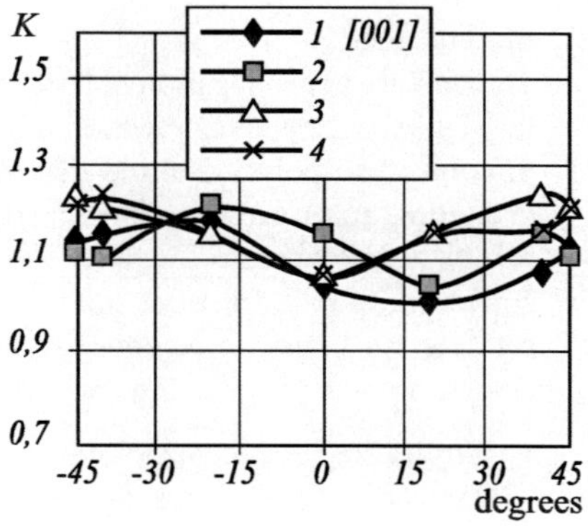

Fig. 4.52 Change in the value of the conversion factor for the maximum reduced shear stresses depending on the azimuthal orientation of the cooled blade. *1* - entrance edge of the root section, *2* - ribs in the inner channel from the side of the entrance edge of the root section, *3* - slotted hole on the trailing edge of the root section, *4* - partitions between slotted holes in the internal channel

The critical shear stress τ_{given} along the slip plane, reduced to the slip direction:

$$\tau_{\text{given}} = \sum_i \sum_j \sigma_{ij} F_{ij}, \tag{4.13}$$

where σ_{ij} is the stress tensor component, F_{ij} is the Schmid factor corresponding to the given component of the stress tensor in the slip system.

The process of fatigue fracture of single-crystal materials, as well as polycrystalline ones, has a pronounced stage character. The stages differ in their duration and destruction mechanisms. As a rule, two main stages of the fatigue fracture process in general are distinguished: the stage of nucleation and the stage of propagation of fatigue macrocracks. The ratio of the number of cycles during nucleation to the number of cycles during crack propagation varies widely depending on the material, loading conditions, and geometry of the object under study.

When analyzing each of the two indicated periods in the process of fracture, characteristic stages can also be distinguished. At the stage of crack propagation, the problem arises of determining the rate of crack growth, as well as determining the direction of crack propagation. Depending on the direction of growth, the following are distinguished: **I stage** (crystallographic growth, shear crack stage); **II stage** (subcritical growth, stage of normal separation crack).

At stage II, various stages of growth are also distinguished, usually due to the peculiarities of the microstructure of the material under consideration:

1. Short cracks

   a. microstructural - short cracks (the length of which is commensurate with the size of the microstructural components of the material);
   b. physically small cracks (the size of which is comparable to the size of the zone of the beginning of the fracture process);
   c. mechanically short cracks (the length of which is comparable to the size of the plastic zone, for example, small cracks that have grown in the plastic zone from the concentrator).

2. Macrocracks
   The process of propagation of fatigue cracks in single-crystal bodies demonstrates the stages listed above, however, it has a number of specific features, in particular, in contrast to polycrystalline materials, a significant proportion of the stage of crystallographic growth is observed. To consider these features, let us turn to the results of fractographic analysis carried out for single-crystal nickel alloys. on the surface of fatigue fractures obtained in tests by the method of pure bending of a rotating specimen, three zones can be distinguished, corresponding to different stages of crack propagation (Shalin et al., 1997):

   - **Crystallographic zone I** corresponds to the stage of formation of stable slip bands along the planes, along which cracks grow.
   - **Noncrystallographic zone II** corresponds to the stage of propagation of the main crack, most often perpendicular to the main normal stresses.
   - Finally, in **zone III**, a one static separation occurs.

The propagation of fatigue cracks in single crystals usually occurs along the planes of the {111} octahedron in the [011] direction. Let us consider the crystallographic features of fatigue fractures of single crystals of various orientations in tests by the method of pure bending with rotation, described in Shalin et al. (1997).

The crystallographic zone of fatigue fracture of the samples has a smooth shear appearance and can occupy up to 80% of the surface. As the stress level increases, the fraction of the area occupied by the crystallographic zone decreases. The propagation of fatigue cracks in zone I occurs along the planes of the {111} octahedron in the [011] direction, which coincides with the usual slip system in single crystals of heat-resistant alloys. On the octahedral cleavage planes, a characteristic "brook" pattern and stripes are formed that coincide with the direction of crack propagation [011].

The rivulet pattern represents micro-fractures that arise when a fatigue crack propagates along parallel planes at different levels. The stripes coinciding with the directions of <011> correspond to the boundaries of the crack propagation regions along the nonparallel octahedral planes. In most cases, the centers of fatigue fracture are located in the locations of carbides or micropores on the surface of the sample or under its surface, determined from the crystallographic point of view. From Table 4.35 (Shalin et al., 1997) it follows that in single crystals of the two investigated orientations [001] and [111], an orientation dependence of the occurrence of fatigue fracture centers is observed.

At room temperature, in single crystals with **orientation [001]**, a fatigue crack propagates in an octahedral plane inclined to the sample axis. The fracture contour acquires the shape of an ellipse, and the fracture site is located at the intersection of the major axis of the ellipse with the lateral surface of the sample. In this case, two cases may occur. The major axis of the ellipse coincides with the crystallographic direction [112], which lies in the octahedral plane. If the fracture source is oriented relative to the directions lying in the azimuthal plane of the specimen, then in the case of crack propagation along one plane {111} the source will be located at the intersection of the [010] direction with the lateral surface. There will be four such intersections, which corresponds to the maximum number of possible octahedral fracture planes in single crystals with the [001] orientation. If the fracture source occurs at the exit point of the [010] direction, then the crack propagates along two of the four possible planes of the {111} octahedron, if the source arises at the intersection of the [110] with a lateral surface, then the crack propagates along one plane of the octahedron (Fig. 4.53).

Table 4.35 Crystallographic elements on the surface of fatigue fracture of single crystals of ZhS32 and ZhS6F alloys

Test temperature, °C	Axial orientation of the sample	Azimuthal orientation of the source of fracture	Observed number of planes in a fracture	Fracture planes indices
20	[001]	[010]	2	{111}, {001}
		[110]	1	-
	[111]	[112]	2...3	{111}, {001}
900	[001]	[110]	1...2	{111}, {001}
	[111]	[112]	1...2	{111}
1000	[001]	[001]	1	{001}
	[111]	-	3	{001}

In single crystals with **orientation [110]**, fatigue cracks can nucleate at two exit points of the [001] directions, and, accordingly, fracture occurs along two octahedral planes, since two other octahedral planes are parallel to the sample axis. In single crystals with the **orientation [111]**, the centers of crack initiation are located mainly at the points where the [112] directions exit onto the surface of the working part of the sample. Fatigue failure occurs along 1. . . 3 octahedron planes inclined to the sample axis. Fatigue failure along the (111) octahedron plane perpendicular to the sample axis is not observed, since the Schmid factor is equal to zero, therefore, there is no plastic deformation preceding fatigue failure.

In single crystals with the [001] orientation, the crack in zone II grows in certain crystallographic planes perpendicular to the axis of the applied stress. It can be either the (001) plane, or the surface formed by the intersecting planes of the octahedron in the form of a "gable" roof (see Fig. 4.53). The lines of intersection of the planes of the octahedron correspond to the [110] direction. Such fracture is typical for specimens with the [001] orientation with a notch, where four centers of fracture appear, corresponding to the exit of the <011> directions onto the specimen surface. As a result, characteristic figures in the form of a cross appear on the surface of destruction. Study of the fracture structure at high magnifications made it possible to reveal fatigue grooves, i.e. crack propagation front, in direction [112]. Based on these data, it can be concluded that the crack simultaneously grows in two directions <112>, which is equivalent to the macroscopic direction <100>.

The fatigue failure of single crystals at **elevated temperatures** has some features associated mainly with the oxidation of the lateral surface. The higher the test temperature and the less heat-resistant the alloy, the more oxidation affects the fatigue strength. As for the structure of fractures at high temperatures, the above-mentioned regularities remain. We only note that for single crystals with orientations <001> and <111> at elevated temperatures, the tendency to fracture along cubic planes {001} increases, and the fraction of the crystallographic break decreases. Accordingly, the zone of a noncrystallographic fracture perpendicular to the sample axis increases; at this stage, the crack grows along the dendritic sections. If high temperature fatigue

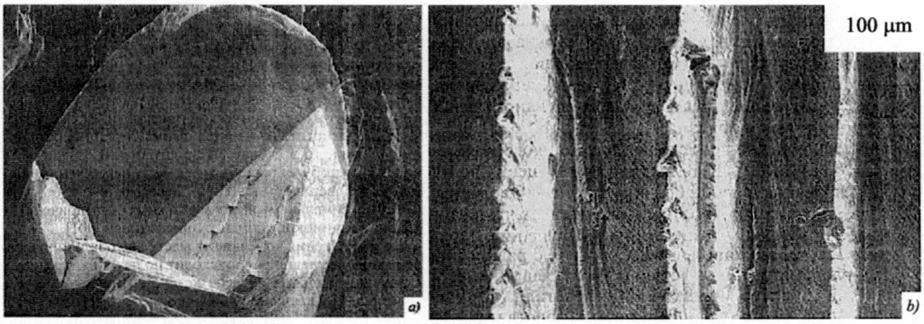

Fig. 4.53 Damaged surface of a single-crystal alloy after high-cycle fatigue test: *a)* - predominant propagation of the fatigue crack along the plane of the octahedron, *b)* - relief in the form of a "gable roof" (Shalin et al., 1997)

tests are carried out on coated specimens, fracture is initiated by cracks in the coating that penetrate deep into the specimen along the interdendritic spaces.

Fatigue fractures **notched specimens** are characterized by multi-site crack initiation, which propagate along the same crystallographic planes as in smooth specimens. For example, in single crystals with the [001] orientation, the fracture consists of four sectors, each of which has its own nucleation site at the intersection of the [001] direction with the lateral surface of the sample. First, the crack grows from the source along one plane of the octahedron, and then passes into two others, intersecting in the form of gable roofs. Fatigue fractures of single-crystal specimens with orientation [111] with a notch in shape resemble a trihedral pyramid, the edges of which consist of steps in the form of gable roofs.

A map of fatigue growth mechanisms was proposed for <001>/<010> PWA1484 single crystals, see Fig. 4.54, obtained on the basis of consideration of thermally activated sliding processes in the plastic zone in the vicinity of the crack tip

$$f\Delta K_e = A' e^{-\frac{Q}{RT}}. \tag{4.14}$$

The transition from the crystallographic growth mode at high temperatures is also facilitated by a decrease in the anisotropy of the elastic and strength characteristics of crystals. Thus, at high temperatures (~ 980°C), the growth rate of fatigue cracks is practically independent (Shalin et al., 1997) on the orientation of single crystals under conditions of uniaxial and complex stress states. Probably, this circumstance explains the absence of anisotropy of high-temperature high-cycle fatigue.

It should be noted that the presence of a dendritic structure also affects the change in the orientation of fatigue crack growth and can cause a transition from one crystallographic plane to another. The experimentally observed alignment in

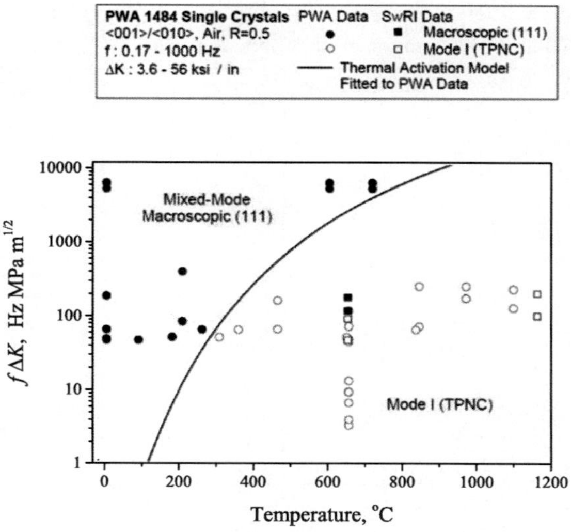

Fig. 4.54 Fatigue crack growth morphology map for PWA 1484 single crystals in air (Telesman and Ghosn, 1996) shows two fracture regimes: mixed-mode macroscopic (111) fracture, and Mode I transprecipitate non-crystallographic (TPNC) fracture. The solid line is the fatigue crack growth morphology transition boundary calculated based on the proposed thermal activation model. Mixed-mode macroscopic (111) fracture is favored at high frequency, high ΔK and low temperature, while TPNC fracture is favored at low frequency, low ΔK, and high temperature.

crack growth rates for single crystals with orientations [111] and [001] is explained by Dul'nev et al. (1988a); Shalin et al. (1997) that as ΔK grows, the fatigue crack path deviates from the (111) plane, perpendicular to the load axis, and transition to the (001) plane, where the interdendritic regions are concentrated, i.e. into the same plane along which the fatigue crack propagates in specimens with orientation [001], see Fig. 4.55.

4.2.11 Oxidation Resistance

Monocrystalline heat-resistant nickel alloys with a low chromium content are oxidized noticeably more intensively than wrought and cast polycrystalline alloys containing 10% Cr or more. This can be illustrated by the following experimental data obtained by V.I. Nikitin (Table 4.36). In this regard, single-crystal alloys are not used for GTU blades without protective coatings. Detailed studies of the corrosion processes of alloys are given in Getsov et al. (2019)

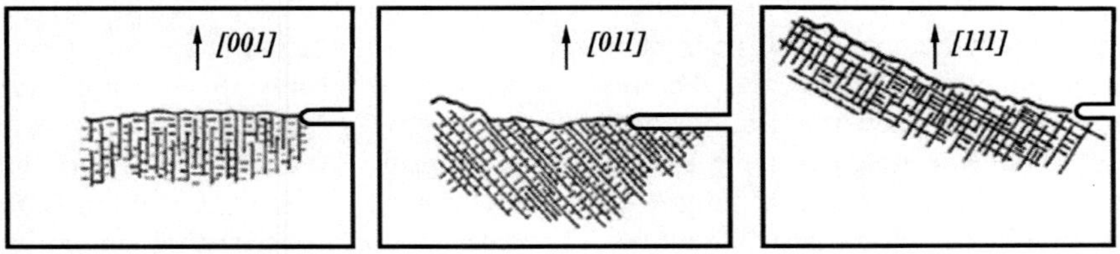

Fig. 4.55 Fracture profiles of specimens of different axial orientations in comparison with their dendritic structure (Shalin et al., 1997)

Table 4.36 Corrosion depth of various alloys

Alloy	Temperature, °C	Corrosion depth (mm) during time (h)	
		1000	2000
ChN60WT	900	0.004	-
	950	0.010	-
EI893(L)	900	0.005	-
	950	0.008	-
EI929	900	0.004	-
	950	0.006	-
ZhS6K	900	0.006	-
	950	0.010	-
ZhS36 mono	900	0.010	0.029
	950	0.044	0.286

4.3 Relationship of Structure and Properties

4.3.1 Influence of Heat Treatment Regime

Heat treatment of heat-resistant alloys is used at various stages of fabrication of blanks and parts, as well as to restore the structure and properties of metal parts during operation. If the modes of heat treatment of workpieces are regulated by the developers of the alloy (they ensure the obtaining of properties guaranteed by the technical conditions and the passport for the alloy), then the modes of recovery heat treatment after coating deposition, modes of heat treatment after one or another hardening treatment, surfacing, welding are usually regulated by the developer of the corresponding process. When determining these modes, the scope of tests confirming the adequacy of the structure and properties after the selected mode is usually much less than the scope of tests performed when developing the mode according to technical specifications. In this regard, the probability of deviations from one or another of the characteristics of the alloy from the passport in these conditions increases sharply.

When setting the modes of heat treatment of new alloys, they are usually guided by the modes adopted for similar materials, sometimes supplementing them with experiments to refine only some parameters. The improvement of the modes of technological processing is carried out practically after the introduction of alloys. In this case, the obtained characteristics of heat resistance, structural stability are usually compared and the features of the structural state are studied. A large number of publications are devoted to a detailed presentation of these questions (see Borzdyka and Zeitlin, 1964; Levin et al., 1975, 1977; Pigrova, 1980; Getsov et al., 1999; Manilova, 2005, among others).

Let us consider some aspects of the influence of heat treatment modes on the structure and properties of austenitic steels and nickel-based alloys: the effect of the hardening temperature and cooling rate, temperature and aging time, double and step hardening. An increase in the quenching temperature and holding time leads to a more complete homogenization of the structure and composition), to the intensification of the processes of recrystallization and grain growth. From Table 4.37 it can be seen that an increase in temperature and holding time noticeably reduces the value of the segregation coefficient to almost 1.0.

As a result, hardness of austenitic steels decreases, long-term strength and creep resistance increase, and ductility decreases. As a very typical example, we can cite the value of the long-term strength of steel 20Ch23N18 at 800°C: after quenching from 930°C σ_{100} = 35 MPa, and after quenching from 1180°C σ_{100} = 70 MPa. A similar picture is observed for nickel-based alloys. However, these influences take place only at temperatures that do not cause the formation of low-melting eutectics. The choice of the optimal temperature regimes for hardening is based on the need to obtain sufficiently high values of both long-term strength and long-term plasticity.

For nickel-based alloys, long-term ductility significantly depends on the cooling rate after homogenization at temperatures of 1100-1200°C, and this effect is

Table 4.37 Influence of the heat treatment regime on the level of dendritic segregation of the modified ZhS6U DC alloy (Paton et al., 1987)

Mode		Coefficient of segregation			
T,°C	t, h	Ti	Nb	W	Hf
1100	50	1.1	1.2	0.83	1.9
1100	60	1.1	1.1	0.9	1.5
1100	80	1.03	1.05	0.93	1.2
1230	3	1.2	1.2	0.72	1.3
1240	3	1.1	1.1	0.72	1.4
1250	3	1.04	1.1	0.9	1.3
1250	6	1.04	1.05	0.9	1.1

Liquation coefficient: the ratio of the concentration of an element in the interdendritic region to the concentration near the axes.

different for alloys containing different amounts of γ'-phase. For nickel-based alloys containing more than 30% γ'-phases (EI929, EP220, ZhS6KP, ZhS6K, etc.), slow cooling leads to an increase in ductility; for materials with less γ-phase, rapid cooling is preferable (Rybnikov, 1979). The cooling rate significantly affects the grain-boundary structure of alloys: with a decrease in the cooling rate, elongated particles are precipitated instead of spherical precipitates of the Me_6C phase, and the filling factor of the boundaries decreases.

When choosing the aging temperature, it is usually assumed that this temperature should be at least 50-100°C higher than the operating temperature of the material. In this case, one can hope that significant changes in the structure will not take place at the initial stages of work. The choice of the optimum aging temperature for the alloy is a multi-parameter task; it depends on the hardening temperature, cooling rate, and in some cases on the melting characteristics of the metal. In this regard, of interest are the results (Rybnikov, 1979), which showed that heat-resistant alloys of the EI893, EI698, EI929 types acquire high values of plasticity and impact toughness at room temperature only in cases when the size of intergranular particles does not exceed 0.3 μm, regardless of the size of intragranular allocations γ'-phases. At the same time, under conditions of prolonged high-temperature loading, the plasticity increases with an increase in the size of intergranular precipitates.

In case of double or step quenching of heat-resistant nickel-based alloys, the second quenching (usually at 1000-1050°C) provides the release of coagulated carbide particles (usually Me_6C), which after aging are surrounded by particles of the γ-phase, which has a positive effect on the properties under high-temperature deformation (Rybnikov, 1979). It should be noted that at the temperature of the second quenching, the γ'-phase also precipitates in the preferred nucleation centers, which are grain boundaries. The rate of the carbide formation process depends on the carbon content.

When choosing the second hardening temperature, it should be borne in mind that the filling factor of the boundaries with carbides at temperatures corresponding to the average solubility temperature of the γ'-phase decreases, and the filling factor of γ' decreases with the phase increases. Special attention should be paid to the influence of the jagged boundaries formed in the process of thermomechanical processing. It was found that such a structure leads to an increase in heat resistance, in particular, associated with inhibition of intergranular slip processes.

It should be noted that, depending on the method of crystallization of cast heat-resistant alloys, various types of heat treatment are used. So, the ZhS6K alloy with an equiaxed structure is sometimes used in the cast state - no heat treatment is carried out; its structure is formed in the process of casting cooling. To reduce the degree of segregation, the cast directional solidification alloys are homogenized. It promotes more complete dissolution of rough precipitates of the γ'-phase, formed in the process of directed crystallization. The result is a slight increase in long-term strength.

Depending on the level of the acting stresses and the operating temperature, the heat treatment mode of some alloys may change. These changes primarily relate to the aging or vacation regimen. A special type of heat treatment is reduction heat treatment, which is used to restore structure and properties after applying a high-temperature coating to an alloy part or after prolonged operation. Section 4.3.3 is devoted to its consideration.

4.3.2 Effects of Long-term Aging

During the operation of parts made of heat-resistant and heat-resistant steels and alloys, changes in their structure can occur, causing a change in properties and the formation of structural deformations. Experimental and computational methods are used to study issues related to changes in the structural state of a material during long-term operation.

Metal intended for long-term operation is subjected to long-term aging in laboratory conditions at various temperatures (with advanced operating time) and examined using optical and electron microscopy, phase and X-ray structural analysis of changes in structure in comparison with the initial state (after technological heat treatment). At the same time, the preference of one or another mode of technological heat treatment is also established.

The characteristic structural transformations in materials of different classes, observed during long-term operation at elevated temperatures and leading to noticeable changes in properties, are presented in the next subsubsecctions based on Getsov et al. (1965, 1999, 2000); Pimenova (1973); Levin et al. (1977); Borzdyka (1990); Rtishchev (1991); Protasova (1993); Manilova (2005); Pigrova (2009).

4.3.2.1 Carbide Transformations

We will follow the carbide transformations that occur in materials of different classes during their long-term operation using the following examples. In Cr-Mo-V alloyed *perlitic steels* with an increase in the operating time, one can observe the direction of carbide reactions from the simplest types of carbides to complex alloyed types $Me_{23}C_6$ and Me_6C: $Me_3C \Rightarrow Me_2C + Me_7C_3 + MeC \Rightarrow Me_{23}C_6 + Me_6C$ (Pigrova, 2009).

During long-term operation at 500°C *high-chromium steel* EP428 (equivalent to 20Ch12WNMF), fine particles of carbo-nitride phases M_2X, MX are formed, which undergo phase-structural transformations according to the reaction (Manilova, 2005): $M_2X \rightarrow M_2X + MX + M_{23}C_6$, where MX and $M_{23}C_6$ are secondary phases. After operating the blades made of EP428 steel for 100000 h at 500°C, optical metallography methods practically did not reveal any noticeable structural changes. The average area of particles in the area of the feather was 11% larger than the average area of particles in the blade joint operating at ~ 400°C. The density of particles and their volume fraction in the metal of the feather was 12% less than in the metal of the blade lock. Some differences were also observed in the average size of carbides at the grain boundaries. In the area of the blade airfoil, there was a slight increase in the size of carbides $(Cr, Fe)_{23}C_6$ along the grain boundaries (by 12.7 %) compared to the size of similar carbides in area of the castle. Thus, after long-term operation of the rotor blades for 100856 h, some coagulation of carbides $(Cr, Fe)_{23}C_6$ along the grain boundaries and in the grain itself was revealed.

The analysis made it possible to establish the sensitivity to the duration of operation of the values of the Cr/V and V/N ratios in the carbo-nitride phases Me_2X, and Cr/Fe in the carbide $(Cr, Fe)_{23}C_6$. The dependence of ω = Cr/Fe on the operating time t can be described by the expression

$$\omega = \omega_0(1 + \alpha t), \tag{4.15}$$

where at 500°C $\omega_0 = 3{,}0$ and $\alpha = 4 \cdot 10^{-6}$.

In the EI481 *austenitic steel* used for gas turbine engine disks, during long-term operation at 600 and especially 650°C, the number of carbide phases increases and their coagulation occurs. The development of similar processes was observed in the *austenitic steel* EI572 (31Ch19N9MWBT), also used for gas turbine disks. It was found that in this steel over time of operation at 600-650°C large particles of σ - needle-shaped phases are formed, which causes a sharp decrease in impact strength and long-term plasticity (Figs. 4.56 and 4.57).

In the blades made of ZhS6KP alloy according to G.P. Okatova (nee Pimenova) during long-term operation, needle-like particles of double carbides $Me_6^m Me_6^n C$ (Fig. 4.58). In the cast *high heat-resistant alloy* ZhS6K at exposure in the range of 1100-1180°C, the formation of a needle-like modification of Me_6C carbide is also observed.

During long-term operation of blades made of EI893 alloy, small changes in the amount of carbide MC and boride M_5B_3 phases were observed, which, nevertheless,

Fig. 4.56 Change in impact toughness of steel EI572 depending on temperature and aging duration: • - 600°C, ○ - 650°C, □ - 750°C

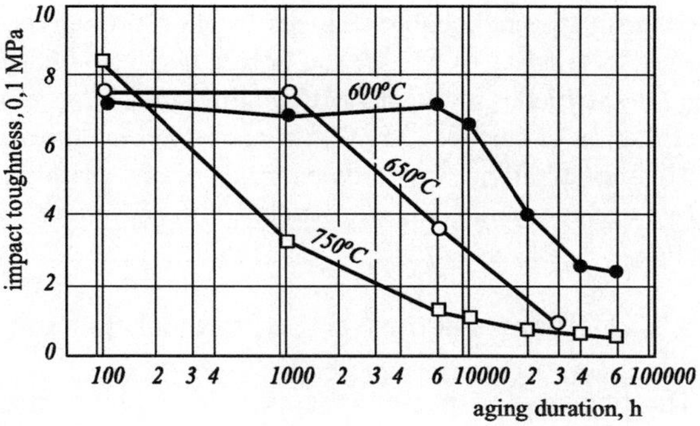

Fig. 4.57 Change in ductility during long-term strength testing of steel EI572 at temperatures of 600 and 650°C: ▲ - tangential and radial samples, • - axial samples, + - aging 700°C, 50 h, × - aging 750°C, 12 h + 800°C, 15 h

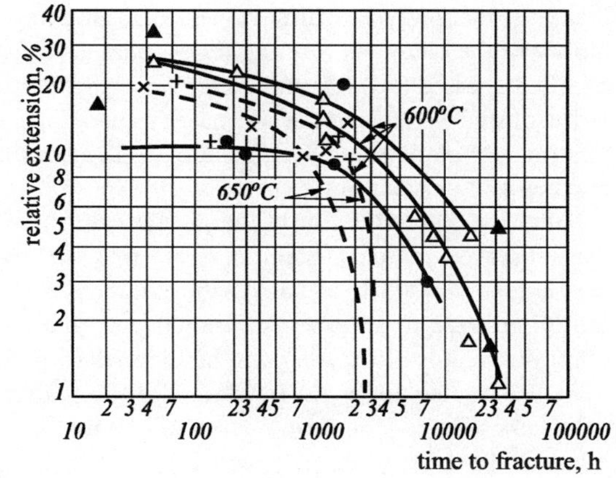

Fig. 4.58 Microstructure of ZhS6KP alloy after standard heat treatment and subsequent long-term exposure to limiting operating temperatures (x500)

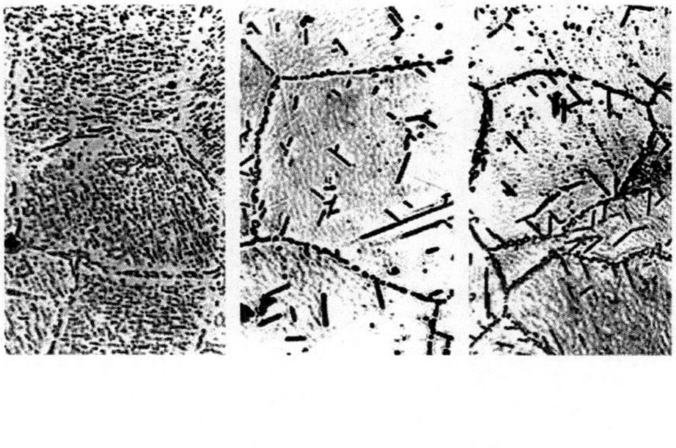

cannot cause noticeable changes in alloy properties. The flow of carbide (MC → $M_{23}C_6$ → M_6C → $M_{12}C$) and boride (M_5B_3 → M_3B_2) reactions.

In the blades made of ZMI3U alloy, after long-term operation, an additional precipitation of carbides $M_{23}C_6$ was observed in comparison with the initial structure. The precipitation of carbide particles occurs mainly along grain boundaries around MC carbides and primary particles of the γ'-phase.

4.3.2.2 Changes in the Number, Composition and Size of Particles of γ'-phase

The amount of γ'-phase in relatively low-alloy deformable nickel-based alloys (containing up to 10-20 % γ'-phases) at long-term aging at temperatures of 500-600°C increases markedly, reaching values corresponding to the values of the limiting solubility of the elements forming the γ'-phase. Particles of the precipitated phase are highly dispersed (≈ 0.03 μm). At higher aging temperatures, the processes of coagulation of the particles of the γ'-phase formed during the previous technological heat treatment, mainly occur. Thus, in the EI698 alloy the particle size after 3000 h aging reaches a size of 0.3 μm.

Additional precipitation of the γ'-phase during prolonged aging at 600 and 650°C was observed in the austenitic steel EI787 (ChN35WTYu). For example, at an exposure of 30000 h at 650°C the amount of γ'-phases increased from 8 % in the initial state to 15.9 %. Additional precipitates of the intermetallic phase were observed mainly along the grain boundaries (Fig. 4.59).

It was carried out by the method of phase analysis study of changes in the amount of γ'- phase after long-term operation in blades of different gas turbines made from alloy EI893 (see Table 4.38 and Fig. 4.60). To determine the features of phase transformations in the alloy, studies of the structure of the alloy were carried out after long holding of the samples in the range of 550-750°C. It was found that at temperatures below 650°C for 10000 h, equilibrium in the separation of the γ'-phase is not achieved. At the same time, it is enriched with tungsten and molybdenum. Long-term aging in the range of 700-750°C practically does not cause changes in

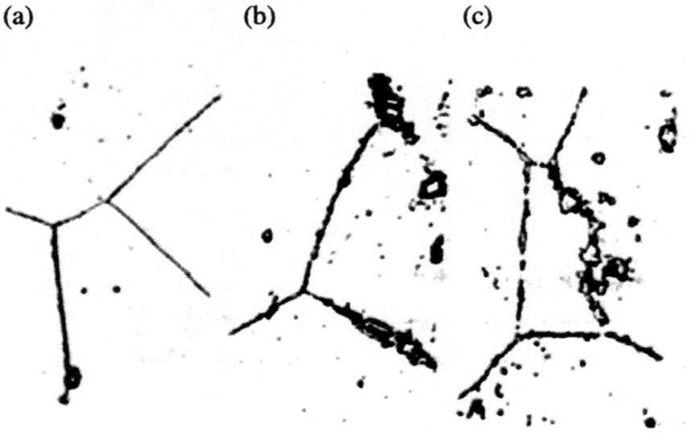

Fig. 4.59 Microstructure of steel EI787 in the initial state (a) and after aging according to the modes: at 650°C, 10000 h (b) and at 650°C, 30000 h (c)

Table 4.38 Change in the amount of γ'-phase during long-term operation of blades made of EI893 alloy

GTE	Lifetime	Amount of γ'-phases in the initial state	Amount of γ'-phases in the 1st stage blade	Amount of γ'-phases in the 2nd stage blade	Relative change of the amount of the γ'-phase	
					1st stage	2nd stage
GT-750-6	22430	9.5	14	-	1.47	
	30165	9	16.6	12	1.84	1.33
GTK-10	11300	11.4	17	-	1.49	
	20500	9,4	15.9	-	1.69	
GTN-9	65000	11.4	16.5	15.8	1.45	

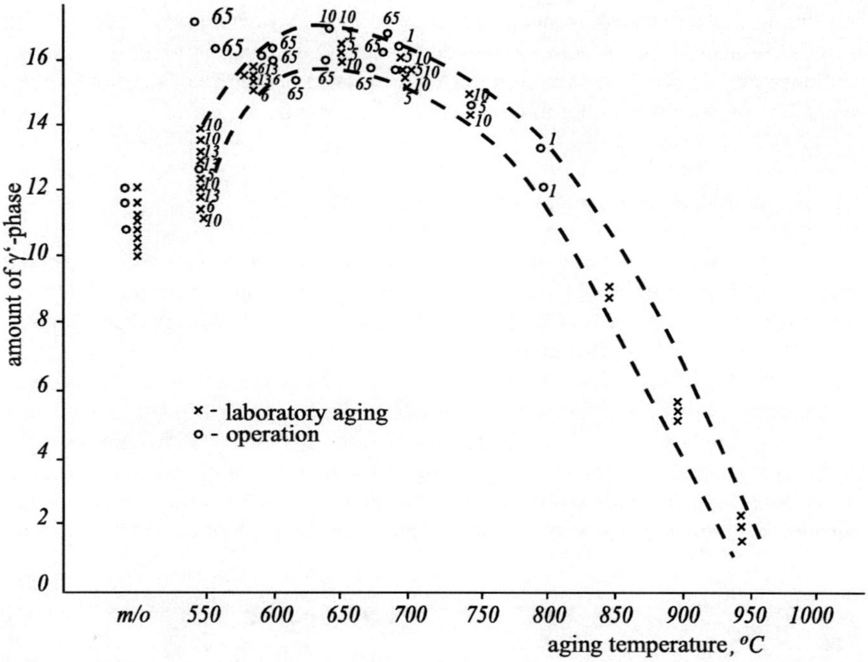

Fig. 4.60 Influence of temperature on the amount of γ'-phase in alloy EI893 after long-term operation (Pigrova, 2006)

the amount of γ'- the phase precipitated at the initial stages of aging after regular heat treatment.

During long-term operation of rotor blades made of alloy ZMI-3, additional release of the γ'-phase (instead of 41-42 % in the initial state, 43-45 % after operation for 16000 and 22000 h) and some coagulation of particles of the γ'- phase was observed. In the course of long-term operation of blades made of alloy EI929, changes in the particle size of the γ'-phase from the temperature and time of alloy operation are observed (Figs. 4.61 and 4.62). The obtained temperature dependencies are non-monotonic. A structural change after prolonged aging was also observed for the heat-resistant alloys EP220 and ZhS6U (Fig. 4.63).

4.3.2.3 Formation of Intermetallic Phases like Ni_3Ti, R-Phases and Laves Phases

During long-term aging at 300-650°C up to 5000 h of maraging steel 0Ch11N10M2T, the formation of various intermetallic phases was observed in its structure: Ni_3Ti type, R-phases and Laves phases (Pigrova, 2009). At temperatures of 400-450°C, the process of matrix separation with the separation of zones of a chromium-based solid solution was also revealed. In steels of the 20Ch13 type with a relatively high carbon content, these processes develop more slowly, because inhibited by carbide reactions. The questions of structural changes during long-term operation of high-chromium steels are considered in detail in Lanskaya (1976); Manilova (2005).

4.3.2.4 Formation of Topologically Close-packed Phases

In high-chromium martensitic steels, during long-term aging, carbide reactions occur with the formation of $Me_{23}C_6$ carbide and the precipitation of intermetallic compounds, a typical representative of which is the TCP Laves phase (Fe_2Me). Thus, it was found in the EI802 alloy after aging 150000 h at 600°C.

In nickel-based alloys, the TCP phases (σ- and μ-phases) are solid solutions based on the known binary and ternary compounds of the Ni-Co-Cr-Mo-W composition. The study of the composition of the phases showed that the σ- and μ-phases differ in the concentration of the elements of the VI group: the σ-phase is enriched in chromium, μ - in the phase with molybdenum and tungsten. These phases are formed during long-term aging at 800 – 900°C. It has been established that TCP-phases of

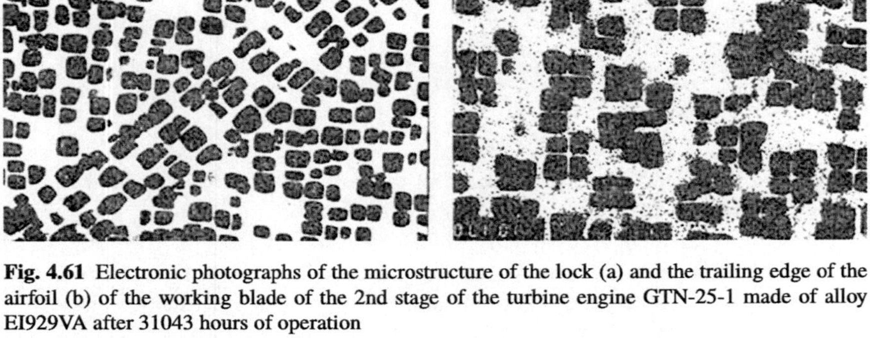

Fig. 4.61 Electronic photographs of the microstructure of the lock (a) and the trailing edge of the airfoil (b) of the working blade of the 2nd stage of the turbine engine GTN-25-1 made of alloy EI929VA after 31043 hours of operation

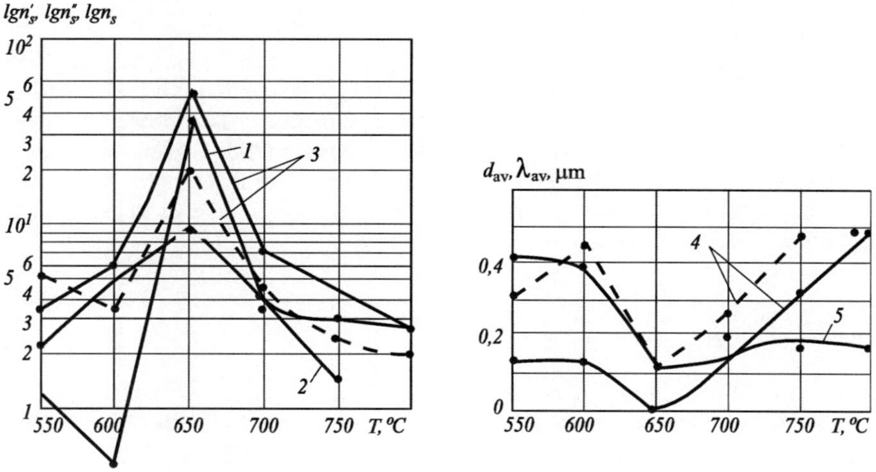

Fig. 4.62 Change in the structural characteristics of the metal of the EI929VA alloy blades depending on the temperature and operating time (5000 h - solid lines, 10000 h - dashed lines). *1, 2* - average surface density of small (n'_S) and large (n''_S) particles of the γ'-phase; *3* - total density n_S; *4* - average size d_{av} of particles of the γ'-phase; *5* - average distance λ_{av} between particles of the γ'-phase

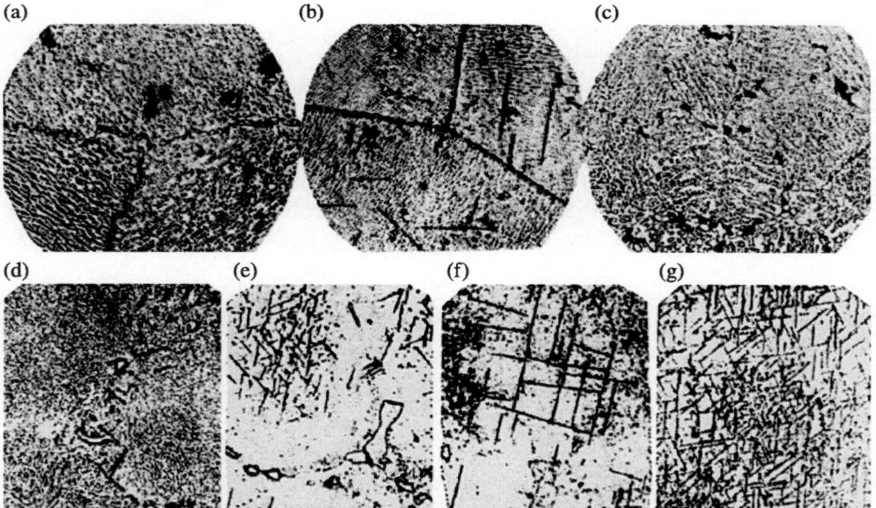

Fig. 4.63 Microstructure of EP220 (a)-(c) and ZhS6U (d)-(g) alloys in the initial state (a), (d) and after aging at 850°C for 10000 h (b), at 900oC - 5000 h (c), at 850°C - 5000 h (f), (g), at 900°C - 2000 h (g)

the type of needle particles σ-phases are formed in the EP99 alloy with an increase in the chromium content in it from 18 to 22 %. The data obtained for a large number of alloys were used to create methods for predicting the conditions for the formation of TCP phases (Fig. 1.58). The kinetics of the formation of TCP phases in EP99, EP958, and EP539 alloys is described in detail in Pigrova (1980). As the experience of operation of blades of the GTK 25 I gas turbine made of IN738 alloy has shown in the conditions of compressor stations (Fig. 4.64), needle-like particles of TCP phases are formed in them.

4.3.2.5 Boundary Structure Changes

For the nickel-based alloy EI698, after prolonged aging at temperatures of 500-600°C along the grain boundaries, in addition to carbide MeC particles, $Me_{23}C_6$ particles appear. After aging at 650°C, precipitates of carbides $Me_{23}C_6$ prevail, while at the aging temperature 750°C particles carbide MeC is completely absent. The size of carbide particles increases from 0.3 to 0.6-0.8 μm.

It should be noted that the rates of aging processes at constant and cyclically varying temperatures differ significantly. Thus, studies of the influence of cyclic thermal elastoplastic stresses on the processes of precipitation aging of the ChN77TYu alloy carried out by V.S. Ermakov, showed that precipitation hardening processes are accelerated tens of times.

There are theoretical methods for determining the structural stability of alloys. In the 60s of the last century, the PHACOMP calculation method was developed (Sims and Hagel, 1972; Pigrova, 1980), which makes it possible to determine with varying accuracy the tendency of a nickel-based alloy of a given composition to form TCP phases — the main cause of embrittlement and a decrease in the performance of alloys. It is also possible to predict the behavior of the alloy on the basis of the diagram developed by Pigrova (1980), shown in Fig. 1.58.

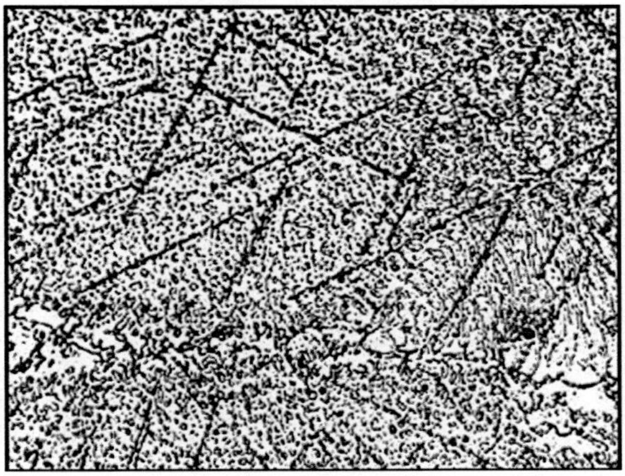

Fig. 4.64 Microstructure of metal at the trailing edge of the 1st stage turbine blades GTK 25 I made of IN 738 alloy after operation for 23000 h (x1000)

As for the determination of quantitative features of the relationship between structure and properties, they are usually determined only experimentally. In this case, we are talking about mechanical properties (σ_B, $\sigma_{0,2}$, δ, ψ, ISM), creep resistance, creep strength, fatigue and thermal fatigue resistance and other characteristics. In well-known handbooks on material properties, only a few data are usually given concerning only changes in short-term mechanical properties, which, as is known, at high temperatures only indicate the operability of the metal in terms of its sensitivity to a notch and say little about the effect on long-term strength and fatigue resistance.

4.3.2.6 Changes in Properties During Long-term Operation

Long-term operation at high temperatures can cause noticeable changes in material properties. In Getsov et al. (1999, 2000) it was proposed to extend the principle of formulation the creep equations based on Rabotnov's structural parameters s_i (Rabotnov, 1969)

$$\dot{p} = F(s_1(t,T), s_2(t,T), s_3(t,T), \ldots, T, \sigma) \tag{4.16}$$

to similar dependencies for other material characteristics:

$$\sigma_B = F(s_1(t,T), s_2(t,T), s_3(t,T), \ldots, T, v) \tag{4.17}$$

$$\sigma_{0,2} = F(s_1(t,T), s_2(t,T), s_3(t,T), \ldots, T, v) \tag{4.18}$$

$$\sigma_{\mathrm{l.t.}} = F(s_1(t,T), s_2(t,T), s_3(t,T), \ldots, T, t) \tag{4.19}$$

$$\delta = F(s_1(t,T), s_2(t,T), s_3(t,T), \ldots, T, v) \tag{4.20}$$

$$\Delta\varepsilon = F(s_1(t,T), s_2(t,T), s_3(t,T), \ldots, T_{\mathrm{max}}, T_{\mathrm{min}}, N, t_{\mathrm{cycl}}) \tag{4.21}$$

$$S_{0,4} = F(s_1(t,T), s_2(t,T), s_3(t,T), \ldots, T, v) \tag{4.22}$$

$$\mathrm{d}l/\mathrm{d}t = F(s_1(t,T), s_2(t,T), s_3(t,T), \ldots, T, K_1) \tag{4.23}$$

Here s_i are structural parameters depending on temperature and time of operation, v is the rate of deformation, $t_{\mathrm{cycl}}, T_{\mathrm{max}}, T_{\mathrm{min}}, N$ is the cycle period, the maximum and minimum cycle temperatures, the number of cycles before the formation of thermal fatigue cracks under thermal cyclic loading, K_1 is the stress intensity factor. Thus, the tensile strength σ_B, yield strength $\sigma_{0,2}$, ductility δ, creep rate $\dot{p}$, creep crack growth rate $\mathrm{d}l/\mathrm{d}t$, resistance to cyclic deformation $S_{0,4}$, thermal fatigue resistance $\Delta\varepsilon$ and long-term strength σ_{lt} depend on the operating conditions in accordance with the ratios (4.17)-(4.23). To assess the stress-strain state and safety margins of parts made of a given alloy, taking into account changes in its structure, it is necessary to know the kinetics of changes in the listed characteristics during a long stay at high temperatures. The studies have shown the feasibility of using the structural state of materials as parameters:

- composition, number and size of particles of carbide phases,
- composition, particle size and amount of intermetallic phases,
- discrepancy between the lattice periods of the precipitated phases and the matrix,

- number, composition and shape of particles of TCP phases,
- phase structure of grain boundaries, and
- composition and amount of the matrix phase.

It is by changes of these parameters that it is advisable to characterize the values $s_1(t,T), s_2(t,T), s_3(t,T), \ldots$ Unfortunately, however, to experimentally determine, in relation to a large resource, the patterns of these microstructural changes in various grades of materials is a task that is practically insolvable, methods for predicting microstructural changes in materials are of great importance. This prediction solves the problem of determining the parameters of the microstructure of the current state and the limiting (stabilized) state of the material with the given parameters of the initial state in relation to a given temperature-time history of operation.

Unfortunately, within the framework of this monograph, it is not possible to illustrate with specific examples all kinds of relationships between the microstructure and various properties, schematically shown in Fig. 4.65 for various materials. It is supposed to consider only individual cases that demonstrate the considered approach to assessing the stress-strain state of parts operating at increased temperatures. How significant in a number of cases the effect of microstructural changes on mechanical properties can be seen from the results of tests carried out with EP99, EI787 and EI481 alloys (Fig. 4.66).

Yield Point Changes

The effect of aging on the resistance to elastoplastic deformation of a number of pearlitic and martensitic steels is clearly illustrated by the data given in Table 4.39

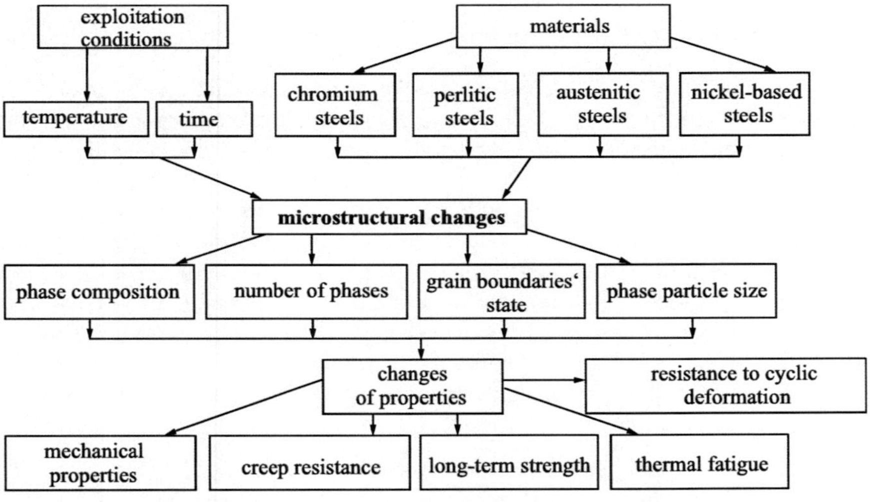

Fig. 4.65 Scheme of the influence of changes in the structure of materials on properties

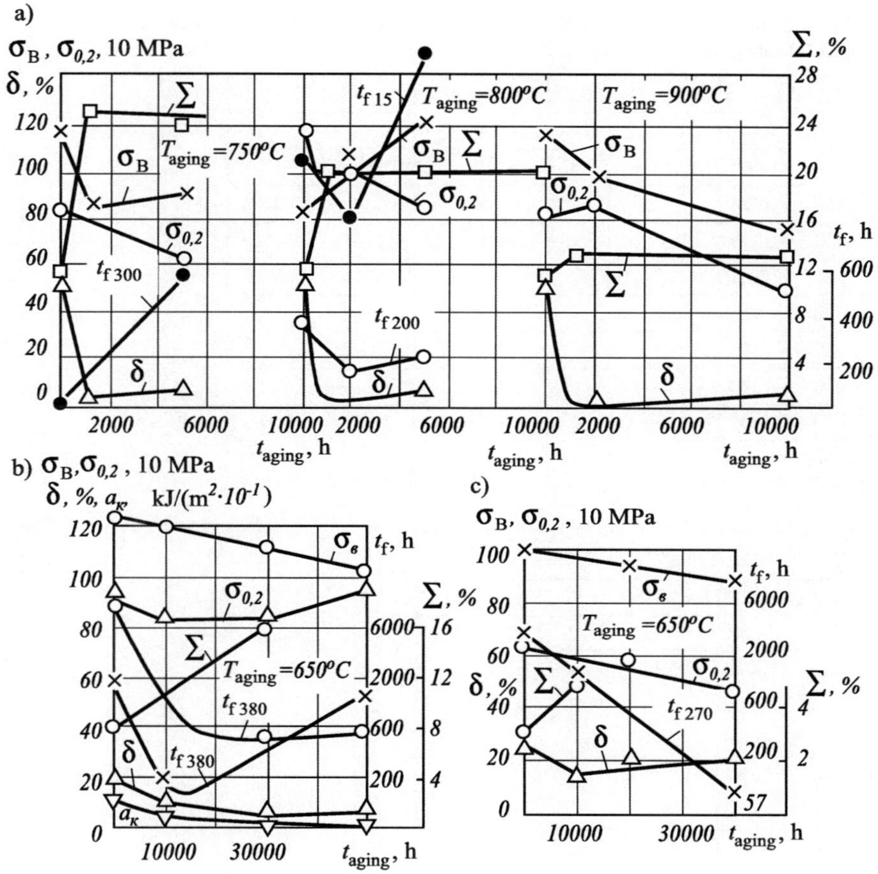

Fig. 4.66 Graphs of changes in mechanical properties and phase composition of ally EP99 (a), steel EI787 (b) and steel EI481 (c) during prolonged aging at various temperatures T_{aging}, Σ is the amount of the hardening phase $t_{f\sigma}^{(\text{main})}$, $t_{f\sigma}^{(\text{n})}$ is the time to failure at stress σ of smooth specimens and specimens with notches; α_K - impact strength

(Maslenkov and Maslenkova, 1991), for austenitic steels in Table 4.40 and nickel-based alloys in Table 4.41 (Getsov, 1996). For rotor pearlitic steel P2MA (Chizhik and Verkina, 1983), long-term aging at 540°C leads to a 10-15 % decrease in the yield strength.

Tables 4.42 and 4.43 show the results of mechanical tests of the metal of blades from various alloys in relative units, characterizing the degree of influence of operating conditions on the characteristics of mechanical properties. Consideration of the results presented in Tables 4.42 and 4.43 shows that during operation the strength properties of all alloys at 20°C either increase or do not change; the ductility of

Table 4.39 Influence of the mode of long-term aging on the resistance to elastic-plastic deformation of pearlitic and martensitic steels ($\sigma_{0.2}^{aging}/\sigma_{0.2}^{initial}$) at 20°C

T_{aging}, °C	t_{agin}, h	15Ch1M1F	EI802	EP752	EP291	15Ch11MF	20Ch13
500	40000	0.88	-	-	-	-	-
550	5000	-	0.92	-	-	-	-
550	10000	-	-	-	-	-	0.86
600	5000	-	0.89	-	0.92	-	-
600	10000	-	-	-	-	0.78	0.75
600	30000	0.78	-	-	-	-	-
620	2500	-	-	0.84	-	-	-
650	2500	-	-	0.78	-	-	-
650	5000	-	-	-	0.86	-	-
650	10000	-	-	-	0.82	-	-

Table 4.40 Influence of the long-term aging regime on the resistance to elastoplastic deformation of austenitic steels ($\sigma_{0.2}^{aging}/\sigma_{0.2}^{initial}$) at 20°C

T_{aging}, °C	t_{agin}, h	20Ch23N18	EI572	EI481	EI787	EI703
600	5000	-	1	1	1.2	-
650	10000	1.13	-	0.83	0.94	-
	20000	-	1	0.87	-	-
	30000	-	-	-	0.91	-
700	5000	-	0.79	0.7	-	-
	10000	1.1	-	-	-	-
750	5000	1	0.75	-	-	0.8

the alloys decreases with rare exceptions. At elevated temperatures, in contrast the strength properties practically do not change, but the ductility increases.

It should be kept in mind that microstructural changes occurring in the alloy at a particular temperature can either be stable and affect the characteristics of the material during its operation at other temperatures, or unstable (for example, upon heating or cooling, the original microstructure can be restored). The study of these features of changes in the structure of heat-resistant materials are devoted Sims and Hagel (1972); Maslenkov and Maslenkova (1991); Getsov (1996); Pigrova (2009); Naumenko et al. (2011); Naumenko and Gariboldi (2014) among others.

Table 4.41 Influence of the mode of long-term aging on the resistance to elastic-plastic deformation of nickel-based alloys ($\sigma_{0.2}^{aging}/\sigma_{0.2}^{initial}$) at 20°C

T_{aging}, °C	t_{aging}, h	EI868	EP99	EP220	EI607	VZh85	EI867	EI437B
700	10000	-	-	-	0.69	-	0.89	-
750	500	1	-	-	-	-	-	1.36
	1000	1	-	-	-	1.03	-	-
	5000	0.9	-	-	-	1.07	-	-
	10000	0.83	-	-	-	-	-	-
800	2000	0.90	1.20	0.9	-	1.03	-	-
	5000	0.87	1.04	0.83	-	0.95	-	-
	16000	0.77	-	0.9	-	-	-	-
900	1000	0.75	-	-	-	1	0.82	-
	2000	0.78	1.05	0.93	-	0.82	-	-
	5000	0.78	-	0.85	-	0.59	-	-
	10000	0.68	0.62	-	-	-	-	-

Table 4.42 Mechanical properties at 20°C in relative coordinates of the blade metal after long-term operation

Alloy	GTU	$t_{operation}$, h	$\sigma_{0.2}^{feather}/\sigma_{0.2}^{shank}$	$\sigma_{B}^{feather}/\sigma_{B}^{shank}$	$\delta^{feather}/\delta^{shank}$	$\psi^{feather}/\psi^{shank}$
IN 738	GTK 25 I	24000	1.07	0.95	0.69	0.48
		30000	1.04	0.94	0.25	0.83
		55000	1.08	1.04	1.15	1.07
EI893	GT 750-6 (1st and 2nd stage)	20500	0.97	0.95	0.64	-
		52000	1.08	1.09	0.75	-
		54601	1.21	1.15	0.56	0.54
			1.09*	1.07*	0.56*	0.57*
EI726	GT 750-6	54601	0.96-1.09	1.02	0.95-1.0	0.91-1.0
CNK7	GTK 25 I	23000	1.08	1.06	0.18	0.67
EI893(L)	GT 6-750	7556	1.23	1.08	0.27	0.68
		14856	1.24	1.20	0.29	0.82

* - 2nd stage

Change in Resistance to Deformation at High Temperatures

We studied the change in the deformation relief at 650 and 750°C in steel EI787 in the initial state (1) and after aging at 650°C for 50000 h (2). Both states were characterized by approximately the same values of the instantaneous yield stress at 20°C (880-900 MPa). Samples in two states were deformed at a rate of 1 %/h to the

Table 4.43 Mechanical properties at 700 and 750°C in relative coordinates of the blade metal after long-term operation

Alloy	GTU	$t_{operation}$, h	T_{test}, °C	$\frac{\sigma_{0.2}^{feather}}{\sigma_{0.2}^{shank}}$	$\frac{\sigma_{B}^{feather}}{\sigma_{B}^{shank}}$	$\frac{\delta^{feather}}{\delta^{shank}}$	$\frac{\psi^{feather}}{\psi^{shank}}$
IN 738	GTK 25 I	55000	750	0.99-1.08	1.0-1.06	1.48-3.0	1.44-1.45
EI893	GT 750-6	54601	750	0.9	0.84-1.0	1.23-1.55	1.3-1.35
CNK7	GTK 25 I	23000	700	1.0	1.0	1.25	2.0

same deformation value of 1.9 % (at 650°C) and 2.2 % (at 750°C). It was found (Table 4.44) that in state 2, compared with state 1, the following properties decrease: resistance to delayed deformation, the amount of intergranular slip and the number of boundaries at which it is observed, the fraction of deformation localized at the grain boundaries, and the tendency to crack initiation.

Ductility Change

Figure 4.66 shows some typical cases of the effect of microstructural changes on the ductility of a number of steels and alloys. However, the formation of complex alloyed carbides such as $Me_{23}C_6$ and Me_6C in pearlite Cr-Mo-V steels does not lead to a decrease in ductility. The reactions MeC⇒ $Me_{23}C_6$ at grain boundaries during prolonged aging of nickel-based alloys (EI698) has a negative effect on ductility characteristics with a simultaneous decrease in strength properties.

The formation of TCP phases leads to a sharp decrease in ductility. So, in the EP99 alloy with 22 % Cr, its catastrophic embrittlement is observed after prolonged aging (1000-5000 h) at 750 – 850°C, while this alloy with 18.5 % Cr has a high level of ductility. This behavior of the alloy is directly related to the formation of topological close-packed phases σ and μ during long-term aging of the alloy. An increase in

Table 4.44 Influence of the structural state of steel EI787 on the features of its deformation

Characteristic	Test temperature			
	650°C		750°C	
	State 1	State 2	State 1	State 2
ε, %	1.9	1.9	2.2	2.2
σ, MPa	730-870	460-470	380-460	280-290
Number of boundaries where slippage is observed, %	13.5	4.5	25	13
Average value of intergranular slip, μm	0.098	0.018	0.326	0.072
Fraction of deformation (%) localized at grain boundaries	3.3	0.5	8.2	1.6
Number of cracks	2	0	10	0

the aging time at 900°C up to 10000 h nevertheless causes a slight increase in the ductility of the alloy.

Allocation at aging temperatures of 500-600°C alloy EI893 highly dispersed γ'-phase causes its embrittlement, which caused the destruction of the blades of the gas turbine blades. This issue will be discussed in more detail below. It should be noted that the nature of the change in the ductility of materials is also largely determined not only by the conditions of long-term operation, but also by their structure obtained as a result of technological heat treatment and, in particular, by the state of grain boundaries.

Similar changes in the structure and mechanical properties were observed after prolonged aging in the EP220, ZhS6U alloys (Fig. 4.63). As shown by the results of X-ray spectral analysis, during aging of these alloys at temperatures up to 850°C, a redistribution of alloying elements occurs between the particles of the strengthening phase (γ'-phase and $Me_{23}C_6$ carbide) and a solid solution, as a result of which a solid the solution is enriched in titanium and chromium and depleted in tungsten and molybdenum.In the alloys, particles of the acicular (film) phase - double carbide Me_6C are precipitated. The change in the shape of the particles of the strengthening phase and their sizes practically does not affect the ductility of these materials.

The appearance of film and cellular formations of $Me_{23}C_6$ carbides observed in the alloys EI437B, EI617, EI826, EI827, Rene-41, Nimonic, Inconel X, Udimet 500 and others, leads to a significant decrease in ductility, long-term strength and impact strength. The formation of a continuous layer of $Me_{23}C_6$ particles along the boundaries leads to the same results. The analysis of the nature of the influence of the processes of changes in the structure during prolonged aging on the properties of various heat-resistant alloys indicates the differences in these influences on the resistance to elastoplastic (instantaneous) deformation.

Change in Creep Resistance

As can be seen from Fig. 4.67, the creep rate of the EI481 austenitic steel at 600°C and $\sigma = 400$ MPa in the state after 10000 hours of aging at 650°C, which caused the coagulation of carbide particles, increased sharply ($480 \cdot 10^{-4}$ %/h instead of $2.9 \cdot 10^{-4}$ %/h in the initial state).

Change in Long-term Strength

The influence of changes in the microstructure that develop during prolonged holding of refractory materials at elevated temperatures also affects the long-term strength curves in the region of large times. In Pigrova (2009) was shown that in the EP539 alloy, after 10000 h aging at 750–800°C, depending on the Al, Ti and Cr content, up to 10 % σ and μ, which leads to a 10fold decrease in the time to failure in long-term strength tests, as well as to a sharp decrease in ductility. A distinction should be made between cases where microstructural changes lead to a hardening of the material (an

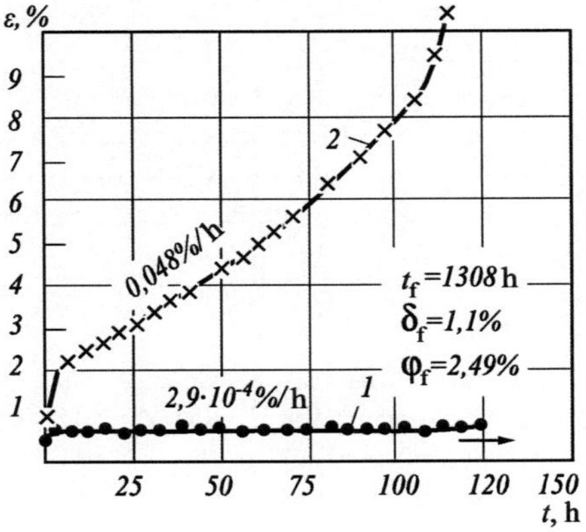

Fig. 4.67 Creep curves of EI481 steel at a temperature of 600°C, $\sigma = 400$ MPa; *1* - initial state; *2* - after 10000 h aging at 650°C

increase in its long-term strength) or to a softening of the material (a decrease in its long-term strength).

Figure 4.68 shows a diagram of the influence of preliminary long aging over time t_{aging1}, t_{aging2} and t_{aging3} for long-term strength of the material. The scheme is based on the assumption that the intensity of the strain aging process and phase transformations does not depend on the stress level. There can be three cases of mutual arrangement of curves for the original and aged metal:

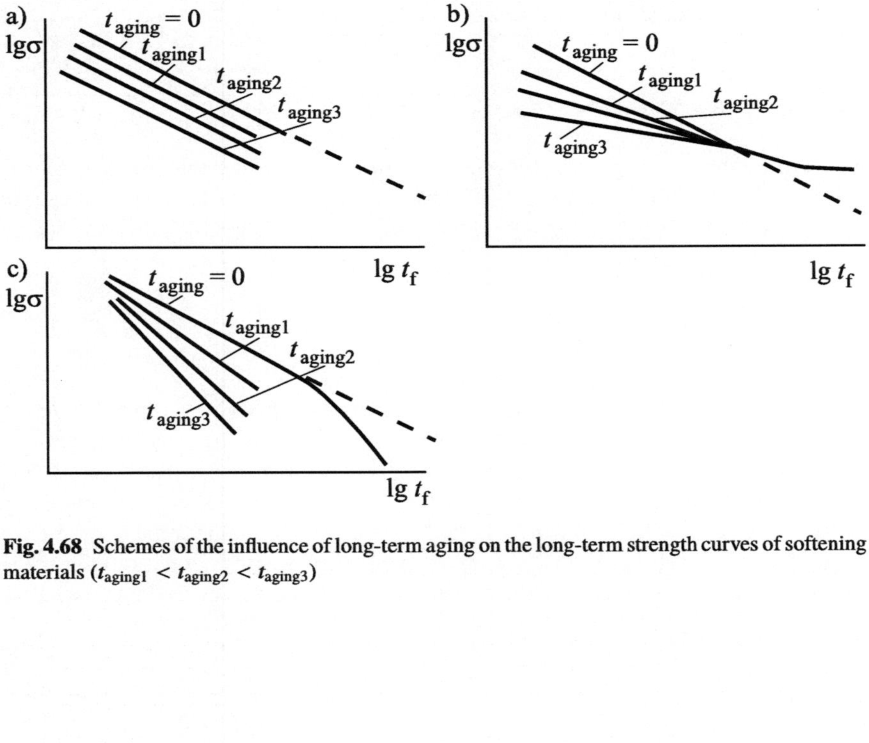

Fig. 4.68 Schemes of the influence of long-term aging on the long-term strength curves of softening materials ($t_{aging1} < t_{aging2} < t_{aging3}$)

1. the curves are located in parallel (see Fig. 4.68a),
2. the curves converge (see Fig. 4.68, b), and
3. the curves diverge (see Fig. 4.68, c).

Cases 1 and 2 correspond to the behavior of the material in which its long-term strength at long service lives is not less than the extrapolated values of the long-term strength of the initial material. In case 3, extrapolation of the results of long-term strength tests of the starting material at long service lives gives overestimated values of the time to failure. Refinement of the slope of the long-term strength curve of the material tested in the initial state can be carried out by the method shown in Fig. 4.69, where the lines *aab, bbc* and *ccd* are parallel to the long-term strength curves aging at the test temperature during $t = t_{aging1}, t = t_{aging2}$ and $t = t_{aging3}$, respectively. The proposed scheme, due to the assumptions made, needs a wider experimental testing; nevertheless, it seems to be useful for evaluating the behavior of unstable materials.

Some typical examples of the negative effect of long-term aging and the resulting changes in the structure of austenitic steels and nickel-based alloys on long-term strength, which are considered above, are shown in Table 4.45. The formation of the EP539 10 % σ-phase leads to a decrease in the level of its long-term strength at 800-850 °C by 25-30 % (in terms of stress). Similar results were obtained for the IN100 alloy (Fig. 4.70).

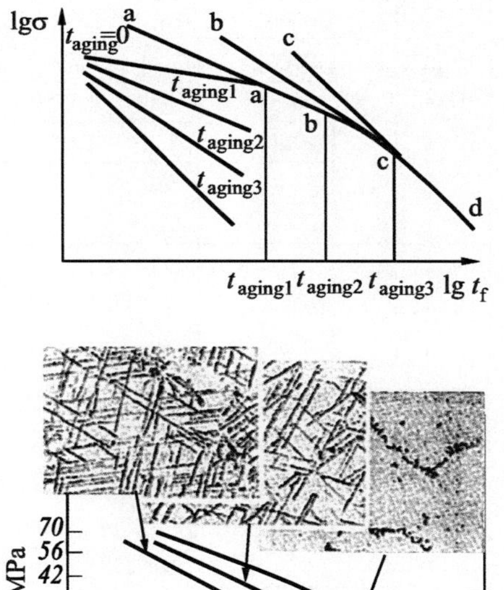

Fig. 4.69 Diagrams illustrating the method of extrapolation of creep rupture strength curves for softening materials

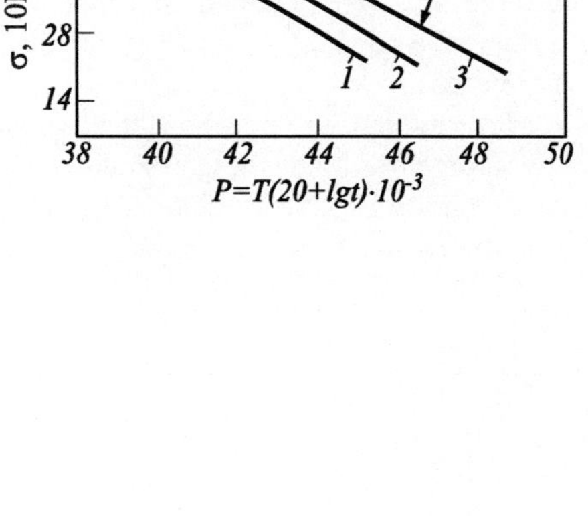

Fig. 4.70 Influence of the previously formed σ-phase on the long-term strength characteristics of the IN100 alloy and the corresponding structures, the properties were determined at 740-910°C after holding for 2500 h at 840°C (Coutsouradis et al., 1994): *1* - a large number of σ-phase; *2* - average; *3* - small (*P* - Larson-Miller parameter, *T* - temperature in K)

Table 4.45 Effect of long-term aging on long-term strength

Material	Aging mode		Test mode		t_f, h	ε_f, %
	T,°C	t, h	T,°C	σ, MPa		
EP99	800	0	800	200	374	24
		2000			147	24,4
		5000			193	20,4
EI703	800	0	800	50	101	78
		2000			94	82
		5000			175	66
EP126	800	0	800	100	684	28,8
		2000			381	50
		5000			151	10
		0	800	80	3480	22
		2000			1871	20
		0	900	50	242	18
		2000			94	75
		5000			65	45
EI787	650	0	650	350	8000	1,6
		30000			428	5,3
		50000			442	15
		0		340	10 000	1,46
		10000		236*	10 056	
		0		380*	1350	
		10000		370*	148	
EI481	650	0	650	270	1000	
		10000			1280	4,14
		40000			72	30

* notched samples

At the same time, it was found that long-term operation of blades made of IN 738 and EI 893 alloys under compressor station conditions did not affect the characteristics of long-term strength of materials. It should, however, be noted that these results were obtained on samples cut from areas remote from the blade airfoil surface; the effect of de-alloying and surface corrosion damage was not modeled in these tests.

Nevertheless, the question of the properties of the EI893 alloy after long-term operation of the blades at temperatures of 550-650°C deserves more attention. Under these conditions, fine particles of the γ'-phase are precipitated, which, in the case of a low hardening temperature and a tempering temperature of 800°C, leads to a sharp increase in hardness and to embrittlement of the alloy. Another factor in the changes in the structure of the alloy during long holdings in the indicated temperature range is the processes of long-range ordering of the γ-phase of the Ni_2Cr type (Pigrova, 2009). In the absence of stress concentrators, the high long-term strength of the alloy in the specified temperature range guarantees the reliable operation of the blades. However, in the presence of concentrators, cracking is possible under conditions of stress relaxation, which is associated with the low long-term ductility of the EI893 alloy under these conditions. Such cracks were found in the area of the holes in the blades for the shroud wire in the blades of the TLP GT-750-100 (Fig. 4.71) (Getsov, 1996), as well as on the lower blades of the old design GTU-750-6 NZL. Therefore, to prevent significant embrittlement of the EI893 alloy blades, when an increase in hardness above 300 HV is detected, they are subjected to a reduction heat treatment.

In Pimenova (1973); Getsov et al. (2003), during X-ray diffraction studies at large angles of metal recording of samples made of high-heat-resistant nickel alloys cut from the "hot" zone of the airfoil after their long-term operation, after long-term strength tests, splitting of the lines of the fcc lattice K_α-doublet (331) and (420). As a result, it was found that the alloys in these states are characterized by the loss of coherent lattices of the γ- and γ'-phases, which leads to a significant decrease in the level of long-term strength. It was shown that an increase in the difference between the lattice parameters γ- and γ'-phases (misfit) leads to softening of the material. On the issue under discussion, the data obtained by V.I. Gladshtein on the very strong influence of operating time on the characteristics of the long-term strength of steel 15Ch1M1F(L) (Fig. 4.72).

The results of tests for the long-term strength of EP428 steel showed that after long-term operation the slope of the curves changes (Fig. 4.73). If we take into account the structural changes occurring in this case, then it is possible to relate the change in the slope of the curves using structural parameters. The established relationship between the Cr/Fe ratio in carbide $Me_{23}C_6$ of EP428 steel and the operating time at a temperature of 500°C allows, based on Eq. (4.15) to determine the type of dependence of long-term strength on the operating time of the blades, namely, to represent them as a fan of curves of the type shown in Fig. 4.68c. For this, we use the power-law dependence of the time to fracture on temperature $t_f = A\sigma^{-m}$, where m is a constant depending on the operating time. Taking into account the linearity of the Eq. (4.15), we obtain the possibility of describing the dependence of m on the operating time: $m = m_0(1 + \alpha t)$. With this in mind, it is advisable to

Fig. 4.71 Crack in the hole for the tie wire

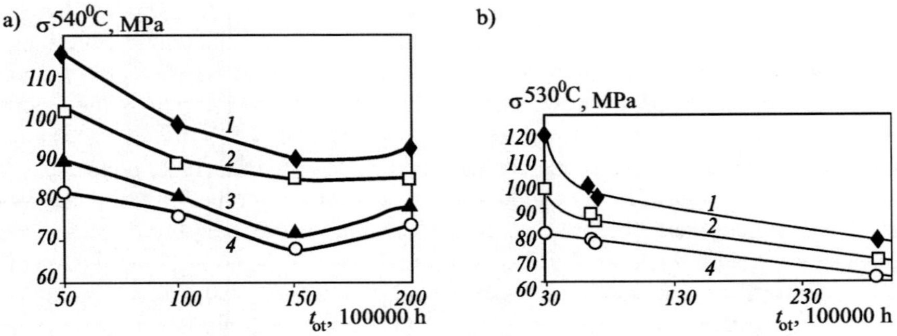

Fig. 4.72 Long-term strength of steel 15Ch1M1F(L) at 540°C (a) and steel 20ChMF(L) at 540°C (b) as a function of operating time t_{ot}: *1* - 50000 h, *2* - 100000 h, *3* - 150000 h, *4* - 200000 h

Fig. 4.73 Long-term strength of metal working blades made of EP428 steel at 500°C:
1 - passport curve of long-term strength of metal with $\sigma_{0.2} = 700 - 750$ MPa,
2 - - - - calculated curve for metal after operation for 60000 h,
3 - ×, • - experimental points for metal after operation for 60000 and 100000 h,
4, *5* - sections of the extrapolated long-term strength curve (*4* parallel *2*, *5* parallel *3*)

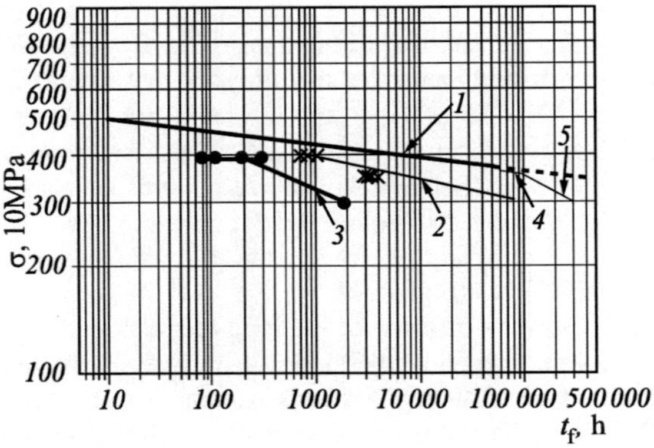

extrapolate the passport curve in Fig. 4.74 not according to a linear law in logarithmic coordinates, but in accordance with the diagram in Fig. 4.69, taking into account the changes in the slope caused by microstructural changes. Long-term strength σ_{200000}

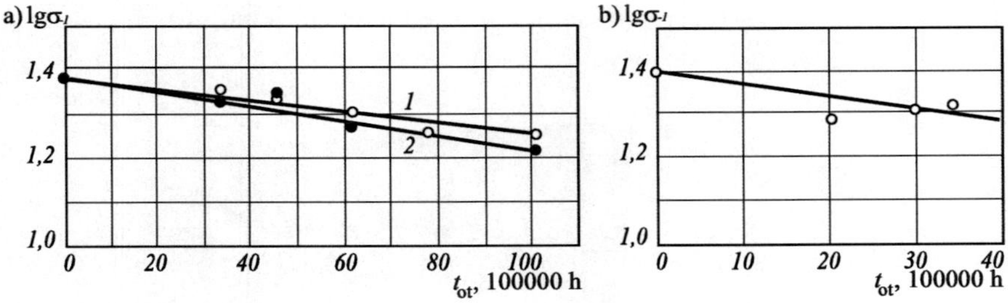

Fig. 4.74 Change in the average endurance limit of rotor blades in the process of operating time. a) GTK-10, EI893, 1st stage HPT, 2nd stage LPT; b) GTK 25 I, CNK7

of EP 428 steel at 500°C in this case will not be 360 MPa, but 320 MPa.

Change in Fatigue Resistance

The effect of long-term aging on the fatigue resistance of the ZhS6K alloy is illustrated by the data given in Table 4.46. It is seen that after a long (1000 h) aging, the fatigue limit of the ZhS6K alloy decreases by about 10 %. Noticeable changes in the fatigue limit of turbine blades of the GTK-10 and GTK 25 I units made of the EI893 and CNK-7 alloys in the process of operating time up to 100000 h were also observed by Ivanov and Rybnikov (2006); Ivanov et al. (2006); Shaydak et al. (2006) (see Fig. 4.74).

Thermal Fatigue Resistance Change

As noted above, after prolonged aging of the EP99 alloy for 1000-5000 h at 750-900°C, a sharp decrease in its ductility occurs: it becomes extremely sensitive to stress concentration under tensile test conditions at 20°C (Fig. 4.75). The brittleness of a long-aged metal is also manifested in thermal fatigue tests: if for the metal in

Table 4.46 Effect of long-term aging on the fatigue resistance of the ZhS6K alloy at 20°C (according to I.A. Ischenko, V.N. Omelchenko, B.N. Sinaisky, etc.)

Aging mode	σ_{-1} for 10^7 cycles, MPa
no aging	285/275
950°C (5 h)	280/270
950°C (1000 h)	225/230

In the numerator σ_{-1} after hardening, in the denominator - without heat treatment.

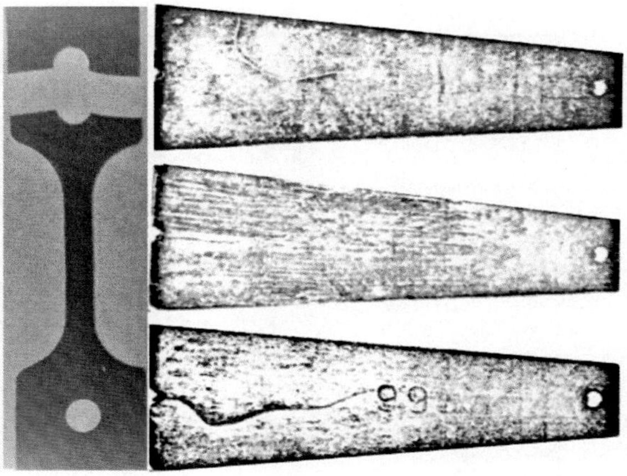

Fig. 4.75 Brittle fractures of EP99 alloy after aging at T_{aging} = 750 and 800°C for 2000-5000 h: during tensile tests at 20°C (left), during thermal shock tests with the mode heating to T_{max} = 800°C, cooling in water (right)

the initial state, the number of cycles before cracks appear in tests according to the 750⇄20°C mode is 90-200, then cracks in metal specimens after prolonged aging at 750-900°C were detected after a single thermal shock (see Fig. 4.75b). Long-term strength of this alloy, which has a structure formed in the process of long aging, as shown above, is also noticeably lower than the long term strength of the parent metal (Table 4.45 and Fig. 4.66).

Studies have shown that in the ZhS6K alloy with an acicular carbide phase (state 2), the durability before cracking and fracture of specimens under cycle conditions of 150-850°C is noticeably reduced in comparison with the alloy in state 1, characterized by the presence of in the structure of globular carbides (Table 4.47).

At the same time, when no serious changes in the metal structure during long-term operation are observed, no noticeable changes in thermal fatigue resistance are observed. So, from the turbine blades GTK 25 I, made of ZMI3U alloy, after 16153 h of operation, flat corset samples were cut out for testing the resistance of metal to thermal fatigue. the results of such tests of the metal of the blades shown in Fig. 4.76 in comparison with the data obtained in the initial state indicate that there are no noticeable differences between them.

The considered examples of the effect on the properties of steels and nickel-based alloys of changes in their structure occurring during a long stay at elevated temperatures indicate the need to take them into account when determining the stress-strain state and safety margins of high-temperature machine parts with long service life. Therefore, the development of methods for predicting changes in properties in the materials of such parts in connection with structural changes during long-term operation, summarized above, deserve closer attention and use in normative documents on strength calculations.

Table 4.47 Thermal fatigue test results for ZhS6K alloy

State	Cycles until cracking	Number of cycles until the destruction of samples
1	30-48	236-516
2	4	15-69

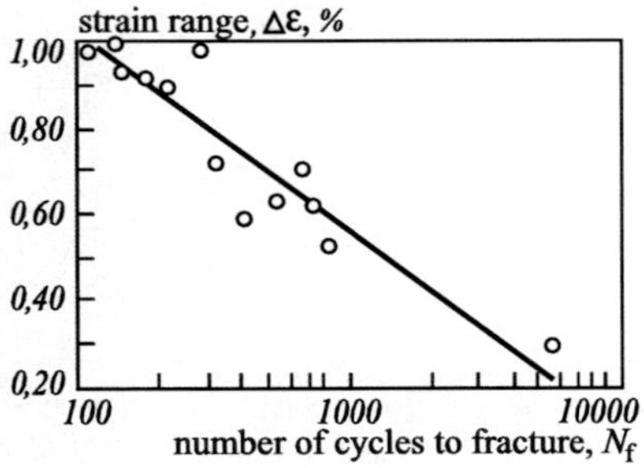

Fig. 4.76 Thermal fatigue curve of metal blades made of ZMI3U alloy in the initial state (straight line) and after operation of the GTK 25 I turbine for 16155 h at the Pervomaiskaya compressor station (points)

4.3.3 Reductive Heat Treatment

One of the ways to increase the resource of the blades is the reduction heat treatment. Such processing leads to restoration of the metal structure and healing of some types of damage. Thus, as early as in Getsov and Taubina (1960); Getsov (1964), it was established that alloys with intermetallic hardening have the ability to regenerate the structure and properties after heating to a temperature lower than the temperature of austenitization (but causing the dissolution of phases) and re-aging. for the EI869 alloy, after intermediate heat treatment with a temperature of 900-950°C, produced during the third stage of creep, the creep rate decreased and the time to fracture increased.

Subsequently, the reductive heat treatment was used to increase the resource of the blades made of the EI893 alloy (Levin et al., 1975; Rybnikov et al., 1984). It was found that in the state of the EI893 alloy after low-temperature quenching, its structure is highly stable, and therefore it is impossible to prevent its embrittlement at operating temperatures of 580-600°C, which occurs due to the release of a large amount (up to 18-20 %) fine γ'-phase. In this state, the hardness of the alloy can reach values of 30-40 HRC, and the ultimate strength at 600°C is reduced to two times. Therefore, in some cases, a significant increase in hardness and a decrease in the ductility of the metal of the blade joint part is observed, which can be the cause of brittle fractures. In this case, the use of a lowered austenitization temperature turns out to be unacceptable - it is necessary to carry out a high temperature heat treatment containing the austenitization operation at a temperature of 1040-1100°C. The analysis of the processes occurring in the process of high-temperature treatment with blades made of EI893 alloy made it possible to introduce this operation to increase the resource of the blades GT-6-750 (GTN-6) UTMZ, GT-750-6 NZL and GTK-10-4 NZL. For example, for GT-6-750 blades made of EI893 alloy, high temperature heat treatment was carried out after 60000 h of operation. It has been established that after high temperature heat treatment the amount of the γ'-phase decreases (9-13 % instead of 13-18 % after operation), the hardness restores its values (220-280 HB instead of 300-450 HB), ductility and impact strength (δ = 30-45 instead of 5-20 %, ψ = 30-50 % instead of 5-25 %, and ISM = 7,0-16,0 instead of 1,5-5,0 kJ/m^2). Similar works related to the development of high temperature heat treatment modes for GTK 10 I blades made of IN738 LC alloy were carried out by Popov et al. (1989). It has been established that after high temperature heat treatment the mechanical properties and long-term strength at 850°C of the metal of the blades with degradation of the structure and properties after 17000 h of operation are restored to the values in the initial state.

The possibility of increasing the service life of GTK 25 I blades made of CNK-7 alloy was also investigated (Dashunin, 2006; Ivanov et al., 2001). As an object of research, we took rotor blades made of CNK-7 alloy after operating time of 30000 hours. It was found that in this state the fatigue and long-term strength have lower values compared to the initial state. The microstructure of the blades is characterized by an increased content of the γ'-phase, an enlargement of its particles, as well as an increase in the number and size of carbide particles $Me_{23}C_6$ around particles of the

MC, primary γ'-phases and eutectics $\gamma+\gamma'$ (Fig. 4.77). As a mode of regenerative heat treatment of blades made of CNK-7 alloy, the heating of the part to 1180°C, holding for 2 hours, cooling in air and subsequent aging at 850°C for 24h It was found that reductive heat treatment improved the structure of the metal of the blades: there was a partial equalization of the chemical composition due to the dissolution of particles of the γ'-phase in the interdendritic layers and the structure of grain boundaries improved - the boundaries became thinner with smaller particles of the γ'-phase and carbide $Me_{23}C_6$. The mechanical properties of the metal are improved (Fig. 4.78), and the time to fracture and long-term ductility increased (Table 4.48).

The nature of the influence of reductive heat treatment on the fatigue resistance of blades made of alloys EI893 and CNK-7 can be seen from the data shown in Figs. 4.79 and 4.80. Obviously, reductive heat treatment also noticeably increases the fatigue strength values.

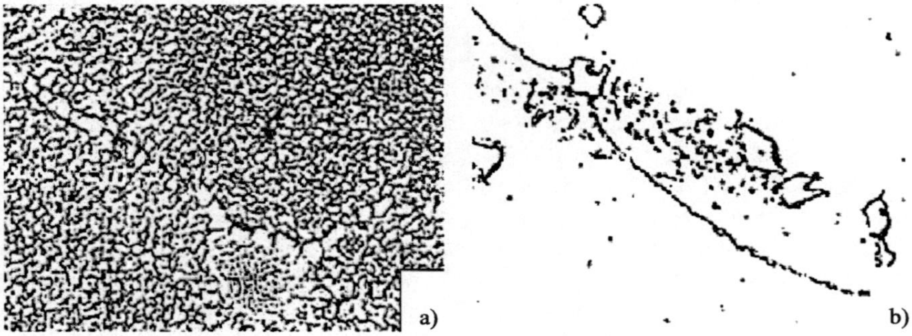

Fig. 4.77 Structure of grain boundaries of metal blades of low pressure turbines GTK 25 I from alloy CNK7: a) - after operating time of 30000 hours, b) - after operating time of 30000 hours and reductive heat treatment

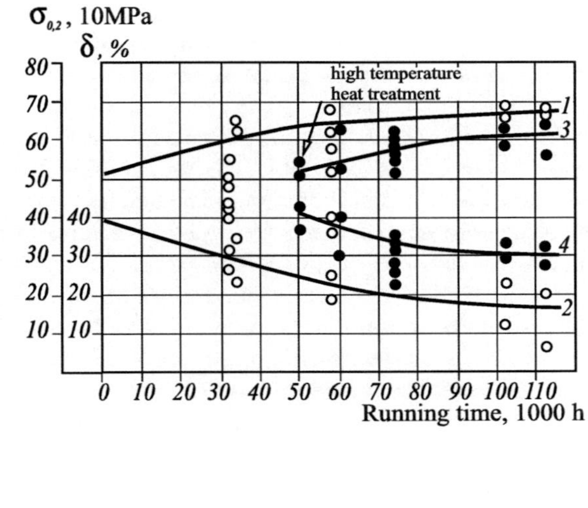

Fig. 4.78 Change in mechanical properties of stamped turbine blades made of EI893 alloy depending on operating time: *1* - $\sigma_{0.2}$ after operating time, *2* - δ after operating, *3* - $\sigma_{0.2}$ after restorative heat treatment, *4* - δ after reductive heat treatment

Table 4.48 Results of long-term strength testing of blade metal made of CNK-7 alloy

State after	Time to fracture (t_f, h) at T = 700°C, σ = 600 MPa	δ_f, %	ψ_f,%
operating time	309-676	1.8-3.0	3.7-5.2
reductive heat treatment	709-804	2.9-4.0	4.2-7.0

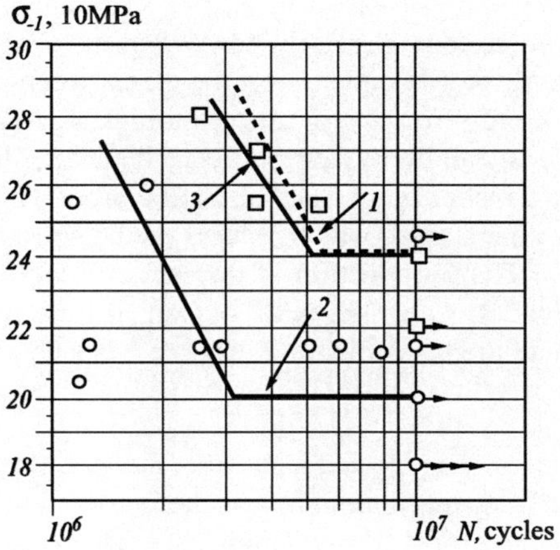

Fig. 4.79 Fatigue resistance of LPT blades made of CNK7 alloy of GTK 25 I unit after 30000 h operating time and restorative heat treatment: *1* - blades before operating time, *2* - blades after operating time 30000 h, *3* - blades after reductive heat treatment

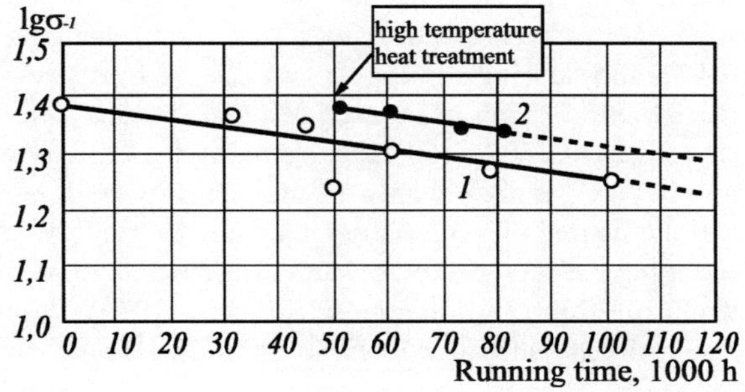

Fig. 4.80 Change in the endurance limit of stamped GTK-10 blades from alloy EI893 depending on the operating time: *1* - blade endurance limit after long-term operation, *2* - blade endurance limit after reductive heat treatment

4.4 Dependence of Properties on the State of the Surface

The aviation industry has accumulated significant experience in establishing the influence of various technological operations used in the processing of GTE parts on their strength. This primarily applies to turbine and compressor blades (Grinchenko, 1971; Sulima and Evstigneev, 1974; Semenchenko and Mirer, 1977).

4.4.1 Effect of Mechanical and Heat Treatment on the State of the Surface Layer

In the process of heating the blanks for stamping and during high-temperature holding during the heat treatment of the stampings of the blades, oxidation and depletion of the surface layer with alloying elements occurs (Table 4.49). Taking into account precisely this size of the depletion layer, allowances for machining are assigned. However, in some cases, when processing the base planes of the blades, due to disruption of the technological process and, in particular, uneven wear of the dies, a uniform distribution of the total allowance along the trough and the back of the blades is not ensured, as a result of which a depleted layer remains in some parts of the feather, which reduces the heat resistance of the metal.

The introduction of any new technological process of finishing machining of blades, as a rule, is accompanied by a study of its effect on the fatigue strength and the residual stress diagram. A decrease in labor intensity due to a decrease in strength is considered unacceptable. The nature and value of residual stresses formed in the surface layer during machining are influenced by cutting conditions (speed, feed), the type and features of the tool, the method and efficiency of cooling the part during machining. When these conditions vary during processing, the depth and degree of work hardening, the distribution and the absolute level of temperatures change and, as a consequence, the values and nature of the distribution of residual stresses change (Fig. 4.81). So, for example, changing the cutting conditions of the EI617 alloy during turning ($s = 0.15 - 0.8$ mm/rev; $t = 0.5\ldots3$ mm) causes a change in the depth of induced residual stresses from 75 to 220 μm and maximum stresses from 220 to 450 MPa; changing the milling modes ($v = 3{,}2\ldots9{,}6$ m/min; $t = 0{,}4\ldots3$ mm) - change in the depth of work hardening from 125 to 250 μm and the degree of work hardening from 25 to 39 %; changing the grinding mode - changing the degree of work hardening from 38 to 55 % at a constant work hardening depth of 20 μm (Sulima and Evstigneev, 1974). When polishing with felt wheels, the work-hardening depth is also 20 μm, while the work-hardening degree does not exceed 20 %.

Table 4.49 Depth of the oxidized and depleted layer (in mm) after annealing at 1200°C for 10 h (according to A.V. Guc)

Material	Annealing in the environment					
	Argon		Vacuum 10^{-5} Torr		Air	
	dross depth	depletion depth	dross depth	depletion depth	dross depth	depletion depth
EI868	0	0.12	0.04	0.03	0.08	0.15
EI826	0	0.14	0.05-0.06	0.70	0.13-0.15	0.98
EP220	0	-	0.05-0.06	-	0.13-0.15	-
ZhS6K	0	0.20	0.05-0.06	0.95	0.13-0.15	1.10

Fig. 4.81 Residual stresses in the surface layer of a blade made of VT3-1 alloy. Samples: 1 - in the initial state, 5, 10- after hardening treatment with balls $d = 1.5$ mm, 9 - with ferro-powder, 11, 12 - after hardening treatment with balls $d = 2.0$ and 3.0 mm

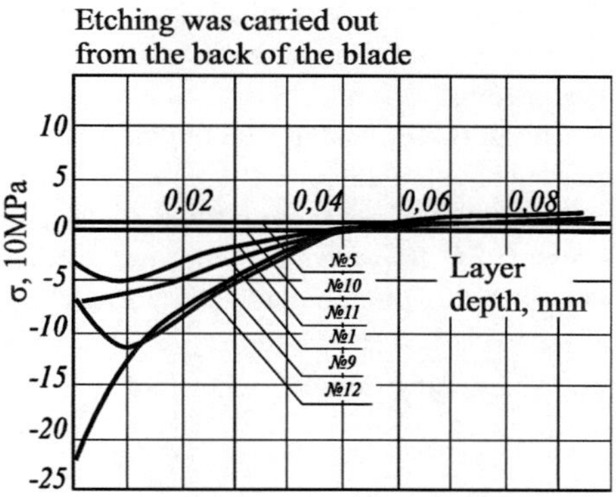

Depending on the method and equipment used during the finishing machining, the nature of the distribution of residual stresses changes; surface stresses can even change sign. For example, grinding both steel 13Ch11N2W2MF and alloy EI617 with abrasive wheels leads to the formation of tensile stresses on the surface, the value of which sharply decreases when grinding with a belt; Vibration-contact polishing with felt discs initiates compressive residual stresses. After vibroabrasive polishing, compressive residual stresses σ_{res} on blades made of EP109 alloy are equal to 500-700 MPa and extend to a depth of more than 140 μm, and after manual polishing $\sigma_{\text{res}} = 650\ldots750$ MPa, but extend to a depth of only 30-50 μm.

The fatigue strength is influenced not only by the degree, but also by the type of plastic deformation. So, when smoothing the bottom of the grooves in specimens with notches made of ZhS6K alloy, causing all-round compression, $\sigma_{-1}^{700°\text{C}} = 242$ MPa, in the case of smoothing the entire groove with the same degree and depth of work hardening, tensile stresses arise at the bottom of the groove and the fatigue limit decreases by 16 %.

In the practice of processing parts made of heat-resistant alloys, heat treatment to relieve residual stresses is widely used as a final operation. When choosing the mode of this processing, certain difficulties arise. So, to significantly reduce the level of residual stresses, the tempering (annealing) temperature should, as a rule, exceed the technological aging temperature[4]. Therefore, the metal controlled by its mechanical properties has to be subjected to tempering, followed by reduction treatment, which usually amounts to repeating the aging operation. However, at present it cannot be considered that there are convincing experimental confirmations of the complete adequacy for various materials of the properties of a metal that has undergone such an additional treatment, with properties after standard heat treatment, regulated by technical conditions. In addition, the reductive heat treatment of the finished blades is carried out at temperatures exceeding the operating temperatures, and therefore

[4] When choosing the optimal tempering (annealing) temperatures, it should be borne in mind that the stress relaxation processes in heat-resistant nickel-based alloys under the influence of preliminary plastic deformation are noticeably intensified (see Sect 2.3).

the use of protective media (usually argon or vacuum) is required. Finally, the last one after recovery heat treatment, the geometric dimensions must remain within the appropriate tolerances.

For all these reasons, the tempering temperature is usually used, which does not guarantee complete removal of residual stresses. When it is known for certain that the residual stresses are compressive, it is usually not sought to completely remove these stresses. So, for example, according to the data (Balashov and Petukhov, 1974), in samples with a notch made of ZhS6K alloy stretched with a work hardening degree of 68 %, before annealing $\sigma_{-1}^{700°C} = 225$ MPa, after annealing, which reduced the work-hardening degree to 35 %, $\sigma_{-1}^{700°C} = 186$ MPa.

4.4.2 Dependence of Properties on Surface Condition

Retention of tensile residual stresses on the surface of finished parts can lead to the formation of surface cracks in the case of insufficient plasticity of the material. Especially dangerous are the cases when the formation of cracks caused by residual stresses occurs during operation due to the fact that the residual tensile stresses are added to the working ones. Thus, Sulima and Evstigneev (1974) describes a number of examples of detecting cracks on the blade tips and edges, caused by the removal of a large uneven allowance with abrasive wheels, as a result of which tensile residual stresses remained on the surface, which were not removed during subsequent polishing.

An enormous amount of research has been devoted to the study of the relationship between fatigue resistance and surface state (Grinchenko, 1971; Sulima and Evstigneev, 1974, among others). An analysis of the results obtained indicates that the fatigue strength of a given material at elevated temperatures is mainly a function of the value and nature of roughness. Given in Table 4.50 data demonstrate that an increase in the degree of roughness both across the specimens and along, causing an increase in stress concentration, leads to a decrease in fatigue strength under flexural vibration.

An increase in the depth and degree of work hardening is accompanied by a decrease in σ_{-1} at high temperatures. The same conclusions are drawn from the analysis of test results at 750°C of samples made of alloy EI437VA-A, processed according to four technological options (Fig. 4.82):

1. electrochemical treatment followed by fine polishing,
2. electrochemical treatment with a left allowance 0.15 mm, hand grinding with an abrasive wheel, hand polishing with a felt wheel with an abrasive grain,
3. the same as in the second version, but without the grinding operation, and
4. milling, hand grinding, hand polishing.

A clearer picture of the influence of the degree of work hardening can be obtained by analyzing the results of fatigue tests of alloys with preliminary plastic deformation by tension ε_{pl} carried out uniformly over the cross section of the samples

Table 4.50 Influence of the surface state on the fatigue resistance of the EI826 alloy at 800°C, $\omega = 5000$ Hz (Sulima and Evstigneev, 1974)

Type of machining	Roughness parameters, μm	Cold work depth, μm	Work hardening degree, %	σ_{res}, 10 MPa		σ_{-1}, 10 MPa, 10^8 cycles	
				after machining	after heat treatment	after machining	after heat treatment
Grinding in a circle along	$R_z = 20\text{-}10$ $R_a = 0.08\text{-}0.16$	50 15	32 26	25-40 3-15	0-2 0	20.1 26.7	22.5 28.3
Grinding in a circle across	$R_z = 20\text{-}10$ $R_a = 20\text{-}10$	50 15	33 29.9	-25 -50	-2 -2	17.6 25.4	20,2 26.8
Roller running	$R_a = 20\text{-}10$	60 225	26.9 37.3	11 35	3 3	26.8 19.0	28.0 21.2

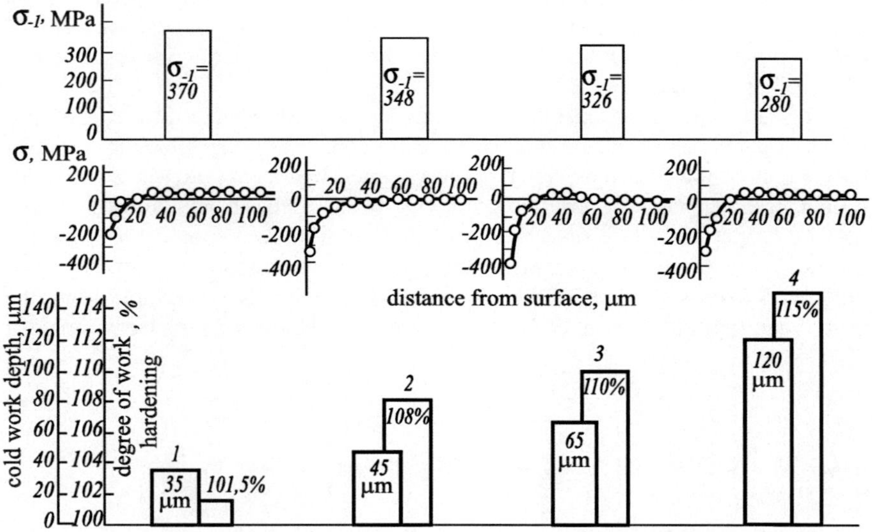

Fig. 4.82 Endurance limits, distribution of residual stresses, depth and degree of work hardening of samples made of alloy EI437B VA, processed according to technological options 1, 2, 3, 4

(Semenchenko and Mirer, 1977). It has been established that each test temperature corresponds to the optimal value of ε_{pl}, which provides the maximum fatigue resistance of the alloy. So, for alloy EI617 at 600, 700 and 800°C the maxima of σ_{-1} take place, respectively, at $\varepsilon_{pl} = 4, \varepsilon_{pl} = 2$ and $\varepsilon_{pl} = 1$ %, this increase is 15-20, 12-15 and 5-8 %. At 900°C, there is a decrease in σ_{-1} for any ε_{pl}.

So, to obtain the maximum fatigue strength of the samples, it is necessary to ensure that the amount of the work-hardening in the surface layer is close to the optimal values. Obviously, this is extremely difficult. The fatigue resistance at elevated temperatures after isothermal heating in vacuum to relieve residual stresses slightly increases in comparison with the σ_{-1} samples without such treatment, regardless of the sign of σ_{res}, which, apparently, is not related to surface effects.

Sulima and Evstigneev (1974) conclude that technological residual macrostresses have practically no effect on fatigue resistance at high temperatures, regardless of their values and sign. This conclusion, however, cannot be extended to the fatigue strength of specimens and blades made of high-temperature and titanium alloys operating at low temperatures. At low temperatures, compressive residual stresses in the surface layer, adding up with stresses from external loads, transform a symmetric cycle into an asymmetric one with an average compressive stress, which leads to an increase in σ_{-1}; tensile residual stresses have a negative effect on fatigue strength.

In recent years, the electron-beam method has been used to perforate holes in cooled blades, combustion chambers, nozzles. As shown by the tests carried out, electron-beam punching of holes does not create additional stresses and does not affect the long-term and fatigue strength. When making holes using pulsed lasers, a heat-affected zone with a depth of 0.1 mm is formed, the degree of negative effect of which on the fatigue and thermal fatigue strength depends on the processing mode (Semenchenko and Mirer, 1977).

In order to mechanize finishing operations for processing blades and reduce labor intensity, a number of methods are used: electrical discharge machining followed by heat treatment and electropolishing; electrochemical treatment followed by vibratory rolling; rolling; magnetic abrasive polishing. The carried out fatigue tests showed that the fatigue resistance of blades processed by these technological processes is practically not lower than after machining and tempering for stress relief. However, this conclusion is valid only in those cases when there is no etching along the grain boundaries and burns do not occur due to violations of the electrochemical processing process (reverse current).

4.4.3 Methods of Surface Hardening of Gas Turbine Parts

At present, a number of different methods of surface hardening of gas turbine parts and mainly turbine and compressor blades are used. However, it should be borne in mind that the efficiency of hardening depends on the temperature and time of operation: for example, for alloy EI929 at 750°C, hardening is maintained only for 1000 h. Hardening of blades from alloy EI437B effective only up to 560°C, from steels of the 15Ch12N2WMF type and titanium alloys (VT3-1, etc.) - up to 300-400°C. The limiting temperature values are related to the stress relaxation resistance of the alloy (Borzdyka and Getsov, 1978).

In Kolotnikova (1981) investigated the processes of relaxation of residual stresses in the surface layer of GTE blades made of ZhS6KP alloy. It was found that at 650°C already after 200 h residual stresses on the surface decrease by 2.5 times, and at

800°C the rate of decrease sharply increases: after 10 h residual stresses decrease by 2.5 times, after 200 h by 6.5 times.

The most widespread types of surface hardening are listed in Table 4.51 according to the data given in Grinchenko (1971); Sulima and Evstigneev (1974); Semenchenko and Mirer (1977). The same books describe the equipment, technology features and the effectiveness of strengthening the blades. These methods are used mainly to increase the fatigue strength of parts both by creating favorable residual stress diagrams and by increasing the surface finish.

Thus, in Balashov and Petukhov (1974), the effect of blowing with microspheres on the fatigue strength of the protrusions of discs made of EI698 alloy under conditions of symmetric bending with static tension was studied (Table 4.52). The results indicate a significant increase in the strength of the protrusions of the discs when they are blown with microballs.

Table 4.51 Types of hardening surface treatment

Purpose	Hardening type	Material
Turbine blade shanks hardening	Roller processing	EI929
	Rotary work hardening	ZhS6K
Reinforcement of the outer and inner surfaces of the airfoil of cooled blades	Thermoplastic hardening	ZS6U; EI598
Reinforcement of turbine disc protrusions	Microbead peening	EI698
Strengthening the airfoil of the compressor blades	Vibrating roller with balls of diameter 2-6 mm	Titanium alloys (VT9, VT20, VT3-1), steel (14Ch17N2, EI802, EI718)
	Hydroblasting with balls of diameter 1.5-2.5 mm	
	Hydrogalting with balls with a diameter of 0.6-0.8 mm	
	Ultrasonic hardening with balls of diameter 0.1 mm	
	Shot-peening hardening with microbeads 0.005-0.2 mm	

Table 4.52 Fatigue strength of disc protrusions at 650°C

Manufacturing technology	Fatigue limit, MPa	
	at σ_m = 100 MPa	at σ_m = 300 MPa
Tooth pulling Δ = 0,02 mm	125	-
Also, with Δ =0,05 mm	105	-
Also, with Δ = 0,08 mm	140	80
Also, followed by blowing with microballs	180	140

In Ivanov (2007) the hardening mode of GTK-10 rotor blades was tested. Figures 4.83 and 4.84 show diagrams of residual stresses depending on the diameter of the balls and the duration of processing. The work carried out made it possible to significantly increase the fatigue resistance of the blades (see Fig. 4.85).

The efficiency of hardening of GTE parts can be illustrated by the results of testing pearlitic steel shafts (Filimonov and Getsov, 2004). The effect of surface hardening by rolling with a roller on the fatigue resistance was studied on specimens with a diameter of 50 mm made of 38ChN3MFA steel with KP90. It was found that the fatigue limit of smooth shafts with a hardened surface in air tests increases by 10-20 %, and for shafts with fitted parts by 2-2.5 times (see Fig. 4.86). In the case of a corrosive environment (water with 3.5 % NaCl), surface hardening is an effective means of increasing fatigue resistance (based on up to 3 10^7 cycles).

Investigations were also carried out on the effect of surface plastic deformation on the rate of propagation of fatigue cracks in heat-resistant and titanium alloys. A significant decrease in speed was found in the presence of residual compressive stresses. When comparing various types of hardening surface treatment, attention is paid not only to the level of the obtained characteristics of mechanical properties, but

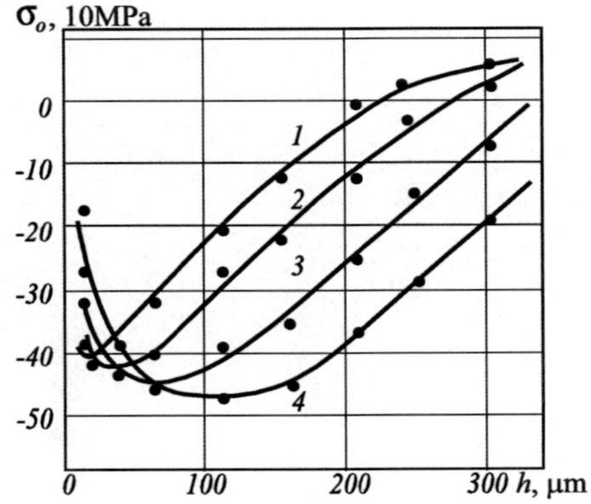

Fig. 4.83 Residual stress diagrams depending on the diameter of the balls $\varnothing_b$.
1 - $\varnothing_b = 0.8$ mm;
2 - $\varnothing_b = 1.3$ mm;
3 - $\varnothing_b = 2.0$ mm;
4 - $\varnothing_b = 2.5$ mm.

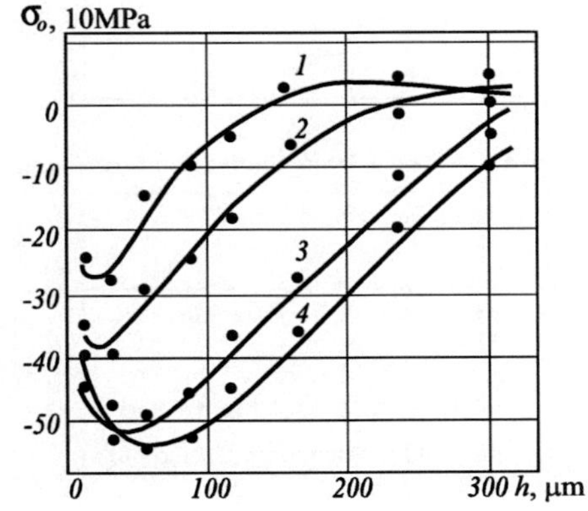

Fig. 4.84 Residual stress diagrams depending on the duration of processing, $\varnothing_b$ = 1 mm. *1* - t = 1 min; *2* - t = 2 min; *3* - t = 4 min; *4* - t = 8 min

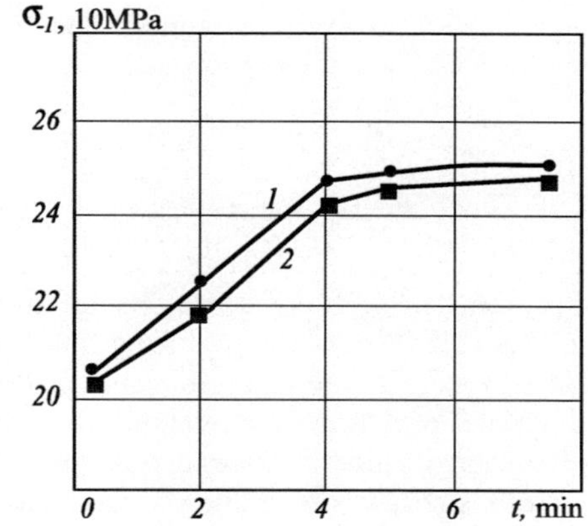

Fig. 4.85 Average endurance limit of blades made of alloy EI893 after hardening depending on the duration of processing: *1* - working blades of the high pressure turbine; *2* - working blades of low pressure turbine

also to their spread. So, for example, the use of ultrasonic deformation hardening of the blade airfoil after vibration abrasive treatment is recognized as inexpedient, since the fatigue strength changes little, and the spread of characteristics for individual

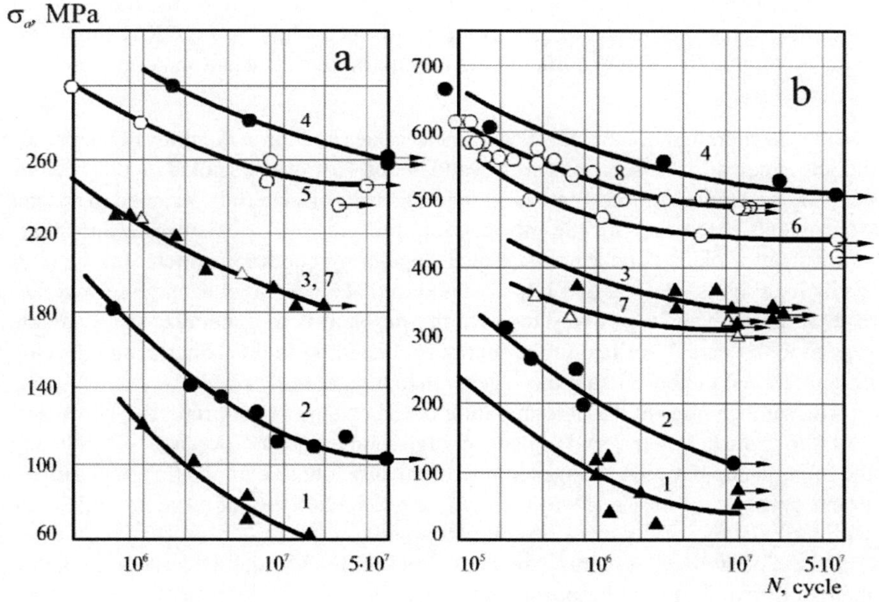

Fig. 4.86 Effectiveness of surface rolling for shafts of different diameter: 1, 2 - diameters 200 and 50, respectively, shafts with fitted sleeves without rolling; 3, 4 - diameters 200 and 50, respectively, shafts with fitted sleeves with rolling; 8, 6, 5, 7 - plain shafts, diameters 8, 40, 50, 200 mm, respectively; a - steel 38Ch2H2MA ($\sigma_{0.2}$ = 570-610 MPa, σ_b = 740-830 MPa), b - steel 38ChN3MFA ($\sigma_{0.2}$ = 930-1080 MPa, σ_b = 1060-1190 MPa)

blades increases. To increase the fatigue resistance of turbine blades, the method of thermoplastic hardening proposed by B.A. Kravchenko.

4.5 Concluding Remarks

Apparently, there is no need to convince the reader that the use of high-plastic heat-resistant single-crystal alloys for gas turbine engine blades is one of the effective ways to increase the parameters and resource, and also that predicting the service properties of materials in relation to conditions of long-term operation is the most reasonable method of ensuring reliability of gas turbine parts in these conditions.

In this chapter, in addition to the mandatory for consideration issues related to metallurgical and technological features of the manufacture of blanks and GTE parts, much attention is paid to:

- on the one hand, the development of issues (which were the subject of many years of research by VIAM and CIAM) related to the experimental determination of the behavior of single-crystal materials and their mathematical description using computational methods for determining the stress-strain state of samples and parts and the conditions of destruction of objects under study, and on the other,
- summarizing the results of research carried out mainly at the CKTI Polzunov, to determine the effect of the microstructure of heat-resistant materials on their properties.

As already mentioned, the relevance of these issues and the difficulties in the practical implementation of solutions are beyond doubt. However, as noted in the previous chapter, the amount of information on the properties of materials obtained in special experiments, necessary for the introduction into practice of calculating the stress-strain state of blades under non-stationary operating conditions, including thermal cyclic loading, is so great that it is obtained for each new alloy is extremely expensive. In this regard, a real way to solve the problem is to generalize the available experimental data described in the literature, including in this chapter, and develop computational methods using these generalizations as basic ones.

The same applies to the determination of the change in the properties of various new alloys during their very long stay at high temperatures. Therefore, establishing the relationship between changes in microstructure, phase composition, and properties is, perhaps, the only real way to predict the characteristics of refractory materials with the existing variety of modes of their operation.

On the other hand, it seems expedient either for the Russian Federation to participate in the work of the European community under the COST program, or to create and implement its own research program with the participation of various domestic scientific organizations.

Unfortunately, the possibilities for theoretical substantiation of the optimization processes of certain technological solutions that allow increasing the resource of parts are very limited. These decisions are predominantly based on purely experimental results and the intuition of the researcher. That is why, despite the vast information contained, for example, in such monographs as the book of Sulima and Evstigneev (1974) and many others, no theory has yet been proposed to predict the effect of certain technological solutions.

Since the last publication of the book in Russian, the author has published new research results, among them are Getsov et al. (2011a,b, 2012a,b, 2014, 2015); Semenov (2014); Semenov et al. (2014, 2015, 2016); Semenov and Getsov (2014b); Grishchenko et al. (2015); Tikhomirova et al. (2017). In addition, recently on the issues discussed in this chapter was published Amaro et al. (2010); le Graverend et al. (2011); Staroselsky and Cassenti (2011); Barker et al. (2013); Adair et al. (2015); Steuer et al. (2015); Hardy et al. (2016).

References

Adair BS, Johnson WS, Antolovich SD, Staroselsky A (2015) Crystallographic orientation and temperature effects on the fatigue crack growth rate and resulting fracture surface morphology in PWA1484 single crystal superalloy. Fatigue & Fracture of Engineering Materials & Structures 38(1):56–68

Amaro RL, Antolovich SD, Neu RW, Staroselsky A (2010) On thermo-mechanical fatigue in single crystal Ni-base superalloys. Procedia Engineering 2(1):815–824

Balashov BF, Petukhov AN (1974) Fatigue strength of heat resistant alloys in relation to stress concentration, cycle asymmetry, and surface cold working. Strength of Materials 6(4):481–481

Barker VM, Steven Johnson W, Adair BS, Antolovich SD, Staroselsky A (2013) Load and temperature interaction modeling of fatigue crack growth in a Ni-base superalloy. International Journal of Fatigue 52:95–105

Belolipetskiy YP (1970) Issledovaniye usloviy formirovaniya i vliyaniya velichiny zerna i raznozernistosti na sluzhebnyye svoystva zharoprochnykh splavov dlya lopatok gazovykh turbin (in Russ., Investigation of the conditions of formation and the effect of grain size and grain size on service properties of heat-resistant alloys for gas turbine blades). Avtoref. diss. cand. tech. sciences, CNIITMASH, Moscow

Betteridge W (1959) The Nimonic Alloys. Edward Arnold, London

Borzdyka AM (1974) Effect of melting procedure on the heat resistance of steels for machine construction. Metal Science and Heat Treatment 16(8):686–688

Borzdyka AM (1990) Yubilei zharoprochnogo splava (K 25-letiyu promyshlennogo primenneniya splava ei893) (in Russ., Anniversary of heat-resistant alloy (On the 25th anniversary of the industrial use of EI893 alloy). Metallovedenie i Termicheskaya Obrabotka Metallov (3):2–10

Borzdyka AM, Getsov LB (1978) Relaksaciya napryazhenii v metallakh i splavakh (in Russ., Stress Relaxation in Metals and Alloys), 2nd edn. Metallurgiya, Moscow

Borzdyka AM, Zeitlin VZ (1964) Termicheskaya obrabotka staley i splavov (in Russ., Heat Treatment of Steels and Alloys). Mashinostroenie, Moscow

Chan KS, Leverant GR (1987) Elevated-temperature fatigue crack growth. Metallurgical and Materials Transactions A 18(4):593–602

Chan KS, Feiger J, Lee YD, John R, Hudak Jr SJ (2005) Fatigue Crack Growth Thresholds of Deflected Mixed-Mode Cracks in PWA1484. Journal of Engineering Materials and Technology 127(1):2–7

Chizhik AA, Verkina LI (1983) Issledovaniye dlitel'noy prochnosti i soprotivlyayemosti polzuchesti rotorov iz stali 25X1M1VA (R2MA) posle razlichnykh srokov ekspluatacii (in Russ., Investigation of long-term strength and creep resistance of rotors made of steel 25X1M1VA (P2MA) after various periods of operation). Trudy CKTI (204):70–74

Chowdhury H, Naumenko K, Altenbach H, Krüger M (2017) Rate dependent tension-compression-asymmetry of ti-61.8 at% al alloy with long period superstructures at 1050° c. Materials Science and Engineering: A 700:503–511

Chowdhury H, Naumenko K, Altenbach H (2018a) Aspects of power law flow rules in crystal plasticity with glide-climb driven hardening and recovery. International Journal of Mechanical Sciences 146-147:486 – 496

Chowdhury H, Naumenko K, Altenbach H, Krüger M (2018b) Critical stresses estimation by crystal viscoplasticity modeling of rate-dependent anisotropy of al-rich tial alloys at high temperature. Archive of Applied Mechanics 88(1-2):65–81

Cook TS, Chatterjee A (1999) High cycle fatigue of single crystal panels containing holes. In: Wu XR, Wang ZG (eds) Fatigue '99: Proc. of the 7th International Fatigue Congress, Higher Education Press, Beijing

Cormier J, Milhet X, Mendez J, Cailletaud G (2009) Modeling the [001] non-isothermal creep behavior of a second generation single crystal Ni–based superalloy submitted to very high temperature overheat. In: Petreniya YK, Getsova LB (eds) Prochnost' materialov i resurs elementov energooborudovaniya, St. Petersburg, no. 296 in Trudy CKTI, p 199–215

Coutsouradis D, Davidson JH, Ewald J, Greenfield P, Khan T, Malik M, Meadowcroft DB, Regis V, Scarlin RB, Schubert F, Thornton DV (eds) (1994) Materials for Advanced Power Engineering 1994, vol I and II. Springer

Cunningharn SE, DeLuca DP, Haake FK (1994) Crack growth and life prediction in single crystal nickel superalloys. WL-TR-94-4089 Vol. I, Pratt and Whitney, West Palm Beach, FL

Dashunin NV (2006) Nadezhnost' metalla i upravleniye resursom lopatochnogo apparata stat-sionarnykh GTU na magistral'nykh gazoprovodakh (in Russ., Metal reliability and resource management of stationary gas turbine blades on mainline gas pipelines). Avtoref. diss. cand. tech. sciences, NPO CKTI, St. Petersburg

Dul'nev PA, Kotov PI (1980) Termicheskaya ustalost' metallov (in Russ., Thermal Fatigue of Metals). Mashinostroeni, Moscow

Dul'nev RA, Svetlov IL, Bychkov NG, Rybina TV, Sukhanov NN, Gordeeva TA, Dobrokhvalova EN, Epishin AI, Krivko AI, Nazarova MP (1988a) An orientation relationship of the thermal fatigue of nickel alloy single crystals. Strength of Materials 20(11):1415–1421

Dul'nev RA, Svetlov IL, Bychkov NG, Rybina TV, Sukhanov NN, Gordeeva TA, Dobrokhvalova EN, Epishin AI, Krivko AI, Nazarova MP (1988b) An orientation relationship of the thermal fatigue of nickel alloy single crystals. Strength of Materials 20(11):1415–1421

Erickson GL, Harris K (1994) Ds and sx superalloys for industrial gas turbines. In: Materials for Advanced Power Engineering 1994, Springer Netherlands, pp 1055–1074, Proceedings of a Conference held in Liège, Belgium, 3–6 October 1994

Filimonov GN, Getsov LB (2004) Fatigue strength of pumps shafts in connection with a scale factor and corrosion environment effect. Revue Internationale d'Ingénierie des Systèmes de Production Mécanique IX (numero special):55–59

Frost HJ, Ashby MF (1982) Deformation-mechanism Maps: the Plasticity and Creep of Metals and Ceramics. Pergamon Press, Oxford

Gabb TP, Gayda J, Miner RV (1986) Orientation and temperature dependence of some mechanical properties of the single-crystal nickel-base superalloy René N4: Part II. Low cycle fatigue behavior. Metallurgical Transactions A 17(3):497–505

Generson IG, Libman PM, Tokarev VN (1975) Issledovaniye kachestva diska iz stali EP428 vakuumno-dugovogo pereplava (in Russ., Investigation of the quality of a disk made of EP428 steel of vacuum-arc remelting). Energomashnostroyeniye (12):18–21

Getsov L, Dobina N, Rybnikov A (2007a) Thermal fatigue of Ni-Based Superalloy Single Crystal. Materiali in tehnologije / Materials and technology 41(2):67–72

Getsov L, Semenov A, Staroselsky A (2008a) A new TMF failure criterion for single crystal superalloys. In: Nauchno-tekhnicheskiye problemy prognozirovaniya nadezhnosti i dolgovech-nosti konstruktsiy i metody ikh resheniya. RELMAS'2008, Publishing house of Polytechnic University, St. Petersburg, vol 1, pp 128–131

Getsov L, Semenov A, Staroselsky A (2008b) A failure criterion for single crystal superalloys during thermocyclic loading. Materiali in tehnologije / Materials and technology 42(1):3–12

Getsov L, Semenov A, Semenov S, Rybnikov A, Tikhomirova E (2015) Thermal fatigue of single-crystal superalloys: experiments, crack-initiation and crack-propagation criteria. Materiali in tehnologije / Materials and technology 49(5):773–778

Getsov LB (1964) Metod otsenki povrezhdayemosti zharoprochnykh materialov (in Russ., Method for assessing the damageability of heat-resistant materials). Zavodskaya Laboratoriya (1):1371–1379

Getsov LB (1981) Heat resistance of macrononuniform materials. Strength of Materials 13(5):564–568

Getsov LB (1996) Materialy i prochnost' detalei gazovykh turbin (in Russ., Materials and Strength of Gas Turbine Parts). Nedra, Moscow

Getsov LB (1997) Effect of crystallographic orientation on mechanical properties of high-temperature alloys. Metal Science and Heat Treatment 39(3):100–102

Getsov LB, Rybnikov AI (1988) Influence of grain size in high-temperature-strength alloys on their properties. Strength of Materials 20(8):1057–1061

Getsov LB, Semenov AS (2009) Kriterii razrusheniya polikristallicheskikh i monokristallicheskikh materialov pri termociklicheskom nagruzhenii (in Russ., Criteria for the failure of polycrystalline and single-crystal materials under thermal cycling loading). In: Petreniya YK, Getsova L (eds) Prochnost' materialov i resurs elementov energooborudovaniya, St. Petersburg, no. 296 in Trudy CKTI, p 83–91

Getsov LB, Taubina MG (1960) Zharoprochnost' austenitnykh splavov pri tsiklicheski izmenyayushchikhsya temperaturakh i napryazheniyakh (in Russ., Heat resistance of austenitic alloys at cyclically varying temperatures and stresses). Izvestiya AN SSSR OTN Metallurgiya i toplivo (5):100–109

Getsov LB, Zhirnov DF, Terekhov KI, Taubina MG (1960) Effect of structure on the hot strength of alloys. Metal Science and Heat Treatment of Metals 2(9):477–480

Getsov LB, Shovkun VE, Filatova MA (1965) Opyt primeneniya splava EI893 dlya lopatok gazovykh turbin (in Russ., Experience of EI893 alloy application for gas turbine blades). Energomashnostroyeniye (2):30–32

Getsov LB, Rybnikov AI, Pigrova GD (1999) Change of steels and alloys structure and properties during long operation at high temperature. J Phys IV France 09:Pr9–105–Pr9–115

Getsov LB, Rybnikov AI, Pigrova GD (2000) Izmeneniye struktury i svoystv staley i splavov v protsesse dlitel'noy ekspluatatsii pri povyshennykh temperaturakh (in Russ., Changes in the structure and properties of steels and alloys during long-term operation at elevated temperatures). Teploenergetika (4):27–33

Getsov LB, Okatova GP, Rybnikov AI, Fedorchenko DG (2003) Strukturnoye sostoyaniye i resurs detaley GTD posle dlitel'noy narabotki (in Russ., Structural condition and resource of GTE parts after long-term operating time). In: Nauchnyye trudy VIgo mezhdunarodnogo seminara *Sovremennyye problemy prochnosti* im. V.I. Likhacheva, Velikii Novgorod, vol 2, pp 204–219

Getsov LB, Rybnikov AI, Dobina NI (2007b) Osobennosti termoustalostnogo razrusheniya monokristallicheskogo zharoprochnogo splava (in Russ., Features of thermo-fatigue fracture of single-crystal heat-resistant alloy). Tyazheloe Mashinostroenie (8):12–15

Getsov LB, Dobina NI, Rybnikov AI, Semenov AS, Starosel'skii A, Tumanov NV (2008c) Thermal fatigue resistance of a single-crystal alloy. Strength of Materials 40(5):538–551

Getsov LB, Rybnikov AI, Semenov AS (2009) Progressiruyushcheye deformirovaniye materialov pri termociklicheskom nagruzhenii (in Russ., Progressive deformation of materials under thermal cycling loading). In: Petreniya YK, Getsova L (eds) Prochnost' materialov i resurs elementov energooborudovaniya, St. Petersburg, no. 296 in Trudy CKTI, p 105–120

Getsov LB, Katanaha NA, Krylov AP, Petrenya YK (2011a) Choice of tempering mode for relieving stresses in parts. Metal Science and Heat Treatment 53(9):491–494

Getsov LB, Rybnikov AI, Dobina NI, Dashunin NV (2011b) Specific features relating to thermal fatigue rupture of the blade materials of gas turbine units. Thermal Engineering 58(9):736–740

Getsov LB, Semenov AS, Grigoriev AV, Tikhomirova EA (2012a) Soprotivleniye deformirovaniyu i razrusheniyu monokristallicheskikh splavov pri staticheskom i termociklicheskom nagruzhenii (in Russ., Resistance to deformation and fracture of single-crystal alloys under static and thermal cyclic loading). Nadezhnost' i Bezopasnost' Energetiki (3):53–62

Getsov LB, Semenov AS, Rybnikov AI, Tikhomirova EA (2012b) Kriterii razrusheniya monokristallicheskikh splavov pri staticheskom i termociklicheskom nagruzhenii (in Russ., Criteria for the destruction of single-crystal alloys under static and thermal cyclic loading). Vestnik Dvigatelestroyeniya (2):255–263

Getsov LB, Semenov AS, Tikhomirova EA, Rybnikov AI (2014) Thermocyclic and static failure criteria for single crystal superalloys of gas turbine blades. Materiali in tehnologije / Materials and technology 48(2):255–260

Getsov LB, Laptev AB, Puzanov AI, Shelyapina NM (2019) Sulfide oxide corrosion of modern heat-resistant alloys. Russian Aeronautics 62(4):689–695

Golubovskiy E, Svetlov I, Nozhnitsky Y (2005a) Relationship of stress rupture and crystallographic orientation for Ni-base superalloys single crystal. In: EUCASS. European Conference for Aerospace SciencesMoskow (CD), Moscow

Golubovskiy ER, Epishin AI, Svetlov IL (2004) Anizotropiya kharakteristik staticheskoy i tsiklicheskoy prochnosti monokristallov litogo nikelevogo zharoprochnogo splava (in Russ., Anisotropy of static and cyclic strength characteristics monocrystals of cast nickel heat-resistant alloy). Vestnik Dvigatelestroyeniya (2):143–147

Golubovskiy ER, Svetlov IL, Khvatskiy KK (2005b) Zakonomernosti izmeneniya aksial'noy i azimutal'noy anizotropii prochnostnykh kharakteristik monokristallov zharoprochnykh nikelevykh splavov dlya lopatok gtd (in Russ., Regularities of changes in axial and azimuthal anisotropy of strength characteristics of single crystals of heat-resistant nickel alloys for gas turbine engine blades). Aviation-space Engineering and Technology (10):50–54

Grinchenko IG (1971) Uprochneniye detaley iz zharoprochnykh i titanovykh splavov (in Russ., Hardening of parts made of heat-resistant and titanium alloys). Mashinostroenie, Moscow

Grishchenko AI, Semenov AS, Getsov LB (2015) Modeling inelastic deformation of single crystal superalloys with account of γ/γ'-phases evolution. Materials Physics and Mechanics 24(3):325–330

Hardy M, Huron E, Glatzel U, Griffin B, Lewis B, Rae C, Seetharaman V, Tin S (eds) (2016) Superalloys. Wiley-TMS, Hoboken, NJ

Ivanov SA (2007) Soprotivleniye ustalosti metalla rabochikh lopatok statsionarnykh GTU v zadachakh prodleniya resursa (in Russ., Resistance to fatigue of metal of rotor blades of stationary gas turbines in problems of resource extension). Avtoref. diss. cand. tech. sciences, NPO CKTI, St. Petersburg

Ivanov SA, Rybnikov AI (2006) Vliyaniye dlitel'noy ekspluatatsii na soprotivleniye ustalosti turbinnykh lopatok (in Russ., Influence of long-term operation on fatigue resistance of turbine blades). Morskoy Vestnik (4):19–22

Ivanov SA, Ivanov AV, Rybnikov AI, Masaleva EN (2001) Vliyaniye vosstanovitel'noy termoobrabotki na sluzhebnyye kharakteristiki metalla rabochikh lopatok gazovykh turbin posle narabotki (in Russ., Influence of recovery heat treatment on service characteristics of the metal working blades of gas turbines after operating time). Turbiny i Kompressory (16-17):34–36

Ivanov SA, Rybnikov AI, Kuznecov VG (2006) Vliyaniye dlitel'noy ekspluatatsii na ustalostnuyu prochnost' lopatok gazovykh turbin (in Russ., Influence of long-term operation on the fatigue strength of gas blades turbines). Metalloobrabotka (5-6):51–53

Jo CY, Kim HM (2003) Effect of recrystallisation on microstructural evolution and mechanical properties of single crystal nickel based superalloy CMSX-2. Part 2 - Creep behaviour of surface recrystallised single crystal. Materials Science and Technology 19(12):1671–1676

Kablov EN (ed) (2006) Litejnye zharoprochnye splavy. Effekt Kishkina (in Russ., Casting Heat-resistant Alloys. Kishkin's Effect). Nauka, Moscow

Khan T (1986) Recent developments and potential of single crystal superalloys for advanced turbine blades. In: Betz W, Brunetaud R, Coutsouradis D, Fischmeister H, Gibbons TB, Kvernes I,

Lindblom Y, Marriott JB, Meadowcroft DB (eds) High Temperature Alloys for Gas Turbines and Other Applications, D. Reidel, Dordrecht, vol I, pp 21–50

Kishkin ST (2006) Sozdaniye, issledovaniye i primeneniye zharoprochnykh splavov. Izbrannyye trudy (in Russ., Creation, Research and Application of Heat-resistant Alloys. Selected Works). Nauka, Moscow

Kolotnikova OV (1981) Issledovaniye relaksatsii napryazheniy kak diagnosticheskogo priznaka ischerpaniya resursa lopatok GTD (in Russ., Investigation of stress relaxation as a diagnostic sign of exhaustion of the resource of GTE blades). Avtoref. dis. cand. tech. sciences, Kuybyshevskiy aviacionnyy institut im. S. P. Koroleva, Kuibyshev

Kulygin GV, Bespyatyi VA, Tuchinskii LI, Samgin VA, Pis'mennyi NN (1971) Mechanical properties of heat-resistant castings. Strength of Materials 3(8):997–999

Lanskaya KA (1976) Vysokokhromistyye zharoprochnyye stali (in Russ., High-chromium Heat-resistant Steels). Metallurgiya, Moscow

le Graverend JB, Cormier J, Caron P, Kruch S, Gallerneau F, Mendez J (2011) Numerical simulation of γ/γ' microstructural evolutions induced by TCP-phase in the MC2 nickel base single crystal superalloy. Materials Science and Engineering: A 528(6):2620–2634

Leverant GR, Gell M (1975) The influence of temperature and cyclic frequency on the fatigue fracture of cube oriented nickel-base superalloy single crystals. Metallurgical Transactions A 6(375)

Levin EE, Pivnik EM, Stanyukovich AV, IG G, VI T (1975) Povysheniye resursa rabochikh lopatok gazovykh turbin iz splava EI893 (in Russ., Increasing the resource of the working blades of gas turbines made of EI893 alloy). Trudy CKTI (130):54–60

Levin EE, Pivnik EM, Generson IG (1977) Opyt primeneniya zharoprochnogo splava EI893 dlya izgotovleniya lopatok gazovykh turbin (in Russ., Experience of using heat-resistant alloy EI893 for the manufacture of gas turbine blades). Progressivnoye formoobrazovaniye, metallovedeniye i termoobrabotka, Leningr. dom nauch.-tekhn. propagandy, Leningrad

Li SX, Smith DJ (1994) An overview of combined fatigue and creep response of single crystal nickel base superalloys. In: Materials for Advanced Power Engineering 1994, Part II, Springer Netherlands, pp 1175–1184, Proceedings of a Conference held in Liège, Belgium, 3–6 October 1994

Manilova EP (2005) Kinetika fazovo-strukturnykh protsessov v usloviyakh dlitel'noy ekspluatatsii v 12% khromistoy stali (EP428) (in Russ., Kinetics of phase-structural processes under conditions of long-term operation in 12% chromium steel (EP428)). Avtoref. dis. cand. tech. sciences, NPO CKTI, St. Petersburg

Marchal N, Forest S, Duvinage LRS (2006) Simulation of fatigue crack growth in single crystal superalloys using local approach to fracture. In: Steglich D, Besson J, Moinereau D (eds) Local Approach to Fracture (9th European Mechanics of Materials Conference, Euromech–Mecamat), Presses de l'Ecole des Mines de Paris, Paris, p 353–358

Maslenkov SB, Maslenkova EA (1991) Spravochnik stali i splavy dlya vysokikh temperatur (in Russ., Handbook Steels and Alloys for Advanced Temperatures), vol 1 & 2. Metallurgiya, Moscow

Miles DE, Mclean M (1977) Anisotropic creep behaviour of (Co, Cr)(Co, Cr)$_7$C$_3$ eutectic composite. Metal Science 11(12):563–570

Naumenko K, Gariboldi E (2014) A phase mixture model for anisotropic creep of forged Al-Cu-Mg-Si alloy. Materials Science and Engineering: A 618:368–376

Naumenko K, Altenbach H, Kutschke A (2011) A combined model for hardening, softening, and damage processes in advanced heat resistant steels at elevated temperature. International Journal of Damage Mechanics 20(4):578–597

Nozhnitsky YA, Golubovsky ER (2009) O prochnostnoy nadezhnosti monokristallicheskikh rabochikh lopatok vysokotemperaturnykh turbin perspektivnykh GTD (in Russ., On the strength reliability of high-temperature single-crystal rotor blades turbines of promising gas turbine engines). In: Petreniya YK, Getsova L (eds) Prochnost' materialov i resurs elementov energooborudovaniya, St. Petersburg, no. 296 in Trudy CKTI, p 74–82

Paton BE, Stroganov GB, Kishkin ST, Bokshtein SE, Logunov AV, Malashenko IS, Movchan BA, Chumakov VA (1987) Zharoprochnost' liteynykh nikelevykh splavov i zashchita ikh ot okisleniya (in Russ., Heat resistance of casting nickel alloys and their protection from oxidation). Naukova dumka, Kiev

Pigrova GD (1980) O metodakh prognozirovaniya TPU faz v splavakh na nikelevoy osnove, primenyayemykh dlya gazovykh turbin (in Russ., On the methods for predicting TPU phases in nickel-based alloys used for gas turbines). Trudy CKTI (172):59–65

Pigrova GD (2006) Diagnostika strukturnykh izmeneniy metalla detaley gtu v protsesse ekspluatatsii metodom fazovogo fiziko-khimicheskogo analiza (in Russ., Diagnostics of structural changes in the metal of gas turbine parts during operation by the method of phase physical and chemical analysis). Trudy CKTI (295):55–64

Pigrova GD (ed) (2009) Fazovyye prevrashcheniya v materialakh energeticheskogo mashinostroyeniya (in Russ., Phase transformations in materials of power engineering). No. 297 in Trudy CKTI, NPO CKTI, St. Petersburg

Pimenova GP (1973) Issledovaniye lopatok iz splava ZhS6KP i puti uvelicheniya ikh resursa (in Russ., Research of blades made of ZhS6KP alloy and ways to increase their resource). Avtoref. diss. cand. tech. sciences, Kazanskiy aviatsionnyy institut, Kazan

Popov VS, Shumikin SA, Tkachenko YM (1989) Issledovaniye vozmozhnosti vosstanovleniya struktury i svoystv metalla rabochikh lopatok TVD GTK-10I posle ekspluatatsii termicheskoy obrabotkoy (in Russ., Study of the possibility of restoring the structure and properties of metal workers turbine blades GTK-10I after operation by heat treatment). In: Novyye konstruktsionnyye stali i splavy i metody ikh obrabotki dlya povysheniya nadezhnosti i dolgovechnosti izdeliy. Tezisy doklada IV Vsesoyuznoy nauchno- tekhnicheskoy konferentsii, Zaporozh'ye, p 9

Pridantsev MV (1968) Zharoprochnyye stareyushchiye splavy na nikelevoy i zheleznoy osnovakh (in Russ., Heat-resistant aging alloys on nickel and iron bases). In: Fiziko-khimicheskiye issledovaniya zharoprochnykh splavov, Nauka, Moscow, pp 88–95

Pridorozhny RP, Sheremetyev AV (2005) Osobennosti vliyaniya kristallograficheskoy oriyentatsii na ustalostnuyu prochnost' monokristallicheskikh lopatok turbin (in Russ., Features of the influence of crystallographic orientation on fatigue strength single-crystal turbine blades). Aviation-space Engineering and Technology (10):55–59

Protasova NA (1993) Rentgenodifraktometricheskoye issledovaniye tonkoy struktury monokristallicheskikh lopatok iz nikelevykh splavov (in Russ., X-ray diffractometric study of the fine structure of single-crystal nickel blades alloys). Avtoref. diss. cand. tech. sciences, VIAM, Moscow

Rabotnov YN (1969) Creep Problems in Structural Members. North-Holland, Amsterdam

Reed PAS, Sinclair I, Wu XD (2000) Fatigue crack path prediction in UDIMET 720 nickel-based alloy single crystals. Metallurgical and Materials Transactions A 31(1):109–123

Rtishchev VV (1991) Legirovaniye i termicheskaya obrabotka zharoprochnykh nikelevykh deformiruyemykh i liteynykh izotropnykh i anizotropnykh lopatochnykh splavov statsionarnykh gazoturbinnykh ustanovok (in Russ., Alloying and heat treatment of heat-resistant nickel wrought and foundry isotropic and anisotropic blade alloys of stationary gas turbine plants). Avtoref. diss. dokt. tech. sciences, Sankt Peterburgskiy politekhnicheskiy universitet, St. Petersburg

Rtishchev VV (1992) Perspektivnye anizotropnye materialy lopatok stacionarnykh GU so stolbchattoi i monokristallicheskimi strukturami (in Russ., Promising anisotropic materials of blades of stationary GU with columnar and single-crystal structures. Trudy CKTI 270:104–119

Rtishchev VV (1994) Anisotropic alloys with columnar and single crystal structures used for blades of stationary gas turbine plants. In: Materials for Advanced Power Engineering 1994, Part II, Springer Netherlands, pp 1135–1144, Proceedings of a Conference held in Liège, Belgium, 3–6 October 1994

Rybnikov AI (1979) Issledovaniye mezhzerennykh prosloyek v zharoprochnykh splavakh na nikelevoy osnove dlya lopatok statsionarnykh gazovykh turbin s tsel'yu ustanovleniya trebovaniy k optimal'noy strukture splavov (in Russ., Investigation of intergranular interlayers in heat-resistant nickel-based alloys for stationary blades gas turbines in order to establish the requirements for the optimal alloy structure). Avtoref. dis. cand. tech. sciences, CKTI, Leningrad

Rybnikov AI, Melnikova IS, Pigrova GD, Kryukov II, Bodrov IS, Levin AE, Pavlov AF (1984) Sposob vosstanovitel'noy termicheskoy obrabotki lopatok iz zharoprochnykh splavov na osnove nikelya (in Russ., Method of recovery heat treatment of blades made of high-temperature alloys based on nickel). Author's certificate No. 1269528

Savin GN (1970) Stress Distribution Around Holes. No. F-607 in NASA Technical Translations, NASA, Washington D.C

Schmid E, Boas W (1935) Kristallplastizität: Mit Besonderer Berücksichtigung der Metalle, Struktur und Eigenschaften der Materie in Einzeldarstellungen, vol 17. Springer, Berlin, Heidelberg

Schubert F, Rieck T, Ennis PJ (2000) The Growth of Small Cracks in the Single Crystal Superalloy CMSX-4 at 750 and 1000°C. In: Pollock TM, Kissinger RD, Bowman RR, Green KA, McLean M, Olson S, Schirra JJ (eds) Proc. Superalloys 2000, TMS (The Minerals, Metals & Materials Society), Philadelphia, pp 341–346

Semenchenko IV, Mirer YG (1977) Povysheniye nadezhnosti lopatok gazoturbinnykh dvigateley tekhnologicheskimi metodami (in Russ., Improving the reliability of gas turbine engine blades by technological methods). Mashinostroenie, Moscow

Semenov AS (2003) PANTOCRATOR – konechno-elementnyy programmnyy kompleks, oriyentirovannyy na resheniye nelineynykh zadach mekhaniki (in Russ., PANTOCRATOR - finite element software package focused on solving non-linear problems of mechanics). In: Trudy V Mezhdunarodnoy konferentsii *Nauchno-tekhnicheskiye problemy prognozirovaniya nadezhnosti i dolgovechnosti konstruktsiy i metody ikh resheniya*, Publishing house SPbSPU, St. Petersburg, pp 466–480

Semenov AS (2014) Identifikatsiya parametrov anizotropii fenomenologicheskogo kriteriya plastichnosti monokristallov na osnove mikromekhanicheskoy modeli (in Russ., Identification of the anisotropy parameters of the phenomenological criterion of plasticity of single crystals based on the micromechanical model). Nauchno-tekhnicheskiye vedomosti SPbGPU Fiziko-matematicheskiye nauki 194(2):15–29

Semenov AS, Getsov LB (2010) Ispol'zovaniye deformatsionnogo kriteriya razrusheniya dlya otsenki termoustalostnoy prochnosti obraztsov i detaley iz monokristallicheskogo splava (in Russ., Using the deformation criterion of fracture to assess the thermal fatigue strength of samples and parts made of a single crystal alloy). In: Prochnost' materialov i elementov konstruktsiy, Kiev, vol 2

Semenov AS, Getsov LB (2014a) Thermal fatigue fracture criteria of single crystal heat-resistant alloys and methods for identification of their parameters. Strength of Materials 46(1):38–48

Semenov AS, Getsov LB (2014b) Thermal fatigue fracture criteria of single crystal heat-resistant alloys and methods for identification of their parameters. Strength of Materials 46(1):38–48

Semenov AS, Getsov LB, Semenov SG, Grishchenko AI (2014) Deformation and fracture strength of single-crystal superalloys under static and cyclic loading. Vestnik Samarskogo gosudarstvennogo aerokosmicheskogo universiteta (5, part 3):66–75

Semenov AS, Semenov SG, Getsov LB (2015) Methods of computational determination of growth rates of fatigue, creep, and thermal fatigue cracks in poly- and monocrystalline blades of gas turbine units. Strength of Materials 47(2):268–290

Semenov SG, Getsov LB, Tikhomirova EA, Semenov AS (2016) Special features of creep and long-term strength of single-crystal refractory nickel-base alloys. Metal Science and Heat Treatment 57(11):731–738

Shalin RE, Bulygin IP, Golubovsky ER (1981) Zharoprochnost' splavov dlya gazoturbinnykh dvigateley (in Russ., Heat Resistance of Alloys for Gas Turbine Engines). Metalurgiya, Moscow

Shalin RE, Svetlov IL, Kachanov EB, Toloraia VN, Gavrilin OS (1997) Monokristally nikelevykh zharoprochnykh splavov (in Russ., Monocrystals of Heat-resistant Nickel Alloys). Mashinostroenie, Moscow

Shaydak BP, Ivanov SA, Rybnikov AI (2006) Vliyaniye strukturnogo faktora na konstruktsionnuyu ustalostnuyu prochnost' rabochikh lopatok promyshlennykh GTU (in Russ., Influence of the structural factor on the structural fatigue strength of workers industrial gas turbine blades). Trudy CKTI (295):79–89

Shpindler SS, Landa MK (1976) Rezhim okhlazhdeniya i kachestvo litykh lopatok gazoturbinnykh dvigateley (in Russ., Cooling mode and quality of cast blades of gas turbine engines. Liteinoe proizvodstvo (9):24–25

Sims CT, Hagel WC (eds) (1972) The Superalloys. Wiley-Interscience, New York

Stanyukovich AV (1967) Khrupkost' i plastichnost' zharoprochnykh materialov (in Russ., Brittleness and Plasticity of Heat-resistant Materials). Metallurgiya, Moscow

Staroselsky A, Cassenti BN (2011) Creep, plasticity, and fatigue of single crystal superalloy. International Journal of Solids and Structures 48(13):2060–2075

Steuer S, Villechaise P, Pollock T, Cormier J (2015) Benefits of high gradient solidification for creep and low cycle fatigue of AM1 single crystal superalloy. Materials Science and Engineering: A 645:109–115

Sulima AM, Evstigneev MI (eds) (1974) Kachestvo poverkhnostnogo sloya i ustalostnaya prochnost' detaley iz zharoprochnykh i titanovykh splavov (in Russ., Surface Layer Quality and Fatigue Strength of Heat-resistant and Titanium Parts Alloys). Mashinostroenie, Moscow

Taylor GI (1938) Analysis of plastic strain in a cubic crystal. In: Contributions to the Mechanics of Solids dedicated to Stephen Timoshenko by his friends on the occasion of his sixties birthday anniversary, MacMillian, New York, pp 218–224

Telesman J, Ghosn LJ (1996) Fatigue crack growth behavior of PWA1484 single crystal superalloy at elevated temperatures. Journal of Engineering for Gas Turbines and Power 118(2)

Tikhomirova EA, Rybnikov AI, Getsov LB (2017) Changes in the structure and properties of single-crystal refractory alloys under long-term operation. Metal Science and Heat Treatment 59(1):34–38

Troshchenko VT, Pokrovskiy VV, V PA (1987) Treshchinostoykost' metallov pri ciklicheskom nagruzhenii (in Russ., Fracture Toughness of Metals Under Cyclic Loading). Naukova Dumka, Kiev

Yagodkin YD, Shulyak VL, Orekhov VB (1987) Mekhanicheskiye svoystva kristallov nikelevogo splava s razlichnoy kristallograficheskoy oriyentatsiyey (in Russ., Mechanical properties of crystals of nickel alloy with different crystallographic orientation). Energomashinostroenie (56):30–35

Zhang GP, Wang ZG (1999) Creep–fatigue crack growth of intermetallic compound Ni_3 Al(B) at elevated temperatures. Materials Science and Technology 15(11):1230–1235

Zhou H, Harada H, Kobayashi T, Koizumi Y (2003) Aspects of thermomechanical fatigue of two single crystal nickel based superalloys. Materials Science and Technology 19(7):847–852

Appendix A
Russian Steels and Alloys for Gas Turbine Units

Below are given a list of Russian steels and alloys discussed in the book. In the first column is presented the material class. In the second column is given the steel or alloy brand designation used in practice in Russia. The basic brand designation according to the state standard GOST is presented in column three. The last two columns contain the numbers of the tables and figures that relate to the steel or alloy brands in columns two and three. Note that the original Russian brands are given mainly in Cyrillic letters - they are presented in the table by Latin letters.

L. B. Getsov, *Materials and Strength of Gas Turbine Parts*, Advanced Structured Materials 150, https://doi.org/10.1007/978-981-16-0534-5

Table A.1: Materials used in gas turbine units in the Russian Federation

Material class	Material	GOST designation	Technical conditions	Table(s)	Figure(s)
Pearlitic steels	R2MA	25Ch1M1FA	TU 108.1029	2.4, B.1, D.4	B.2, D.1
	EI723	25Ch2M1F	GOST 20072-74	2.16, 2.19, 3.6, 3.11, 3.29, C.1	2.9, C.1, C.2
	26ChN3M2FAA	26ChN3M2FAA	-	1.7	
	EI415	20Ch3MBF	GOST 20072-74	2.4, 3.2, 3.11, 3.12, 4.1, B.1	3.4, 3.15, 4.1, B.4
	EP182	20Ch1M1F1TR			2.36
	25Ch2M2FA	25Ch2M2FA			2.8
	15Ch1M1F	15Ch1M1F	GOST 5632-72	2.4, 4.39, B.1	4.72, B.1
Martensitic, ferritic, ferritic-martensitic austenitic steels	2Ch13*	20Ch13		1.8, 1.9, 2.4, 3.20, 4.39, D.1, D.4	2.8, 3.3, 3.92, D.5, D.17
	15Ch11MF	15Ch11MF		4.39	
	EP291	18Ch11MFB		4.39, B.1	B.3
	EI802	15Ch12WNMF		2.4, 2.5, 3.6, 3.31, 4.39, B.1, C.1	2.9, 2.28, 3.67, B.5, C.3, C.4
	EI961	13Ch11N2W2MF		1.9, 3.20, 3.31, D.1, D.4	3.92, D.8
	EI56	09Ch16N4B		1.9	
	EP718	ChN45MBTYuBR		1.9	
	EP428	20Ch12BNMF			4.73
	08Ch17N6T	08Ch17N6T		1.8, 1.9, 2.4, 3.31, D.1, D.4	3.92, D.9
	EI268*	14Ch17N2		1.8, 1.9, 2.4, 3.20, 3.31, 4.51, D.1, D.4	3.92, D.7

Continuation on next page

Table A.1: Continued

Material class	Material	GOST designation	State Standard	Table	Figure
Austenitic steels	Ch18N10T*	12Ch18N10T			2.5, 2.27, 2.31, 2.32, 3.57, 3.83
	EI417*	20Ch23N18		2.5, 2.7, 4.40, C.1	2.9, 3.67, C.5
	EI703	ChN38T		3.15, 4.40, 4.45	
	EI696	0Ch11N20T3R			2.20
	EI481	37Ch12N7G8MFB		1.3, 2.4, 2.6, 2.7, 2.16, 2.19, 3.12, 3.14, 3.15, 3.20, 3.29, 4.40, 4.45, B.1	1.22, 1.23, 1.25, 1.33, 2.33, 3.5, 3.15-3.17, 3.29, 3.57, 4.66, 4.67, B.6
	EI572	31Ch19N9MWBT		1.3, 4.40	1.29, 4.56, 4.57
	EI726	09Ch14N18W2BR1		4.42	1.13
	EP126	ChN28WMAB		3.15,4.45	
	EI612	ChN35WT		1.3, 2.4, 2.19, 3.3, 3.14, 3.29, B.1	1.9, 1.23, 1.26, 1.27, 3.5, 3.6, 3.15, 3.16, 3.33, 3.76, 3.79, B.7
	EI692	ChN35WMT		3.12	3.26
	EI787	ChN35WTYu		1.3, 3.14, 3.15, 4.40, 4.44, 4.45	1.5, 1.29, 4.59, 4.66
Deformed heat-resistant nickel-based alloys	EI607	ChN80TBYu		4.41	
	EI607A	ChN80T1BYu		2.4, 2.6, 3.3, 3.11, 3.14, 3.18, 3.24, 3.25, 3.29, B.1	1.55, 2.9, 3.15, 3.28, 3.49, 3.55, 3.56, 3.66, 3.67, 3.79, 3.82, B.8
	EI437B	ChN77YuR		3.14, 3.17, 3.18, 4.41, D.3	1.56, 1.57, 1.60, 3.44, 3.57, 3.75, 4.82

Continuation on next page

Table A.1: Continued

Material class	Material	GOST designation	State Standard	Table	Figure
	EI617	ChN70WMTYu		3.18, 3.21	1.57, 1.60
	EI765	ChN70WMYuT		2.5, 2.16, 2.19, 3.6-3.8, 3.12, 3.14, 3.24, 3.29, B.1, C.1	1.7, 1.11, 1.12, 1.55, 1.60, 2.2, 2.8, 2.9, 2.14, 2.19, 2.22, 2.24, 2.28, 2.30, 2.41, 3.28, 3.35, 3.43, 3.56, 3.58, 3.66, 3.67, 3.68, 3.70, 3.72, 3.76, 3.77, 4.2, B.9, C.6
	EI893	ChN65WMTYu		3.17, 3.18, 3.20, 3.25, 4.2, 4.38, 4.42, 4.43, B.1	1.8, 1.20, 1.59, 1.60, 3.51, 4.60, 4.74, 4.78, 4.80, 4.85, B.12
	EI826	ChN70WMTYuF		2.5, 2.6, 2.16, 2.18, 2.19, 3.7, 3.8, 3.18, 3.25, 4.4, 4.5, 4.6, 4.49, 4.50, B.1, C.1	1.1, 1.14, 1.57, 1.60, 2.29, 3.28, 3.43, 3.44, 3.56, 3.58, 3.59, 3.67, 3.70, 3.72, 3.76, 3.77, 4.3, 4.4, B.10, C.7
	EI827	ChN75WMYu		2.7, 2.16, 2.19, 3.3, 3.7, 3.15, 3.18, 3.24, B.1	1.55, 1.57, 1.60, 2.11, 2.33, 3.28, 3.35, 3.36, 33.43, 3.56, 3.66, 3.70, 3.76, B.11
	EI661	-	-	3.18	2.33, 3.36
	EK78	ChN65KWMYuTB	TU 14-1-4026-85		1.60
	EP800	ChN65KMWYuB	TU 14-1-4834-90		1.60

Continuation on next page

Table A.1: Continued

Material class	Material	GOST designation	State Standard	Table	Figure
	EP957	ChN60KWKMB	TU 14-1-4827-90		1.60, 3.65
	EP539	ChN60MYuWT	GOST 23705-79	3.18, 3.25	1.60, 3.63
	ZhS6KP	-	-	D.3, D.4	1.57, 3.22, 3.314.58, D.14
	EI929	ChN55WMTKYu	GOST 5632-72	2.9, 3.25, 4.4, 4.36, B.1	1.57, 1.60, 3.62, 4.61, 4.62, B.13
	EI867	ChN62MWKYu		1.60, 3.17, 3.21, 4.41	
	EP109	ChN56WMKYu			1.57, 1.60
	EP220	ChN51WMTYuKFR	GOST 23705-79	2.5, 2.19, 3.22, 3.24, 3.25, 4.4, 4.6, 4.41, 4.49, B.1, C.1	1.15, 1.55, 1.57, 1.60, 2.27, 2.28, 2.32, 3.45, 3.58, 3.60, 3.52, 3.66, 3.70, 4.63, B.14, C.9
	EP238	ChN58WMKYuR			1.57
	EI698	ChN73MBTYu	TU 14-1-1466-75	3.12, 3.13, 3.14, 4.52, D.4	1.25, 2.9, 2.25, 3.67, D.12
	EP741	ChN60KMWTYu	TU 14-1-131-369-77	D.11	1.60
	EP742	ChN62BMKTYu	TU14-1-3998-85	3.32, D.2	
	EP962	ChN60KMYuBWTF	STU 14-1-2345-78	D.2	
	EK79 (EP742U)	EK79	TU 14-131-561-83	1.13, D.3, D.4	D.13
	EK151	EK151	-	1.13	
	EK152	EK152	-	1.13	
	EP975 (VDS-75)	EP975	TU 14-131-568-84	1.13	
Cast heat-resistant steels and nickel-based	EI893(L)	EI893L	TU 108.1109-82	1.10, 3.51, 4.36, 4.42	1.2,
	ZMI-2	ZMI-2		3.25	
	ZMI-3, ZMI3U	ChN64WMKYuT	-	1.5, 3.26	1.3, 1.60, 3.64, 4.76

Continuation on next page

Table A.1: Continued

Material class	Material	GOST designation	State Standard	Table	Figure
alloys	ChC70VI	ChS70VI	TU 1-809-1025-98	1.5, 3.28, B.1	1.60, 2.17
	ChS104	ChS104		1.5	1.60, 2.16
	SN35	SN35	-	1.5	1.60
	CNK7	CNK7	-	1.5, 2.1, 3.27, 4.42, 4.43	1.60, 3.64, 4.10, 4.20, 4.74, 4.77, 4.79
	EP539LM	EP958	-	1.5	1.60, 3.53
	ZhS6K	Zh6K	-	1.5, 3.3, 3.17, 3.18, 3.25, 4.3, 4.36, 4.46, 4.49, B.1	1.1, 1.6, 2.15, 3.41, 3.50, 4.16
	ZhS6F	ZhS6F	-	2.16, 2.19, 4.12, 4.34, 4.35, D.3	1.60, 4.11, 4.17, 4.29, 4.31, 4.33, 4.35
	VZhL12U	VZhL12U		1.5, 3.18	1.60, 3.41
	ZhS6U	ZhS6U	TU 1-331514	1.5, 3.5, 3.24, 4.15, 4.37, D.3	1.60, 3.13, 3.41, 3.42, 3.66, 3.86, 4.27, 4.63
	ZhS26	Zh26	-	1.5	1.60, 4.22
	ZhS32	Zh32		1.5, 2.1, 4.11, 4.13, 4.16, 4.34, 4.35	1.60, 4.5, 4.12, 4.17
	ZhS36	Zh36		1.5, 2.4, 4.11, 4.17, 4.19, 4.34, 4.36, B.1	1.60, 2.18, 4.17, 4.19, 4.25, 4.49
	ZhS40	Zh40		1.5, 4.11, 4.34	1.60, 1.63
	ZhS47	ZhS47		1.5, 1.11, 1.12, 4.10, 4.17	
	ZhS49	ZhS49		1.5	
Heat-resistant	VZh85 (EP847)	ChN70W	-	1.14, 4.41	
	EI435	ChN78T	GOST 5632-72	1.14	1.57

Continuation on next page

Table A.1: Continued

Material class	Material	GOST designation	State Standard	Table	Figure
nickel-based alloys	EI868 (VZh98)	ChN60WT	-	1.14, 2.5, 3.24, 3.25, 4.36, 4.41, 4.49, C.1	1.5, 1.38, 3.54, 3.57, C.8
	EP99	ChN50MWKTYuR		1.14, 4.41, 4.45	4.66, 4.75
	EI602	ChN75MBTYu		1.14	
	EI962	11Ch11N2W2MF		D.2	
	EP914	ChN65WMBYu	TU 14-1-2689-79	1.14	
Titanium alloys	OT4-0	-	OST1 90013-71	1.18	1.64
	OT4-1	-		1.18	1.64
	OT4	-		1.18	1.64
	VT4	-		1.18, 1.19	1.64, 1.64
	VT5	-		1.18, 1.19	1.64
	VT5-1			1.18	1.64
	VT6C	-		1.18	1.64
	VT18U	-		1.20	
	VT36			1.20	
	VT22M	-		1.20	1.64
	VT1-0			1.19	
	VT20			1.19	1.64
	VT6			1.18, 1.19	1.64
	VT6L			1.18	
	VT6C				1.64
	VT14			1.18	1.64
	VT14L			1.18	

Continuation on next page

Table A.1: Continued

Material class	Material	GOST designation	State Standard	Table	Figure
	VT15			1.18	
	VT18				1.64
	VT20	-		1.18, 1.19	1.64
	VT20L	-		1.18	
	VT21L	-		1.18	
	VT21L	-		1.18	
	VT21L	-		1.18	
	VT22	-			1.64
	VT23	-		1.18	1.64
	TC-5	-	-	1.19	
	AT-6			1.19	
	VT3-1		OST 90013-71e	1.19, 3.31, 3.32, D.1, D.4	1.64, 4.81, D.10
	VT8	-		1.19	1.64
	VT9			1.18, 1.19, D.2	1.64

*Old designation according to GOST

Technical conditions: documents containing information about the chemical composition

The following abbreviations are used (Russian designation): Yu - aluminum Al, G - manganese Mn, R - boron B, C - silicon Si, A - nitrogen N, B - niobium Nb, W - tungsten W, D - copper Cu, M - molybdenum Mo, N - nickel Ni, F - vanadium V, K - cobalt Co, Ch - chromium Cr, T - titanium Ti, VI- vacuum induction alloying, VT - high strength titanium

End of the table

Appendix B
Isochronous Creep Curves

B.1 Materials and Test Conditions

Table B.1 Materials and test conditions

No.	Material	Foreign designation	Temperature, °C	Test duration, h	Figure
1.	15Ch1M1V	15Cr1Mo1V	450	25000	B.1
2.	R2MA	25Cr1Mo1VN	500, 525, 550	200000	B.2
3.	EP291	18Cr12MoNiVNb	500, 525, 550	200000	B.3
4.	EI415	20Cr3MoWV	550	20	B.4
5.	EI802	15Cr12WNiMoV	500, 550	25	B.5
6.	EI481	37Cr12Ni7Mn8MoVNb	550, 600, 650	2000	B.6
7.	EI612	CrNi35WTi	600, 650	50	B.7
8.	EI607A	CrNi80TiNbAl	600, 650, 700, 750, 800	1000-2000	B.8
9.	EI765	CrNi70WMoAlTi	600, 700, 750, 800	2000	2.9, B.9
10.	EI826	CrNi70WMoTiAlV	600, 700, 750, 800, 850, 900	2000	B.10
11.	EI827	CrNi75WMoAl	600, 700, 750, 800, 850, 900	2000	B.11
12.	EI893	CrNi65WMoTiAl	650, 750	100000 calculation from relaxation curves	B.12
13.	EI929	CrNi55WMoTiCoAl	850	500	B.13
14.	EP220	CrNi51WMoTiAlCoVNb	850, 900, 950, 1000	2-1000	B.14
15.	ChS70VI	-	700, 750	5000	2.17
16.	ChS104	-	700, 750	5000	2.16
17.	ZhS6K	-	850, 950, 1000	5000	2.15
18.	ZhS36	-	850, 900, 1000	300	2.18

L. B. Getsov, *Materials and Strength of Gas Turbine Parts*, Advanced Structured Materials 150, https://doi.org/10.1007/978-981-16-0534-5

B.2 Diagrams

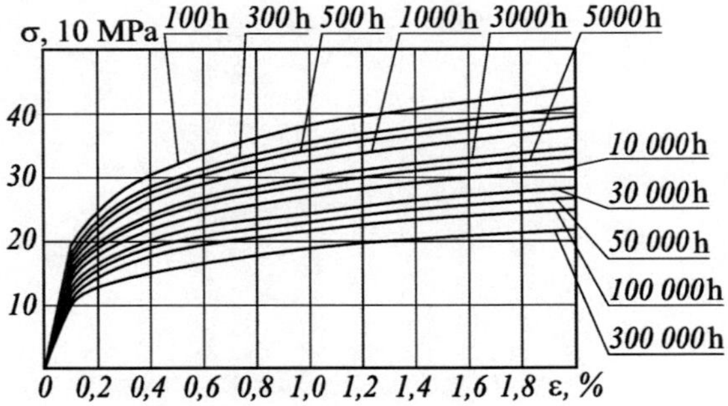

Fig. B.1 Isochronous creep curves of steel 15Ch1M1F, 450°C, $\sigma_{0,2}^{20} = 440$ MPa

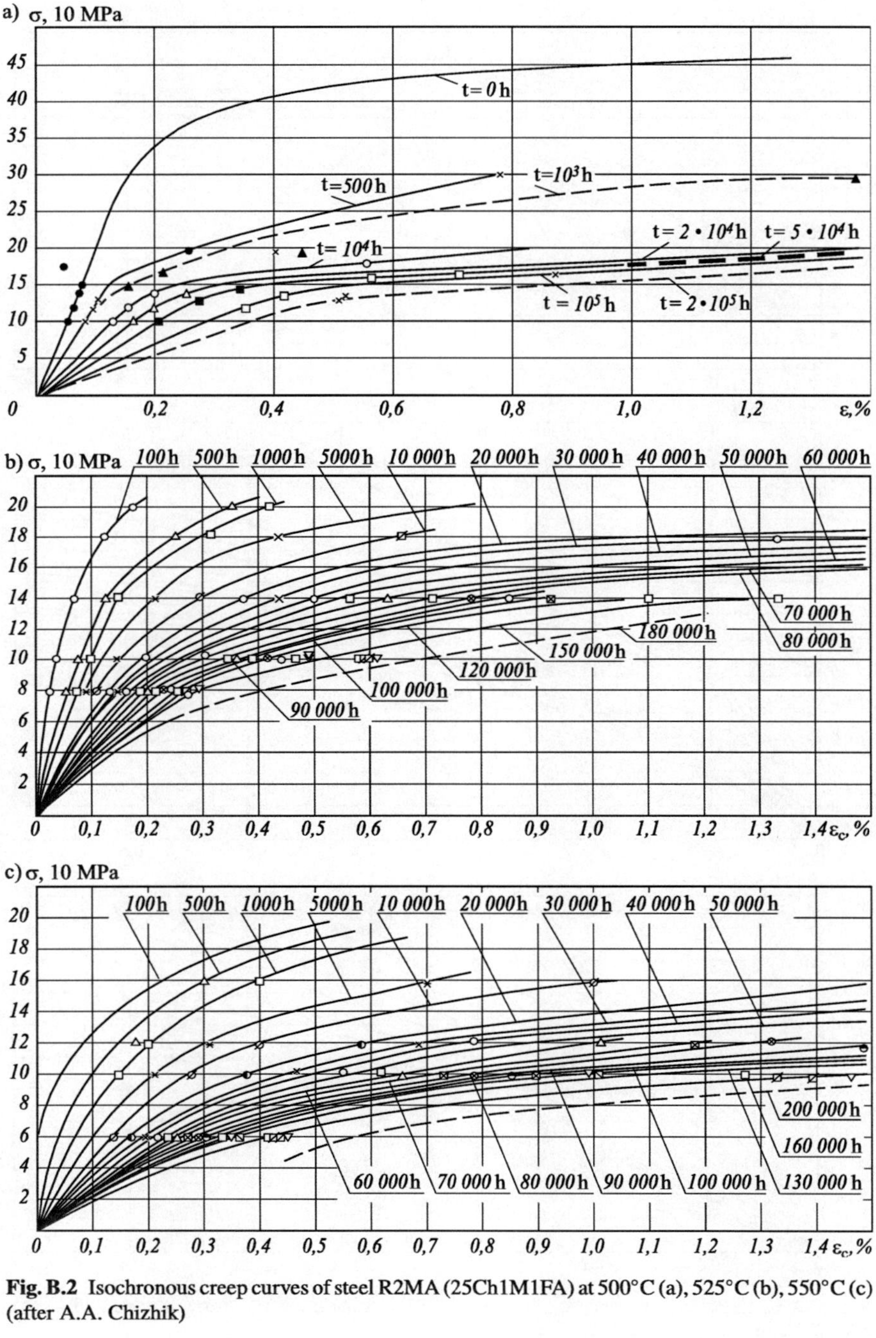

Fig. B.2 Isochronous creep curves of steel R2MA (25Ch1M1FA) at 500°C (a), 525°C (b), 550°C (c) (after A.A. Chizhik)

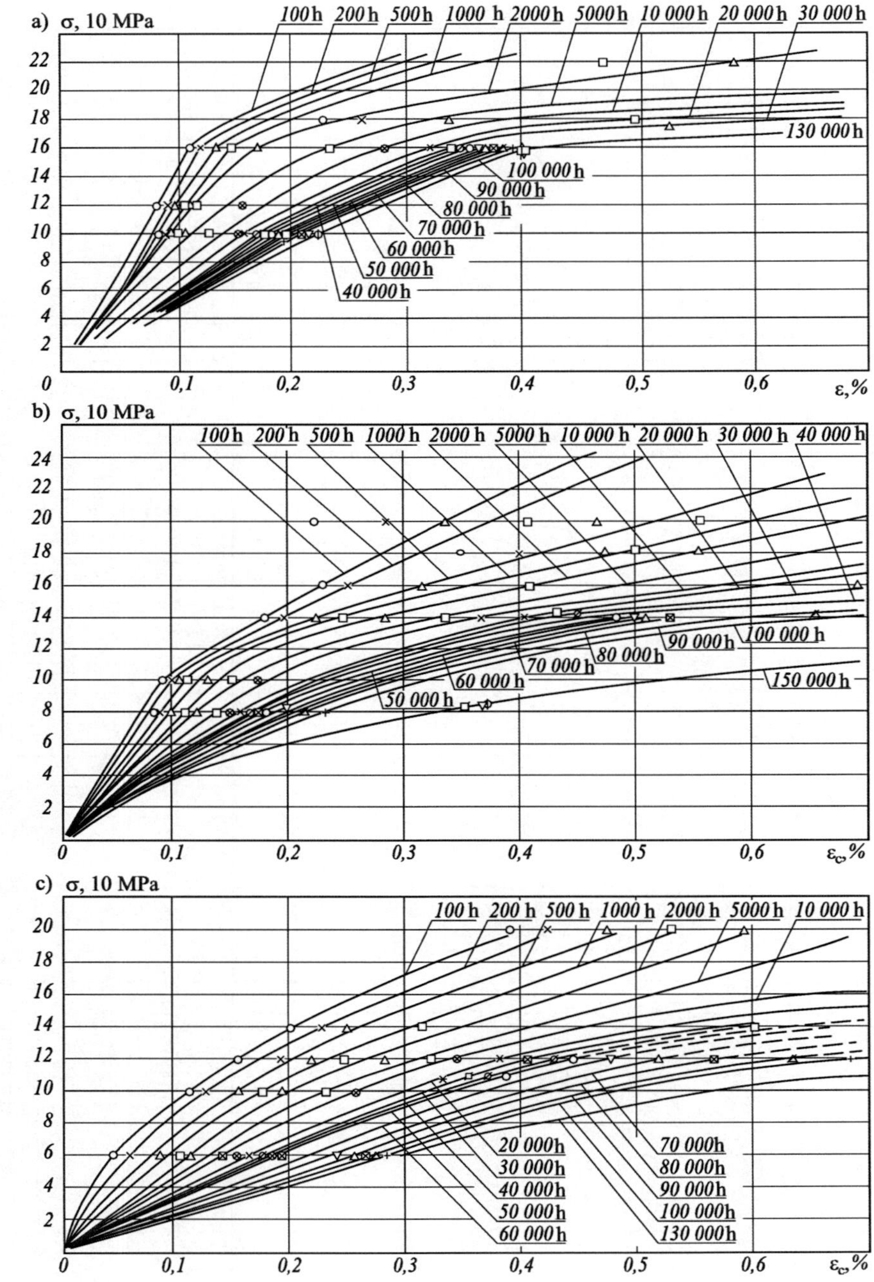

Fig. B.3 Isochronous creep curves of steel EP291 (18Ch12MNFB) at 500°C (a), 525°C (b), 550°C (c) (after A.A. Chizhik)

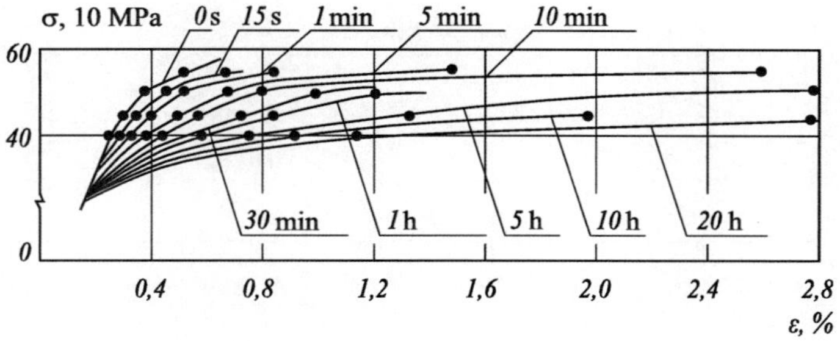

Fig. B.4 Deformation curves of EI415 (20Ch3MWF) steel at temperature 550°C

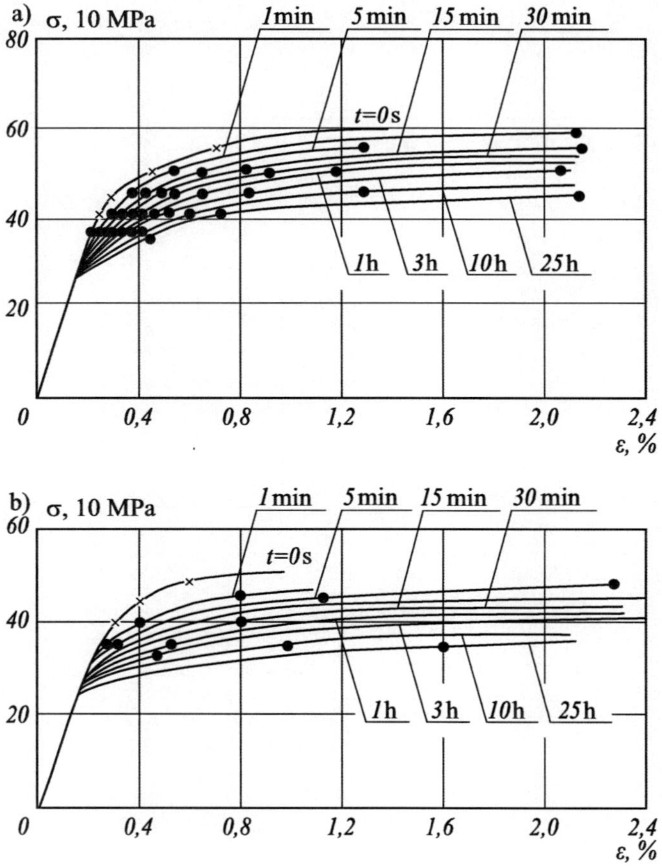

Fig. B.5 Deformation curves of EI802 (15Ch12WNMF) steel at temperature 500°C (a), 550°C (b)

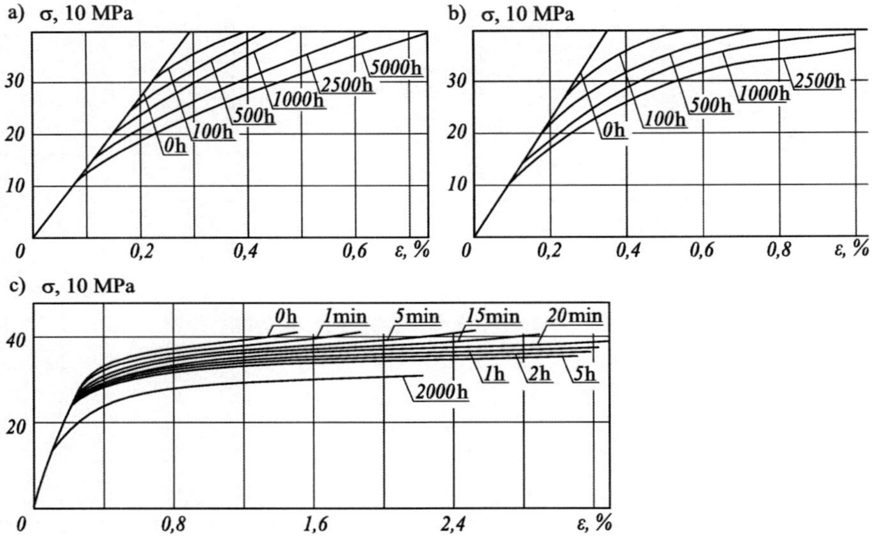

Fig. B.6 Deformation curves of EI481 (37Ch12N7G8MFB) steel at temperature 550°C (a), 600°C (b), 600°C (c)

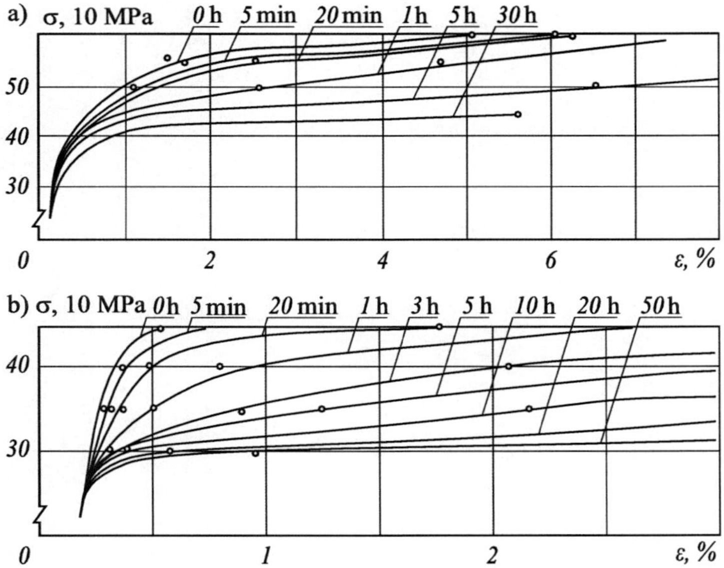

Fig. B.7 Isochronous creep curves of steel EI612 (ChN35WT) steel at temperature 600°C (a), 650°C (b)

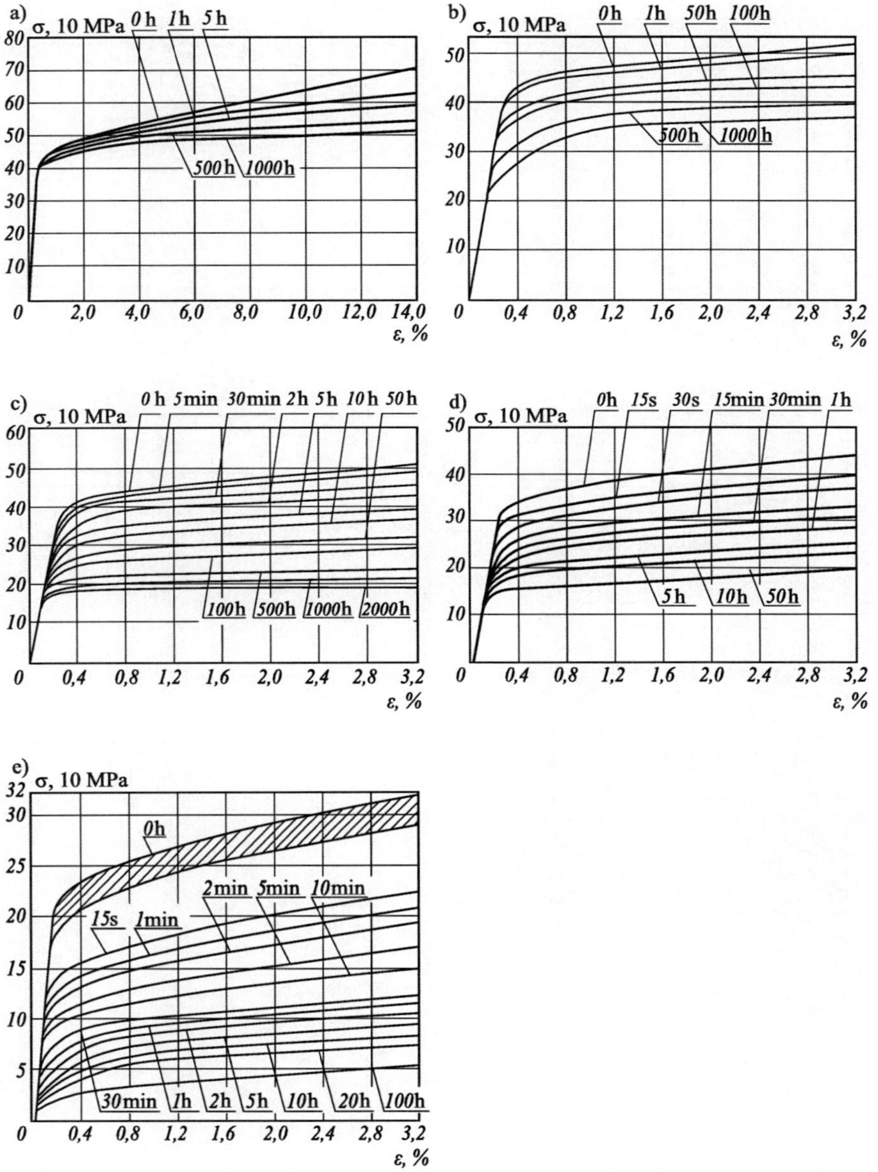

Fig. B.8 Deformation curves of EI607A (ChN80T1BYu) alloy at temperature 600°C (a), 650°C (b), 700°C (c), 750°C (d), 800°C (e)

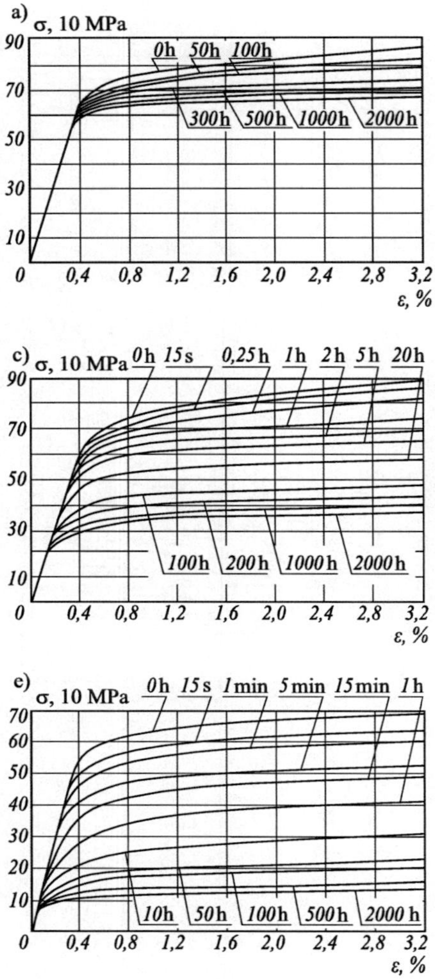

Fig. B.9 Deformation curves of EI765 (ChN70WMYuT) alloy at temperature 600°C (a), 650°C (b), 700°C (c), 750°C (d), 800°C (e)

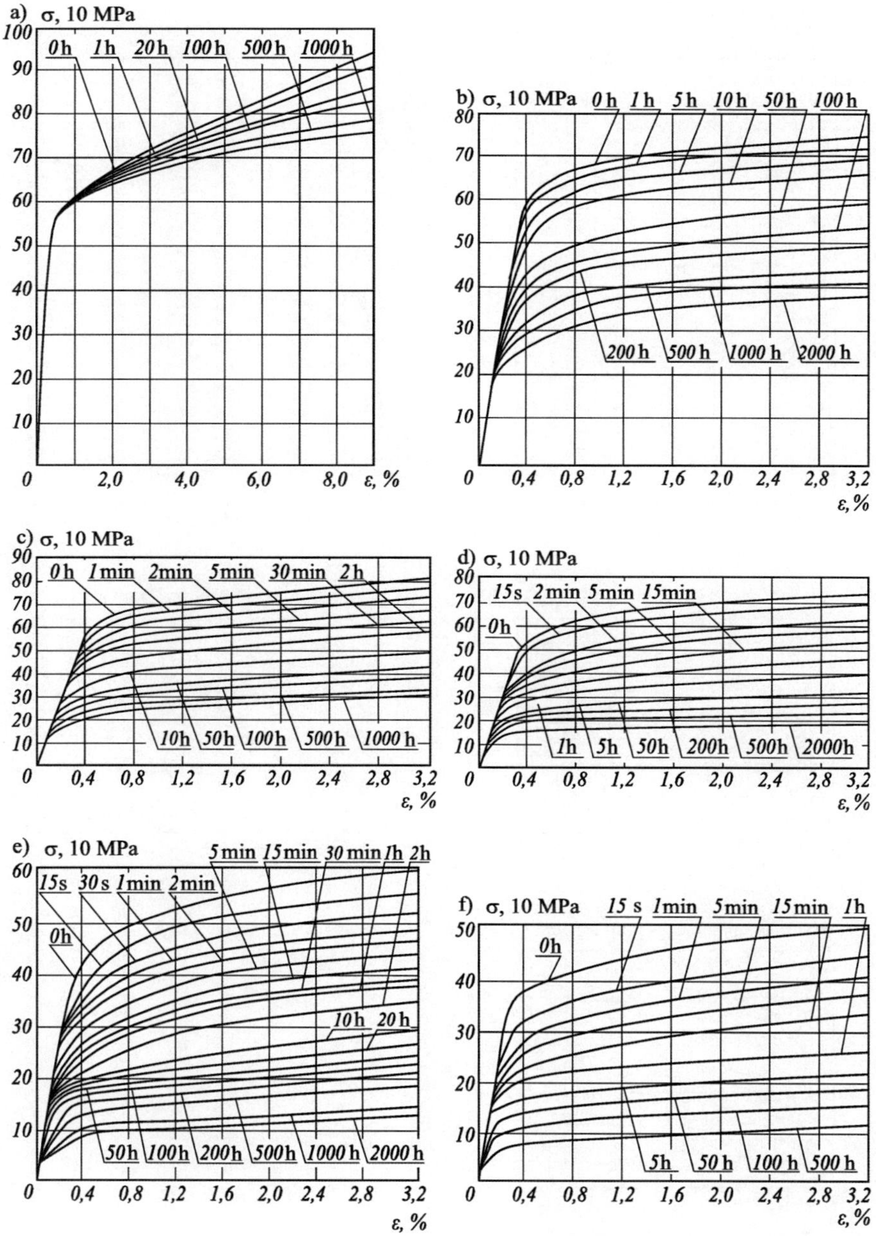

Fig. B.10 Deformation curves of EI826 (ChN70WMTYuF) alloy at temperature 600°C (a), 700°C (b), 750°C (c), 800°C (d), 850°C (e), 900°C (f)

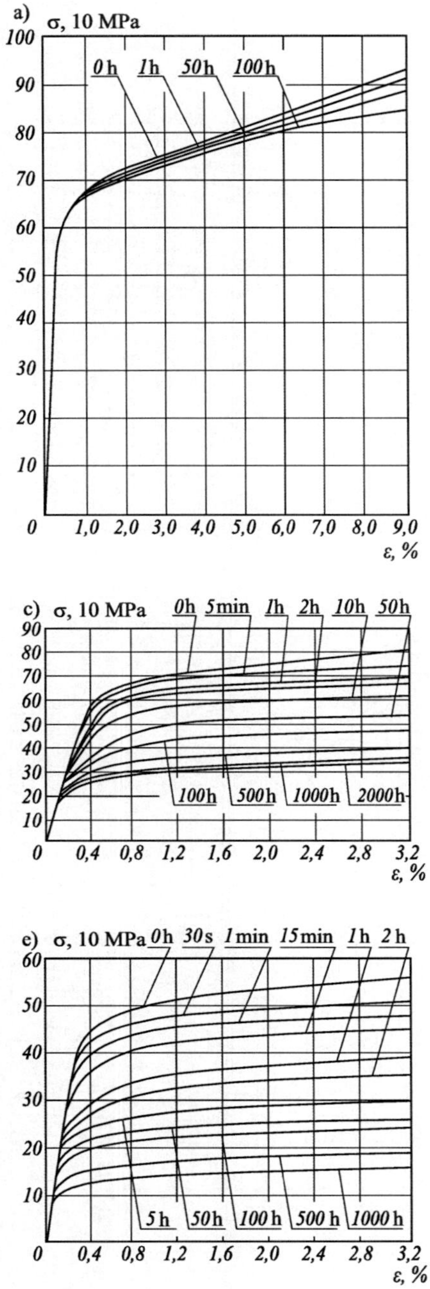

Fig. B.11 Deformation curves of alloy EI827 (ChN75WMYu) at temperature 600°C (a), 700°C (b), 750°C (c), 800°C (d), 850°C (e)

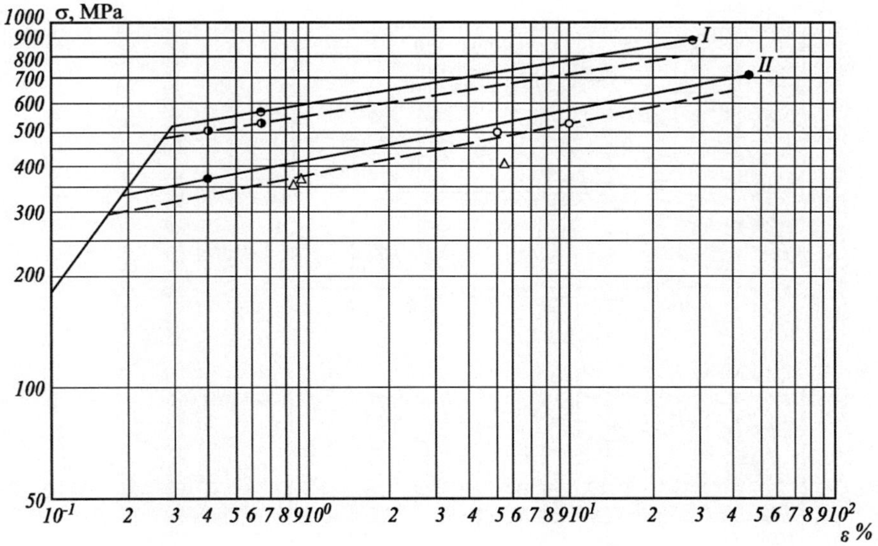

Fig. B.12 Isochronous curves of the EI893 (ChN65WMTYu) alloy at temperature of 650°C, obtained by calculation for 100000 h. *I* - 1st mode heat treatment, *II* - 2nd mode heat treatment

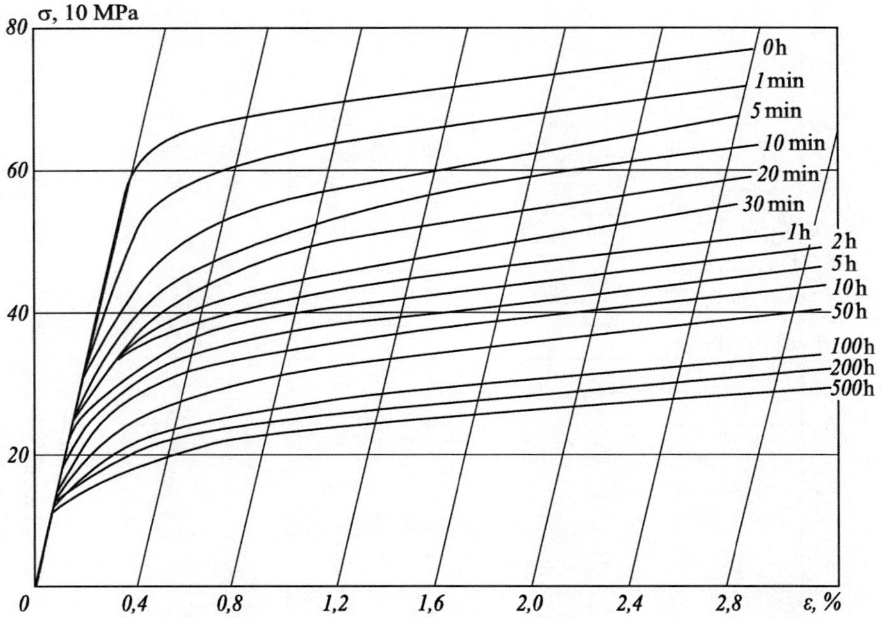

Fig. B.13 Deformation curves of alloy EI929 (ChN55WMTKYu) at a temperature of 850°C

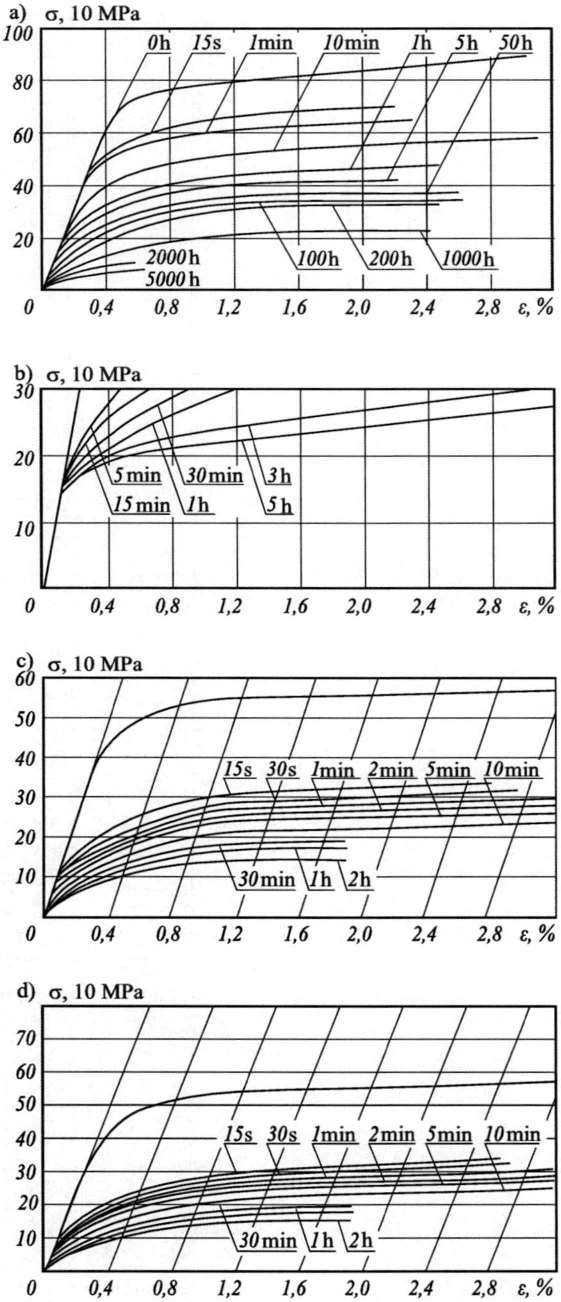

Fig. B.14 Deformation curves of alloy EP220 (ChN51WMTYuKFR) at a temperature of 850°C (a), 900°C (b), 950°C (c), 1000°C (d)

Appendix C
Cyclic Creep Curves

C.1 Materials and Test Conditions

Below are the curves of cyclic creep under alternating torsion loading of heat-resistant nickel-based steels and alloys (γ_c - creep shear strain, t_c - time, numbers on the curves - half-cycle number, τ_c - shear stress).

Table C.1 Materials and test conditions

No.	Material	Foreign brand	Temperature, °C	Figure
1.	EI723	25Cr1Mo1V	550	C.1
2.			600	C.2
3.	EI802	15Cr12WNiMoV	500	C.3
4.			550	C.4
5.	EI417	20Cr23Ni18	550	C.5
6.	EI765	CrNi70WMoAlTi	700	C.6
7.	EI826	CrNi70WMoTiAlV	700	C.7
8.	EI868	CrNi60WTi	600, 700, 800	C.8
9.	EP220	CrNi51WMoTiAlCoVB	900	C.9

L. B. Getsov, *Materials and Strength of Gas Turbine Parts*, Advanced Structured Materials 150, https://doi.org/10.1007/978-981-16-0534-5

C.2 Diagrams

Figures C.1-C.9 shows the curves of cyclic creep under alternating loading under the temperature conditions specified in Table C.1.

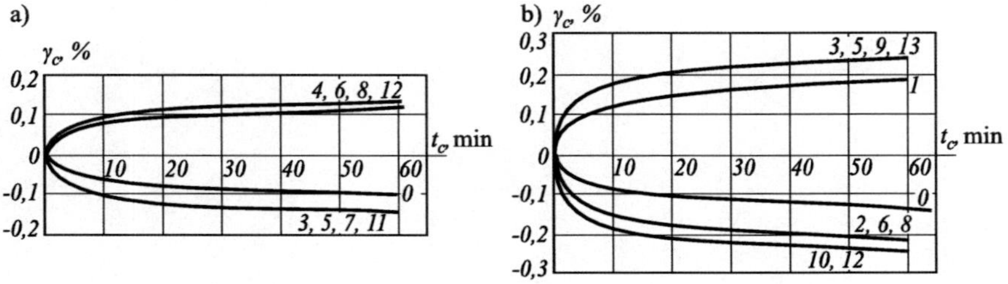

Fig. C.1 Cyclic creep curves of steel EI723 (25Ch1M1F) at 550°C, $\tau_c = \pm 324$ MPa (a), $\tau_c = \pm 366$ MPa (b)

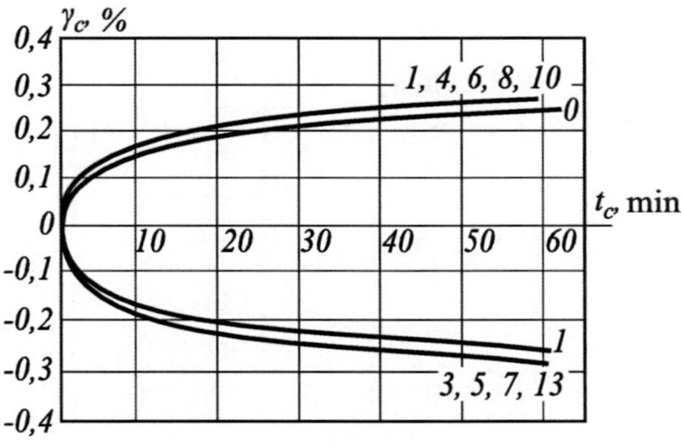

Fig. C.2 Cyclic creep curves of EI723 steel at 600°C, $\tau_c = \pm 232$ MPa

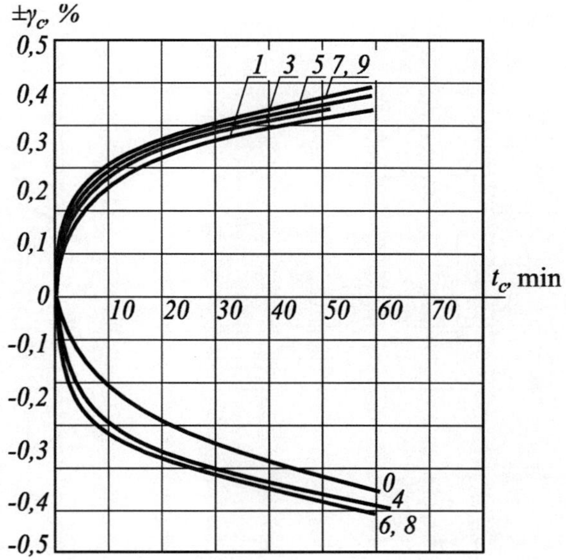

Fig. C.3 Cyclic creep curves of EI802 steel at 500°C, $\tau_c = \pm 258$ MPa

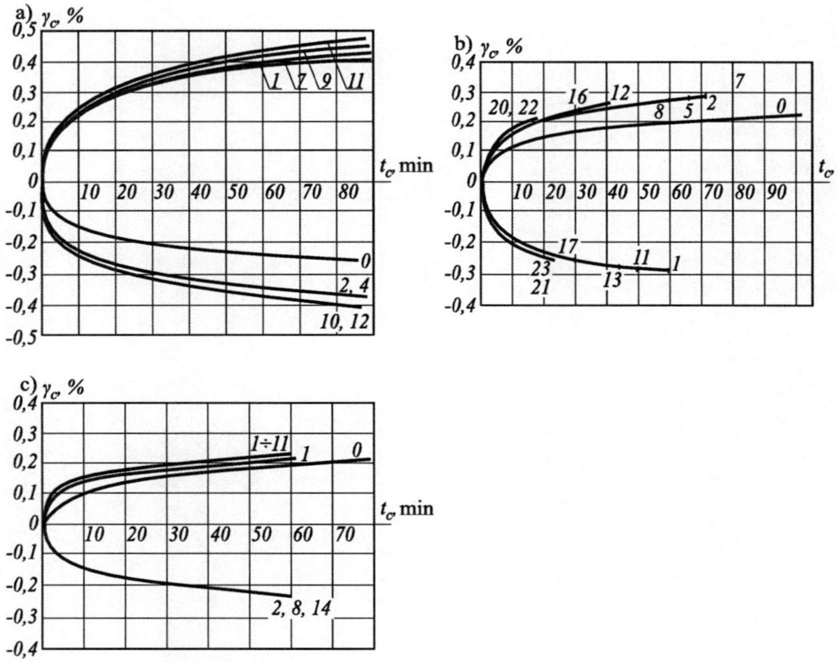

Fig. C.4 Cyclic creep curves of EI802 steel at 500°C, $\tau_c = \pm 236$ MPa (a), $\tau_c = \pm 195$ MPa (b), $\tau_c = \pm 157$ MPa (c)

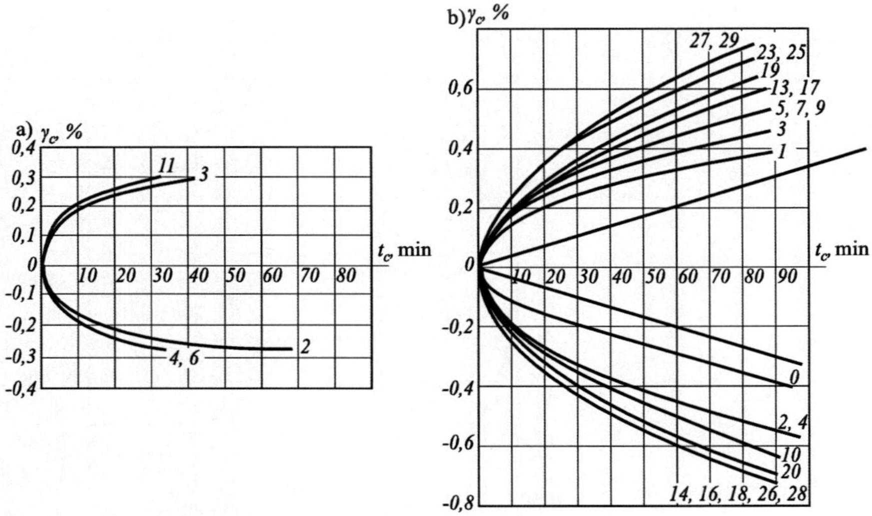

Fig. C.5 Cyclic creep curves of EI417 (20Ch23N18) steel at 550°C, $\tau_c = \pm 135$ MPa (a) and at 700°C, $\tau_c = \pm 67$ MPa (b)

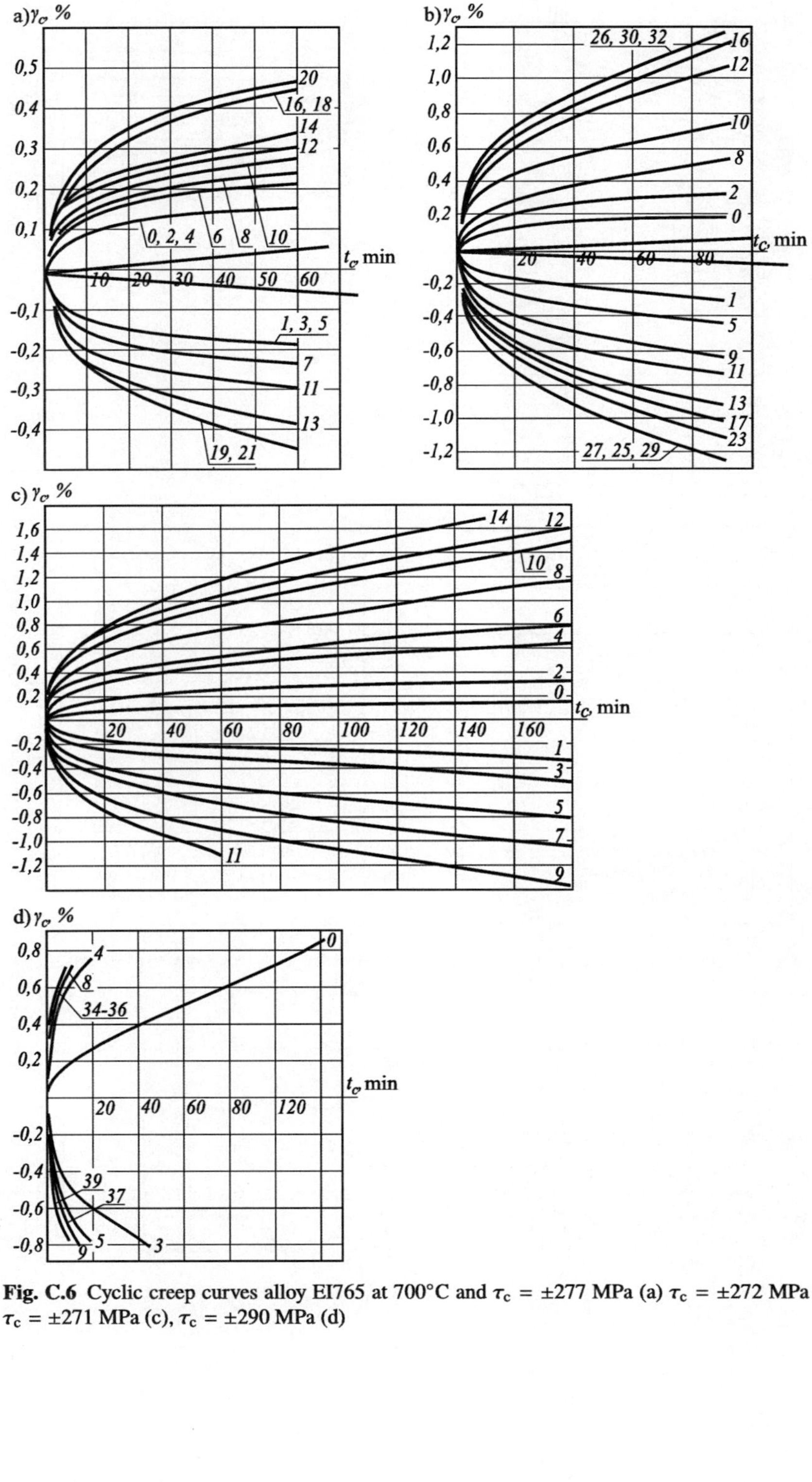

Fig. C.6 Cyclic creep curves alloy EI765 at 700°C and $\tau_c = \pm 277$ MPa (a) $\tau_c = \pm 272$ MPa (b), $\tau_c = \pm 271$ MPa (c), $\tau_c = \pm 290$ MPa (d)

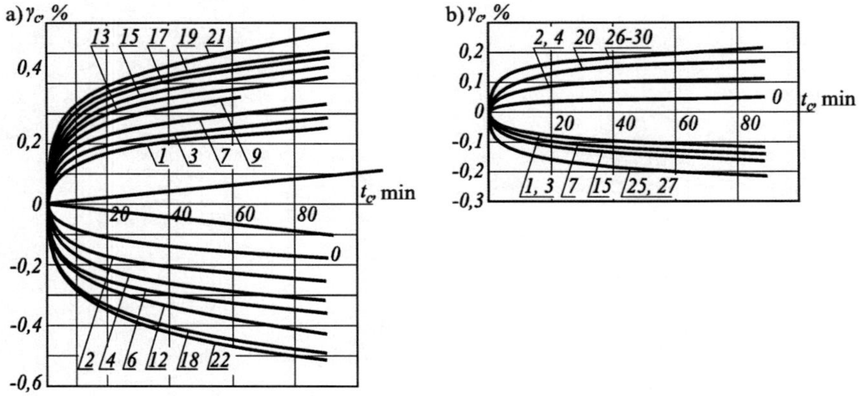

Fig. C.7 Cyclic creep curves alloy EI826 at 700°C and $\tau_c = \pm 314$ MPa (a) $\tau_c = \pm 270$ MPa (b)

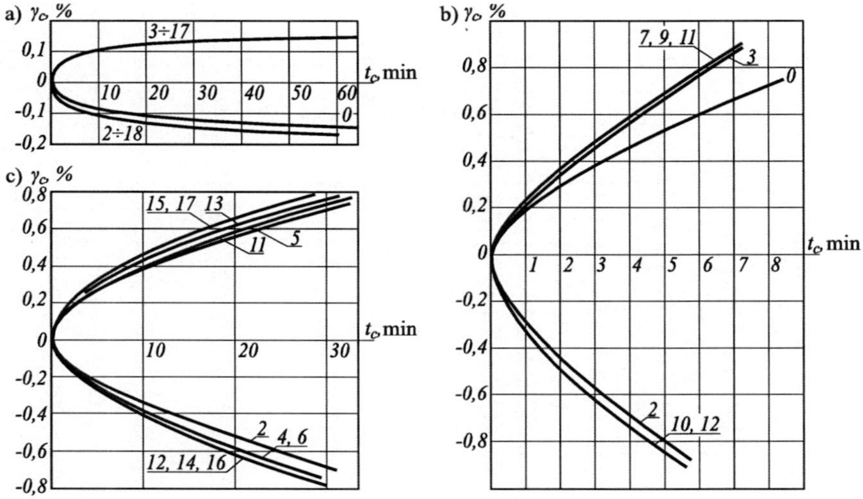

Fig. C.8 Cyclic creep curves alloy EI868 at 600°C and $\tau_c = \pm 163$ MPa (a), at 700°C and $\tau_c = \pm$ 120 MPa (b), at 800°C and $\tau_c = \pm 126$ MPa (c), $\tau_c = \pm 290$ MPa (d)

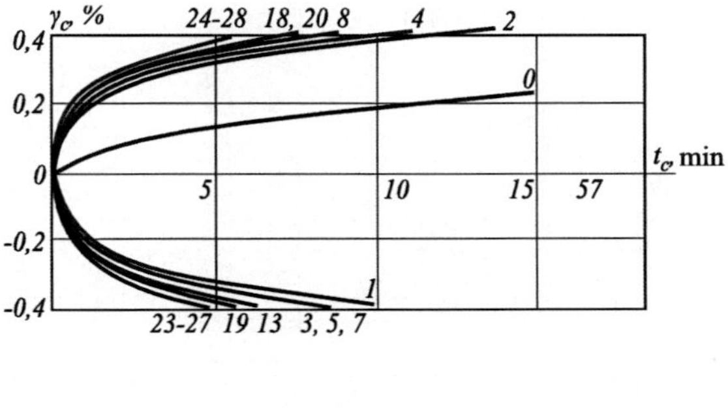

Fig. C.9 Cyclic creep curves at cyclic alternating loading EP220 alloy at 900°C, $\tau_c = \pm$ 135 MPa

Appendix D
Fracture Toughness Characteristics at Cyclic Load

D.1 Materials and Their Parameters

Figures D.1-D.4 present the rate of growth of fatigue cracks in pearlitic steels used for rotors as a function of the stress intensity factor. Cyclic fracture toughness characteristics of materials used for the manufacture of compressor blades are shown in Fig. 3.92, in Table D.1 (Pokrovskii et al., 1994a,b) and in Figs. D.5-D.10; for disc materials in Figs. 3.93, D.11-D.13; for materials of blades made of heat-resistant alloys in Figs. D.14-D.16; in Fig. D.17 - for the material of the pump body. Figures D.18-D.20 show the characteristics of the cyclic fracture toughness of a number of foreign alloys. Tables D.2 and D.3 show the threshold values of the stress intensity factor range for materials of discs and blades (Troshchenko et al., 1987; Prokopenko et al., 1984; Pokrovskii et al., 1994a,b; Pavlenko et al., 2004). Here A and m are the parameters of the Paris equation (3.140), where the crack growth rate has the

Table D.1 List of modern heat-resistant alloys

Material	test environment	$\sigma_{0.2}$ MPa	σ_B MPa	δ %	ψ %	K_{th} MPa$\sqrt{m}$	m	A	$\Delta K_{th}^C/\Delta K_{th}^B$
Steel 20Ch13	air	691	841	23,6	61	5,47	4,161	9,826 10^{-11}	0,77
	sea fog					4,2	4,412	4,959 10^{-11}	
Steel 14Ch17N2	air	782	943	17,5	60,8	6,1	4,866	7,595 10^{-11}	0,65
	sea fog					3,98	4,950	1,918 10^{-11}	
Steel 13Ch11N2W2MF	air	885	1015	16,2	57	6,16	4,559	4,585 10^{-11}	0,59
	sea fog					3,64	3,793	1,545 10^{-11}	
Steel 08Ch17N6T	air	840	895	17,3	55,8	5,54	3,686	2,046 10^{-11}	0,77
	sea fog					4,23	4,032	8,237 10^{-11}	
Ti-alloy VT3-1	air	968	1032	15,5	42,4	5,69	4,732	6,428 10^{-11}	0,55
	sea fog					3,08	3,824	1,192 10^{-9}	

L. B. Getsov, *Materials and Strength of Gas Turbine Parts*, Advanced Structured Materials 150, https://doi.org/10.1007/978-981-16-0534-5

Table D.2 Cyclic fracture toughness parameters of disk materials

Material	T, °C	σ_B, MPa	$\sigma_{0.2}$, MPa	δ, %	f, Hz	m	A
Ti-alloy VT9	20	1150	1000	10	0,5	3,18	9,1 10^{-12}
					40	4,2	7,3 10^{-13}
	500	810	660	8	0,5	3,35	8,7 10^{-13}
					t = 20 s	3,7	5,45 10^{-12}
					t = 40 s	3,8	2,2 10^{-12}
Ni-alloy EP742	20	1230	880	21	0,5	5,1	6 10^{-16}
					35	4,2	6,25 10^{-15}
	500	1170	820	20	0,5	3,8	1 10^{-11}
	700	940	760	16	0,5	2,7	1,4 10^{-11}
Ni-alloy EP962	20	1500	1150	10	25	4,66	9,7 10^{-15}
	600	1410	1050	8	25	2,41	3,9 10^{-11}
					t = 20 s	6,39	1,7 10^{-17}
	750	1200	1010	11	25	3,16	7,5 10^{-12}
					t = 20 s	6,39	1,7 10^{-17}

dimension of *m*/cycle. The condition for non-propagation of cracks in the blades is determined by the formula $\sigma_{th} = K_{th}/Y(c)\sqrt{c}$, where $Y(c)$ is determined experimentally based on the fact that the dependence of crack growth rate (CGR) on the stress intensity factor (SIF) for blades and samples from the same materials coincide (K-calibration).

Table D.3 Threshold ranges of stress intensity factor values and parameters of the Paris equation

Material	T, °C	R	ΔK_{th}, MPa$\sqrt{m}$	m	A	Application limits for ΔK_I, MPa$\sqrt{m}$
EK79	20	-1	10,55	12,34	5,93 10^{-20}	
	500	-1	7,43	6,81	7,08 10^{-13}	
EI437B	20	-1	9,08	-	-	
	700	-1	6,47	-	-	
ZhS6F [111]	20	-1	8,4	4,42	1,62 10^{-11}	≤40
		0	6,96	4,4	2,87 10^{-11}	≤20
		0,5	5,97	4,78	6,06 10^{-11}	≤20
	800	-1	7,65	4,38	3,2 10^{-11}	≤30
		0	6,47	4,36	8,85 10^{-11}	≤25
		0,5	5,47	4,92	1,18 10^{-10}	≤20
	1000	-1	5,57	4,88	8,31 10^{-11}	≤16
		0	4,73	4,98	1,78 10^{-10}	≤16
		0,5	3,82	6,08	2,23 10^{-10}	≤9
ZhS6F [001]	20	-1	5,75	4,21	9,10 10^{-11}	≤30
		0	3,99	4,12	4,92 10^{-10}	≤30
		0,5	2,78	3,88	3,33 10^{-9}	≤20
	800	-1	4,73	3,85	8,92 10^{-10}	≤25
		0	3,45	9,92	3,03 10^{-9}	≤16
		0,5	2,38	4,01	1,63 10^{-8}	≤10
ZhS6U	20	-1	8,71	-	-	
	600	-1	7,27	-	-	
	800	-1	6,47	-	-	
ZhS6KP	20	-1	5,44	-	-	
	600	-1	5,35	-	-	
	800	-1	3,17	-	-	

D.2 Cyclic Fracture Toughness Curves

In Figs. D.1-D.20 the curves of the cyclic fracture toughness of steels and alloys listed in Table D.4 are given.

Table D.4 List of materials and conditions of their tests for crack resistance under vibration loading

Figure	Material designation	Test temperature, °C	Reference
D.1	R2MA (25CrMo1VA)	20, 550	according to Shlyannikov et al. (2015)
D.2	20Ch2WNMF	20	according to A.A. Chizhik
D.3	27Ch3M2FA	20	according to A.A. Chizhik
D.4	27Ch3Mo2VA	20	according to A.A. Chizhik
D.5	20Ch13	20	according to Prokopenko et al. (1984)
D.6	13Ch12N2W2MF	20	according to Troshchenko et al. (1987)
D.7	14Ch17N2	20	according to Troshchenko et al. (1987)
D.8	13Ch11N2W2MF	20	according to Troshchenko et al. (1987)
D.9	08Ch17N6T	20	according to Troshchenko et al. (1987)
D.10	VT3-1	20	according to Troshchenko et al. (1987)
D.11	EP741P	20-700	according to Pokrovskii et al. (1994a,b)
D.12	EI698	20-600	according to Pokrovskii et al. (1994a,b)
D.13	EK79	20, 500	according to Pavlenko et al. (2004)
D.14	ZhS6KP	20-1000	according to Troshchenko et al. (1987)
D.15	ZhS6F	20	according to E.R. Golubovsky
D.16	ZhS6F	750, 900, 1000	according to E.R. Golubovsky
D.17	15Ch5MF, EK-32, 20Ch13, 08Ch13, CT-28		
D.18	IN718	650	according to Sims et al. (1987)
D.19	Rene88	593	according to Sims et al. (1987)
D.20	Rene95, NASA 11B-7, IN100, Astroloy, Waspaloy, MERL 76	649	according to Sims et al. (1987)

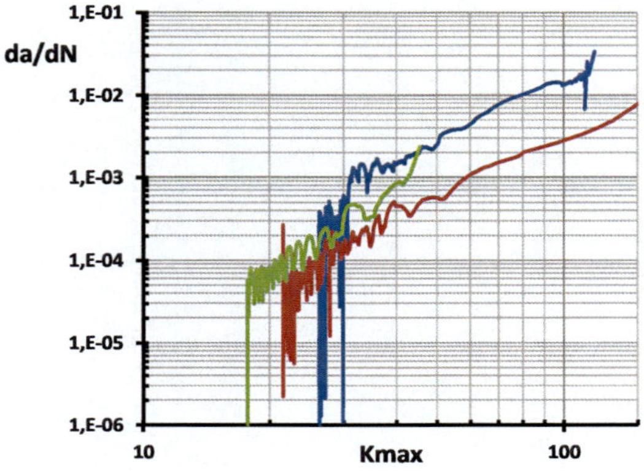

Fig. D.1 Rate of growth of fatigue cracks in steel R2MA (25Ch1M1FA) at 20°C (red line - 1), 550°C (green line - 2), and cracks under fatigue conditions with exposure times in a cycle of 2 min (fatigue + creep) at 550°C (blue line - 3) depending on the stress intensity factor (Shlyannikov et al., 2015); the tests were realized w.r.t. ASMT E647 (1, 2) and ASMT E2760 (3)

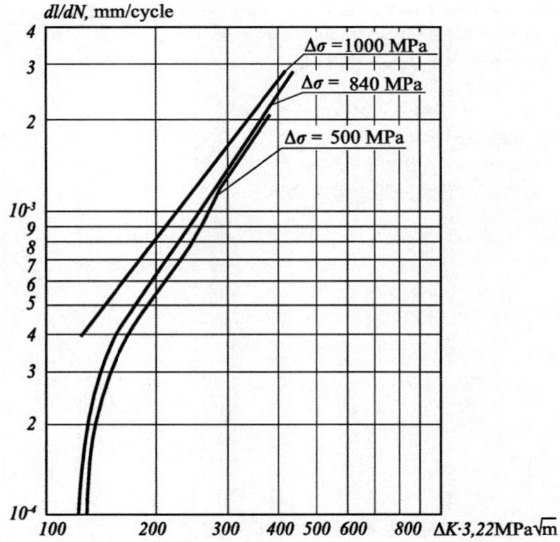

Fig. D.2 Dependence of crack growth rate on stress intensity factor for steel 20Ch2WNMF and different stress amplitudes (according to A.A. Chizhik)

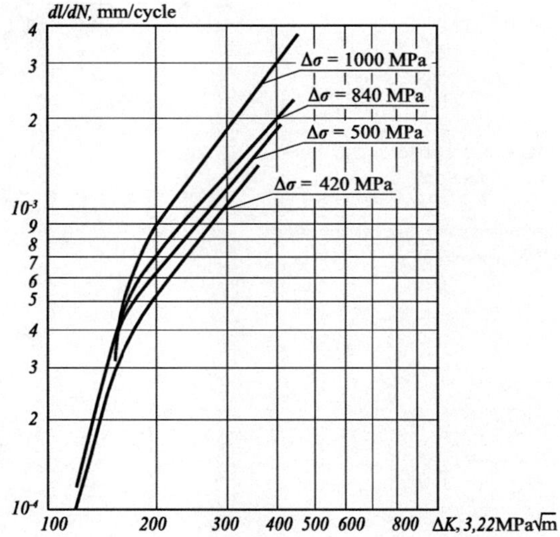

Fig. D.3 Dependence of crack growth rate on stress intensity factor for steel 27Ch23M2FA and different stress amplitudes (according to A.A. Chizhik)

Fig. D.4 Dependence of crack growth rate on stress intensity factor for steel 27Ch3M2FA (melted in different ways: 1 - electroslag melting, 2 - open arc melting, 3 - vacuum arc melting) and different stress amplitudes (according to A.A. Chizhik)

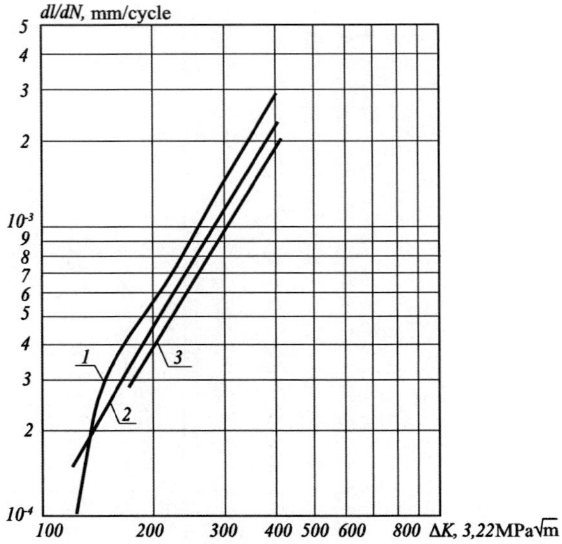

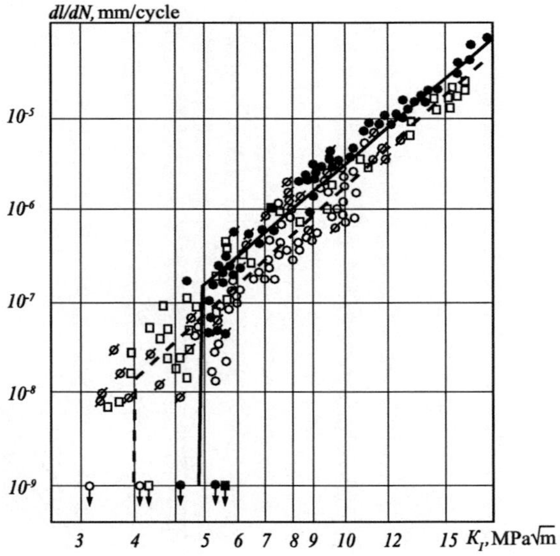

Fig. D.5 Dependence of crack growth rate on stress intensity factor at air for steel 20Ch13 specimens (•) and blades (○) (Prokopenko et al., 1984)

Fig. D.6 Dependence of crack growth rate on stress intensity factor at air for steel 20Ch12N2WMF specimens (•) and blades (○) (Troshchenko et al., 1987)

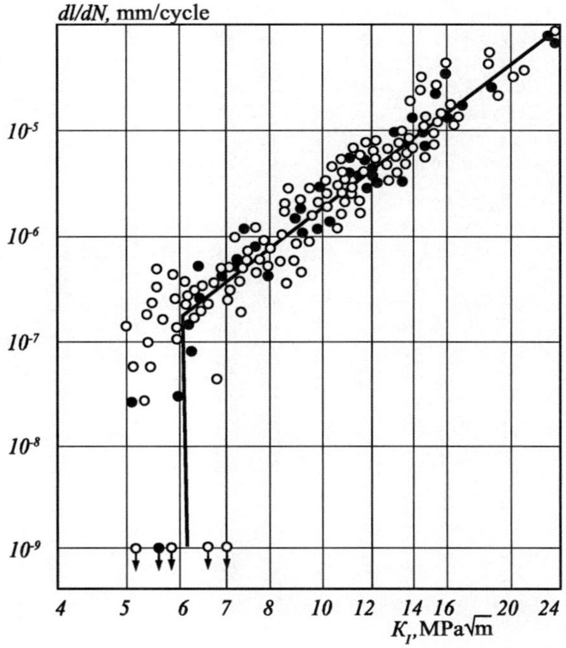

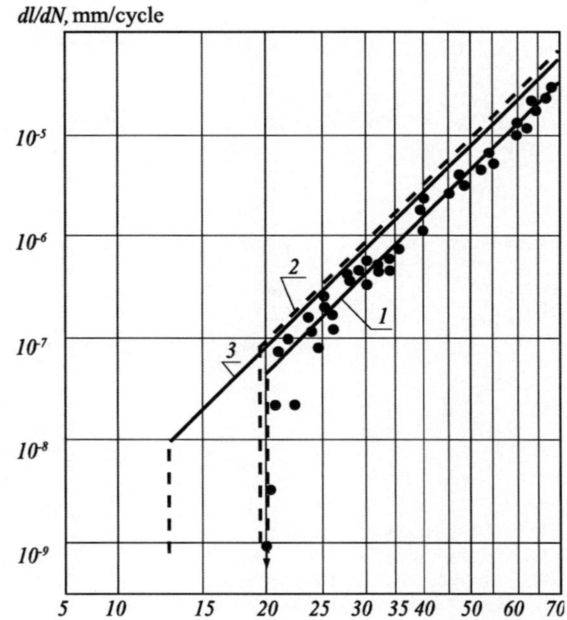

Fig. D.7 Dependence of crack growth rate on stress intensity factor for steel 14Ch17N2 specimens: 1 - with protector, 2 - at air, 3 - in contact with drops of sea water (Prokopenko et al., 1984)

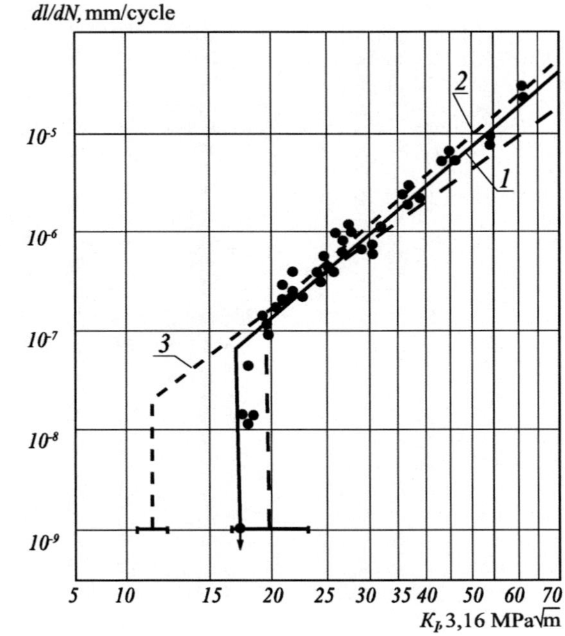

Fig. D.8 Dependence of crack growth rate on stress intensity factor for steel 13Ch11N2W2MF specimens: 1 - with protector, 2 - at air, 3 - in contact with drops of sea water (Prokopenko et al., 1984)

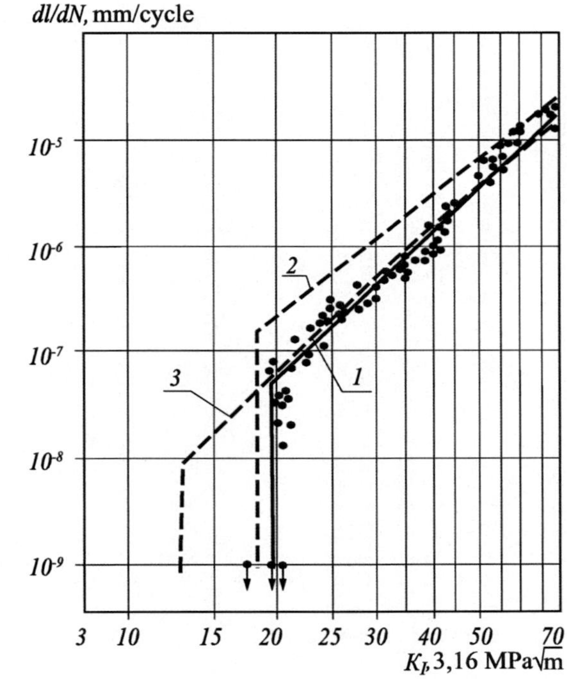

Fig. D.9 Dependence of crack growth rate on stress intensity factor for steel 08Ch17N6T specimens: 1 - with protector, 2 - at air, 3 - in contact with drops of sea water (Prokopenko et al., 1984)

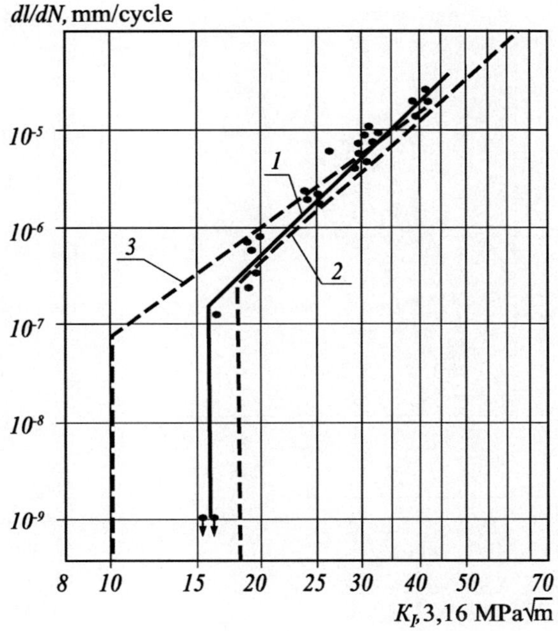

Fig. D.10 Dependence of crack growth rate on stress intensity factor for steel VT3-1 specimens: 1 - with protector, 2 - at air, 3 - in contact with drops of sea water (Prokopenko et al., 1984)

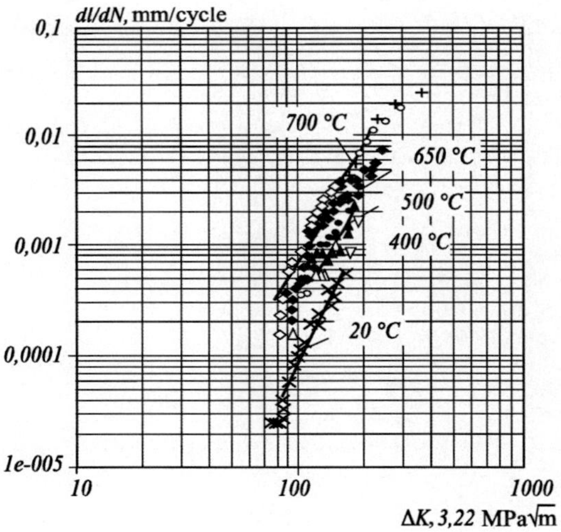

Fig. D.11 Fracture toughness EP741P (Pokrovskii et al., 1994a,b)

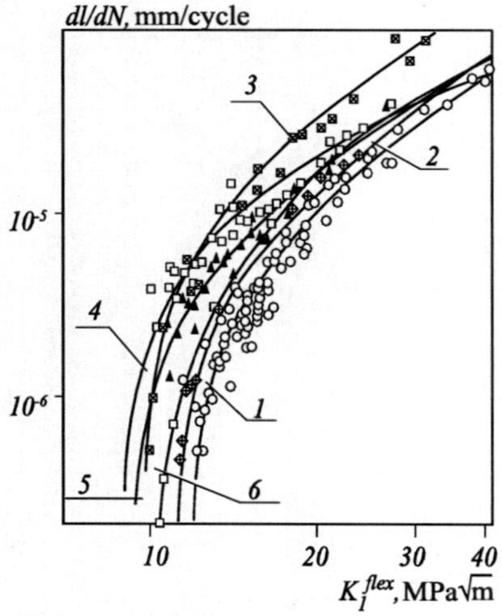

№	C	n	K_{th}
1	$4.4\ 10^{8}$	1.96	11.7
2	$5.0\ 10^{-8}$	1.98	11.0
3	$1.6\ 10^{7}$	1.87	9.6
4	$1.4\ 10^{6}$	1.07	8.9
5	$1.0\ 10^{-7}$	1.80	9.2

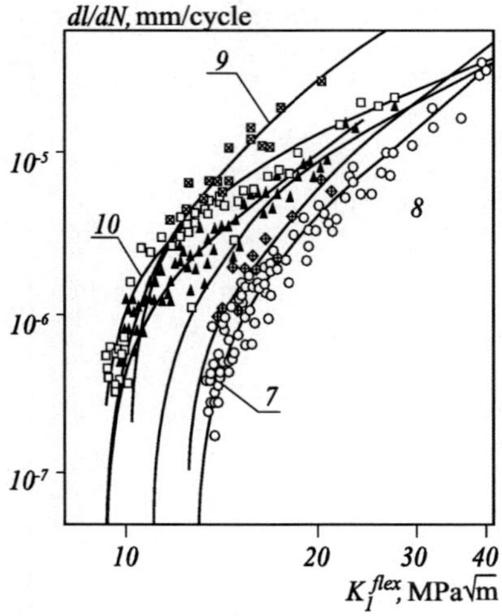

№	C	n	K_{th}
1	$6.0\ 10^{-10}$	3.00	13.0
2	$9.5\ 10^{10}$	2.96	12.4
3	$9.0\ 10^{9}$	2.698	9.97
4	$3.8\ 10^{7}$	1.30	9.0
5	$5.5\ 10^{8}$	1.78	9.2

Fig. D.12 Dependence of crack growth rate on stress intensity factor for alloy EI698 (Pokrovskii et al., 1994a,b)

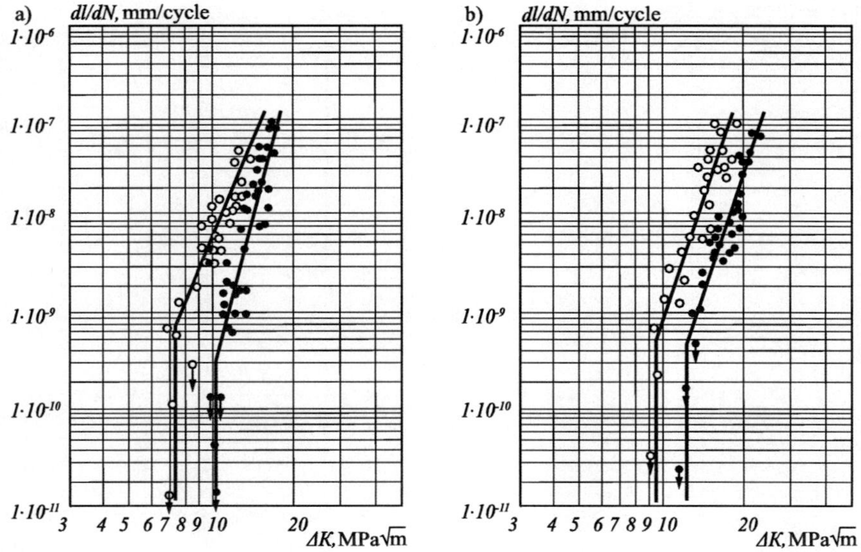

Fig. D.13 Dependence of the crack growth rate of fatigue cracks in the alloy EK79 in the initial state (a) and after surface hardening (b) at 20°C (•) and 500°C (○) (Pavlenko et al., 2004)

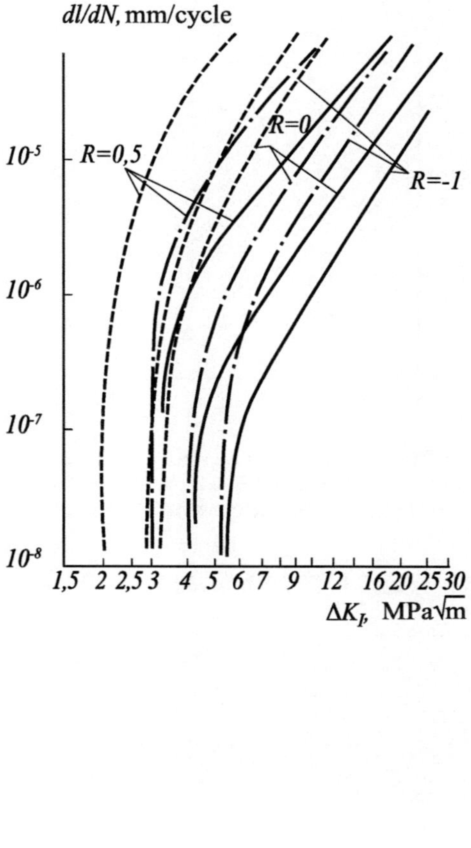

Fig. D.14 Dependence of the crack growth rate on ΔK_{I} of the ZhS6KP alloy (Troshchenko et al., 1987) at f = 400 Hz, solid lines - for T = 20°C, dash-dotted - for T = 800°C, dash-lines for T = 1000°C

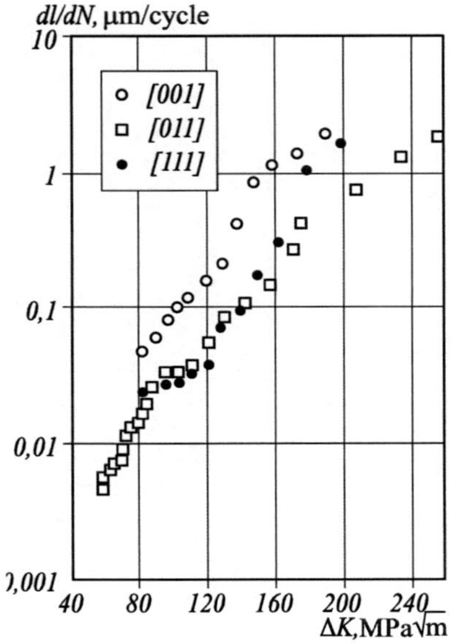

Fig. D.15 Dependence of the fatigue crack growth rate in the single-crystal alloy ZhS6F on ΔK (according to E.R. Golubovsky)

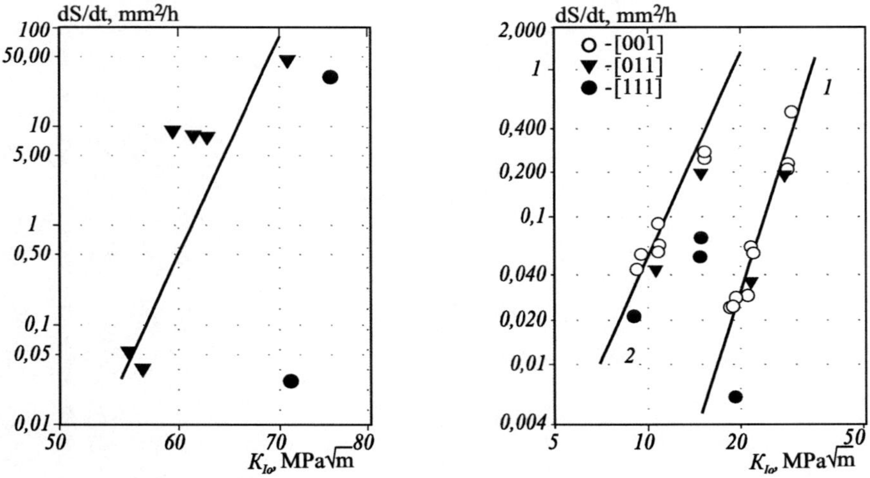

Fig. D.16 Dependence of the creep crack growth rate of the ZhS6F alloy on the initial value of K_0 (according to the data of E.R. Golubovsky). S is the crack area. a - T = 750°C (curve 1 - [011], curve 2 - [111]), plane deformation b - [001], 1 - T = 900°C, 2 - T = 1000°C, plane stress state

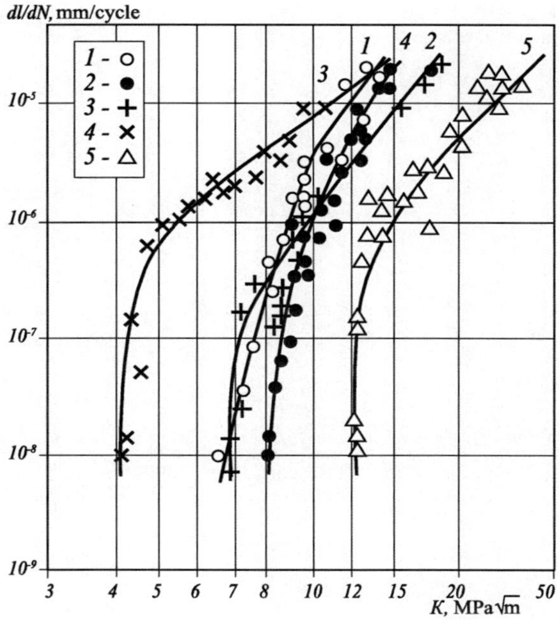

Fig. D.17 Crack toughness characteristics of steels in seawater: 1 - steel 15Ch5MF (ΔK_{th} = 6.53 MPa$\sqrt{m}$), 2 - steel cladding EK-32 (ΔK_{th} = 6, 84 MPa$\sqrt{m}$), 3 - surfacing made of steel 20Ch13 (ΔK_{th} = 4, 1 MPa$\sqrt{m}$), 4 - on melting from steel 08Ch13 (ΔK_{th} = 8, 1 MPa$\sqrt{m}$), 5 - steel surfacing CT-28 (ΔK_{th} = 12, 0 MPa$\sqrt{m}$)

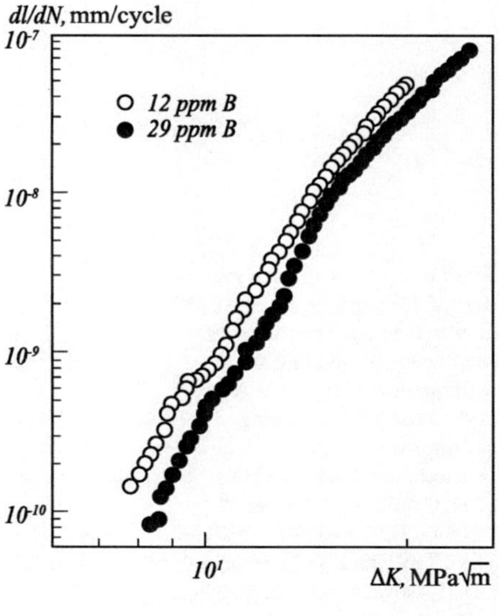

Fig. D.18 The effect of the boron content in the IN718 alloy on the crack growth rate

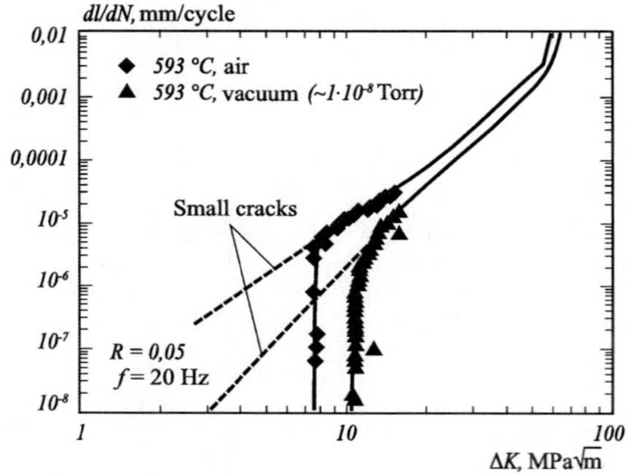

Fig. D.19 Fatigue crack growth rate in Rene 88 alloy

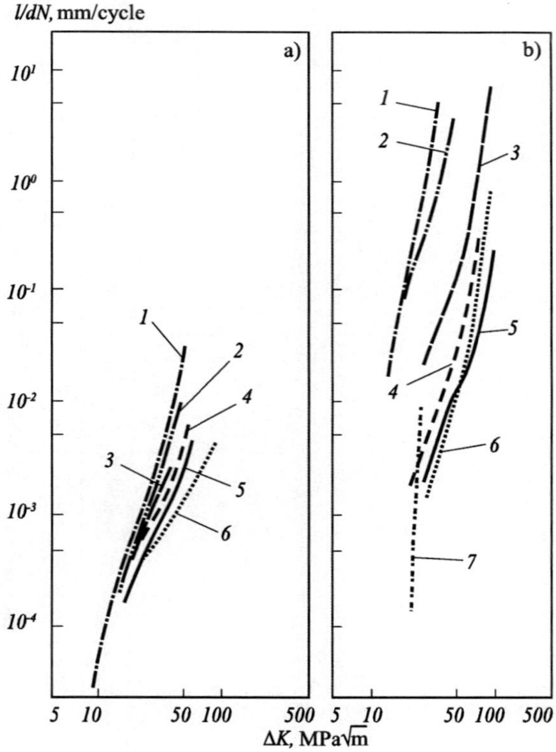

Fig. D.20 Crack growth rate dl/dN for some superalloys at 649°C under conditions of regular cyclic loading with a frequency of 0.33 Hz (a) and under cyclic loading conditions with 15 min delays at maximum load (b) (Sims et al., 1987): 1 - Rene 95; 2 - NASA 11B-7 and HIP MERL 76; 3 - IN100; 4 - Astroloy; 5 - HIP Astroloy; 6 - Waspaloy; 7 - MERL 76

References

Pavlenko DV, Gryaznov BA, Yatsenko VK, Ezhov VN, Orlov MP (2004) Cyclic crack growth resistance of EK79-ID alloy specimens hardened by surface plastic deformation. Strength of Materials 36(3):327–331

Pokrovskii VV, Troshchenko VT, Tseitlin VI, Ezhov VN, Zamotaev VS, Sidyachenko VG (1994a) Estimating working life for aircraft gas turbine engine disks at the stage of fatigue crack development. Part 1. Strength of Materials 26(11):796–800

Pokrovskii VV, Troshchenko VT, Tseitlin VI, Ezhov VN, Zamotaev VS, Sidyachenko VG, Samuleev VV (1994b) Estimating working life for aircraft gas turbine engine disks at the stage of fatigue crack development. Part 2. Strength of Materials 26(12):861–871

Prokopenko AV, Torgov VN, Getsov LB (1984) Osobennosti rasprostraneniya ustalostnykh treshchin v lopatkakh kompressorov (in Russ., Features of distribution of fatigue cracks in compressor blades). Energomashinostroenie (4):20–21

Shlyannikov VN, Tumanov AV, Boychenko NV (2015) A creep stress intensity factor approach to creep–fatigue crack growth. Engineering Fracture Mechanics 142:201–219

Sims CT, Stoloff NS, Hagel WC (eds) (1987) Superalloys II. John Wiley & Sons, New York

Troshchenko VT, Pokrovskiy VV, V PA (1987) Treshchinostoykost' metallov pri tsiklicheskom nagruzhenii (in Russ., Fracture Toughness of Metals Under Cyclic Loading). Naukova Dumka, Kiev